Selected Titles in This Series

287 **Marlos A. G. Viana and Donald St. P. Richards, Editors,** Algebraic methods in statistics and probability, 2001

286 **Edward L. Green, Serkan Hoşten, Reinhard C. Laubenbacher, and Victoria Ann Powers, Editors,** Symbolic computation: Solving equations in algebra, geometry, and engineering, 2001

285 **Joshua A. Leslie and Thierry P. Robart, Editors,** The geometrical study of differential equations, 2001

284 **Gaston M. N'Guérékata and Asamoah Nkwanta, Editors,** Council for African American researchers in the mathematical sciences: Volume IV, 2001

283 **Paul A. Milewski, Leslie M. Smith, Fabian Waleffe, and Esteban G. Tabak, Editors,** Advances in wave interaction and turbulence, 2001

282 **Arlan Ramsay and Jean Renault, Editors,** Groupoids in analysis, geometry, and physics, 2001

281 **Vadim Olshevsky, Editor,** Structured matrices in mathematics, computer science, and engineering II, 2001

280 **Vadim Olshevsky, Editor,** Structured matrices in mathematics, computer science, and engineering I, 2001

279 **Alejandro Adem, Gunnar Carlsson, and Ralph Cohen, Editors,** Topology, geometry, and algebra: Interactions and new directions, 2001

278 **Eric Todd Quinto, Leon Ehrenpreis, Adel Faridani, Fulton Gonzalez, and Eric Grinberg, Editors,** Radon transforms and tomography, 2001

277 **Luca Capogna and Loredana Lanzani, Editors,** Harmonic analysis and boundary value problems, 2001

276 **Emma Previato, Editor,** Advances in algebraic geometry motivated by physics, 2001

275 **Alfred G. Noël, Earl Barnes, and Sonya A. F. Stephens, Editors,** Council for African American researchers in the mathematical sciences: Volume III, 2001

274 **Ken-ichi Maruyama and John W. Rutter, Editors,** Groups of homotopy self-equivalences and related topics, 2001

273 **A. V. Kelarev, R. Göbel, K. M. Rangaswamy, P. Schultz, and C. Vinsonhaler, Editors,** Abelian groups, rings and modules, 2001

272 **Eva Bayer-Fluckiger, David Lewis, and Andrew Ranicki, Editors,** Quadratic forms and their applications, 2000

271 **J. P. C. Greenlees, Robert R. Bruner, and Nicholas Kuhn, Editors,** Homotopy methods in algebraic topology, 2001

270 **Jan Denef, Leonard Lipschitz, Thanases Pheidas, and Jan Van Geel, Editors,** Hilbert's tenth problem: Relations with arithmetic and algebraic geometry, 2000

269 **Mikhail Lyubich, John W. Milnor, and Yair N. Minsky, Editors,** Laminations and foliations in dynamics, geometry and topology, 2001

268 **Robert Gulliver, Walter Littman, and Roberto Triggiani, Editors,** Differential geometric methods in the control of partial differential equations, 2000

267 **Nicolás Andruskiewitsch, Walter Ricardo Ferrer Santos, and Hans-Jürgen Schneider, Editors,** New trends in Hopf algebra theory, 2000

266 **Caroline Grant Melles and Ruth I. Michler, Editors,** Singularities in algebraic and analytic geometry, 2000

265 **Dominique Arlettaz and Kathryn Hess, Editors,** Une dégustation topologique: Homotopy theory in the Swiss Alps, 2000

264 **Kai Yuen Chan, Alexander A. Mikhalev, Man-Keung Siu, Jie-Tai Yu, and Efim I. Zelmanov, Editors,** Combinatorial and computational algebra, 2000

263 **Yan Guo, Editor,** Nonlinear wave equations, 2000

(Continued in the back of this publication)

Algebraic Methods in Statistics and Probability

CONTEMPORARY MATHEMATICS

287

Algebraic Methods in Statistics and Probability

AMS Special Session on Algebraic Methods in Statistics
April 8–9, 2000
University of Notre Dame
Notre Dame, Indiana

Marlos A. G. Viana
Donald St. P. Richards
Editors

American Mathematical Society
Providence, Rhode Island

This volume contains the proceedings of the AMS Special Session on Algebraic Methods and Statistics held at The University of Notre Dame, Notre Dame, Indiana on April 8–9, 2000.

2000 *Mathematics Subject Classification.* Primary 05B20, 60F05, 62A01, 62C10, 62H15; Secondary 13P10, 15A52, 20B30, 30E20, 62G08.

Library of Congress Cataloging-in-Publication Data

AMS Special Session on Algebraic Methods and Statistics (2000 : University of Notre Dame)
Algebraic methods in statistics and probability : AMS Special Session on Algebraic Methods and Statistics, April 8–9, 2000, University of Notre Dame, Notre Dame, Indiana / Marlos A. G. Viana, Donald St. P. Richards, editors.
p. cm. — (Contemporary mathematics, ISSN 0271-4132 ; 287)
Includes bibliographical references.
ISBN 0-8218-2687-5 (alk. paper)
1. Mathematical statistics—Congresses. 2. Probabilities—Congresses. 3. Algebra—Congresses. I. Viana, Marlos A. G. II. Richards, Donald St. P. III. Title. IV. Contemporary mathematics (American Mathematical Society) ; v. 287.

QA276.A1 A48 2000
519.5—dc21   2001045884

---

10 9 8 7 6 5 4 3 2 1   06 05 04 03 02 01

Contents

Preface ix

Simplicial Inference
JOHN AITCHISON 1

A Note on Nyman's Equivalent Formulation of the Riemann Hypothesis
JEAN-FRANÇOIS BURNOL 23

On the Construction of Linear Orthogonal Arrays by Extension
DOMINIQUE COLLOMBIER AND ASTRID JOURDAN 27

A Coordinate-free Approach to Multivariate Exponential Families
ALESSANDRO DI BUCCHIANICO AND DANIEL E. LOEB 37

Best Invariant Predictive Distributions
MORRIS L. EATON AND WILLIAM D. SUDDERTH 49

A Family of Probability Densities Related to the Riemann Zeta Function
WERNER EHM 63

Local Field U-Statistics
STEVEN N. EVANS 75

Krawtchouk Matrices from Classical and Quantum Random Walks
PHILIP FEINSILVER AND JERZY KOCIK 83

Some Rank-based Hypothesis Tests for Covariance Structure
and Conditional Independence
YONGHONG GAO AND JOHN I. MARDEN 97

Gaussian Measures as Limits on Irreducible Symmetric Spaces and Cones
PIOTR GRACZYK 111

The Covariance Structure of the Multivariate Liouville Distributions
RAMESHWAR D. GUPTA AND DONALD ST. P. RICHARDS 125

Reduction of Regression Models under Symmetry
INGE S. HELLAND 139

Deconvolution Density Estimation on Compact Lie Groups
PETER T. KIM AND DONALD ST. P. RICHARDS 155

On Efficiency of Indirect Estimation of Nonparametric Regression Functions
CHRIS A. J. KLAASSEN, EUN-JOO LEE AND FRITS H. RUYMGAART 173

Patterned Matrices Treated via Linear Spaces
TÕNU KOLLO AND DIETRICH VON ROSEN 185

Random Walks on Regular Languages and Algebraic Systems of Generating Functions
STEVEN P. LALLEY 201

The Normal Quasi-Wishart Distribution
GÉRARD LETAC AND HÉLÈNE MASSAM 231

Rank Score Statistics for Spherical Data
TERESA NEEMAN AND TED CHANG 241

Graphical Model Search via Essential Graphs
MICHAEL D. PERLMAN 255

Computational Commutative Algebra in Discrete Statistics
GIOVANNI PISTONE, EVA RICCOMAGNO AND HENRY P. WYNN 267

Maximum Covariance Difference Test for Equality of Two Covariance Matrices
AKIMICHI TAKEMURA AND SATOSHI KURIKI 283

The Covariance Structure of Random Permutation Matrices
MARLOS A. G. VIANA 303

The Extendibility of Statistical Models
ERNST WIT AND PETER MCCULLAGH 327

Preface

The role of algebraic methods and arguments in statistics and probability is well-known, from Gauss' least squares principle through R. A. Fisher's method of variance decomposition. The work of A. James [**Jam57**], according to E. Hannan [**Han65**, p.65], appears to be among the first describing the group-theoretic nature of Fisher's argument, giving meaning to the notion of *relationship algebra*, or the commuting algebra of a representation of the group of symmetries of an experimental design. During the same period, U. Grenander [**Gre63**] showed the effectiveness of harmonic analysis techniques to extend the classical limit theorems to algebraic structures such as locally compact groups, Banach spaces and topological algebras. Two decades later, the relevance of group invariance and group representation arguments in statistical inference would become evident in the works of, among others, P. Diaconis [**Dia88**], M. Eaton [**Eat89**], R. Farrell [**Far85**] and R. Wijsman [**Wij90**]. The integral of Haar, as didactically presented in L. Nachbin's monograph [**Nac65**], became a familiar tool among statisticians, and S. Andersson's group symmetry covariance models are recognized (e.g., [**Per87**]) now as a landmark concept extending and setting definitive boundaries to the standard of multivariate statistical analysis, as known in the tradition of T.W. Anderson. The contributions of S. Lauritzen [**Lau96**] on modelling complex stochastic systems using graph theory, and the monograph by G. Pistone, E. Riccomagno and H. Winn [**PRW00**] on computational commutative algebra in statistics are examples of recent new complements to the seminal work of Alan James.

The opportunity present to all of us statisticians and probabilists today is in the formulation and application of comprehensive concepts and language capable of capturing in their essence the research questions posed by those with whom we collaborate – reasoning tools which represent the broad physical, biological, behavioral, social interface between phenomena and data. The challenge is in representing and including the interface, the context, in our proposed methods of scientific explanation. As recently observed by V. Lakshminarayanan [**LSJ98**], group-theoretical analysis may lead to prediction of new measurable effects or allow newer methods of analysis of data and offer greater insight into [specific] research hypotheses. This remark exemplifies the many opportunities ahead.

The present volume was originally proposed by the editors during the AMS Sectional Meeting, in Montreal, 1997. A later sectional meeting in South Bend, Indiana, in the Spring of 2000, gave us the opportunity to continue with the project. The papers appearing in the *South Bend* volume result from presented talks and subsequent invited contributions.

We are grateful to all who assisted us during the refereeing process, namely J. Aitchison, S. Andersson, J-F. Burnol, P. Chan, T. Chang, D. Collombier,

D. Dawson, I. Dinwoodie, C. Dunkl, W. Ehm, S. Evans, J. Faraut, U. Franz, D. Geiger, S. Hedayat, I. Helland, T. Kollo, S. Lalley, H. Massam, B. McNeney, R. Mukerjee, H. Neudecker, M. Perlman, F.'Ruymgaart, and P. Sarnak. We also acknowledge the support of the Series managing editor, Dennis DeTurck. Our sincere appreciation goes to the authors, for their timely cooperation during the editorial process.

We feel that the twenty-three contributions that follow give an excellent overview of current opportunities and directions in algebraic methods in statistics and probability.

References

[Dia88] P. Diaconis, *Group representation in probability and statistics*, IMS, Hayward, California, 1988.

[Eat89] M. L. Eaton, *Group invariance applications in statistics*, IMS-ASA, Hayward, California, 1989.

[Far85] R. H. Farrell (ed.), *Multivariate calculation*, Springer-Verlag, New York, NY, 1985.

[Gre63] U. Grenander, *Probabilities on algebraic structures*, Wiley, New York, NY, 1963.

[Han65] E. J. Hannan, *Group representations and applied probability*, Journal of Applied Probability **2** (1965), 1–68.

[Jam57] A. T. James, *The relationship algebra of an experimental design*, Annals of Mathematical Statistics **28** (1957), 993–1002.

[Lau96] S. L. Lauritzen, *Graphical models*, Oxford University Press, New York, NY, 1996.

[LSJ98] V. Lakshminarayanan, R. Srudhar, and R. Jagannathan, *Lie algebraic treatment of dioptric power and optical aberrations*, J. Optical Society of America, A **15** (1998), 2497–2503.

[Nac65] L. Nachbin, *The Haar integral*, Van Nostrand, Princeton, N.J., 1965.

[Per87] M. D. Perlman, *Group symmetry covariance models*, Statistical Science **2** (1987), 421–425.

[PRW00] G. Pistone, E. Riccomagno, and P. Wyinn, *Algebraic statistics: Computational commutative algebra in statistics*, Chapman & Hall/CRC, Boca Raton, FL, 2000.

[Wij90] R. A. Wijsman, *Invariant measures on groups and their use in statistics*, IMS, Hayward, California, 1990.

Marlos A. G. Viana
Chicago, Illinois

Donald St. P. Richards
Charlottesville, Virginia

July 26, 2001

Contemporary Mathematics
Volume **287**, 2001

Simplicial Inference

John Aitchison

ABSTRACT. The statistical analysis of data in a simplex sample space has been a source of great confusion and misinterpretation in both the theoretical and applied literature. Yet the concepts and principles at the basis of such problems are simple and lead, by a sequence of logical necessities, to simple forms of statistical methodology. This paper is to some extent pedagogical in that it sets out these simplicial concepts and principles in logical sequence; connects the resulting theory to statistical practice through easily applied techniques of log ratio transformations; and presents a concise guide to a number of promising new techniques for the investigation of more complex problems involving such data.

1. Introduction

There are many practical situations in statistical analysis where the unit simplex is the natural sample space. We distinguish between two main types of data which fall into this mode. Compositions, such as major oxide compositions of rocks in geology and time budgets in sociological studies, take the form of a vector $x = (x_1, \dots, x_D)$ of positive proportions, or *components*, of the D *parts* with unit sum $x_1 + \cdots + x_D = 1$. Similarly, any probability statement about a finite number D of possible states or hypotheses takes a similar form with the components $x_1, \dots, x_D$ being the probabilities assigned to the D hypotheses, such as when a clinician assigns diagnostic probabilities to a set of mutually exclusive and exhaustive disease types. The appropriate sample space for such vector data is the d-dimensional unit simplex

$$S^d = \{(x_1, \dots, x_D) : x_i > 0 \ (i = 1, \dots, D),\ x_1 + \cdots + x_D = 1\}, \tag{1.1}$$

where $d = D - 1$. In this paper we bring together in one framework the concepts and principles which provide a sound basis for inference from such data.

Despite the obvious fact that the simplex S^d is radically different from R^d, it is still quite common to see analyses which ignore the constrained nature of the vector x and apply standard multivariate analysis, with consequent misinterpretations of the nature of the compositional variability. Two such recent examples are to be

1991 *Mathematics Subject Classification.* Primary 62F30; Secondary 62A05.

Key words and phrases. Compositional data; Differential perturbation process; Log contrasts; Logistic normal distributions; Mellin transform; Perturbation; Power transformation; Probability statement data.

found in [**Wo1**], [**Wo2**]. For a detailed account of the history of such difficulties and misinterpretations, see [**A1**], [**A2**], [**A5**], [**A12**], [**A13**].

2. The nature of compositional and probability statement problems: scale invariance

When we say that a problem is compositional we are recognizing that the *sizes* of our specimens are irrelevant. This trivial admission has far-reaching consequences. A simple example can illustrate the argument. Consider two specimen vectors $w = (1.6, 2.4, 4.0)$ and $W = (3.0, 4.5, 7.5)$ in R_+^3 representing the weights of the three parts (a, b, c) of two specimens of total weight 8 gm and 15 gm, respectively. If we are interested in compositional problems, we recognize that these are of the same composition, the difference in weight being taken account of by the scale relationship $W = (15/8)w$. More generally two specimen vectors w and W in R_+^D are compositionally equivalent, written $W \sim w$, when there exists a positive proportionality constant p such that $W = pw$. The fundamental requirement of compositional data analysis can then be stated as follows: any meaningful function f of a specimen vector must be such that $f(W) = f(w)$ when $W \sim w$, or equivalently

$$f(pw) = f(w) \text{ for every } p > 0. \tag{2.1}$$

In other words, the function f must be *invariant under the group of scale transformations.* Since any group invariant function can be expressed as a function of any maximal invariant h and since

$$h(w) = (w_1/w_D, \ldots, w_{D-1}/w_D)$$

is such a maximal invariant we have the following important consequence.

> *Any meaningful (scale-invariant) function of a composition can be expressed in terms of ratios of the components of the composition.*

Note that there are many equivalent sets of ratios which may be used for the purpose of creating meaningful functions of compositions. For example, a more symmetric set of ratios such as $w/g(w)$, where $g(w) = (w_1 \ldots w_D)^{1/D}$ is the geometric mean of the components of w, would equally meet the scale-invariant requirement.

The choice of the unit simplex (1.1) as sample space is justified by the requirement of scale invariance since clearly the choice of units of measurement is irrelevant in a compositional problem. Statisticians will readily recognize the above arguments in relation to probabilistic statements. It is clear that the standard practice of measuring probabilities on the scale of 0 to 1 is merely a convention and that any meaningful probabilistic statement can be expressed in terms of ratios, equivalently odds.

3. Marginal concepts in the simplex: subcompositional and conditional coherence

The marginal or projection concept for simplicial data is slightly more complex than that for unconstrained vectors in R^D, where a marginal vector is simply a subvector of the full D-dimensional vector. For example, a geologist interested only in the parts (Na_2O, K_2O, Al_2O_3) of a ten-part major oxide composition of a rock commonly forms the subcomposition based on these parts. Formally, the subcomposition based on parts $(1, 2, \dots, C)$ of a D-part composition $(x_1, \dots, x_D)$ is the $(1, 2, \dots, C)$-subcomposition $(s_1, \dots, s_C)$ defined by

$$(s_1, \dots, s_C) = (x_1, \dots, x_C)/(x_1 + \dots + x_C). \tag{3.1}$$

This operation $\mathcal{C} : S^d \to S^c$, where $c = C - 1$, is commonly referred to as the *closure* operation by geologists and allows (3.1) to be expressed as

$$(s_1, \dots, s_C) = \mathcal{C}(x_1, \dots, x_C). \tag{3.2}$$

Less familiar is another logical necessity of compositional analysis, namely subcompositional coherence. Consider two scientists A and B interested in soil samples, which have been divided into aliquots. For each aliquot A records a 4-part composition (animal, vegetable, mineral, water); B first dries each aliquot without recording the water content and arrives at a 3-part composition (animal, vegetable, mineral). Let us further assume for simplicity the ideal situation where the aliquots in each pair are identical and where the two scientists are accurate in their determinations. Then clearly B's 3-part composition (s_1, s_2, s_3) for an aliquot will be a subcomposition of A's 4-part composition (x_1, x_2, x_3, x_4) for the corresponding aliquot related as in (3.2) above with $C = 3$, $D = 4$. It is then obvious that any compositional statements that A and B make about the common parts, animal, vegetable and mineral, must agree. This is the nature of subcompositional coherence.

The ignoring of this principle of subcompositional coherence has been a source of great confusion in compositional data analysis, The literature, even currently, is full of attempts to explain the dependence of components of compositions in terms of product moment correlation of raw components. Consider the simple data set:

Full compositions (x_1, x_2, x_3, x_4)	**Subcompositions** (s_1, s_2, s_3)
(0.1, 0.2, 0.1, 0.6)	(0.250, 0.500, 0.250)
(0.2, 0.1, 0.1, 0.6)	(0.500, 0.250, 0.250)
(0.3, 0.3, 0.2, 0.2)	(0.375, 0.375, 0.250)

Scientist A would report the correlation between animal and vegetable as $\text{corr}(x_1, x_2) = 0.5$, whereas B would report $\text{corr}(s_1, s_2) = -1$. There is thus incoherence of the product-moment correlation between raw components as a measure of dependence. Note, however, that the ratio of two components remains unchanged when we move from full composition to subcomposition: $s_i/s_j = x_i/x_j$, so that, as long as we work with scale invariant functions, or equivalently express all our statements about compositions in terms of ratios, we shall be subcompositionally coherent.

Similar considerations apply to probabilistic statements. A clinician may be faced with a differential diagnostic problem among five forms (1, 2, 3, 4, 5) of which 1, 2, 3, are malignant and 4, 5 benign. At a stage in the diagnostic process the clinician, having ruled out the benign forms 4 and 5, may wish to make a conditional probabilistic statement involving only the malignant states 1, 2, 3. The process of moving from the full probabilistic statement to the conditional probability statement is exactly analogous to the closure operation of forming a subcomposition from a full composition. Moreover, clearly there is also a principle of conditional coherence, analogous to the subcompositional coherence principle, that must apply here.

4. Group operations in the simplex

4.1. The role of group operations in statistics. For every sample space there are basic group operations which, when recognized, dominate clear thinking about data analysis. In R^D the two operations, translation (or displacement) and scalar multiplication, are so familiar that their fundamental role is often overlooked. Yet the change from y to $Y = y + t$ by the translation t or to $Y = ay$ by the scalar multiple a are at the heart of statistical methodology for R^D sample spaces. For example, since the translation relationship between y_1 and Y_1 is the same as that between y_2 and Y_2 if and only if Y_1 and Y_2 are equal translations t of y_1 and y_2, any definition of a difference or a distance measure must be such that the measure is the same for (y_1, Y_1) as for $(y_1 + t, Y_1 + t)$ for every translation t. Technically this is a requirement of invariance under the group of translations. This is the reason, though seldom expressed because of its obviousness in this simple space, for the use of the mean vector $\mu = E(y)$ and the covariance matrix $\Sigma = V(y) = E\{(y - \mu)(y - \mu)^T\}$ as meaningful measures of central tendency and dispersion. Recall also, for further reference, two basic properties: for a fixed translation t,

$$E(y + t) = E(y) + t, V(y + t) = V(y). \tag{4.1}$$

The second operation, that of scalar multiplication, also plays a substantial role in, for example, linear forms of statistical analysis such as principal component analysis, where linear combinations $a_1 y_1 + \cdots + a_D y_D$ with certain properties are sought. Recall, again for further reference, that for a fixed scalar multiple,

$$E(ay) = aE(y),\ V(ay) = a^2 E(y). \tag{4.2}$$

Similar considerations of groups of fundamental operations are essential for other sample spaces. For example, in the analysis of directional data, as in the study of the movement of tectonic plates, it was recognition that the group of rotations on the sphere plays a central role and the use of a satisfactory representation of that group that led [C] to the production of the essential statistical tool for spherical regression.

4.2. Perturbation: a fundamental group operation in the simplex. By analogy with the group operation arguments for R^D, the obvious questions for a simplex sample space are whether there is an operation on a composition x, analogous to translation in R^D, which transforms it into X, and whether this can be used to characterize 'difference' between compositions. The answer is to be found in the perturbation operator as defined in [**A5**, Section 2.8]. If we define a perturbation p as a differential scaling operator $p = (p_1, \ldots, p_D) \in S^d$ and denote by $\circ$ the perturbation operation, then we can define the perturbation operation in the following way. The perturbation $p = (p_1, \ldots, p_D)$ applied to the composition $x = (x_1, \ldots, x_D)$ produces the composition

$$p \circ x = (p_1 x_1, \ldots, p_D x_D)/(p_1 x_1 + \cdots + p_D x_D). \tag{4.3}$$

Note that because of the nature of the scaling in this relationship it is not strictly necessary for the perturbation p to be a vector in S^d.

In mathematical terms the set of perturbations in S^D form a group with the identity perturbation $e = (1/D, \ldots, 1/D)$ and the inverse of a perturbation p being the closure $C(p_1^{-1}, \ldots, p_D^{-1})$. The relation between any two compositions x and X

can always be expressed as a perturbation operation $X = (X \circ x^{-1}) \circ x$, where $X \circ x^{-1}$ is a perturbation in the group of perturbations in the unit simplex S^d. The change from X to x is simply the inverse perturbation defined by $(X \circ x^{-1})^{-1} = x \circ X^{-1}$. Thus any measure of difference between compositions x and X must be expressible in terms of one or other of these perturbations. A consequence of this is that if we wish to define any *scalar measure of distance* between two compositions x and X, say $\Delta(x, X)$, then we must ensure that it is a function of the ratios $x_1/X_1, \ldots, x_D/X_D$. As we shall see later, this, together with attention to the need for scale invariance, subcompositional coherence and some other simple requirements, has led [**A11**] to advocate the use of

(4.4)
$$\Delta(x, X) = \{\sum_{i<j}\{\log(x_i/x_j) - \log(X_i/X_j)\}^2\}^{1/2}$$

as a simplicial metric, reinforcing an intuitive equivalent choice in [**A5**, Section 8.3].

In relation to probability statements the perturbation operation is a standard process. Bayesians perturb the prior probability assessment x on a finite number D of hypotheses by the likelihood p to obtain the posterior assessment X through the use of Bayes' formula (4.3). Again, in genetic selection, the population composition x of genotypes of one generation is perturbed by differential survival probabilities represented by a perturbation p to obtain the composition X at the next generation, again by the perturbation probabilistic mechanism (4.3). In certain geological processes, such as metamorphic change, sedimentation, crushing in relation to particle size distributions, change may be best modeled by such perturbation mechanisms, where an initial specimen of composition x_0 is subjected to a sequence of perturbations $p_1, \ldots, p_n$ in reaching its current state x_n:

$$x_1 = p_1 \circ x_0, x_2 = p_2 \circ x_1, \ldots, x_n = p_n \circ x_{n-1},$$

so that

(4.5)
$$x_n = (p_1 \circ p_2 \circ \cdots \circ p_n) \circ x_0.$$

It is clear that in (4.5) we have the makings of some form of central limit theorem but we delay consideration of this until we have completed the more mathematical aspects of the simplex sample space.

A further role which perturbation plays in simplicial inference is in characterizing imprecision or error. A simple example will suffice for the moment. In the process of replicate analyses of aliquots of some specimen in an attempt to determine its composition, ξ, we may obtain different compositions $x_1, \ldots, x_N$ because of the imprecision of the analytic process. In such a situation we can model by setting

(4.6)
$$x_r = \xi \circ p_r \quad r = 1, \ldots, N,$$

where the p_r are independent error compositions characterizing the imprecision.

4.3. Power transformation: another fundamental group operation in the simplex. First we define the power operation and then consider its relevance in simplicial inference. For any positive scalar a and any composition $x \in S^d$, we define

(4.7)
$$a \cdot x = \mathcal{C}(x_1^a, \ldots, x_D^a)$$

as the a-power transform of x. Such an operation $\cdot$ arises in simplicial inference in two distinct ways. First it may be of relevance directly because of the nature of the sampling process. For example, in grain size studies of sediments, sediment samples may be successively sieved through meshes of different diameters and the weights of these successive separations converted into compositions based on proportions by weight. Thus, though separation is based on the linear measurement *diameter*, the composition is based essentially on a *weight*, or equivalently, a *volume* measurement, with a power transformation being the natural connecting concept. More indirectly, the power transformation can be useful in describing regression relations for compositions. For example, the finding of [**A5**, Section 7.7] of the relationship of a (sand, silt, clay) sediment x to depth d can be expressed in the form

$$x = \xi \circ (\log d \cdot \eta) \circ p, \tag{4.8}$$

where η is a composition playing the counterpart of regression coefficients and p is a perturbation playing the role of error in more familiar regression situations.

It must be clear that together the operations perturbation $\circ$ and power $\cdot$ play roles in the geometry of S^d analogous to translation and scalar multiplication in R^D and indeed can be used to define a vector space in S^d. We can now complete this geometric picture by considering how to arrive at a suitable metric.

5. A simplicial metric

For the simplex S^d any metric $\Delta : S^d \times S^d \to R_+$ must satisfy not only the usual axioms that $\Delta(x,y) \geq 0$, $\Delta(x,y) = 0$ if and only if $x = y$, $\Delta(x,y) = \Delta(y,x)$ (symmetry), $\Delta(x,y) + \Delta(y,z) \geq \Delta(x,z)$ (the triangular inequality) and $\Delta(a\cdot x, b\cdot y) = ab\,\Delta(x,y)$ (the scalar relationship), but also certain important logical requirements associated with compositional and probabilistic statement analysis. For example, since, as we have seen above, the change from composition x to y is the same as from similarly perturbed compositions $x \circ p$ and $y \circ p$, namely $y \circ x^{-1}$, we require the metric to be perturbation invariant in the sense that $\Delta(x \circ p, y \circ p) = \Delta(x,y)$ for every x, y, p. This perturbation invariance along with scale invariance and the obvious requirement of permutation invariance led [**A11**] to the adoption of the following simple and equivalent forms for a simplicial metric:

$$\Delta(x,X) = \left[\sum_{i=1}^{D}\left\{\log\frac{x_i}{g(x)} - \log\frac{X_i}{g(X)}\right\}^2\right]^{\frac{1}{2}} \tag{5.1}$$

$$= \left[\frac{1}{D}\sum_{i<j}\left\{\log\frac{x_i}{x_j} - \log\frac{X_i}{X_j}\right\}^2\right]^{\frac{1}{2}} \tag{5.2}$$

$$= [(y-Y)^T H^{-1}(y-Y)]^{\frac{1}{2}}, \tag{5.3}$$

where $y_i = \log(x_i/x_D)$, $Y_i = \log(X_i/X_D)$, $i = 1,\ldots,d$, $H = [h_{ij}]$, with $h_{ij} = 2$ for $i = j$ and $h_{ij} = 1$ for $i \neq j$, as in [**A5**, Section 4.7], and $g(x)$ is the geometric mean of the components of x. We may note here that this metric also satisfies another obvious requirement that the distance between any two compositions must be at least as great as the distance between any corresponding subcompositions of the compositions. The metric also has an associated norm $\|x\|$ and inner product $\langle x, X\rangle$ defined by

$$\|x\| = \Delta(x,e),\ \langle x, X\rangle = \frac{1}{2}(\|x\|^2 + \|X\|^2 - \Delta^2(x,X)), \tag{5.4}$$

where $e = (1/D)(1,\dots,1)$ is the identity of the perturbation group. The statistician can then rest assured that with the natural and meaningful operations of perturbation and power and this simple metric there is available a relevant sample space within which to develop a statistical methodology for the analysis of data in the form of compositions or probabilistic statements.

Note that for probability statement data, the perturbation invariance requirement is appropriate since the difference between two probability statements must remain unaltered if each is updated Bayesianly by the same likelihood. For a study of the failure of other simplicial metrics, see [**Mar-B-P**], and for an account of the confusion in determining an appropriate simplicial metric, see [**A7**], [**A8**], [**A10**], [**A11**], [**W1**] [**W2**], [**W-P**].

6. Distributional concepts in the simplex

Suppose that we have a distribution of the unit of probability spread over the simplex S^d by a density function $f(x)$ $(x \in S^d)$. Before we investigate specific useful forms of density functions we consider the customary tasks of finding suitable definitions for central and dispersion characteristics of any simplicial distribution. There are two related questions in the characterization of the variability of vectors. How can we describe characteristics which in meaningful ways define (1) a center around which the variability takes place, and (2) measures of dispersion around this center. Within (2) we include measures of the dependence between the various components of the composition.

6.1. Distributional characteristics: measure of central tendency. It is worth recalling the arguments which determine sensible centers and dispersions in R^D. In such a sample space, in which ideas of Euclidean distance dominate, it is claimed to be sensible to consider as center the μ which minimizes the average squared distance $E(\|y-\mu\|^2)$, and this turns out to be simply $E(y)$. For compositions and the simplex we have the metric (5.1-5.3), and can use it in a similar way. More specifically we can define as center $\xi = \text{cen}(x)$ the composition ξ which minimizes $E\{\Delta^2(x,\xi)\}$ subject to the condition that $\xi \in S^d$. This simple optimization problem leads to the following definition:

$$\xi = \text{cen}(x) = \mathcal{C}\{\exp(E(\log x)\}. \tag{6.1}$$

This may seem at first sight a very unfamiliar object until we realize that for any positive random variable z the formal definition of the geometric mean is $\exp\{E(\log z)\}$. Note here that although in (6.1) it seems that we have abandoned, in the use of $\log(x)$, our scale-invariant directive to use only ratios, the complete expression for $\text{cen}(x)$ involves a closure operation $\mathcal{C}$ which ensures ratios. Indeed an alternative and equivalent definition of center could be used involving ratios at the first stage of the computation, namely

$$\text{cen}(x) = \mathcal{C}\{\exp(E(\log x/g(x))\}, \tag{6.2}$$

where $g(x)$ denotes the geometric mean of the components of x.

There is an analogy here with the use of the lognormal distribution $\Lambda(\mu, \Sigma)$ to describe the pattern of variability of a positive quantity and the use of the geometric mean $\exp\{E(\log z)\} = \exp(\mu)$, dating back to [**Mc**]. We shall refer to ξ as the *geometric center* and note that, for any fixed perturbation p, $\text{cen}(p \circ x) = p \circ \text{cen}(x)$, in analogy with $E(t+y) = E(y)+t$ in (4.2) for unconstrained variability in R^D. Note also that with this definition we have a power result $\text{cen}(a \cdot x) = \mathcal{C}\{a\ \text{cen}(x)\}$ in analogy with $E(ay) = aE(y)$ for unconstrained variability. Note also that $\text{cen}(x \circ y) = \text{cen}(x) \circ \text{cen}(y)$, whether or not x and y are independent, in conformity with a similar R^D result. We may digress here to note the practical implications of this simple choice of center. For a compositional data set

$$\mathcal{D} = \{x_r = (x_{r1}, \dots, x_{rD}) : r = 1, \dots, N\}, \tag{6.3}$$

standard practice seems to be to take the arithmetic center

$$\bar{x} = (x_{\cdot 1}, \dots, x_{\cdot D}) \text{ where } x_{\cdot i} = N^{-1} \sum_{r=1}^{N} x_{ri}. \tag{6.4}$$

A consequence of the above analysis is the advocacy of

$$\mathcal{C}(g_1, \dots, g_D) = (g_1, \dots, g_D)/(g_1 + \dots + g_D) \tag{6.5}$$

as center of the compositional data set, where $g_i = (\Pi_r x_{ri})^{1/N}$ is the geometric mean of the ith component over all N cases. There can be a substantial difference in the use of these different centers. For such examples, see [**A6**], [**A12**].

6.2. Distributional characteristics: measures of dispersion and dependence. There are a number of criteria which dictate the choice of any measure $V(x)$ of dispersion and dependence, which forms the basis of characteristics of compositional variability in terms of second-order moments.

(a) Interpretability in relation to the specific hypotheses and problems of interest in fields of application.
(b) Conformability with the definition of center as defined in (6.2).
(c) Invariance under the group of perturbations. Can we ensure that $V(p \circ x) = V(x)$ for every constant perturbation p? (Recall the result in (4.2) that for $y \in R^D$ the covariance matrix V is invariant under translation: $V(t+y) = V(y)$.)
(d) Satisfaction of the power transformation relationship $V(a \cdot x) = a^2 V(x)$, in a way similar to $V(ay) = a^2 V(y)$ for R^D variability.
(e) Mathematical tractability.

Criterion (a) clearly requires that we work in terms of ratios of the components of compositions to ensure scale invariance. At first, thought, this might suggest the use of variances and covariances of the form $\text{var}(x_i/x_j)$ and $\text{cov}(x_i/x_j, x_k/x_l)$. These, however, are mathematically intractable since there is no exact or even simple approximate relationship between $\text{var}(x_i/x_j)$ and $\text{var}(x_j/x_i)$. Fortunately criterion (b) suggests that logarithms of ratios are more appropriate for our purpose to conform with the definition of the geometric center. We are thus led to consider the use of such dispersion characteristics as

$$\text{var}\{\log(x_i/x_j)\}, \ \text{cov}\{\log(x_i/x_j),\ \log(x_k/x_l)\}. \tag{6.6}$$

Obvious advantages of this are simple relationships such as $\text{var}\{\log(x_i/x_j)\} = \text{var}\{\log(x_j/x_i)\}$ and $\text{cov}\{\log(x_i/x_j),\ \log(x_k/x_l)\} = \text{cov}\{\log(x_j/x_i),\ \log(x_l/x_k)\}$.

There are a number of useful and equivalent ways [**A5**, Chapter 4] in which to summarize such a *sufficient* set of second-order moment characteristics: the *log ratio covariance matrix*

$$\Sigma(x) = [\sigma_{ij}] = \text{cov}\{\log(x_i/x_D), \log(x_j/x_D)], \tag{6.7}$$

using only the final component x_D as the common ratio divisor; the *centered log ratio covariance matrix*

$$\Gamma(x) = [\text{cov}\{\log(x_i/g(x)\}, \log\{x_j/g(x)\}]; \tag{6.8}$$

or the variation matrix

$$T(x) = [\tau_{ij}] = [\text{var}\{\log(x_i/x_j)\}], \tag{6.9}$$

with elements involving only pairs of components. These three dispersion characteristics are equivalent: each can be derived from any other by simple matrix operations [**A5**, Chapter 4]. Note that T is symmetric, has zero diagonal elements, and cannot be expressed as the standard covariance matrix of some vector, but has the advantage of considering components only two at a time. We shall see later in Section 8 that it has an important restricted negative definite property.

A first reaction to this variation matrix characterization is often surprise because it is defined in terms of variances only. The simplest statistical analogue is in the use of a completely randomized block design in, say, an industrial experiment From such a situation, information about $\text{var}(y_i - y_j)$ for all i, j is a sufficient description of the variability for purposes of inference.

Note also that if x and y are independent compositions then, for any form V of the dispersion characteristic $V(x \circ y) = V(x) + V(y)$, meeting, in particular, the perturbation criterion (c), and $V(a{\cdot}x) = a^2V(x)$, meeting the power transformation criterion (d).

Relative variances such as $\text{var}\{\log(x_i/x_j)\}$ provide some compensation for deprivation of the meaningless product-moment correlation interpretations. For example, $\tau_{ij} = 0$ means a perfect relationship between x_i and x_j in the sense that the ratio x_i/x_j is constant, replacing the unusable idea of perfect positive correlation between x_i and x_j by one of perfect proportionality. Again, the larger the value of τ_{ij} the more the departure from proportionality with $\tau_{ij} = \infty$ replacing the unusable idea of zero correlation or independence between x_i and x_j. Note that if we are really interested in hypotheses of independence these are most appropriately expressed in terms of independence of subcompositions. For example independence of the (1,2,3)- and (4,5)-subcompositions would be reflected in the following statements:

$$\text{cov}\{\log(x_1/x_3), \log(x_4/x_5)\} = 0, \;\; \text{cov}\{\log(x_2/x_3), \log(x_4/x_5)\} = 0. \tag{6.10}$$

In the study of unconstrained variability in R^D it is often convenient to have available a measure of total variability, for example in principal component analysis and in biplots. For such a sample space the trace of the covariance matrix is the appropriate measure. Here we might consider $\text{trace}\{\Gamma(x)\}$, the trace of the symmetric centered log ratio covariance matrix $\Gamma(x)$. Equally, we might argue on common sense grounds that the sum of all the possible relative variances in $T(x)$, namely $\sum_{i<j} \text{var}\{\log(x_i/x_j)\}$, would be equally good. These two measures indeed differ only by a constant factor and so we can define $\text{totvar}(x)$, a measure of total

variability, as

$$\text{totvar}(x) = \text{trace}\{(\Gamma(x)\} = (1/D)\sum_{i<j} \text{var}\{\log(x_i/x_j)\}. \tag{6.11}$$

We may also note here that the simplicial metric (5.1)-(5.3) is compatible with the above definitions of covariance analogous to the compatibility of Euclidean distance with the covariance matrix of an unconstrained vector. As an illustration of this consider how we might construct a measure of the total variability for a compositional data set $\mathcal{D}$ as at (6.3). The definition at (6.9) suggests that we may obtain such a total measure, totvar_1 say, by replacing each $\text{var}\{\log(x_i/x_j)\}$ in (6.11) by its standard estimate. An alternative intuitive measure of total variation is surely the sum of all the possible distances between the N compositions, namely $\text{totvar}_2 = \sum_{r<s} \Delta^2(x_r, x_s)$. The easily established proportional relationship $\text{totvar}_1 = [D/\{N(N-1)\}]\text{totvar}_2$ confirms the compatibility of the defined covariance structures and scalar measures of distance for compositional variability.

6.3. Generating functions for simplicial distributions. The characteristic and moment generating functions for distributions in R^d are familiar useful tools of distributional analysis. It is relatively easy to design the analogous tools for the study of simplicial distributions in S^d. The transform which seems to be most suited to this purpose is a multivariate adaptation of the Mellin transform. Let

$$U^d = \{(u_1, \ldots, u_D) : u_1 + \cdots + u_D = 0\}. \tag{6.12}$$

For a composition $x \in S^d$ with density function $p(x)$, define its Mellin generating function $M_x : U^d \to R_+$ by the relationship

$$M_x(u) = \int_{S^d} x_1^{u_1} \cdots x_D^{u_D} p(x) dx. \tag{6.13}$$

Note that the restriction of the vector u to the hyperplane U^d rather than R^D is dictated by the need to meet the requirement of scale invariance, here ensured by the fact that integrand is expressible in terms of ratios of the components of x.

The Mellin generating function has perturbation, power and limit properties similar to additive and scale properties of characteristic and moment generating functions for distributions in R^D.

Property M1: $M_x(0) = 1$.

Property M2: If x and y are independent compositions then

$$M_{x \circ y}(u) = M_x(u) M_y(u).$$

Property M3: If a is a positive scalar, then $M_{a \cdot x}(u) = M_x(au)$.

Property M4: If b is a fixed perturbation, then

$$M_{b \circ x}(u) = b_1^{u_1} \cdots b_D^{u_D} M_x(u).$$

Property M5: Combining M2 and M3, if x and y are independent compositions, then

$$M_{a \cdot x \circ b \cdot y}(u) = M_x(au) M_y(bu).$$

Property M6: Moment generating properties. In a manner similar to the use of moment-generating functions in R^D, we can obtain expansions which produce moments of any order:

$$\log M_x(u) = \sum_{i=1}^{D} u_i \log \xi_i - \frac{1}{4} uTu^T + \text{terms of higher order},$$

where ξ and T are the center and variation matrix of the distribution. Moments can also be found through a differentiation process. If we denote by $\partial/\partial u_i \setminus u_j$ differentiation with respect to u_i, after u_j is expressed as $-\Sigma_{k\neq l} u_k$, then we have the following useful results:

$$\partial \log M_x(0)/\partial u_i \setminus u_j = \log(\xi_i/\xi_j),\ \partial^2 \log M_x(0)/\partial(u_i \setminus u_j)^2 = \text{var}\{\log(x_i/x_j)\}.$$

M7: A limit property. Let $\{x_n\}$ be a sequence of compositions with density functions $\{f_n\}$ and Mellin transforms $\{M_n\}$. If $M_n(u) \to M(u)$ and $M(u)$ is the Mellin transform of $f(x)$, then f_n converges in distribution to f.

7. Central limit theorem for compositions

An obvious question to ask about compositional variability is whether there is an analogue of the well-known limiting results for sequences of additive and multiplicative changes leading to normal and lognormal variability through the central limit theorems. As we have already noted the relationship (4.5) depicts the result of a sequence of independent perturbations. In exactly the same way as moment generating functions can be used to establish central limit theorems in R^D so we could use the above properties of the Mellin generating function to establish a similar result for x_n in (4.5). A simple version for the case where p_r $(r = 1, 2, \dots)$ are independently and identically distributed with $\text{cen}(p_r) = \xi = (\xi_1, \dots, \xi_D)$ and $T(p_r) = T(r = 1, \dots, n)$ leads to the following limiting Mellin generating function for $y_n = n^{\frac{1}{2}} \cdot x_n$:

$$M(u) = \exp\left(\sum_{i=1}^{D} u_i \log \xi_i - \frac{1}{4} uTu^T\right). \tag{7.1}$$

Alternatively we can very simply relate (4.5) to an additive central limit theorem by rewriting it in terms of log ratios:

$$\log(x_{ni}/x_{nD}) = \{\log(p_{1i}/p_{1D}) + \cdots + \log(p_{ni}/p_{nD})\} + \log(x_{0i}/x_{0D}),$$

for $i = 1, \dots, d$. If the perturbations are random then the sum within the brackets will, under certain regularity conditions which need not divert us here, tend for large n towards a multivariate normal pattern of variability. It is a simple application of distribution theory to deduce the form of the probability density function $f(x)$ on the unit simplex as

$$\begin{aligned} f(x) &= \det(2\pi\Sigma)^{-\frac{1}{2}}(x_1 \cdots x_d)^{-1} \\ &\quad \exp\left[-\frac{1}{2}\{\log(x_{-D}/x_D) - \mu\}\Sigma^{-1}\{\log(x_{-D}/s_D) - \mu\}^T\right], \end{aligned} \tag{7.2}$$

for $x \in S^D$, where $x_{-D} = (x_1, \dots, x_{D-1})$, μ is a d row vector and Σ is a positive definite square matrix of order d. This is the parametric class of additive logistic normal distributions $L^d(\mu, \Sigma)$ described by [**A-S**]. This result differs from the Mellin transform result only in the parameterization of the parameters. We use the notation $L^D(\xi, T)$ to denote the distribution (7.3) in this parameterization.

8. Log contrasts

In unconstrained multivariate analysis with sample space R^D, substantial use is made of properties of linear combinations (transformations) of the components of vector observations, for example in all techniques involving eigen-analysis. Inspection of the forms involved in the definitions of geometric center, dispersion matrices, Mellin generating function, and the distribution emerging from the central limit theorem suggest that the simplex analogue of a linear combination is a log contrast [**A3**] of a composition x defined by

$$a_1 \log x_1 + \cdots + a_D \log x_D, \tag{8.1}$$

where $a_1 + \cdots + a_D = 0$. Just as linear combinations can be used to define subspaces of the vector space R^D by way of null spaces or range spaces, so log contrasts can be used to identify subspaces of the already identified vector space S^d. We shall see later the role that such log contrasts play in statistical analysis. The main distributional result for log contrasts can be expressed as follows.

Property L1: If composition x has geometric center ξ and variation matrix T, then the vector $\ell = (\ell_1, \ldots, \ell_C) \in R^C$, where

$$l_r = \sum_{i=1}^{D} a_{ri} \log x_i, \quad r = 1, \ldots, C$$

has moment generating function $G(t)$, where $t = (t_1, \ldots, t_C)$, given by $G_l(t) = M_x(tA^T)$, where $A = [a_{ri}]$. A corollary of this result is that, if x follows a $\mathcal{L}^D(\xi, T)$ distribution, then ℓ follows a $\mathcal{L}^C(A \log \xi, -\frac{1}{2}ATA^T)$ distribution.

We may comment here on the negative signs that appear in this last result and, for example, in the result (7.1). This is because of the nature of the variation matrix T. This can easily be shown to have a restricted form of negative definiteness in the sense that, for any $u \in U^d$ as defined by (6.12), $uTu^T \leq 0$, so that the covariance matrix $-\frac{1}{2}ATA^T$ in the above result is positive definite.

9. Parametric classes of distributions

The emergence of the logistic normal distribution $L^{D-1}(\mu, \Sigma)$ or $L^D(\xi, T)$ in a central limit theorem ensures for this parametric class of distributions a central role in the study of distributions on the simplex in a way similar to the multivariate normal and lognormal distributions in R^D and R^D_+. In particular, in addition to simple logistic normal subcompositional and conditional properties, this class of distributions has the essential and useful properties of being closed under the basic simplex operations of perturbation and power: see [**A5**, Chapter 6] for details. The popular Dirichlet class $\mathrm{Di}(\alpha)$ on the simplex with density function

$$f(x) \propto x_1^{\beta_1 - 1} \cdots x_D^{\beta_D - 1} \quad (x \in S^d)$$

has so many drawbacks that it has virtually no role to play in simplicial inference. For example, it has no simple perturbation or power transformation properties and so is ill-suited to the basic group operations of the simplex. Moreover, it has so many built-in independence properties that, apart from being a model of extreme independence, it has almost no role to play in the investigation of the nature of the dependence structure of compositional variability.

There are other classes of distributions on S^d. The fact that the $L^d(\mu,\Sigma)$ class is simply related to the $N^d(\mu,\Sigma)$ class in R^d by way of the log ratio transformation led [**A5**] to consider other transformations from S^d to R^d and defining corresponding logistic-normal classes of distributions, the multiplicative and partitioned classes, which are directed at specific practical problems in compositional data analysis. Also [**A4**], [**A5**, Section 13.4] extends the $L^d(\mu,\Sigma)$ class by the introduction of a single parameter to produce a generalization which includes both the Dirichlet class and the $L^d(\mu,\Sigma)$ logistic normal class. While this is a useful extension, it is somewhat restricted by computational problems involving multiple integrals. A more promising generalization, which is simpler computationally, is an extension based on the recently introduced multivariate skew normal class of distributions [**Az-D**]. In terms of a class of distributions on the simplex, a composition x can be said to have a logistic skew normal distribution if the log ratio vector $[\log(x_1/x_D),\dots,\log(x_d/x_D)]$ has a multivariate skew normal distribution. For recent applications of this class to compositional data problems, see [**Mat-B-P**], [**A-B**].

For comparison with the fitting of parametric distributions to simplicial data or for use when there is no satisfactory parametric class, resort may be made to a non-parametric approach through kernel density estimation [**A-L**].

10. Inferential principles: from theory to practice

So far in this paper we have concentrated on providing a sound logical basis for study of data within the simplex, first by demonstrating that certain fundamental aspects of compositional data analysis—scale invariance, subcompositional coherence, and the central roles of the perturbation and power operations—shape the nature of the appropriate algebra and geometry of the simplex sample space. Then we have identified within the consequent study of probabilistic aspects, promising parametric classes of distributions on the simplex. While, for this development of the logical sequence of arguments, we have confined ourselves to developments within the simplex it must be clear to statisticians, particularly in the relationship of log ratio vectors to the multivariate normal distribution in the last section, that a promising approach to compositional data analysis must be through the use of transformation techniques along the same lines as logarithmic transformations or Box-Cox transformations. This approach has formed the basis of what has come to be known in some circles as log ratio analysis.

Basically the philosophy here is as follows. Any problem concerning compositions can be expressed in terms of log ratios of the components of the composition. Any convenient sufficient set of log ratios can be used. For example the set of final divisor log ratios

$$y_i = \log(x_i/x_D), \quad i = 1,\dots,d, \tag{10.1}$$

can be used with inverse transformation

$$\begin{aligned} x_i &= \exp(y_i)/\{\exp(y_1)+\cdots+\exp(y_d)+1\}, \quad i=1,\dots,d, \\ x_D &= 1/\{\exp(y_1)+\cdots+\exp(y_d)+1\}. \end{aligned} \tag{10.2}$$

Note that transformations (10.1) and (10.3), which are asymmetric in the components of the composition, form a mapping between the unit simplex S^d and

D-dimensional real space R^D. If we insist on a symmetric set of log ratios then we may take

$$z_i = \log\{(x_i/g(x)\}, \quad i = 1, \ldots, D, \tag{10.3}$$

with inverse

$$x_i = \exp(z_i)/\{\exp(z_1) + \cdots + \exp(z_D)\}, \quad i = 1, \ldots, D, \tag{10.4}$$

where $g(x)$ is the geometric mean of the components of x. This is a transformation between the unit simplex S^d and the hyperplane $z_1 + \cdots + z_D = 0$ in d-dimensional real space R^d. The new constraint on the transformed composition is not a transfer of the so-called constant-sum constraint but a penalty for the insistence on a symmetric treatment of the components of the composition. It is linked to the use of the singular centered log ratio covariance matrix $\Gamma(x)$ at (6.8). In practice this singularity causes no interpretational or computational problem.

There are essentially four steps in any log ratio analysis of compositional data.

(1) Reformulate the compositional problem in terms of log ratios of the components.
(2) Transform the compositional data set into compatible log ratio vectors.
(3) Since the log ratio vectors are in real space and free of the constant sum constraint simply apply the appropriate multivariate methodology associated with unconstrained vectors.
(4) Reinterpret the inference from the statistical analysis of the log ratios into terms of the compositions.

A wide variety of compositional problems which can be studied through the above log ratio transformation techniques is described in Aitchison [**A5**]. These include tests of distributional form, log linear modeling to take account of experimental design and concomitant factors, testing various forms of pseudo-independence, discriminant analysis, log contrast principal component analysis. Moreover the link to the multivariate normal allows simple Bayesian analysis including the use of predictive distributions. A question that often arises in the use of form (10.1) of the log ratio transformation is whether the inference is sensitive to the choice of divisor. Aitchison [**A5**] demonstrates that all these procedures are invariant under the group of permutations of the components, and so in particular of the choice of divisor. Rather than reiterate these procedures we concentrate on some more recent developments.

11. Graphical display of compositional data

The biplot [**G1**], [**G2**] is a well-established graphical aid in other branches of statistical analysis. Its adaptation for compositional and probability statement data is simple and can prove a useful exploratory and expository tool. For the compositional data set (6.3) the biplot is based on a singular value decomposition of the doubly centered log ratio matrix $Z = [z_{ri}]$, where

$$z_{ri} = \log\{x_{ri}/g(x_r)\} - N^{-1}\sum_{r=1}^{N}\log\{x_{ri}/g(x_r)\},$$

$i = 1, \ldots, D$, $r = 1, \ldots, N$. Let $Z = U\,\mathrm{diag}(k_1, \ldots, k_R)V^T$ be the singular value decomposition, where R is the rank of Z, in practice usually $R = d$, and where the singular values $k_1, \ldots, k_R$ are in descending order of magnitude. The biplot

(Figure 1) then converts the second-order approximation to Z given by the singular value decomposition into a graphical display. As for standard multivariate biplots the quality of the approximations is measured in terms of the proportion

$$\frac{k_1^2 + k_2^2}{k_1^2 + \ldots + k_R^2}$$

of the total variability of the compositional data set retained by Z^2, with $k_2 >> k_3$ an obvious requirement.

Figure 1 consists of an *origin* O which represents the center of the compositional data set, a *vertex* at position $(k_1 v_{i1}, k_2 v_{i2})/(N-1)^{\frac{1}{2}}$ for each of the parts, labeled $1, \ldots, D$, and a *case marker* at position $(N-1)^{\frac{1}{2}}(u_{r1}, u_{r2})$ for each of the N cases, labeled $c_1, \ldots, c_N$. We term the join of O to a vertex i the *ray* Oi and the join of two vertices i and j the *link* ij. These features constitute a biplot with the following main properties for the interpretation of the compositional variability.

Links, rays and covariance structure. The links and rays provide information on the covariance structure of the compositional data set, in the sense that

$$|ij|^2 \approx \text{var}\{\log(x_i/x_j)\}, \tag{11.1}$$

$$|Oi|^2 \approx \text{var}[\log\{x_i/g(x)\}], \tag{11.2}$$

$$\cos(iOj) \approx \text{corr}[\log\{x_i/g(x)\}, \log\{x_j/g(x)\}]. \tag{11.3}$$

It is tempting to imagine that (11.2) can be used to replace discredited $\text{corr}(x_i, x_j)$ as a measure of the dependence between two components. Unfortunately, this measure does not have subcompositional coherence.

A more useful result is the following. If links ij and kl intersect in M, then

$$\cos(iMk) \approx \text{corr}[\log(x_i/x_j), \log(x_k/x_l)], \tag{11.4}$$

A particular case of this is when the two links are at right angles so that $\cos(iMk) \approx 0$, implying that there is zero correlation of the two log ratios. This is useful in investigation of subcompositions for possible independence.

Subcompositional analysis. The center O is the centroid (center of gravity) of the D vertices $1, \ldots, D$. Since ratios are preserved under formation of subcompositions it follows that the biplot for any subcomposition s is simply formed by selecting the vertices corresponding to the parts of the subcomposition and taking the center of the subcompositional biplot as the centroid of these vertices.

Coincident vertices. If vertices i and j coincide or nearly so this means that var(log(xi/xj) is zero or nearly so, so that the ratio x_i/x_j is constant or nearly so.

Collinear vertices. If a subset of vertices, say $1, \ldots, C$ is collinear, then we know from our comment on subcompositional analysis that the associated subcomposition has a biplot that is one-dimensional, and then a technical argument leads us to the conclusion that the subcomposition has one-dimensional variability. Technically, this one-dimensionality is described by the constancy of $(C-2)$ log contrasts of the components $x_1, \ldots, x_C$. Inspection of these constant log contrasts may then give further insights into the nature of the compositional variability.

Case markers and recovery of data. Such markers have the easily established property that the scalar or dot product $Oc_n \cdot ji$ represents the departure of $\log(x_i/x_j)$ for case c_n from the average of this log ratio over all the cases. Let P and P_n in Figure 1 denote the projections of the center O, and the compositional marker c_n on the possibly extended link ji. Then $Oc_n \cdot ji = \pm|PP_n||ji|$, where the positive sign is taken if the directions of PP_n and ji are the same, otherwise the negative sign is taken. A simple interpretation can be obtained as follows. Consider the extended line ji as divided into positive and negative parts by the point P, the positive part being in the direction of ji from P. If P_n falls on the positive (negative) side of this line then the log ratio of $\log(x_{ni}/x_{nj})$ of the nth composition exceeds (falls short of) the average value of this log ratio over all cases and the further P_n is from P the greater is this exceedance (shortfall); if P_n coincides with P, then the compositional log ratio coincides with the average. In Figure 1, the nth composition clearly has a log ratio $\log(x_{ni}/x_{nj})$ which falls short of the overall average of this log ratio. A

FIGURE 1. Components of a compositional relative variation diagram (a) and interpretation of markers in relation to log ratio $\log(x_i/x_j)$ (b) .

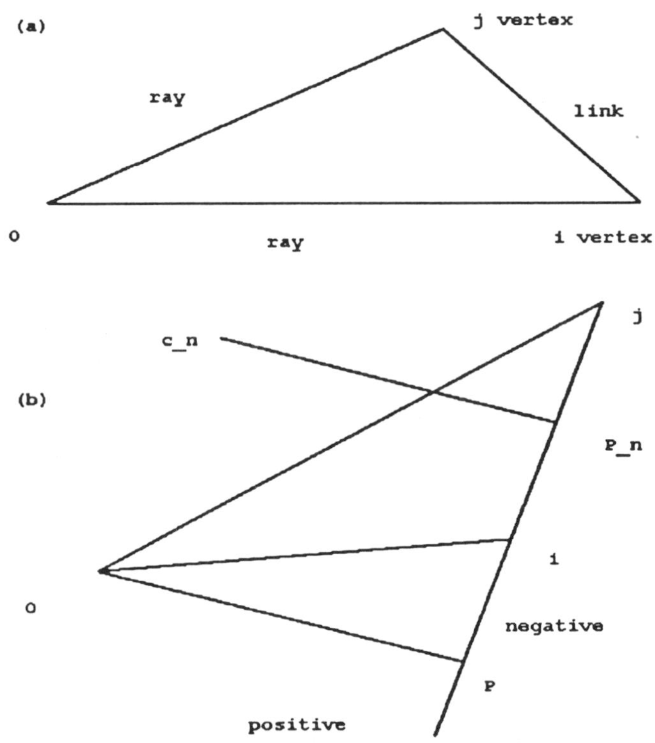

similar form of interpretation can be obtained from the fact that $Oc_n \cdot O_i$ represents the departure of the centered log ratio $\log\{x_{ni}/g(x_n)\}$ of the nth composition from the average of this centered log ratio over all replicates.

It must be clear from the above aspects of interpretation that the fundamental elements of a compositional biplot are the links, not the rays as in the case of variation diagrams for unconstrained multivariate data. The complete set of links, by specifying all the relative variances, determines the compositional covariance structure and provides direct information about subcompositional variability and independence. It is also obvious that interpretation of the relative variation diagram is concerned with its internal geometry and would, for example, be unaffected by any rotation or indeed mirror-imaging of the diagram.

Another fundamental difference between the practice of biplots for unconstrained and compositional data is in the use of data scaling. For unconstrained data, if there are substantial differences in the variances of the components, biplot approximation may concentrate its effort on capturing the nature of the variability of the most variable components and fail to provide any picture of the pattern of variability within the less variable components. Since such differences in variances may simply arise because of scales of measurement, a common technique in such biplot applications is to apply some form of individual scaling to the components of the unconstrained vectors prior to application of the singular value decomposition. No such individual scaling is necessary for compositional data when the analysis involves log ratio transformations. Indeed, since for any set of constants $(c_1, \ldots, c_D)$, we have

$$\operatorname{cov}\{\log(c_i x_i / c_j x_j), \log(c_k x_k / c_l x_l)\} = \operatorname{cov}\{\log(x_i/x_j), \log(x_k/x_l)\},$$

it is obvious that the covariance structure and therefore the compositional biplot are unchanged by any differential scaling or perturbation of the compositions. Only the centering process is affected by such differential scaling. Moreover any attempt at differential scaling of the *log ratios* of the components would be equivalent to applying differential power transformations to the *components* of the compositions, a distortion which would prevent any compositional interpretation from the resulting diagram.

For some applications of biplots to compositional data in a variety of geological contexts, see [**A9**], and for applications in other disciplines and to extensions to conditional biplots, see [**A14**].

12. Subcompositional analysis

A common problem in compositional data analysis appears to be marginal analysis in the sense of locating subcompositions of greatest or of least variability. For this purpose, the measure of total variation discussed in Section 6.2 provides for any subcomposition s of a full compositions x the estimate of the ratio trace $\Gamma(s)/\text{trace}\Gamma(x)$ as the proportion of the total variation explained by the subcomposition. In such forms of analysis it should be noted that a $(1, \ldots, C-1)$-subcomposition is a set of $C-1$ particular log contrasts and so the variability explained by a C-part subcomposition can also be compared with that achieved by the first $C-1$ principal log contrasts.

Another interesting form of subcompositional analysis is where the composition plays the role of regressor, for example in categorical regression, where we wish to examine the extent to which, for example, type of rock depends on full major oxide composition or some subcomposition. For binary regression a sensible approach is to set the conditional model of type t, say 0 and 1, for given composition x as follows:

$$(12.1)\quad pr(t=1|x) = 1 - pr(t=0|x) = F\left(a_0 + \sum_{i=1}^{D} a_i \log x_i\right), \text{ where } \sum_{i=1}^{D} a_i = 0.$$

Hypotheses that the categorization depends only on a subcomposition, for example on the subcomposition formed from parts $1,\dots,C$ is then simply specified by $a_{C+1} = \cdots = a_D = 0$, and so the whole lattice of subcompositional hypotheses can be readily and systematically investigated.

A striking example of the use of this technique is to be found in discriminating between two types of limestone. Thomas and Aitchison [**T-A**] show that of the 17-part (major-oxide, trace element) composition a simple major-oxide subcomposition (CaO, Fe_2O_3, MgO) provides excellent discrimination, equal to that of the full composition.

13. Convex linear mixing problems

By a convex linear mixing problem we mean one in which some target D-composition y is visualized as arising from some convex linear combination $\pi = (\pi_1,\dots,\pi_c)$ of C *source* or end member compositions $x_1,\dots,x_c$, as

$$(13.1)\qquad y = \pi_1 x_1 + \cdots + \pi_c x_c.$$

One such problem is where information is available only on the target in the form of a compositional data set and the objective is to attempt to find *fixed* end member source compositions from which the target compositions could have arisen through varying mixtures of these source compositions. For successful approaches to, and algorithms for, this difficult problem, see [**R**] and [**We**].

A completely different problem arises if we have some information about the sources. If there are C sources with precise compositions $x_1,\dots,x_C$ the problem commonly posed is to what extent a target composition y can be expressed as a convex linear combination (13.1) of the source compositions. The problem is often made more complicated by the following features which increase substantially its statistical nature.

(a) The source compositions are not precise but are defined only as an observed cluster of compositions.
(b) The target composition is not precise but is defined only as a cluster of compositions.

If we denote by y and $x_1,\dots,x_C$ generic target and source compositions, then the question of interest is the extent to which the data may support a relationship of the form (13.1). A reasonable technique for resolving this problem [**A-B**] is to use the ith source samples to assess in terms of $\mathcal{L}^D(\xi_i, T_i)$ the compositional variability of the ith source. Although the distributional problem of determining the consequent variability of y in (13.1) cannot be resolved exactly, there are excellent approximations in terms of logistic normal and skew logistic normal distributions which allow the evaluation of the likelihood function for π and hence the testing of various hypotheses concerning π. Aitchison and Bacon-Shone [**A-B**] also consider other models of convex linear mixing, for example where π may be variable and where there may extra variability characterized by a perturbation of (13.1). Such models are of interest in investigating the sources of pollution; see [**A-B**] for one such application.

14. Perturbation analysis

14.1. Perturbation analysis of simple change. A common application in compositional data analysis is to investigate the way in which a specimen in an initial state with composition x is altered by some process to a specimen in a final state with composition X. As we have seen such a change may be characterized by an unknown perturbation π, so that $X = x \circ \pi$. There are two distinct forms of data sets which may allow the investigation of such perturbations.

Paired compositions. When each x_r is paired with an associated X_r ($r = 1, \ldots, N$), for example when each x_r and X_r, refer to the same physical unit, before and after treatment. Then we have essentially a set of observed perturbations

$$P_r = X_r \circ x_r^{-1} \quad (r = 1, \ldots, N),$$

and so we can analyze these as a compositional data set, for example, by fitting a $\mathcal{L}^D(\xi, T)$ distribution.

Separate data sets. Here we have a set of specimens with compositions $\{x_r : r = 1, \ldots, N_I\}$ in a state regarded as initial, and also a completely separate set of specimens with compositions $\{X_r : r = 1, \ldots, N_F\}$ in a state regarded as final. Modeling here may suppose that the x_r are identically and independently distributed as $\mathcal{L}^D(\xi, T)$. Then, a generic X may be supposed to be an unknown perturbation π of a generic x, say $X = x \circ \pi$, so that the X_r are identically and independently distributed as $\mathcal{L}^D(\xi \circ \pi, T)$. With this modeling we can construct the likelihood of (ξ, T, π) for a given data set, and so carry out any inference on π.

14.2. Differential perturbation processes. Consider a process which results in an observable D-part composition $x(t) = (x_1(t), \ldots, x_D(t))$ which varies with some ordered variable such as time t, although other variables such as temperature or depth may be equally appropriate. Since processes are commonly assumed to take place continuously over time we can attempt to describe such a process in a differential way by relating the composition $x(t + dt)$ at time $t + dt$ to the composition $x(t)$ at previous time t in terms of a small perturbation, say $dp(t)$. Since such an infinitesimal perturbation will be a slight departure from the identity perturbation $(1/D, \ldots, 1/D)$, it is most simply expressed in the form

$$dp(t) = (1/D)(1 + \delta_1(t)dt, \ldots, 1 + \delta_D(t)dt), \tag{14.1}$$

where, for the sake of generality, we have allowed the perturbation to depend on t. Sometimes it is convenient to assume that such a perturbation is in the D-simplex but since the perturbation operation is invariant with respect to scale there is strictly no need for such a requirement. As long as the perturbation densities $\delta_i(t)$ ($i = 1, \ldots, D$) are small enough to ensure that the elements of $dp(t)$ are positive, and since we are dealing with differential perturbations this will always be the case.

Thus we define a differential perturbation process for the composition $x(t)$ as

$$x(t + dt) = x(t) \circ (1/D)(1 + \delta_1(t)dt, \ldots, 1 + \delta_D(t)dt), \tag{14.2}$$

from which we obtain, in terms of log ratios,

$$\log\left(\frac{x_i(t + dt)}{x_j(t + dt)}\right) = \log\left(\frac{x_i(t)}{x_j(t)}\right) + \log\left(\frac{1 + \delta_i(t)}{1 + \delta_j(t)}\right). \tag{14.3}$$

In the limit, as $dt \to 0$, the differential perturbation process (14.2) is equivalent to the set of differential equations

$$\frac{d}{dt}\log\frac{x_i(t)}{x_j(t)} = \delta_i(t) - \delta_j(t) \quad (i < j). \tag{14.4}$$

There is redundancy in this set of $\frac{1}{2}D(D-1)$ differential equations; the reduced set

$$\frac{d}{dt}\log\frac{x_i(t)}{x_D(t)} = \delta_i(t) - \delta_D(t), \quad i = 1,\dots,d, \tag{14.5}$$

is sufficient to describe the process. Given the initial composition $x(0)$ at time 0 we can readily obtain the composition $x(t)$ at time t for the process by solving the differential equations

$$x(t) = x(0) \circ P(t), \tag{14.6}$$

where the ith component $P_i(t)$ of the overall perturbation $P(t)$ is

$$P_i(t) = \exp\left\{\int_0^t \delta_i(u)du\right\}. \tag{14.7}$$

We note that (14.6) and (14.7) can be expressed in terms of log ratios:

$$\log\{x_i(t)/x_D(t)\} = \log\{x_i(0)/x_D(0)\} + \delta_i(t) - \delta_D(t), \quad i = 1,\dots,d, \tag{14.8}$$

and so we may anticipate that log ratio analysis of compositional data sets will have a central role to play in the investigation of differential perturbation processes. For purposes of exposition we have restricted our use of differential perturbation processes to the case of a single continuous process variable t. No such restrictions are necessary. There may be a number of subprocesses, each with its own driving continuous variable, in which case the differential equations defining the process will involve partial derivatives. For compositional data sets with no continuous variable defined, a log contrast principal component analysis will determine the dimensionality of any such underlying continuous vector. Investigation of the constant log contrast principal components is then the first step in any attempt to identify the underlying processes. For some simple applications to sedimentary and olivine compositions, see [**A-T**].

15. Discussion

The history of simplicial inference abounds with all sorts of devices purporting to overcome the difficulties associated with the special nature of the simplex sample space. These range from imagining that there is no problem, ignoring the constraint and inappropriately applying standard unconstrained R^D multivariate analysis; 'opening out' the compositions into imaginary vectors in R^D_+; recognizing that $\text{corr}(x_i, x_j) = 0$ does not mean that components x_i and x_j are independent but attempting to discover a non-zero 'null correlation' which characterizes independence in imagined open vectors; studying compositional pathology in the sense of identifying what goes wrong in applying standard multivariate analysis to simplicial data; transforming from the simplex to the positive orthant of the sphere and imagining that distributions over the whole sphere are appropriate for further analysis; attempting to rid the Dirichlet class of its inherent independence properties. See [**A12**] for further discussion of these attempted solutions.

We hope that the above account of how the nature of compositional and probability statement problems leads to requirements of scale, perturbation and permutation invariance and subcompositional coherence which dictate the form of a meaningful and appropriate methodology, will encourage the statistician, faced with simplicial data, to apply these statistical techniques. Whether the statistician opts to perform log ratio analysis and interpret the results in the transformed space R^d or to transform back into the simplex S^d, or indeed never to leave the simplex S^d, is clearly a matter of individual choice, possibly depending on the mathematical skills of the statistician's client.

References

[A1] Aitchison, J. *A new approach to null correlations of proportions,* Math. Geology **13** (1981), 175–189.

[A2] ——— *The statistical analysis of compositional data (with discussion),* J. R. Statist. Soc. **B44** (1982), 139–177.

[A3] ——— *Principal component analysis of compositional data,* Biometrika **70** (1983), 57–65.

[A4] ——— *A general class of distributions on the simplex,* J. R. Statist. Soc. **B47** (1985), 136–146.

[A5] ——— *The Statistical Analysis of Compositional Data,* Chapman and Hall, London, 1986.

[A6] ——— *Letter to the Editor. Measures of location of compositional data sets,* Math. Geology **21** (1989), 787–790.

[A7] ——— *Comment on "Measures of variability for geological data," by D. F. Watson and G. M. Philip,* Math. Geology. **22** (1990), 223–226.

[A8] ——— *Letter to the Editor. Delusions of uniqueness and ineluctability,* J. Math Geology **23** (1990), 275–277.

[A9] ——— *Relative variation diagrams for describing patterns of variability of compositional data,* Math. Geology **22** (1990), 487–512.

[A10] ——— *A plea for precision in Mathematical Geology,* Math. Geology **23** (1991), 1081–1084.

[A11] ——— *On criteria for measures of compositional differences,* Math. Geology **24** (1992), 365–380.

[A12] ——— *The one-hour course in compositional data analysis or compositional data analysis is easy.* In: Vera Pawlowsky Glahn, Ed., Proceedings of the Third Annual Conference of the International Association for Mathematical Geology, CIMNE, Barcelona, 1997, 3–35.

[A13] ——— *Logratios and natural laws in compositional data analysis,* Math. Geology **31** (1999), 563–580.

[A14] ——— *Biplots for compositional data,* available electronically from author, 1999.

[A-B] Aitchison, J. and J. Bacon-Shone, *Convex linear combinations of compositions,* Biometrika **86** (1999), 351–364.

[A-L] Aitchison, J. and I. J. Lauder, *Kernel density estimation for compositional data,* Applied Statistics **34** (1985), 129–137.

[A-S] Aitchison, J. and S. M. Shen, *Logistic-normal distributions: some properties and uses,* Biometrika **67** (1980), 261–272.

[A-T] Aitchison, J. and C. W. Thomas, *Differential perturbation processes: a tool for the study of compositional processes.* In: A. Buccianti, G. Nardi and R. Potenza, Eds., Proceedings of IAMG'98, The Fourth Annual Conference of the International Association for Mathematical Geology, De Frede, Naples, 1998, 499–504.

[Az-D] Azzalini, A. and A. Dalla Valle, *The multivariate skew-normal distribution,* Biometrika **83** (1996), 715–26.

[C] Chang, T. C., *Spherical regression,* Ann. Statist. **14** (1988), 907–924.

[G1] Gabriel, K. R., *The biplot-graphic display of matrices with application to principal component analysis,* Biometrika **58** (1971), 453–467.

[G2] ——— *Biplot display of multivariate matrices for inspection of data and diagnosis*, In: V. Barnett, Ed., Interpreting Multivariate Data, Wiley, New York, 1981, 147–173.

[Mc] McAlister, D., *The law of the geometric mean*, Proc. Roy. Soc. **29** (1879), 367–375.

[Mar-B-P] Martin-Fernández, J. A., C. Barceló-Vidal, and V. Pawlowsky-Glahn, *Measures of difference for compositional data and hierarchical clustering methods*, In: A. Buccianti, G. Nardi and R. Potenza, Eds., Proceedings of IAMG'98, The Fourth Annual Conference of the International Association for Mathematical Geology, De Frede, Naples, 1998, 526–531.

[Mat-B-P] Mateu-Figueras, G., C. Barceló-Vidal and V. Pawlowsky-Glahn, *Modeling compositional data with multivariate skew-normal distributions*, In: A. Buccianti, G. Nardi and R. Potenza, Eds., Proceedings of IAMG98, The Fourth Annual Conference of the International Association for Mathematical Geology, De Frede, Naples, 1998, 532–537.

[R] Renner, R. M., *The resolution of a compositiional data set into mixtures of fixed source components*, Appl. Statist. **42** (1993), 615–631.

[T-A] Thomas, C. W. and J. Aitchison, *The use of logratios in subcompositional analysis and geochemical discrimination of metamorphosed limestones from the northeast and central Scottish Highlands*, In: A. Buccianti, G. Nardi and R. Potenza, Eds., Proceedings of IAMG98, The Fourth Annual Conference of the International Association for Mathematical Geology, De Frede, Naples, 1998, 549–554.

[W1] Watson, D. F., *Reply to Comment on "Measures of variability for geological data" by D. F. Watson and G. M. Philip*, Math. Geology **22** (1990), 227–231.

[W2] Watson, D. F., *Reply to "Delusions of uniqueness and ineluctability" by J. Aitchison*, Math. Geology **23** (1991), 279.

[W-P] Watson, D. F. and G. M. Philip, *Measures of variability for geological data*, Math. Geology **21** (1989), 233–254.

[We] Weltje, G. J., *End-member modeling of compositional data: numerical-statistical algorithms for solving the explicit mixing problem*, Math. Geology **29** (1997), 503–549.

[Wo1] Woronow, A., *The elusive benefits of logratios*, In: V. Pawlowsky-Glahn, Ed., Proceedings of IAMG97, The Third Annual Conference of the International Association for Mathematical Geology, CIMNE, Barcelona, 1997, 97–101.

[Wo2] ——— *Regression and discrimination analysis using raw compositional data - is it really a problem?* In: V. Pawlowsky-Glahn, Ed., Proceedings of IAMG97, The Third Annual Conference of the International Association for Mathematical Geology, CIMNE, Barcelona, 1997, 157–162.

Department of Statistics, University of Glasgow, Glasgow G12 8QQ, U.K.

Current address: Department of Statistics, University of Glasgow, Glasgow G12 8QQ, U.K.

E-mail address: John.Aitchison@btinternet.com

Contemporary Mathematics
Volume **287**, 2001

A Note on Nyman's Equivalent Formulation of the Riemann Hypothesis

Jean-François Burnol

ABSTRACT. A certain subspace of $L^2((0,1),dt)$ has been considered by Nyman, Beurling, and others, with the result that the constant function **1** belongs to it if and only if the Riemann Hypothesis holds. I show in this note that the product $\prod_{\zeta(\rho)=0,\mathrm{Re}(\rho)>\frac{1}{2}}\left|\frac{1-\rho}{\rho}\right|$ is the norm of the projection of **1** to this subspace. This provides a quantitative refinement to Nyman's theorem.

1. Nyman's criterion for the Riemann Hypothesis

We first recall a theorem of Nyman [**15**]. We write $\{u\}$ for the fractional part of the real number u. To each α in $]0,1[$ we associate the function ρ_α of the positive real number t defined by $\rho_\alpha(t) = \left\{\frac{\alpha}{t}\right\} - \alpha\left\{\frac{1}{t}\right\}$. Of course ρ_α vanishes identically on $]1,+\infty[$, and is bounded on $]0,1]$. We let **NB** be the closed subspace of $\mathbf{H} = L^2((0,1),dt)$ spanned by the functions ρ_α for $0<\alpha<1$. We consider both **NB** and **H** as closed subspaces of $L^2((0,\infty),dt)$. We let **1** be the step function with value 1 for $0<t\le 1$, and 0 for $t>1$. We note that both **H** and **NB** are invariant under the semi-group of unitary contractions $U(\lambda): f(t)\mapsto \frac{1}{\sqrt{\lambda}}f(\frac{t}{\lambda})$, $0<\lambda\le 1$. Indeed $U(\lambda)\cdot\rho_\alpha = \frac{1}{\sqrt{\lambda}}(\rho_{\alpha\lambda}-\alpha\rho_\lambda)$. So **NB** contains the constant function **1** if and only if it contains all step functions if and only if it actually coincides with all of **H**.

THEOREM 1.1 (Nyman, 1950 [**15**]). *The Riemann Hypothesis holds if and only if the function* **1** *belongs to* **NB** *(hence if and only if* **NB** = **H**.*)*

The paper by Balazard and Saias ([**3**]) is an exposition of the use of the theory of Hardy spaces and especially of fundamental results on invariant subspaces by Beurling ([**8**]) and Lax ([**14**]) in this context. Nyman's criterion has stimulated various developments in recent years, theoretical as well as numerical (see the above cited [**3**] as well as [**1**], [**2**], and references therein; see also the author's [**9**] and [**10**]). We will work exclusively in an Hilbert space setting but should not fail to mention the extension by Beurling ([**7**]) of Nyman's work to an L^p-setting, leading to a characterization of the absence of zeros with real part (strictly) greater than

1991 *Mathematics Subject Classification.* Primary 11M26; Secondary 30D50.

$\frac{1}{p}$. Bercovici and Foias ([**5**], [**6**]) have shown how the geometry of the contraction semi-group $U(\lambda)$, $\lambda \leq 1$ can also be the basis of a proof in the L^p-case. We prove:

THEOREM 1.2. *Let* **P** *be orthogonal projection onto* **NB**. *We have (counting the zeros with their multiplicities):*

$$\|\mathbf{P}(\mathbf{1})\| = \prod_{\zeta(\rho)=0,\ \mathrm{Re}(\rho)>\frac{1}{2}} \left|\frac{1-\rho}{\rho}\right|.$$

Clearly this implies Nyman's criterion. The proof requires some classical results of harmonic analysis (Hardy spaces, the factorization theorem, the Beurling–Lax description of invariant subspaces) for which accessible reviews are given in the books by Dym–McKean ([**11**]) and Hoffman ([**13**]). Specifically we will put to use an old general argument of Beurling ([**8**, Theorem II]) (originating in the study of the Hardy space of the unit disc). It is natural to discuss **1.2** together with a proof (whose idea is given in [**9**]) of the elegant following result:

THEOREM 1.3 (Balazard-Saias-Yor, [**4**]). *One has:*

$$\frac{1}{2\pi}\int_{\mathrm{Re}(w)=\frac{1}{2}} \frac{\log|\zeta(w)|}{|w|^2}\,|dw| = \sum_{\zeta(\rho)=0,\ \mathrm{Re}(\rho)>\frac{1}{2}} \log\left|\frac{\rho}{1-\rho}\right|.$$

2. Proofs of 1.1, 1.2 and 1.3

As in the papers cited above on this subject the main tool is the Fourier Transform, here in its multiplicative version:

$$f(t) \in L^2((0,\infty), dt) \mapsto \widehat{f}(s) = \int_0^\infty f(t)\, t^{s-1}\, dt.$$

The integral is to be understood in the L^2-sense, with $\mathrm{Re}(s) = \frac{1}{2}$. Indeed we really want to look at $\sqrt{t}f(t)$ in $L^2((0,\infty), \frac{dt}{t})$ and at its transform in the dual group $\int_0^\infty \sqrt{t}f(t)\, t^{i\tau}\, \frac{dt}{t}$, $\tau \in \mathbb{R}$. With $s = \frac{1}{2} + i\tau$ we end up with the formula above. If $f(t)$ actually belongs to $L^2((0,1), dt)$, then $\widehat{f}(s) = \int_0^\infty f(t)\, t^{s-1}\, dt$ makes sense as an analytic function in the half-plane $\mathrm{Re}(s) > \frac{1}{2}$ and the set of such $\widehat{f}$ is characterized by the Paley–Wiener Theorem ([**11**, Chapter 3]) as the Hardy space $\mathbb{H}^2$ of analytic functions $h(s)$ whose L^2-norms on vertical lines $\frac{1}{2\pi}\int_{\mathrm{Re}(s)=\sigma>\frac{1}{2}} |h(s)|^2\,|ds|$ are bounded independently of σ. Such an analytic function $h(s)$ has (pointwise almost everywhere) non-tangential limits $h(\frac{1}{2} + i\tau)$ which are also obtained as the Fourier–Mellin Transform of $f \in L^2((0,1), dt)$. Furthermore, there is a Cauchy formula:

$$\mathrm{Re}(s) > \frac{1}{2} \quad \Rightarrow \quad h(s) = \frac{1}{2\pi}\int_{-\infty}^{+\infty} \frac{h(\frac{1}{2} + i\tau)}{s - \frac{1}{2} - i\tau}\, d\tau. \tag{2.1}$$

We note in passing that $\widehat{\mathbf{1}}(s) = \frac{1}{s}$. The zeta-function appears in this story thanks to

$$\int_0^1 \left\{\frac{1}{t}\right\} t^{s-1}\, dt = \frac{1}{s-1} - \frac{\zeta(s)}{s}.$$

Multiplication by $\frac{s-1}{s}$ leaves invariant $\mathbb{H}^2$, so $\frac{(s-1)\zeta(s)}{s^2}$ belongs to $\mathbb{H}^2$ ([**9**], [**12**]). As reviewed in [**11**], [**13**] a general fact following from this is the absolute convergence for any s in the half-plane $\mathrm{Re}(s) > \frac{1}{2}$ of the Blaschke product

$$B(s) = \prod_{\zeta(\rho)=0,\ \mathrm{Re}(\rho)>\frac{1}{2}} \frac{s-\rho}{s-(1-\bar{\rho})} \frac{1-\bar{\rho}}{\rho} \left| \frac{\rho}{1-\rho} \right| .$$

It is known that such a product built from the zeros of an element of $\mathbb{H}^2$ is an *inner function*, that is an analytic function bounded by 1 in the half-plane whose non-tangential limits on the critical line have modulus 1 (almost everywhere). The expression as an infinite product might cease to make sense (pointwise) for the boundary values, but in the case at hand it is absolutely convergent for all s in $\mathbb{C}$, except at the possible poles $1-\bar{\rho}$. A general fact is Norbert Wiener's theorem on the gain of a causal filter: it implies that the integral involved in **1.3** is absolutely convergent. Furthermore there is a factorization formula ([**13**, chap. 8]):

$$\frac{(s-1)\zeta(s)}{s^2} = c \cdot A^{s-\frac{1}{2}} \cdot \exp\left(\frac{1}{2\pi} \int_{\mathrm{Re}(w)=\frac{1}{2}} \log \left| \frac{\zeta(w)}{w} \right| \frac{s+w-2sw}{s-w} \frac{|dw|}{|w|^2} \right) \cdot B(s),$$

where $|c| = 1$, $0 < A \leq 1$. There could have been a further term associated to a singular measure on the critical line, but the analytic continuation of $\zeta(s)$ implies its non-existence. A value of A strictly less than 1 would imply a much too quick decrease of $\zeta(s)$ when s is real and goes to $+\infty$, so $A = 1$ (see [**5**], also [**9**]). Looking at real values of s we see that c is real and positive so $c = 1$. We thus have:

$$\frac{(s-1)\zeta(s)}{s^2} = \exp\left(\frac{1}{2\pi} \int_{\mathrm{Re}(w)=\frac{1}{2}} \log \left| \frac{\zeta(w)}{w} \right| \frac{s+w-2sw}{s-w} \frac{|dw|}{|w|^2} \right) \cdot B(s).$$

We note that the same applies to $\frac{1}{s}$ and gives

$$\frac{1}{s} = \exp\left(-\frac{1}{2\pi} \int_{\mathrm{Re}(w)=\frac{1}{2}} \log|w| \frac{s+w-2sw}{s-w} \frac{|dw|}{|w|^2} \right),$$

and we finally conclude (for $\mathrm{Re}(s) > \frac{1}{2}$):

$$\frac{s-1}{s}\zeta(s) = \exp\left(\frac{1}{2\pi} \int_{\mathrm{Re}(w)=\frac{1}{2}} \log|\zeta(w)| \frac{s+w-2sw}{s-w} \frac{|dw|}{|w|^2} \right) \cdot B(s),$$

from which the Balazard-Saias-Yor formula **1.3** follows by specializing to $s = 1$. One has furthermore

$$\widehat{\rho_\alpha}(s) = \frac{\alpha - \alpha^s}{s} \zeta(s)$$

and the Beurling-Lax (see [**13**, chap. 7]) general description of invariant subspaces of the Hardy spaces then leads to the conclusion that the Mellin transform of **NB** exactly coincides with $B(s)\mathbb{H}^2$. Indeed on one hand the "common denominator" of the Blaschke products associated to the functions $\widehat{\rho_\alpha}(s)$ is $B(s)$ and on the other hand none of these functions has a singular inner factor, nor a special inner factor $A^{s-\frac{1}{2}}$.

LEMMA 2.1. *The orthogonal projection of $\frac{1}{s}$ on $B(s)\mathbb{H}^2$ is $B(1)\frac{B(s)}{s}$.*

PROOF. The argument is essentially taken from Beurling: see [**8**, Theorem II]. We check that $\frac{1}{s} - B(1)\frac{B(s)}{s}$ is perpendicular to $B(s)h(s)$, for any $h(s) \in \mathbb{H}^2$:

$$\begin{aligned} & \frac{1}{2\pi}\int_{\mathrm{Re}(w)=\frac{1}{2}} \overline{\left(\frac{1}{w} - B(1)\frac{B(w)}{w}\right)} B(w)h(w)\,|dw| \\ = & \frac{1}{2\pi}\int_{\mathrm{Re}(w)=\frac{1}{2}} \frac{(B(w)-B(1))\,h(w)}{\overline{w}}\,|dw| \\ = & \frac{1}{2\pi}\int_{\mathrm{Re}(w)=\frac{1}{2}} \frac{(B(w)-B(1))\,h(w)}{1-w}\,|dw| \\ = & (B(1)-B(1))\,h(1) = 0, \end{aligned}$$

where $|B(w)| = 1$ on $\mathrm{Re}(w) = \frac{1}{2}$ and then (**2.1**) were used. □

Hence the norm of the projection of **1** on **NB** is $B(1)$ as stated in **1.2** (we note that $\frac{B(s)}{s}$ is of norm 1 as $B(s)$ is of modulus 1 on the critical line). We can also interpret Theorem **1.2** as a statement involving a conditional expectation for a certain Gaussian system (work in progress).

References

[1] L. Báez-Duarte, *On Beurling's real variable reformulation of the Riemann hypothesis*, Adv. in Maths. **101** (1993), 10–30.

[2] L. Báez-Duarte, M. Balazard, B. Landreau and E. Saias, *Notes sur la fonction ζ de Riemann, 3*, Adv. in Maths. **149** (2000), 130–144.

[3] M. Balazard, E. Saias, *The Nyman–Beurling equivalent form for the Riemann hypothesis*, Expo. Math. **18** (2000), 131–138.

[4] M. Balazard, E. Saias, M. Yor, *Notes sur la fonction ζ de Riemann, 2*, Adv. in Math. **143** (1999), 284–287.

[5] H. Bercovici, C. Foias, *A real variable restatement of Riemann's Hypothesis*, Israel J. of Math. **48** (1984), 57–68.

[6] H. Bercovici, C. Foias, *On the Zorn spaces in Beurling's approach to the Riemann hypothesis*, Analysis and topology, 143–149 (1998), World Sci. Publishing, River Edge, NJ.

[7] A. Beurling, *A closure problem related to the Riemann Zeta-function*, Proc. Nat. Acad. Sci. **41** (1955), 312–314.

[8] A. Beurling, *On two problems concerning linear transformations in Hilbert space*, Acta Mathematica **81** (1949), 239–255.

[9] J.-F. Burnol, *An adelic causality problem related to abelian L-functions*, J. Number Theory **87** (2001), no 2, 253–269.

[10] J.-F. Burnol, *A lower bound in an approximation problem involving the zeros of the Riemann zeta function*, manuscript, March 2001, 18pp., `http://arXiv.org/abs/math/0103058`.

[11] H. Dym, H. P. McKean, *Fourier Series and Integrals*, Academic Press (1972).

[12] W. Ehm, *A family of probability densities related to the Riemann zeta function*, contribution to this volume.

[13] K. Hoffman, *Banach spaces of analytic functions*, Prentice-Hall Inc. 1962, Dover Publication 1988.

[14] P. Lax, *Translation invariant subspaces*, Acta Mathematica **101**, (1959), 163–178.

[15] B. Nyman, *On the One-Dimensional Translation Group and Semi-Group in Certain Function Spaces*, Thesis, University of Uppsala, 1950, 55pp.

UNIVERSITÉ DE NICE SOPHIA-ANTIPOLIS, LABORATOIRE J.-A. DIEUDONNÉ, PARC VALROSE, F-06108 NICE CEDEX 02, FRANCE

E-mail address: `burnol@math.unice.fr`

Contemporary Mathematics
Volume **287**, 2001

On the Construction of Linear Orthogonal Arrays by Extension

Dominique Collombier and Astrid Jourdan

ABSTRACT. Hedayat, Stufken and Su (1996b) have introduced a method of constructing an orthogonal array of strength t from on orthogonal array of strength t-1, based on difference schemes. We extend this result and construct an orthogonal array of strength $t+m$, $m \geq 0$, from an orthogonal array of strength $t-1$. We study, especially, the case of linear orthogonal arrays and their parity check matrices.

1. Introduction

Orthogonal arrays are a very useful branch of combinatorics. They find applications in many fields like statistical designs or coding theory. They were first introduced by Rao (1947). The construction of orthogonal arrays has been studied by various authors, especially for arrays of strength 2. The usual method of construction was introduced by Bose and Bush (1952), and is based on difference schemes of strength 2. Recently Hedayat *et al.*(1992, 1996b) presented a generalization of this method to any strength $t \geq 1$. They showed how to construct an orthogonal array of strength t from an orthogonal array of strength $t-1$, which is also a difference scheme of strength t. This method of extending orthogonal arrays was also studied by Hill (1999), using techniques of linear codes.

In this paper we construct an orthogonal array of strength $t+m$, with $m \geq 0$, from an orthogonal array of strength $t-1$, extending the results of Hedayat *et al.*(1996b). We will pay particular attention to linear orthogonal arrays because they form a useful class of regular factorial designs. In particular, we give the parity check matrix corresponding to the constructed linear orthogonal array. This matrix determines the annihilator (contrasts of definition) of the regular factorial design, and allows the detection of confounded effects.

Section 2 will be devoted to basic definitions and properties. Section 3 will briefly present the method of construction introduced by Hedayat *et al.*(1996b). In Section 4 we will generalize the results presented in Section 3 to linear orthogonal arrays of strength t+m.

1991 *Mathematics Subject Classification.* Primary 05B15, 62K15 Secondary 94B05.

Key words and phrases. Linear orthogonal arrays, difference schemes, parity check matrices, regular factorial designs.

2. Linear orthogonal arrays and difference schemes

DEFINITION 2.1. *A $N \times d$ matrix T, with entries from a set of integers $\{0, \ldots, p-1\}$, is called an orthogonal array of strength t, $0 \leq t \leq d$, denoted $OA(N, d, p, t)$, if every $N \times t$ submatrix of T contains all the t-tuples based on $\{0, \ldots, p-1\}$ λ times each.*

Moreover, if $\{0, \ldots, p-1\}$ represents the elements of the additive group of the Galois field $GF(p)$, with p prime, and if the rows of T are all distinct and form a vector subspace of $GF(p)^d$, then T is called a linear orthogonal array.

The weight of a vector $v \in GF(p)^d$ is the number on non-zero coordinates of v, and is denoted $wt(v)$. For every $r \times d$ matrix P over $GF(p)$, the weight of P is

$$wt(P) = min\ (wt(\lambda P), \forall \lambda \in GF(p)^r : \lambda \neq 0).$$

REMARK 2.1. (a) N is called the size of the array, d the number of factors, p the number of symbols and λ the index of the array; (b) Necessarily, $\lambda = N/p^t \in \mathbb{N}^* = \mathbb{N}\backslash\{0\}$, and if the array is linear, $\lambda = p^k$ with $k \in N$.

Let $\mathcal{E}_T$ denote the subspace of $GF(p)^d$ formed by the rows of T, which is a $N \times d$ linear orthogonal array with $N = p^n$, $n \in \mathbb{N}^*$. As $\mathcal{E}_T$ is a vector subspace of $GF(p)^d$, the array T is entirely defined:

- either by n independent vectors. The $n \times d$ matrix formed by this vectors is called the generator matrix of the array;
- or *via* $(\mathcal{E}_T)^\perp$, the orthogonal subspace of $\mathcal{E}_T$, by P, the $r \times d$ generator matrix of $(\mathcal{E}_T)^\perp$. In this case $rgP = r = d - n$ and the rows of T are the solutions of $Px = 0$.

In coding theory language, P is called the *parity check matrix.* Such a terminology is adopted in this paper. This matrix will receive our attention because of the two following properties.

First, it is well known that the strength of the array T depends on the weight of P. Actually $wt(P) = t + 1$ (see Bailey (1990) and Collombier (1996)).

Moreover P is crucial to determine confounded effects, especially with Fourier (trigonometric) regression models (see Kobilinsky (1990) and Jourdan (2000)).

Thus, the aim of this paper is not only to construct linear orthogonal arrays but also to provide their parity check matrices. The method used to answer both objectives is based on difference schemes.

DEFINITION 2.2. *A $N \times d$ matrix D based on $GF(p)$ is called a difference scheme of strength t, $0 \leq t \leq d$, denoted $D(N, d, p, t)$, if in every $N \times t$ submatrix of D the cosets of the subgroup $\{(g, ..., g), \forall g \in GF(p)\} \subseteq GF(p)^t$ are represented with multiplicity λ.*

REMARK 2.2. (a) This definition of difference schemes of strength t (due to Seiden (1954)) is a generalization of difference schemes of strength 2 introduced by Bose and Bush (1952); (b) In this paper the entries of D are from a Galois field $GF(p)$, because D is used to construct a linear orthogonal array. More generally, however, difference schemes can be defined on Abelian groups.

Before we describe Hedayat *et al.*(1996b) method, we review the basic properties of difference schemes.

LEMMA 2.1. *Let $D = (d_{ij})$ be a $N \times d$ difference scheme of strength t based on $GF(p)$.*

1. For every $h \in \{1, ..., d\}$ *the* $N \times d$ *matrix* $D^* = (d^*_{ij})$, *where* $d^*_{ij} = d_{ij} - d_{ih}$, *is also a difference scheme of strength* t. *Moreover the submatrix formed by the columns of* D, *except the column indexed by* h, *is an orthogonal array of strength* $t-1$.

2. For every $g \in GF(p)$, *the* $N \times d$ *matrix obtained by adding* g *to each entry of* D, *denoted* $g + D$, *is a difference scheme of strength* t *based on* $GF(p)$.

3. If $D_1, ..., D_s$ *are difference schemes of strength* t *based on* $GF(p)$, *then the matrix*

$$D = \begin{pmatrix} D_1 \\ \hline \cdots \\ \hline D_s \end{pmatrix}$$

is also a difference scheme of strength t.

REMARK 2.3. From Property 1 in Lemma 2.1 we may assume, without loss of generality, that a difference scheme has the form

$$D = (0 \,|\, T),$$

where T is both an orthogonal array of strength $t-1$ and a difference scheme of strength t.

3. Construction of linear orthogonal arrays of strength t.

The method of construction introduced by Hedayat *et al.*(1996b) is based on the following property. If D is a difference scheme $D(N, d, p, t)$, then the matrix

$$T = \begin{pmatrix} g + D \\ \hline \cdots \\ \forall g \in GF(p) \end{pmatrix}$$

is an $OA(pN, d, p, t)$. Here, and in the remaining of the paper, we use the above notation to indicate a matrix made up (in this case) of p submatrices $g + D$, for all g in $GF(p)$, positioned one under the other. Given this, the construction of an orthogonal array is divided into two steps: First, conditions to obtain the difference scheme used in the preceding property are specified. From Remark 2.3, the construction of this difference scheme $D(N, d, p, t)$ is equivalent to the construction of an array which is both a difference scheme of strength t and an orthogonal array of strength $t-1$; Finally, the number of factors of T is increased. If $t = 2$, we can always add one or more columns to the matrix T and keep an orthogonal array of strength 2 (see Bose and Bush (1952)). But if $t > 2$, adding columns is not simple. That is why the notion of amicable matrices is introduced.

DEFINITION 3.1. *Let* $D = (d_{ij})$ *be a difference scheme* $D(N, d_1, p, t)$ *and let* $B = (b_{ij})$ *be an* $OA(N, d_2, p, min\ (d_2, t))$. *For a positive integer* v *such that* $v \leq min\ (t-1, d_2)$, *let* $J_v \subseteq \{1, ..., d_2\}$, *with* $|J_v| = v$, $z \in G^v$ *and*

$$D = (d_{ik} |\ i \in I_{B[J_v],z} \text{ and } k = \{1, \ldots, d_1\})$$

with $I_{B[J_v],z} = (i \in \{1, \ldots, N\} : (b_{ij} |\ j \in J_v) = z)$. *We will say that matrices* D *and* B *are amicable if* $\forall v \in \mathbb{N}^* : v \leq min\ (t-1, d_2)$, $\forall J_v \subseteq \{1, ..., d_2\} : |J_v| = v$, *and* $\forall z \in G^v$, *there is a permutation matrix* S *such that, for* $M = SB$, $D_{I_{M[J_v],z}}$ *is a difference scheme of strength t-v.*

Hedayat *et al.*(1996b) established that if matrices B and D are amicable, the array

$$T = \left(\frac{g + D \mid M}{\dots \atop \forall g \in GF(p)} \right)$$

is an $OA(pN, d_1 + d_2, p, t)$. Combining these two results, Hedayat *et al.*(1996b) introduced a method of building an orthogonal array of strength t from an orthogonal array of strength $t-1$ (which is also a difference scheme of strength t), thus increasing the number of factors,

$$OA(N, \mathbf{d}, p, \mathbf{t-1}) \longrightarrow OA(N', \mathbf{d+s+1}, p, \mathbf{t}).$$

We notice that if the array of strength $t-1$ is linear, the constructed array is also linear and its parity check matrix is known. What we present in the next section is a generalization of the above method to the strength $t+m, m \geq 0$.

$$OA(N, \mathbf{d}, p, \mathbf{t-1}) \longrightarrow OA(N', \mathbf{d+s+1}, p, \mathbf{t+m}).$$

4. Construction of linear orthogonal arrays of strength t+m, m≥0

The main objective in this section is the construction of a linear orthogonal array of strength $t+m, m \geq 0$, from a linear orthogonal array of strength $t-1$. The different steps in the construction are similar to those used by Hedayat *et al.*(1996b). The first proposition is a generalization to the class of linear orthogonal array of a result due to Adhikary and Das (1992).

NOTATION 4.1. $\mathbf{1}_d$ is the $d \times 1$ vector with all elements equal to 1 and v^t is the transpose of the vector v.

PROPOSITION 4.1. *Let T be a linear orthogonal array of strength $t-1$ based on $GF(p)$, $OA(N, d, p, t-1)$, such that rank $(T) = d - r$. The array*

$$T' = \left(\frac{g + T}{\dots \atop \forall g \in GF(p)} \right)$$

is a linear orthogonal of strength $t+m \geq 0$ if the matrix $M = \left(\frac{T}{\mathbf{1}_d^t} \right)$ satisfies

$$\begin{cases} \text{1. rank } (M) = d - r + 1, \\ \text{2. every } (t+m) \text{ columns of } M \text{ are linearly independent.} \end{cases}$$

We need the next result to establish the proof of Proposition 4.1. This lemma is due to Hedayat *et al.*(1996a).

LEMMA 4.1. *Let T be a $p^{d-r} \times d$ linear orthogonal array. The strength of T is $t \leq d$ if and only if any t columns of T are linearly independent.*

PROOF OF PROPOSITION 4.1. Every row of T' belongs to $\mathcal{R}(T^t \mid \mathbf{1}_d)$ (the range space). As rank $(T) = d - r$, if rank $(M) = d - r$, then a row of T is equal to $\mathbf{1}_d^t$. Therefore T' is formed by p repetitions of T and its rows do not form a vector subspace of $GF(p)^d$. Hence, T' is a linear orthogonal array if rank $(M) = d-r+1$. Moreover, if every $(t+m)$ columns of M are linearly independent, then the corresponding $(t+m)$ columns of T are linearly independent. Therefore, the two conditions of Lemma 4.1 are satisfied, and it follows that T' is a linear orthogonal array of strength $t+m$. □

These conditions are only necessary. If we modify the second condition we obtain a sufficient and necessary condition. More specifically,

PROPOSITION 4.2. *Let T, T' and M be as in the Proposition 4.1. For every $J \subseteq \{1, ..., d\}$, let p_J denote the canonical projector from $GF(p)^d$ to $GF(p)^{|J|}$, i.e.*

$$p_J : x = (x_1, ..., x_d) \in GF(p)^d \to (x_j |\, j \in J).$$

Let $\mathcal{E}_T$ be the subspace of $GF(p)^d$ formed by the rows of T and $p_J(\mathcal{E}_T)$ its image by p_J. Suppose that rank $(M) = d - r + 1$. Then T' is a linear orthogonal array of strength $t+m$, $m \geq 0$, if and only if $\forall J \subseteq \{1, ..., d\} : |J| = t+m$,
either $\mathbf{1}_{t+m} \in p_J(\mathcal{E}_T)$ and the $t+m$ columns of T indexed by J are independent,
or $\mathbf{1}_{t+m} \notin p_J(\mathcal{E}_T)$ and the $t+m$ columns of M indexed by J are independent.

PROOF. As rank $(M) = d - r + 1$, $\mathcal{E}_{T'}$, the subspace formed by the rows of T', is a subspace of $GF(p)^d$. Moreover $J \subseteq \{1, ..., d\} : |J| = t+m$

$$p_J(\mathcal{E}_{T'}) = \begin{cases} p_J(\mathcal{E}_T) \; if \; \mathbf{1}_{t+m} \in p_J(\mathcal{E}_T) \\ p_J(\mathcal{E}_T) \oplus \langle \mathbf{1}_{t+m} \rangle \; if \; \mathbf{1}_{t+m} \notin p_J(\mathcal{E}_T) \end{cases}$$

where $\langle v \rangle$ is the subspace generated by v. Then in view of Lemma 4.1, T' is a linear orthogonal array of strength $t+m$ if only if

$$p_J(\mathcal{E}_{T'}) = GF(p)^{t+m}.$$

But if $\mathbf{1}_{t+m} \in p_J(\mathcal{E}_T)$, then $p_J(\mathcal{E}_{T'}) = p_J(\mathcal{E}_T) = GF(p)^{t+m}$. That is, the $(t+m)$ columns of T indexed by J are independent.

If $\mathbf{1}_{t+m} \notin p_J(\mathcal{E}_T)$, then $p_J(\mathcal{E}_{T'}) = p_J(\mathcal{E}_T) \oplus \langle \mathbf{1}_{t+m} \rangle = p_J(\mathcal{E}_M) = GF(p)^{t+m}$. That is, the $(t+m)$ columns of M indexed by J are independent. □

We can infer, from the properties of difference schemes, that T' is an orthogonal array of strength $t+m$, if and only if T is a difference scheme of strength $t+m$. Therefore, the array T' used in the preceding propositions, is both a difference scheme of strength $t+m$ and a linear orthogonal array of strength $t-1$. The following theorem specifies the necessary and sufficient conditions to obtain such an array.

THEOREM 4.1. *Let P be a $r \times d$ parity check matrix based on $GF(p)$, with rank $(P) = r$. We denote T the $p^{d-r} \times d$ array whose rows are the solutions of $Px = 0$. Let t and m be integers such that $0 \leq t \leq d$, $0 \leq m < d - t$. For every $J \subseteq \{1, ..., d\}$, we define the subspace of $GF(p)^d$,*

$$\mathcal{S}_J(P) = \{ y \in R(P^t) : supp\ (y) \subseteq J \},$$

where supp (y) is the support of y, that is the set of non-zero entries in y. Then T is both a linear orthogonal array of strength $t-1$ and a difference scheme of strength $t+m$ if and only if

1. *$wt(P) \geq t$*
2. *$\forall J \subseteq \{1, ..., d\} : |J| = t+m$, either $\mathcal{S}_J(P) = 0$,*
   or $dim(\mathcal{S}_J(P)) = 1$ and $\mathcal{S}_J(P)$ is not orthogonal to $\mathbf{1}_d$.

EXAMPLE 4.1. Let $d = 8, p = 2$ and $t = 4$. Hedayat et al. (1996b) constructed an array T which was both an $OA(2^2, 8, 2, 3)$ and $D(2^6, 8, 2, 4)$. This array T is defined by its 2×8 parity check matrix P based on $GF(2)$:

$$P = \left(\begin{array}{cc|cccccc} 1 & 0 & 1 & 1 & 1 & 1 & 0 & 0 \\ 0 & 1 & 1 & 1 & 0 & 0 & 1 & 1 \end{array} \right)$$

If we apply Theorem 4.1 with $m = 1$, we can easily verify that the array T is in fact a difference scheme of strength 5.

EXAMPLE 4.2. Let $d = 6$ and $p = 3$. Let P be the 2×6 matrix based on $GF(3)$

$$P = \left(\begin{array}{cc|cccc} 1 & 0 & 1 & 0 & 2 & 1 \\ 0 & 1 & 1 & 2 & 0 & 0 \end{array} \right).$$

The assumptions of Theorem 4.1 are satisfied with $t = 3$ and $m = 2$. Hence, the array T defined by its parity check matrix P is both an $OA(3^4, 6, 3, 2)$ and $D(3^4, 6, 3, 5)$.

PROOF OF THEOREM 4.1. Hedayat *et al.*(1996b) proved that condition 1 in the statement of Theorem 4.1 is necessary and sufficient to have a linear orthogonal array of strength $t - 1$. We still have to demonstrate that T is also a difference scheme of strength $t + m$.

Sufficient condition : Suppose that T is not a difference scheme of strength $t+m$, then T', as defined in Proposition 4.1, is not a linear orthogonal array of strength $t + m$. Then, there is $J \subseteq \{1, ..., d\} : |J| = t + m$ such that the $(t + m)$ columns of T' indexed by J are not independent. Let T'_j (resp. T_j) be the jth column of T' (resp. T). We have $\sum_{j \in J} \lambda_j T'_j = 0$ for $\lambda_1, \ldots, \lambda_{t+m} \in GF(p)$ with $\lambda_j \neq 0$ for some $j \in J$. Then, by construction of T',

$$\begin{cases} \sum_{j \in J} \lambda_j = 0, \\ \sum_{j \in J} \lambda_j T_j = 0. \end{cases}$$

Thus, the vector $y = (y_1, ..., y_d) \in GF(p)^d$ defined by

$$\begin{cases} y_j = \lambda_j \; j \in J, \\ y_j = 0 \quad j \notin J, \end{cases}$$

satisfies supp $(y) \subseteq J$ and $y \in \mathcal{R}(P^t)$, because $y \in \ker T$, and therefore $y \in \mathcal{S}_J(P)$. Moreover $y\mathbf{1}_d = 0$ with $y \neq 0$, hence y is orthogonal to $\mathbf{1}_d$. This clearly contradicts assumption 2 in the statement of the theorem, and establishes the sufficient condition.

Necessary condition : Suppose that condition 2 is not satisfied, then

1. either $dim(\mathcal{S}_J(P)) > 1$,
2. or $dim(\mathcal{S}_J(P)) = 1$ and $\mathcal{S}_J(P) \perp \mathbf{1}_d$ for $J \subseteq \{1, ..., d\} : |J| = t + m$.

We will first show that $\forall J \subseteq \{1, ..., d\} : |J| = t + m$,

$$p_J(\mathcal{E}_T) = [p_J(\mathcal{S}_J(P))]^\perp$$

• By definition we have $\mathcal{S}_J(P) \subseteq \mathcal{E}_T^\perp$, so $(x \mid y) = 0$, $\forall x \in \mathcal{S}_J(P)$ and $\forall y \in \mathcal{E}_T$. Moreover for every $y \in \mathcal{S}_J(P)$, supp $(y) \subseteq J$, hence $(x \mid y) = (p_J(x) \mid p_J(y))$. It follows that $p_J(\mathcal{E}_T)$ is orthogonal to $p_J(\mathcal{S}_J(P))$, therefore

$$p_J(\mathcal{E}_T) \subseteq [p_J(\mathcal{S}_J(P))]^\perp .$$

• Suppose now, without loss generality, that $J = \{1, ..., t+m\}$. Let $\overline{J} = \{t+m+1, \ldots, d\}$. Then $\mathcal{E}_T^\perp = \mathcal{S}_J(P) \oplus \mathcal{S}_{\overline{J}}(P)$. Consider

$$(x, y) \in \left([p_J(\mathcal{S}_J(P))]^\perp_{GF(p)^{t+m}}, [p_{\overline{J}}(\mathcal{S}_{\overline{J}}(P))]^\perp_{GF(p)^{d-t-m}} \right).$$

It then follows that $\forall (a, b) \in \mathcal{E}_T^\perp$, $(a, b) = (v, 0) + (0, s)$, where $(v, 0) \in \mathcal{S}_J(P)$ and $(0, s) \in \mathcal{S}_{\overline{J}}(P)$, and consequently

$$((x, y) \mid (a, b)) = ((x, y) \mid [(v, 0) + (0, s)]) = (x \mid v) + (y \mid s).$$

$(v, 0) \in \mathcal{S}_J(P) \Rightarrow v \in p_J(\mathcal{S}_J(P))$ then $(x \mid v) = 0$ since $x \in [p_J(\mathcal{S}_J(P))]^\perp_{GF(p)^{t+m}}$,
$(0, s) \in \mathcal{S}_{\overline{J}}(P) \Rightarrow s \in p_{\overline{J}}(\mathcal{S}_{\overline{J}}(P))$ then $(y \mid s) = 0$ since $y \in [p_{\overline{J}}(\mathcal{S}_{\overline{J}}(P))]^\perp_{GF(p)^{d-t+m}}$.

So, $\forall (a, b) \in \mathcal{E}_T^\perp$, $((x, y) \mid (a, b)) = 0$, then $(x, y) \in \left(\mathcal{E}_T^\perp\right)^\perp = \mathcal{E}_T$. Therefore we have

$$p_J\left([p_J(\mathcal{S}_J(P))]^\perp_{GF(p)^{t+m}}, [p_{\overline{J}}(\mathcal{S}_{\overline{J}}(P))]^\perp_{GF(p)^{d-t-m}} \right) = [p_J(\mathcal{S}_J(P))]^\perp \subseteq p_J(\mathcal{E}_T).$$

Consider now the two cases: (1) If $dim(\mathcal{S}_J(P)) > 1$ then $dim\,(p_J(\mathcal{S}_J(P))) > 1$. Thus it follows from the definition of $\mathcal{S}_J(P)$, and from $p_J(\mathcal{E}_T) = [p_J(\mathcal{S}_J(P))]^\perp$ that $dim(p_J(\mathcal{E}_T)) < t+m-1$; (2) If $dim(\mathcal{S}_J(P)) = 1$ and $\mathcal{S}_J(P) \perp \mathbf{1}_d$, then $dim(p_J(\mathcal{E}_T)) = t+m-1$ and $\mathbf{1}_d \in p_J(\mathcal{E}_T)$. In both cases,

$$p_J(\mathcal{E}_{T'}) = p_J(\mathcal{E}_T) + \langle \mathbf{1}_d \rangle,$$

which is a subspace strictly included in $GF(p)^{t+m}$. According to Lemma 4.1, T' is not an linear orthogonal array of strength $t+m$, because some columns of T' are linearly dependent. By construction of T', it follows that T is not a difference scheme of strength $t+m$. Therefore, the necessary condition is proved by contradiction. □

REMARK 4.1. If $wt(P) \geq t$, then $\forall y \in \mathcal{R}(P^t)$, $|\text{supp }(y)| \geq t$. Moreover if $m = 0$, then from the second condition of the theorem, it follows that for $y \in \mathcal{R}(P^t)$ such that supp $(y) \subseteq J$ with $|J| = t$,

$$\text{supp }(y) = J \Leftrightarrow wt(y) = t \text{ and } y\mathbf{1}_t \neq 0.$$

In this case our theorem is equivalent to Theorem 4.3 of Hedayat et al.(1996b).

Hedayat *et al.*(1996b) identify situations where columns can be added to an orthogonal array in order to increase the number of factors without loss the strength of the array. If we apply their results (§4) using difference schemes constructed *via* parity check matrices given in Theorem 4.1, we provide a method to build linear orthogonal array of strength $t+m$ and $d+s+1$ factors ($s \geq 0$), $OA(N, d+s+1, p, t+m)$, from a linear orthogonal array of strength $t-1$ and d factors, $OA(N, d, p, t-1)$.

THEOREM 4.2. *Let $P, T, \mathcal{S}_J(P)$ $J \subseteq \{1, ..., d\}$, t and m be as in Theorem 4.1. Suppose P satisfies conditions of Theorem 4.1, then $D = (0 \mid T)$ is a difference scheme of strength $t+m$ over $GF(p)^{d+1}$. Further let $v_1, ..., v_s$, be $1 \times d$ vectors based on $GF(p)$. For every $I \subseteq \{1, ..., s\}$, we define the $(r+|I|) \times d$ matrix*

$$Q_I = \left(\frac{P}{v_i,\ i \in I} \right).$$

If the following properties are satisfied, the $p^{d-r} \times s$ array B=TV with $V = (v_1^t, \ldots, v_s^t)$ is orthogonal of strength $min\,(t+m, s)$, amicable with D.

1. $\forall I \subseteq \{1, ..., s\} : 1 \leq |I| \leq min\,(t-1, s)$

(a) $wt(Q_I) \geq t - I$,
(b) $\forall J \subseteq \{1, ..., d\} : |J| = t + m - |I|$, *either* $\mathcal{S}_J(Q_I) = 0$, *or* $dim(\mathcal{S}_J(Q_I)) = 1$ *and* $\mathcal{S}_J(Q_I)$ *is not orthogonal to* $\mathbf{1}_d$.

2. $\forall I \subseteq \{1, ..., s\} : min\ (t - 1, s) < |I| \leq min\ (t + m - 1, s)$
(a) $wt(Q_I) \geq t + m - |I|$,
(b) $\forall y \in \mathcal{R}(Q_I^t)$ *such that* $wt(y) = t + m - |I| \Rightarrow y\mathbf{1}_d \neq 0$.

3. If $s \geq t + m$, $wt(Q_I) \geq 1$.

REMARK 4.2. If $m \geq t - 1$ then 2-(a) $\Rightarrow$1-(a).

PROOF OF THEOREM 4.2. The proof of Theorem 4.2 is divided into 2 parts. We have to demonstrate (a) that B is an orthogonal array of strength min $(t+m, s)$, and (b) that B is amicable with D.

(a) B is an orthogonal array of strength $w =$ min $(t+m, s)$, if every $N \times w$ submatrix of B contains the vectors of $GF(p)^w$, p^{d-r-w} times each. As $B = ((x_i \mid v_j))_{ij}$ with $Px_i = 0$, it is equivalent to prove that for every $J \subseteq \{1, \ldots, s\} : |J| = w$ and every $z \in GF(p)^w$, there are p^{d-r-w} solutions to the equation

$$Q_J x = \left(\frac{0}{z}\right). \tag{4.1}$$

If $w = s$ then $|J| = s \leq t + m$, and $|J| \leq$ min $(t + m - 1, s)$. Then, either $|J| < t$ and, from assumption 1-(a), $wt(Q_J) \geq t - |J| > 1$, or $t \leq |J| \leq$ min $(t + m - 1, s)$ and, from assumption 2-(a), $wt(Q_J) \geq t + m - |J| \geq 1$.
If $w = t + m$ then $s \geq t + m$ and from assumption 3, $wt(Q_J) \geq 1$.

So, $\forall J \subseteq \{1, ..., s\} : |J| = w$, $wt(Q_J) \geq 1$. It means that the rows of Q_J are linearly independent. As the number of rows is $r + w$, there are p^{d-r-w} solutions to (4.1) $\forall z \in GF(p)^w$

(b) D is a $p^{d-r} \times (d+1)$ matrix such that $D = (0 \mid T)$. The ith row of D is denoted $(0 \mid x_i)$ with $Px_i = 0$. For every integer $v \leq$ min $(t + m - 1, s)$, let $J \subseteq \{1, ..., s\}$ with $|J| = v$, $z = (z_1, ..., z_v) \in GF(p)^v$, and

$$D_{I_{B[J],z}} = (d_{ik} |\ i \in I_{B[J],z} \text{ and } k = \{0, \ldots, d\})$$

where $I_{B[J],z} = (i \in \{1, ..., p^{d-r}\} : (x_i \mid v_j) = z_j\ \forall j \in J)$. In order to demonstrate that D and B are amicable, we must show that $D_{I_{B[J],z}}$ is a difference scheme of strength $t + m - |J|$. Let $D'_{J,z}$ be the array constituted of the solutions of

$$Q_J x = \left(\frac{0}{z}\right), \tag{4.2}$$

then $D_{I_{B[J],z}} = (0 \mid D'_{J,z})$. For every $J \subseteq \{1, ..., s\}$, $|J|$ varies between 1 and min $(t + m - 1, s)$. This interval is divided into 2 parts. If $1 \leq |J| \leq$ min $(t - 1, s)$, then from condition 1, Q_J verifies the assumptions of Theorem 4.1 with $t - |J|$. Therefore $D'_{J,z}$ is a linear orthogonal array of strength $t - |J|$ and, above all, a difference scheme of strength $t + m - |J|$. If min $(t - 1, s) < |J| \leq$ min $(t + m - 1, s)$, from condition 2, Q_J verifies the assumptions of Theorem 4.1 with $t + m - |J|$. Therefore $D'_{J,z}$ is also a difference scheme of strength $t + m - |J|$. Thus $\forall J \subseteq \{1, ..., s\}$, $1 \leq |J| \leq$ min $(t + m - 1, s)$, $D'_{J,z}$ is a difference scheme of strength $t + m - |J|$ and D and B are amicable. □

The orthogonal array constructed with amicable matrices B and D is linear and its parity check matrix is given in the next corollary.

COROLLARY 4.1. *With P and $V = (v_1^t, \dots, v_s^t)$ as in the preceding theorem, a linear orthogonal array of strength $t+m$ and $d+s+1$ factors can be constructed and its parity check matrix is*

$$P' = \left(\begin{array}{c|c|c} -P\mathbf{1}_d & P & 0 \\ \hline -V\mathbf{1}_d & V & -I_s \end{array} \right).$$

PROOF. The linear orthogonal array of strength $t+m$ with $d+s+1$ factors constructed from the amicable matrices B and D defined in Theorem 4.2 is

$$T' = \left(\begin{array}{c} g\mathbf{1}_{p^{d-r}} \mid g+D \mid B \\ \hline \dots \\ \forall g \in GF(p) \end{array} \right)$$

where $D = (d_{ij})$ with $(d_{ij} \mid j = 1, ..., d) = x_i^t$ and $B = ((x_i \mid v_j))_{ij}$ with $Px_i = 0$.

• The matrix P' defined in Corollary 4.1 satisfies $P'(T')^t = 0$ because $PT^t = 0$ and $VD^t = B^t$. Then the rows of T' form a vector subspace of ker (P').

• As rank $(P) = r$, the rank of the submatrix of P',

$$\left(\begin{array}{c|c} P & 0 \\ \hline V & -I_s \end{array} \right),$$

is $r+s$. Moreover the size of P' is $(r+s) \times (1+d+s)$. Then rank $(P') = r+s$.

• As T is a linear array, the rows of $D = (0 \mid T)$ are distinct, then there are p^{d-r+1} separate rows in T'. Therefore the rows of T' are the vectors of ker (P'). □

EXAMPLE 4.3. Let $d = 7$ and $p = 3$. Let P be the 2×7 parity check matrix based on $GF(3)$

$$P = \left(\begin{array}{cc|ccccc} 1 & 0 & 0 & 0 & 1 & 1 & 2 \\ 0 & 1 & 1 & 1 & 0 & 0 & 2 \end{array} \right).$$

The assumptions of Theorem 4.1 are satisfied with $t = 3$ and $m = 1$. If we apply Theorem 4.2 with the vectors

$$v_1 = (0,0,1,2,1,0,1), \ v_2 = (1,1,2,0,1,2,0) \text{ and } v_3 = (2,1,2,0,1,2,0),$$

we can construct an $OA(3^6, 7+3+1, 3, 3+1) = OA(729, 11, 3, 4)$, which is defined by its 5×11 parity check matrix

$$P' = \left(\begin{array}{c|ccccccc|ccc} 1 & 1 & 0 & 0 & 0 & 1 & 1 & 2 & 0 & 0 & 0 \\ 1 & 0 & 1 & 1 & 1 & 0 & 0 & 2 & 0 & 0 & 0 \\ \hline 1 & 0 & 0 & 1 & 2 & 1 & 0 & 1 & 2 & 0 & 0 \\ 2 & 1 & 1 & 2 & 0 & 1 & 2 & 0 & 0 & 2 & 0 \\ 1 & 2 & 1 & 2 & 0 & 1 & 2 & 0 & 0 & 0 & 2 \end{array} \right).$$

EXAMPLE 4.4. Let $d = 4$ and $p = 5$. Let P be the 2×4 parity check matrix based on $GF(5)$

$$P = \left(\begin{array}{cc|cc} 1 & 0 & 2 & 0 \\ 0 & 1 & 0 & 1 \end{array} \right).$$

The assumptions of Theorem 4.1 are satisfied with $t = 2$ and $m = 1$. If we apply the preceding theorem with the vectors

$$v_1 = (1,2,2,0) \text{ and } v_2 = (1,1,2,0),$$

we can construct an $OA(5^3, 4+2+1, 5, 2+1) = OA(125, 7, 5, 3)$ (index $\lambda = 1$), which is defined by its 4×7 parity check matrix

$$P' = \left(\begin{array}{c|cccc|cc} 2 & 1 & 0 & 2 & 0 & 0 & 0 \\ 3 & 0 & 1 & 0 & 1 & 0 & 0 \\ \hline 0 & 1 & 2 & 2 & 0 & 4 & 0 \\ 1 & 1 & 1 & 2 & 0 & 0 & 4 \end{array} \right).$$

References

[1] B. Adhikary and P. Das, *Multiple fold-over and its application to orthogonal arrays*, Sankhyā, special volume **54** (1992), 21-24.

[2] R.A. Bailey, *Cyclic designs and factorial designs*, In Proceedings of the R.C. Bose symposium, Delhi 1988, on probability, statistics and design of experiments (R.R. Bahadur, K. Sen, Eds.), New York, (1990), 45-54.

[3] R.C. Bose and K.A. Bush, *Orthogonal arrays of strength two and three*, Ann. Math. Statist. **23** (1952), 508-524.

[4] D. Collombier, *Plans d'expérience factoriels*, Springer-Verlag, Berlin, 1996.

[5] A.S. Hedayat, K. Pu and J. Stufken, *On the construction of asymmetrical orthogonal arrays*, Ann. Statist. **20** (1992), 2142-2152.

[6] A.S. Hedayat, N.J.A. Sloane and J. Stufken, *Orthogonal arrays : Theory and applications*, Springer-Verlag, New York, 1996a.

[7] A.S. Hedayat, J. Stufken and G. Su, *On difference schemes and orthogonal arrays of strength t*, J. Statist. Plann. Inference **56** (1996b), 307-324.

[8] R. Hill, *An extension theorem for linear codes*, Designs, Codes and Cryptography **17** (1999), 151-157.

[9] A. Jourdan, *Analyse statistique et échantillonnage d'expériences simulées*, Ph.D. Thesis, Université de Pau et des Pays de l'Adour, France, 2000.

[10] A. Kobilinsky, *Complex linear models and cyclic designs*, Linear Algebra Appl. **127** (1990), 227-282.

[11] C.R. Rao, *Factorial experiments derivable from combinatorial arrangements of arrays*, J. Roy. Statist. Soc. (suppl.) **9** (1947), 128-139.

[12] E. Seiden, *On the problem of construction of orthogonal arrays*, Ann. Math. Statist. **25** (1954), 151-156.

Département de Mathématiques, Université Louis Pasteur, Strasbourg, France
Current address: I.R.M.A., Université Louis Pasteur, F-67084 Strasbourg Cedex, France
E-mail address: `collombier@math.u-strasbg.fr`

Département de Mathématiques, Université de Pau et des Pays de l'Adour, Pau, France
Current address: L.M.A., Université de Pau et des Pays de l'Adour, F-64000 Pau, France
E-mail address: `astrid.jourdan@univ-pau.fr`

Contemporary Mathematics
Volume **287**, 2001

A Coordinate-free Approach to Multivariate Exponential Families

Alessandro Di Bucchianico and Daniel E. Loeb

This paper is dedicated to the memory of Gian-Carlo Rota.

ABSTRACT. We apply the invariant umbral calculus to obtain a coordinate-free approach to multivariate exponential families.

1. Introduction

In [**20**] Morris showed that there are six classes of natural exponential families on the real line with quadratic variance function (i.e., where the variance is a polynomial of degree at most two). It is interesting to note that the Morris classification was discovered a few years earlier in approximation theory by May ([**19**] and for generalizations [**12**]; see also [**7**, Section 5] for the exact relation between natural exponential families and approximation theory). In [**7**], we studied natural exponential families in light of polynomial expansions of their density function. These polynomials belong to the class of Sheffer polynomials. In the present paper, we study these polynomials and their relation to natural exponential families from an umbral calculus view point. In particular, the polynomial expansions associated with a natural exponential family are Sheffer and their delta operator is related to their variance function. Using slightly different terminology, Feinsilver proved [**10**, Chapter 4] that a natural exponential family has a quadratic variance function if and only if the corresponding Sheffer polynomials are orthogonal. This result immediately follows from our umbral calculus approach to natural exponential families.

The multivariate classification problem of natural exponential families with quadratic variance function was initiated by Letac (see [**16**]). A partial solution for so-called simple quadratic variance function has been found by Casalis ([**3, 4, 5**]). For interesting related papers on multivariate exponential families, see [**15, 22, 23, 24**]. The results in [**13, 14, 25, 26**] are relevant for the present paper, because they present multivariate forms of the Favard characterization of orthogonal polynomials. These results require fixed coordinate systems and therefore, are subject to the

1991 *Mathematics Subject Classification.* Primary 05A40, 33C45, 60E05, 62E10.
Key words and phrases. Umbral calculus, natural exponential family, variance function, Sheffer sequence, orthogonal polynomials.

observation made by several authors [**15, 24, 25**] that no coordinate system is preferable to another when dealing with multivariate orthogonal polynomials and natural exponential families. In this paper we extend the umbral calculus setting of [**7**] to an invariant setting. This new setting provides a way to treat these topics without fixing a basis and may lead to future progress in the classification problem for multivariate natural exponential families.

This paper is organized as follows. In Section 2 we review the results of [**7**]. An introduction to invariant umbral calculus is given in Section 3, which is based on [**8**] and [**9**]. Then we proceed in Section 4 to give a treatment of natural exponential families in the spirit of Section 3. Finally, generalizations of the results of [**7**] are given in Section 5.

2. Review of univariate umbral approach to exponential families

In this section we summarize the results of [**7**]. We begin by recalling the basic notions of natural exponential families, in particular the variance function which plays a central role. Our notation closely follows [**18**]. Let ν be a measure on the real line that is not concentrated in one point and such that $\int_{-\infty}^{\infty} x\,d\nu(x) = 0$. Let the Laplace transform of ν be given by

$$L(\theta) = \int_{-\infty}^{\infty} e^{x\theta}\,d\nu(x).$$

We define Θ to be the interior of the set $\{\theta \in \mathbb{R} \,|\, L(\theta) < \infty\}$. If Θ is non-empty, then the natural exponential family generated by ν is the set of probability densities of the form

$$P_\theta(A) = \int_A e^{x\theta - k(\theta)}\,d\nu(x), \tag{2.1}$$

where $k(\theta) = \log L(\theta)$ is the cumulant of ν and $\theta \in \Theta$. Moreover,

$$k'(\theta) = \int_{-\infty}^{\infty} x\,dP_\theta(x)\ ,\ k''(\theta) = \int_{-\infty}^{\infty} (x - k'(\theta))^2\,dP_\theta(x). \tag{2.2}$$

Let $M_\nu = k'(\Theta)$ indicate the range of k'. Since k is strictly convex on Θ by the Hölder inequality, $k' : \Theta \longrightarrow M_\nu$ is a bijection. Its inverse will be denoted by

$$\psi : M_\nu \to \Theta.$$

This means that we may reparametrize the densities with respect to ν in (2.1) as

$$\varphi(m, x) = e^{x\psi(m) - k(\psi(m))}, \tag{2.3}$$

which leads to the following important definition.

Definition 2.1. Let $\{P_\theta \,|\, \theta \in \Theta\}$ be the natural exponential family generated by a measure ν. The function $V_\nu : M_\nu \to \mathbb{R}$ defined by $V_\nu(m) = \int_{-\infty}^{\infty} (x - m)^2\,\varphi(m, x)\,d\nu(x)$ is called the **variance function** of $\{P_\theta \,|\, \theta \in \Theta\}$.

A natural exponential family is uniquely determined by its variance function together with the domain of the variance function ([**20**]). This characterization also holds in the multivariate case ([**17**]). This fact is central to our approach. Note that an exponential family is not uniquely determined by the measure ν. In particular, we may always assume without loss of generality that $\int_{-\infty}^{\infty} x\,d\nu(x) = 0$.

By well-known properties of Laplace transforms, k and ψ are analytic functions in a neighborhood of zero. Hence, we may expand (2.3) into a power series in $m \in M_\nu$. The associated Sheffer polynomials of a natural exponential family are the polynomials $(s_n)_{n\geq 0}$ defined by

$$\varphi(m,x) = \sum_{n=0}^{\infty} s_n(x)\, m^n,$$

where φ is defined by (2.3). It follows from (2.1), (2.2), $\int_{-\infty}^{\infty} x\, d\nu(x) = 0$ and the fact that ν is not concentrated in one point, that $k'(0) = 0$ and $k''(0) \neq 0$. Thus, $\psi(0) = 0$ and $\psi'(0) \neq 0$, which implies that s_n is a polynomial of degree exactly n.

The delta operator of a Sheffer sequence is the linear operator that takes s_n to s_{n-1}. Any delta operator admits a power series expansion in the derivative operator D. If $Q = q(D)$ is a delta operator, then the Pincherle derivative is the operator $q'(D)$. It is shown in [**7**] that $q'(D) = V_\nu(D)$. This relation is used to prove Feinsilver's Theorem: a natural exponential family has quadratic variance function if and only if the associated Sheffer sequence is orthogonal. The proof is based on the Favard's Theorem for orthogonal polynomials which says that orthogonal polynomials are characterized by a three-term recurrence relation. For Sheffer sequences, this recurrence relation can be shown to correspond to a differential equation (in the Pincherle's sense) for its delta operator. Details can be found in [**7**].

3. Introduction to invariant umbral calculus

In this section we give an introduction to a theory of multivariate polynomials that does not require a fixed coordinate system. A first exposition of this theory can be found in [**8**]. A more extensive treatment including Sheffer polynomials is contained in [**9**].

All elements belong to a real Hilbert space $\mathcal{H}$ with inner product $\langle .|. \rangle$. Although for our application to natural exponential families we only need a finite-dimensional setting, the results of this section are presented for Hilbert spaces with a countably infinite dimension.

We first need to define a notion of set where elements have multiplicities. Given some *universe* of discourse $\mathcal{U}$, a *(finite) multiset* is a function with finite support from $\mathcal{U}$ to the nonnegative integers $\mathbf{N}$. We say that $R : \mathcal{U} \to \mathbf{N}$ is a multiset *on* the universe $\mathcal{U}$ and write $R \subset \mathcal{U}$. The value of a multiset R at a point $h \in \mathcal{U}$ is denoted R_h and called the *multiplicity* of h in R. If $R_{h'} = 0$ for all $h' \neq h$ and $R_h = k$, then we denote R by $\{h^k\}$. For an ordinary set, $R_h = 0$ (if $h \notin R$) or $R_h = 1$ (if $h \in R$).

If $R_h \geq S_h$ for all $h \in \mathcal{U}$, we write $S \subset R$ and say S is a *submultiset* of R. If $S \subset R$, then their *difference* $R \setminus S$ is defined to be the multiset with multiplicities $(R \setminus S)_h = R_h - S_h$ for all $h \in \mathcal{U}$. Similarly, the *union* $R \cup S$ of any two multisets R and S is defined to be the multiset with multiplicities $(R \cup S)_h = R_h + S_h$ for all $h \in \mathcal{U}$. We define the *factorial* $R!$, and the *lower factorial* $(R)_S$, as follows:

$$R! = \prod_{h\in\mathcal{U}} R_h!, \qquad (R)_S = \frac{R!}{(R \setminus S)!}.$$

The *binomial coefficients* are defined as

$$\binom{R}{S} = \frac{R!}{S!\,(R \setminus S)!}.$$

Note that the notions of factorial and binomial coefficient reduce to the standard notions when we consider multisets of the form $R = \{h^k\}$.

We now are ready to define our general notion of monomial. Let $R \subset \mathcal{H}$ be a finite multiset. The *monomial* x^R is defined by the product $\prod_{r \in R} \langle x|r \rangle$. The multiset R is said to be the *exponent* of the monomial x^R. In particular, $x^\emptyset$ equals the constant 1 identically, and the map $\mathbf{x} : r \mapsto x^{\{r\}} = \langle x|r \rangle$ is linear. Note that if $\dim \mathcal{H} = n < \infty$ with basis $\{e_1, \ldots, e_n\}$, then we may identify the classical variables $x_1, \ldots, x_n$ with the monomials $\langle x|e_1 \rangle, \ldots, \langle x|e_n \rangle$. Thus, the classical notion of a polynomial in the variables $x_1, x_2, \ldots$ is contained in our setup.

The *degree* of x^R is defined to be $|R|$. A *polynomial* is a finite linear combination of monomials over the real numbers. Its *degree* is the largest degree of its terms. The degree of the zero polynomial being taken to be -1. It can be shown that degree is a well-defined notion which does not depend on the representation of the polynomial. The set of all polynomials is denoted $\mathcal{P} = \mathcal{P}(\mathcal{H})$ and is called the *Hilbert space polynomial ring*. The space $\mathcal{P}$ is called the *Bose algebra* and denoted $\Gamma_0 \mathcal{H}$ by Nielsen [**21**, p. 4]. The completion of $\mathcal{P}$ (with respect to a certain natural inner product, e.g., [**8**, Section 3.2]) is called the exponential space of $\mathcal{H}$ in [**1**, pp. 157-242] and [**2**]. It is called the symmetric Hilbert space $S\mathcal{H}$ in [**11**, Defn. 2.1].

We now define linear operators on $\mathcal{P}$. For $h \in \mathcal{H}$, the *translation* or *shift operator* by h is defined by

$$E^h p(x) = p(x + h).$$

For $h \in \mathcal{H}$, let the *derivative operator* D_h be the derivation of $\mathcal{P}$ defined by $D_h \langle x|r \rangle = \langle h|r \rangle$. Thus, by the definition of derivation,

$$D_h\, x^R = \sum_{r \in R} \langle h|r \rangle\, x^{R \setminus \{h\}}.$$

If u is a unit vector, then D_u can be thought of as the derivative in the direction u. Otherwise, $h = cu$ and $D_h = cD_u$. In particular, D_0 is the zero operator. We will write $D^R = \prod_{r \in R} D_r$. Note that this product is well-defined, since all differentiation operators commute.

A linear operator θ on $\mathcal{P}$ is said to be *shift-invariant* if $\theta E^a = E^a \theta$ for all $a \in \mathcal{H}$. Any shift-invariant operator T admits an expansion of the form

$$T = \sum_{R \subset B} c_R\, D^R, \tag{3.1}$$

where B is an orthonormal basis of $\mathcal{H}$, and $c_R = \left[T x^R / R!\right]_{x=0}$. The indicator of a shift-invariant operator is the formal power series in the formal variable t given by

$$\operatorname{Ind}(T) = \sum_{R \subset B} \left[T x^R / R!\right]_{x=0} t^R.$$

The delta operator is a shift-invariant operator of order one, i.e., an operator such that (3.1) has no constant term and at least one nonzero term with $|R| = 1$. Define a *delta set* to be a set of shift-invariant operators such that Q_h is a delta operator for all $h \neq 0$ in $\mathcal{H}$, the map $h \mapsto Q_h$ is linear, and furthermore the ground field $\mathbf{R}$ is the intersection of their null-spaces

$$\mathbf{R} = \bigcap_{h \in \mathcal{H}} \ker Q_h. \tag{3.2}$$

Condition (3.2) ensures that a "delta set" such as $Q_{b_i} = D_{b_{i+1}}$ is not allowed where $B = \{b_1, b_2, \dots\}$ is an orthonormal basis of $\mathcal{H}$, for in this case $\langle x|b_1\rangle$ would not be in the intersection of the kernels. We will write $Q^R = \prod_{r \in R} Q_r$. Note that this product is well-defined, since all shift-invariant operators can be shown to commute. The notion of indicator as define above allows us to define the *generating function* of a delta set. Let B be an orthonormal basis of $\mathcal{H}$. Then

$$\operatorname{Gen}(\mathbf{Q}) = \sum_{b \in B} \langle x|b\rangle \operatorname{Ind}(Q_b).$$

It can be shown that $\operatorname{Gen}(\mathbf{Q})$ does not depend on the choice of basis B. Note, for example, that $\operatorname{Gen}(D) = \langle x|t\rangle$.

We now return to polynomials. Define an *umbral set* $\mathbf{q}$ of polynomials to be a linear map from $\mathcal{H}$ to $\mathcal{P}$. For example, $\mathbf{x} : r \mapsto \langle x|r\rangle$ is an umbral set of polynomials. Define an *umbral sequence* $\mathbf{q}$ of polynomials to be a map from finite multisets R on $\mathcal{H}$ to polynomials, denoted q_R, which is k-linear when R is considered as a k-tuple $R = (r_1, r_2, \dots, r_k)$. For example, $\mathbf{x} : R \mapsto x^R$ is an umbral sequence. These monomials obey the binomial theorem

$$(x+y)^R = \sum_{S \cup T = R} \binom{R}{S} x^S\, y^T.$$

Similarly, a sequence of polynomials $\mathbf{q}$ obeying the identity

$$q_R(x+y) = \sum_{S \cup T = R} \binom{R}{S} q_S(x)\, q_T(y)$$

is said to be a *sequence of binomial type* if the map $\tilde{q} = \mathbf{q} \circ \mathbf{x} : r \mapsto q_{\{r\}}$ from $\mathcal{H}$ to the space $\mathcal{P}_1$ of homogeneous polynomials of degree 1 is linear and bijective, where $\mathbf{x}$ is defined as $\mathbf{x} : r \mapsto x^{\{r\}} = \langle x|r\rangle$. A *diagonal* set of binomial type is a set $\mathbf{q}$ which is binomial and satisfies the additional property $q_{\{r\}} = \langle x|r\rangle$.

The following theorem links sequences of binomial type to delta sets.

THEOREM 3.1. *Let* $\mathbf{q}$ *be a (diagonal) sequence of binomial type. Then*

(a) *The maps* Q_h *defined by*

$$Q_h\, q_R = \sum_{r \in R} \langle r|h\rangle\, q_{R \setminus \{r\}}$$

form a (diagonal) delta set,

(b) $q_\emptyset$ *is identically one, and*

(c) $q_R(0) = 0$ *for* $R \neq \emptyset$.

PROOF. Part a) follows from a straightforward calculation. Parts b) and c) follow by induction on $|R|$. □

Let $\mathbf{q}$ be an umbral sequence (not necessarily of binomial type). When $q_R(0) = 0$ for all $R \neq \emptyset$, we define the *umbral composition* of $p = \sum_{R \subset \mathcal{H}} c_R \, x^R$ with $\mathbf{q}$ to be $p(\mathbf{q}) = \sum_{R \subset \mathcal{H}} c_R \, q_R$.

Let $\mathbf{q}$ be a (diagonal) sequence of binomial type with delta set $\mathbf{Q}$. A sequence of *nonzero* polynomials $\mathbf{s}$ obeying the identity

$$s_R(x+y) = \sum_{S \cup T = R} \binom{R}{S} s_T(x) \, q_S(y) \tag{3.3}$$

is said to be a *(diagonal) Sheffer sequence* with respect to $\mathbf{q}$ or equivalently, to $\mathbf{Q}$.

The following theorem shows two other important characterizations of Sheffer sequences in terms of shift-invariant operators.

THEOREM 3.2. *Let $\mathbf{q}$ be a sequence of binomial type.*

(a) *A sequence $\mathbf{s}$ is (diagonal) Sheffer with respect to $\mathbf{Q}$ if and only if*

$$Q_h s_R = \sum_{r \in R} \langle r|h \rangle \; s_{R \setminus \{r\}}.$$

(b) *A sequence $\mathbf{s}$ is Sheffer with respect to $\mathbf{q}$ if and only if the map $A\colon s_R \mapsto q_R$ is an invertible shift-invariant operator.*

PROOF. Part a) follows from (3.2) and a straightforward adaptation of the proof of [**6**, Th. 2.2.17]. In order to prove b), we rewrite the left-hand side of (3.3) as $E^y S_R(x)$ and apply shift-invariance together with Theorem 3.1. □

THEOREM 3.3 (Operator Expansion). *Let $\mathbf{s}$ be a Sheffer sequence with invertible operator A and delta set $\mathbf{Q}$. A shift-invariant operator T can be expressed as*

$$T = \sum_{R \subset B} c_R A Q^R,$$

where R is a submultiset of some orthonormal basis B and $c_R = [T s_R(x)/R!]_{x=0}$.

PROOF. This follows by applying both sides to the basis $\mathbf{s}$ and using Theorem 3.2. □

THEOREM 3.4 (Polynomial Expansion). *Let $\mathbf{s}$ be a Sheffer sequence with invertible operator A and delta set $\mathbf{Q}$. A polynomial p can be expressed as*

$$p = \sum_{R \subset B} c_R \, s_R,$$

where R is a submultiset of some orthonormal basis B and $c_R = \left[A Q^R \, p(x)/R!\right]_{x=0}$.

PROOF. Apply Theorem 3.3 to $T = I$. □

We now define the generating function of an umbral sequence $\mathbf{p}$ as follows:

$$\operatorname{Gen}(p) = \sum_{R \subset B} p_R \, t^R / R!,$$

where B is an orthonormal basis. It can be shown that $\operatorname{Gen}(p)$ is a formal power series in the formal variable $\langle x|t \rangle$ that does not depend on the choice of basis B. The generating function of the monomials x^R is easily seen to be $\operatorname{Gen}(\mathbf{x}) = \exp(\langle x|t \rangle)$. In a similar way we may define umbral composition for the generating function of delta set.

THEOREM 3.5. *Let* $\mathbf{q}$ *be a sequence of binomial type with delta set* $\mathbf{Q}$. *Let* $\mathbf{Q}^{(-1)}$ *be the inverse of* $\mathbf{Q}$ *in the monoid of umbral composition. Then*

$$\text{Gen}\,(\mathbf{q}) = \exp\left(\text{Gen}\left(\mathbf{Q}^{(-1)}\right)\right).$$

A sequence $\mathbf{s}$ *is Sheffer (with respect to* $\mathbf{Q}$*) if and only if*

$$\text{Gen}\,(\mathbf{s}) = f(t)\,\text{Gen}\,(\mathbf{q}).$$

PROOF. The first formula follows by noticing that the constant terms on both sides are equal and by showing that Q_h for all $h \in \mathcal{H}$ acts the same way when applied to both sides. The second formula follows by combining (3.3) with the generating function of $\mathbf{q}$. □

We conclude this section with the notion of Pincherle derivative. The Pincherle derivative in the direction h of a linear operator T on $\mathcal{P}$ is given is given by the commutator of θ and the operator X_h of multiplication by $\langle x|h\rangle$:

$$T^{(h)} = TX_h - X_hT. \tag{3.4}$$

As an example, we note that it is easy to compute that $D_k^{(h)} = \langle k|h\rangle\, I$. A more elegant equivalent definition is as follows (see [**9**]). The Pincherle derivative in the direction h of a shift-invariant operator is the shift-invariant operator whose indicator equals $D_h\text{Ind}\,(T)$. The operator D_h is defined term-wise on the coefficients of the formal power series $\text{Ind}\,(T)$.

THEOREM 3.6. *If* T *and* U *are shift-invariant operators and* $h \in \mathcal{H}$, *then* $(TU)^{(h)} = T^{(h)}U + UT^{(h)}$.

PROOF. By (3.4), we have

$$\begin{aligned}
(T\eta)^{(h)} &= TUX_h - X_hTU \\
&= (TUX_h - TX_hU) + (TX_hU - X_hTU) \\
&= T^{(h)}U + TU^{(h)}.
\end{aligned}$$

□

4. Invariant natural exponential families

In this section we generalize the notion of natural exponential families to an invariant setting in the spirit of the previous section.

Let ν be a measure on a *finite dimensional* Hilbert space $\mathcal{H}$. We assume that ν is not concentrated on a hyperplane and that $\int_{\mathcal{H}} x\,d\nu(x) = 0$. Let the Laplace transform of ν be given by

$$L(\theta) = \int_{\mathcal{H}} e^{\langle x|\theta\rangle}\,d\nu(x).$$

We define Θ to be the interior of the set $\{\theta \in \mathcal{H} \,|\, L(\theta) < \infty\}$. If Θ is non-empty, then the *natural exponential family* generated by ν is the set of probability distributions of the form

$$P_\theta(A) = \int_A e^{\langle x|\theta\rangle - k(\theta)}\,d\nu(x), \tag{4.1}$$

where $k(\theta) = \log L(\theta)$ is the cumulant of ν and $\theta \in \Theta$. Since $k(\theta) = \log L(\theta)$, we have

$$e^{k(\theta)} = \int_{\mathcal{H}} e^{\langle x|\theta\rangle}\, d\nu(x). \tag{4.2}$$

It follows by differentiating (4.2) with respect to θ in the direction $h \in \mathcal{H}$, that

$$D_h k(\theta) = \int_{\mathcal{H}} \langle x|h\rangle\, dP_\theta(x). \tag{4.3}$$

Now let B be an arbitrary orthonormal basis of $\mathcal{H}$. Since $x = \sum_{b\in B} \langle x|b\rangle\, b$, it follows from (4.3) that

$$\sum_{b\in B} [D_b k(\theta)]\, b = \int_{\mathcal{H}} x\, dP_\theta(x). \tag{4.4}$$

We write $k'(\theta) = \sum_{b\in B} [D_b k(\theta)]\, b$. This is the gradient in the classical sense, because $k'(\theta)$ does not depend on the choice of B. Differentiating (4.2) with respect to θ in the directions h and a, and using (4.3), we obtain

$$D_h D_a k(\theta) = \int_{\mathcal{H}} (x - D_h k(\theta))\,(x - D_a k(\theta))\, dP_\theta(x). \tag{4.5}$$

Let M_ν be the range of k', i.e., $M_\nu = k'(\Theta)$. Since k is strictly convex on Θ by the Hölder inequality, $k' : \Theta \longrightarrow M_\nu$ is a bijection. Its inverse will be denoted by

$$\psi : M_\nu \to \Theta.$$

This means that we may reparametrize the densities with respect to ν in (4.1) as

$$\varphi(m,x) = e^{\langle x|\psi(m)\rangle\; -\; k(\psi(m))}. \tag{4.6}$$

Using the reparametrization of (4.6), we now come to the following important definition.

DEFINITION 4.1. Let $\{P_\theta \mid \theta \in \Theta\}$ be the natural exponential family generated by a measure ν. The bilinear function defined by

$$V_\nu(m)(h,a) = \int_{\mathcal{H}} \langle x - m|h\rangle\, \langle x - m|a\rangle\, \varphi(m,x)\, d\nu(x)$$

is called the *variance function* of $\{P_\theta \mid \theta \in \Theta\}$.

By well-known properties of Laplace transforms, k and ψ are analytic functions in a neighborhood of zero. Hence, we may expand (4.6) into a power series in $m \in M_\nu$. The associated Sheffer sequence of a natural exponential family is the sequence $\mathbf{s}$ defined by

$$\varphi(m,x) = \sum_{R\subset B} s_R(x)\, m^R, \tag{4.7}$$

where φ is defined by (4.6). It follows from (4.1), (4.3) and $\int_{\mathcal{H}} x\, d\nu(x) = 0$ that $k'(0) = 0$. Moreover, since ν is not concentrated on a hyperplane, it follows from (4.5) that $k''(0) \neq 0$. Thus, $\psi(0) = 0$ and $\psi'(0) \neq 0$, which implies that s_R is a polynomial of degree exactly $|R|$. The fact that ν is not concentrated on a hyperplane also implies that $s_{\{r\}}$ spans $\mathcal{H}$ for $r \in B$.

5. Main results

In this section we show the relations between natural exponential families, Sheffer polynomials, orthogonal polynomials, delta sets and quadratic variance functions. In this section $\mathcal{H}$ is a finite dimensional real Hilbert space.

DEFINITION 5.1. Let Λ be a linear functional on $\mathcal{P}$. An umbral sequence $\mathbf{p}$ is orthogonal with respect to Λ if $\Lambda(p_S p_R) = 0$ when $R \setminus S \perp S$ and $S \setminus R \perp R$.

The following theorem is an coordinate-free version of the multivariate Favard-type recurrence characterization of orthogonal polynomials (cf. [**13, 14, 25, 26**].

THEOREM 5.2. *Let* $\mathbf{p}$ *be an umbral sequence. Then* $\mathbf{p}$ *is an orthogonal sequence with respect to a functional* Λ *if and only if*

$$\langle x|h\rangle \, p_R = \sum_{\substack{T\subset B\\ |T|=|R|+1}} c_T p_T + \sum_{\substack{U\subset B\\ |U|=|R|}} d_U p_U + \sum_{\substack{V\subset B\\ |V|=|R|-1}} e_V p_V .$$

PROOF. The proof goes along the classical lines as in [**14, 25, 26**]. □

With suitable conditions on the coefficients, the functional Λ admits an integral representation (see e.g. [**13**]). The following theorem shows that a Sheffer sequence satisfies a Favard-type recurrence if and only if its delta set and invertible operator satisfy certain differential equations (in the sense of Pincherle derivatives).

THEOREM 5.3. *Let* $\mathbf{s}$ *be a Sheffer sequence with delta set* $\mathbf{Q}$ *and invertible operator* A. *Let* B *be an orthonormal basis of the finite dimensional Hilbert space* $\mathcal{H}$. *Then*

$$\langle x|h\rangle \, s_R = \sum_{\substack{T\subset B\\ |T|=|R|+1}} c_T s_T + \sum_{\substack{U\subset B\\ |U|=|R|}} d_U s_U + \sum_{\substack{V\subset B\\ |V|=|R|-1}} e_V s_V .$$

if and only if for all $k, h \in \mathcal{H}$ *we have*

$$\begin{aligned} Q_k^{(h)} = \left(\sum_{t\in B} c_{\{t\}} \langle k|t\rangle + d_\emptyset\right) I + e_\emptyset \, Q_{\sum_{b\in B} b} + \sum_{u\in B} d_{\{u\}} \langle k|u\rangle \, Q_u \\ + \sum_{v.w\in B} e_{\{v\}} \langle k|v\rangle \, Q_v Q_w \end{aligned} \tag{5.1}$$

and

$$A^{(h)} = A\left(d_\emptyset I + e_\emptyset Q_{\sum_{b\in B} b}\right).$$

PROOF. "Only if" It follows from the Polynomial Expansion Theorem that

$$\begin{aligned} \langle x|h\rangle \, s_R &= \sum_{G\in B} \left[AQ^S \langle x|h\rangle \, s_R\right]_{x=0} s_G = \sum_{G\in B} \left[\left(AQ^S\right)^{(h)} s_R\right]_{x=0} s_G \\ &= \sum_{G\in B} \left[\left(A^{(h)}Q^S + A\left(Q^S\right)^{(h)}\right) s_R\right]_{x=0} s_G \\ &= \sum_{G\in B} \left[\left(A^{(h)}Q^S + A\left(\sum_{s\in S} \langle s|h\rangle \, Q^{S\setminus s} Q_s^{(h)}\right)\right) s_R\right]_{x=0} s_G \\ &= \sum_{G\in B} \left[\left(A\left(d_\emptyset I + e_\emptyset Q_{\sum_{b\in B} b}\right)\right) s_R\right]_{x=0} s_G . \end{aligned}$$

Conversely, we apply the Operator Expansion Theorem to the operator A, the invertible operator of the Sheffer sequence, and AQ_k, where $\mathbf{Q}$ is the delta set of the Sheffer sequence. We denote the basic set of $\mathbf{s}$ by $\mathbf{q}$. Recall that $q_R(0) = 0$ if $R \neq \emptyset$. For the former operator we obtain

$$\begin{aligned}
A^{(h)} &= \sum_{R \subset B} \left[A^{(h)} s_R\right]_{x=0} A \frac{Q^R}{R!} = \sum_{R \subset B} \left[A \langle x|h \rangle \, s_R\right]_{x=0} A \frac{Q^R}{R!} \\
&= \sum_{R \subset B} \left(\sum_{\substack{T \subset B \\ |T|=|R|+1}} c_T q_T(0) + \sum_{\substack{U \subset B \\ |U|=|R|}} d_U q_U(0) + \sum_{\substack{V \subset B \\ |V|=|R|-1}} e_V q_V(0) \right) A \frac{Q^R}{R!} \\
&= d_\emptyset q_\emptyset(0) A Q^\emptyset + e_\emptyset \sum_{b \in \mathcal{B}} Q_b = A \left(d_\emptyset I + e_\emptyset Q_{\sum_{b \in \mathcal{B}} b} \right).
\end{aligned}$$

In the same way we expand $(AQ_k)^{(h)}$ as

$$\begin{aligned}
&\sum_{R \subset B} \left[(AQ_k)^{(h)} s_R\right]_{x=0} A \frac{Q^R}{R!} = \sum_{R \subset B} \left[A Q_k \langle x|h \rangle \, s_R\right]_{x=0} A \frac{Q^R}{R!} \\
&= \sum_{R \subset B} Q_k \left(\sum_{\substack{T \subset B \\ |T|=|R|+1}} c_T q_T(0) + \sum_{\substack{U \subset B \\ |U|=|R|}} d_U q_U(0) + \sum_{\substack{V \subset B \\ |V|=|R|-1}} e_V q_V(0) \right) A \frac{Q^R}{R!} \\
&= \sum_{R \subset \mathcal{B}} \left(\sum_{\substack{T \subset B \\ |T|=|R|+1}} c_T \sum_{t \in T} \langle k|t \rangle \, q_{T \setminus t}(0) + \sum_{\substack{U \subset B \\ |U|=|R|}} d_U \sum_{u \in U} \langle k|u \rangle \, q_{U \setminus u}(0) \right. \\
&\qquad\qquad \left. + \sum_{\substack{V \subset B \\ |V|=|R|-1}} e_V \sum_{v \in V} \langle k|v \rangle \, q_{V \setminus v}(0) \right) \\
&= \sum_{t \in B} c_{\{t\}} \langle k|t \rangle \, q_\emptyset(0) A Q^\emptyset + \sum_{u \in B} d_{\{u\}} \langle k|u \rangle \, q_\emptyset(0) A Q_u + \sum_{v,w \in B} e_{\{v\}} \langle k|v \rangle \, q_\emptyset(0) A Q_v Q_w \\
&= \sum_{t \in B} c_{\{t\}} \langle k|t \rangle \, A + \sum_{u \in B} d_{\{u\}} \langle k|u \rangle \, A Q_u + \sum_{v,w \in B} e_{\{v\}} \langle k|v \rangle \, A Q_v Q_w.
\end{aligned}$$

Hence,

$$\begin{aligned}
AQ_k^{(h)} &= (AQ_k)^{(h)} - A^{(h)} Q_k \\
&= \sum_{t \in B} c_{\{t\}} \langle k|t \rangle \, A + \sum_{u \in B} d_{\{u\}} \langle k|u \rangle \, A Q_u + \sum_{v,w \in B} e_{\{v\}} \langle k|v \rangle \, A Q_v Q_w \\
&\qquad - A \left(d_\emptyset + e_\emptyset Q_{\sum_{b \in \mathcal{B}} b} \right).
\end{aligned}$$

Since A is an invertible operator, we obtain

$$\begin{aligned}
Q_k^{(h)} &= \sum_{t \in B} c_{\{t\}} \langle k|t \rangle + \sum_{u \in B} d_{\{u\}} \langle k|u \rangle \, Q_u + \sum_{v,w \in B} e_{\{v\}} \langle k|v \rangle \, Q_v Q_w \\
&\qquad - \left(d_\emptyset I + e_\emptyset Q_{\sum_{b \in B} b} \right).
\end{aligned}$$

This completes the proof. □

We now give an explicit example of a recurrence formula. The invariant Hermite polynomials are defined in [**8**, Section 5.3] by

$$H_R(x) = \prod_{r\in R} (X_r - D_r)1,$$

where X_r is the multiplication by $\langle x|r\rangle$ operator. The Hermite polynomials are an Appell set, i.e., is Sheffer with delta set $\mathbf{D}$. The invertible operator is $\exp \mathcal{L}/2$, where

$$\mathcal{L}\,x^S = \sum_{r\in S}\sum_{s\in S\setminus\{r\}} \langle r|s\rangle\; x^{S\setminus\{r,s\}}.$$

Using the Appell property, we obtain the following recursion:

$$\begin{aligned}\langle x|h\rangle\; H_R(x) &= X_h \prod_{r\in R}(X_r - D_r)1 \\ &= (X_h - D_h + D_h)\prod_{r\in R}(X_r - D_r)1 \\ &= \prod_{r\in R\cup\{h\}}(X_r - D_r)1 + D_h\prod_{r\in R}(X_r - D_r)1 \\ &= H_{R\cup\{h\}} + \sum_{r\in R}\langle h|r\rangle\; H_{R\setminus\{r\}}.\end{aligned}$$

We now return to natural exponential families.

THEOREM 5.4. *Let $\{P_\theta \,|\, \theta \in \Theta\}$ be the natural exponential family generated by a measure ν and let* $\mathbf{s}$ *be the associated Sheffer sequence with delta set* $\mathbf{Q}$ *and invertible operator A. Then* $\operatorname{Ind}\left(Q_h^{(a)}\right) = V_F(t)(h,a)$.

PROOF. Since $\operatorname{Gen}(\mathbf{Q}) = \psi^{-1}(x) = k'(x)$, it follows that $\operatorname{Ind}\left(Q_h^{(a)}\right) = D_h D_a k(t)$ $= V_F(t)(h,a)$. □

We conclude with an invariant setting of a theorem due to Feinsilver.

THEOREM 5.5. *A natural exponential family has orthogonal Sheffer polynomials if and only its variance function is a quadratic bilinear function.*

PROOF. By Theorem 5.4, the form of the variance function is determined by the derivative of the generating function of the delta set of the associated Sheffer sequence. The quadratic form is a consequence of Theorem 5.3, in particular (5.1). □

Acknowledgment. We thank a referee and the editors for corrections and insightful remarks that improved our paper.

References

[1] H. Araki and E. J. Woods, *Complete Boolean Algebras of Type I Factors*, volume 2 of Lecture Notes in Mathematics, R.I.M.S. Kyoto Univ., Berlin, Heidelberg, New York, 1966.

[2] P. Cartier, *Processus aléatoires généralisés*, Séminaire Bourbaki 272, 1963/4.

[3] M. Casalis, *Les familles exponentielles sur* $\mathbf{R}^2$ *de fonction-variance* $V(m) = am \otimes m + B(m) + C$, C. R. Acad. Sci. Paris Sér. I Math. **314(8)** (1992) 635–638.

[4] M. Casalis, *Les* $2n+4$ *types de familles exponentielles naturelles sur* $\mathbf{R}^n$ *à fonction-variance quadratique simple*, C. R. Acad. Sci. Paris Sér. I Math. **318(3)** (1994), 261–264.

[5] M. Casalis, The $2d + 4$ simple quadratic natural exponential families on $\mathbf{R}^d$, Ann. Statist. **24(4)** (1996), 1828–1854.
[6] A. Di Bucchianico, Probabilistic and analytical aspects of the umbral calculus, volume 119 of *CWI Tract Series*, CWI, Amsterdam, 1997.
[7] A. Di Bucchianico and D.E. Loeb. Natural exponential families and umbral calculus. In B.E. Sagan and R.P. Stanley, editors, *Mathematical essays in honor of Gian-Carlo Rota*, pages 43–60. Birkhäuser, 1998.
[8] A. Di Bucchianico, D.E. Loeb, and G.-C. Rota. Umbral calculus in Hilbert space. In B.E. Sagan and R.P. Stanley, editors, *Mathematical essays in honor of Gian-Carlo Rota*, pages 1–26. Birkhäuser, 1998.
[9] A. Di Bucchianico, D. E. Loeb, and G.-C. Rota, Umbral calculus in vector spaces, in progress.
[10] P. J. Feinsilver, Special functions, probability semigroups, and Hamiltonian flows, Springer, Berlin, 1978.
[11] A. Guichardet, Symmetric Hilbert Spaces and Related Topics, volume 261 of *Lecture Notes in Mathematics*, Springer-Verlag, Berlin, Heidelberg, New York, 1972.
[12] M. E. H. Ismail and C. P. May, On a family of approximation operators, J. Math. Anal. Appl. **63(2)** (1978), 446–462.
[13] M. A. Kowalski, Orthogonality and recursion formulas for polynomials in n variables, SIAM J. Math. Anal. **13(2)**(1982), 316–323.
[14] M. A. Kowalski, The recursion formulas for orthogonal polynomials in n variables. SIAM J. Math. Anal. **13(2)** (1982), 309–315.
[15] E. Labeye-Voisin and D. Pommeret, Polynômes orthogonaux associés aux familles exponentielles naturelles sur $\mathbf{R}^d$, C. R. Acad. Sci. Paris Sér. I Math. **320(1)** (1995), 79–84.
[16] G. Letac, Le problème de la classification des familles exponentielles naturelles de $\mathbf{R}^d$ ayant une fonction variance quadratique. In *Probability measures on groups, IX (Oberwolfach, 1988)*, pages 192–216. Springer, Berlin, 1989.
[17] G. Letac, Lectures on Natural Exponential Families and Their Variance Functions, volume 50 of Monografias de Matemática. Instituto de Matemática Pura e Aplicada, Rio de Janeiro, 1992.
[18] G. Letac and M. Mora, Natural real exponential families with cubic variance functions, Ann. Statist. **18(1)** (1990), 1–37.
[19] C. P. May, Saturation and inverse theorems for combinations of a class of exponential-type operators, Canad. J. Math. **28(6)** (1976), 1224–1250.
[20] C. N. Morris, Natural exponential families with quadratic variance functions, Ann. Statist. **10(1)** (1982), 65–80.
[21] T. T. Nielsen, Bose Algebras: The Complex and Real Wave Representation, Lecture Notes in Mathematics. Springer-Verlag, 1991.
[22] G. Pistone and H. P. Wynn, Finitely generated cumulants. Statist. Sinica **9(4)** (1999), 1029–1052.
[23] D. Pommeret, Natural exponential families and Lie algebras, Exposition. Math. **14(4)** (1996), 353–381.
[24] D. Pommeret, Orthogonal polynomials and natural exponential families, Test **5(1)** (1996), 77–111.
[25] Y. Xu, On multivariate orthogonal polynomials. *SIAM J. Math. Anal.*, **24(3)** (1993), 783–794.
[26] Y. Xu, Recurrence formulas for multivariate orthogonal polynomials, Math. Comp. **62**(1994), 687–702.

EURANDOM AND FACULTY OF TECHNOLOGY MANAGEMENT, EINDHOVEN UNIVERSITY OF TECHNOLOGY, P.O. BOX 513, 5600 MB EINDHOVEN THE NETHERLANDS
E-mail address: `A.d.Bucchianico@tm.tue.nl`

DANIEL H. WAGNER ASSOCIATES, 40 LLOYD AVENUE, MALVERN, PA 19355, USA
E-mail address: `loeb@pawagner.com`

Contemporary Mathematics
Volume **287**, 2001

Best Invariant Predictive Distributions

Morris L. Eaton and William D. Sudderth

ABSTRACT. Best invariant decision rules and predictors are found for statistical decision problems relatively invariant under the action of a group. Under an assumption of a proper group action on the sample space, a basic disintegration formula is used to show that the best invariant rules are formal Bayes for an improper prior induced by a right Haar measure on the group.

1. Introduction

Consider a statistical decision problem in which, after observing data $X \in \mathcal{X}$, a statistician must either estimate a parameter $\theta \in \Theta$ or predict a future observation $Z \in \mathcal{Z}$. In each case, suppose that the statistician must select an action $a \in \mathcal{A}$ based on the observation x thereby incurring a loss $L(a, \theta, x)$ in the estimation problem or $L(a, \theta, x, z)$ in the prediction problem. By taking $\mathcal{A} = \Theta$ or $\mathcal{A} = \mathcal{Z}$, we can treat problems of point estimation or prediction, respectively. For problems in which the statistician is required to specify a probability distribution for the unknown parameter θ or the future observation z, we can take $\mathcal{A}$ to be the set of probability measures on Θ or $\mathcal{Z}$, respectively.

It is assumed throughout the paper that a group G acts on each of the spaces $\Theta, \mathcal{X}, \mathcal{A}$, and $\mathcal{Z}$, and that the statistical model $\{P_\theta, \theta \in \Theta\}$ for the observations is invariant. It is also assumed that the loss function L is relatively invariant under the action of G. The major results of the paper, proved under a number of technical assumptions, are characterizations of a best invariant decision rule for both the case of estimation and that of prediction. These results imply that a best invariant rule can be viewed as a formal Bayes rule for an improper prior on Θ. This improper prior is induced by a right Haar measure on the group G and the multiplier for the relatively invariant loss function.

A central unifying theme in our approach is a basic disintegration formula that enables us to write the risk function of an invariant decision rule as a double integral – first over the group G with respect to right Haar measure and then over the quotient space $\mathcal{X}/G$ with respect to an induced quotient measure. Proofs of special cases of our results are in Berger (1985) and Eaton (1989) among others. Other authors have long since observed the key role of right Haar measure – for

1991 *Mathematics Subject Classification.* Primary 62C10, 62C05.

Research partially supported by NSF Grants DMS96-26601 and DMS97-03285.

example, Hora and Buehler (1966) and Zidek (1969). What is new here is the treatment of relatively, rather than fully invariant, loss functions and our inclusion of prediction problems. However, the most attractive feature for most readers is, we think, our use of the disintegration formula and the resulting simplicity of our arguments.

Here is an outline of the paper. The next section is an exposition of the basic disintegration formula. Section 3 formulates the class of invariant estimation problems we will treat and Section 4 finds a best invariant decision rule for such problems. In Section 5 is the formulation of invariant prediction problems and Section 6 finds best invariant predictors. Section 7 is about invariant prediction problems in which the statistician must specify a predictive distribution for a future observation based on the data. It is shown that a certain predictive distribution, corresponding to a formal posterior for a prior induced by right Haar measure, is best invariant for every invariant fair Bayes loss function as defined in Eaton (1982).

2. The basic disintegration formula

Our main purpose in this section is to explain the formula that is the key to the rest of the paper. The formula is quite natural, but the explanation is fairly technical. A more leisurely and more detailed exposition is given in Eaton (1989).

Suppose λ is a measure on a space $\mathcal{X}$ and that the group G acts on the space $\mathcal{X}$ and has a right Haar measure ν_r. In later sections we will exploit a disintegration of the form

$$\int_{\mathcal{X}} f(x)\lambda(dx) = \int_{\mathcal{X}/G} \left[\int_G f(gx)\nu_r(dg) \right] d\beta, \tag{2.1}$$

where β is a measure on the quotient space $\mathcal{X}/G$. Under regularity conditions to be discussed below, equation (2.1) is established in the works of Bourbaki and has been used by Andersson (1982) to calculate distributions of maximal invariants. (See Andersson (1982) for detailed references to the appropriate sections of Bourbaki.) Here we will show how the formula leads to a simple proof in a general setting of the fact that best invariant decision rules are formal Bayes rules for a prior induced by right Haar measure.

In order that formula (2.1) make sense and be valid, a number of assumptions are needed. First we assume that the space $\mathcal{X}$ is a locally compact, complete separable metric space and that G is another such space that is also a group acting continuously on $\mathcal{X}$. Thus there is a continuous mapping $(g, x) \longrightarrow gx$ from $G \times \mathcal{X}$ to $\mathcal{X}$ such that $(g_1 g_2)x = g_1(g_2 x)$ for all $g_1, g_2 \in G$ and $x \in \mathcal{X}$.

The locally compact group G has a *left Haar measure* ν_l. This means that ν_l is a non-trivial measure defined on the Borel subsets of G such that

$$\int_G f(\hat{g}g)\nu_l(dg) = \int_G f(g)\nu_l(dg)$$

for all $\hat{g} \in G$ and all nonnegative measurable functions $f : G \longrightarrow \mathbf{R}$. (By "measurable" we always mean "Borel measurable.") The *modular function* Δ of the group G is a homomorphism of G into the multiplicative group of positive real numbers whose defining property is that

$$\int_G f(g\hat{g}^{-1})\nu_l(dg) = \Delta(\hat{g}) \int_G f(g)\nu_l(dg). \tag{2.2}$$

A *right Haar measure* ν_r on G can be obtained from ν_l via the formula

$$\nu_r(dg) = \Delta(g^{-1})\nu_l(dg).$$

Furthermore

$$\int_G f(g^{-1})\nu_l(dg) = \int_G f(g)\nu_r(dg), \tag{2.3}$$

and

$$\int_G f(\hat{g}^{-1}g)\nu_r(dg) = \Delta^{-1}(\hat{g})\int_G f(g)\nu_r(dg). \tag{2.4}$$

(These properties of the Haar integral are available in Nachbin (1976).)

We now know how to interpret the inner integral on the right side of (2.1) as an integral with respect to right Haar measure. To begin to make sense of the outer integral, consider the quotient space $\mathcal{X}/G$ whose points are the *orbits* $\pi(x) = \{gx \mid g \in G\}$, $x \in \mathcal{X}$. The function $\pi : \mathcal{X} \to \mathcal{X}/G$ is the *natural projection*, and $\mathcal{X}/G$ is given the quotient topology wherein a set $U \subseteq \mathcal{X}/G$ is open if and only if $\pi^{-1}(U)$ is open in $\mathcal{X}$. In order to guarantee that $\mathcal{X}/G$ is a nice space, we assume that G acts *properly* on $\mathcal{X}$; that is, under the mapping

$$(g, x) \longrightarrow (gx, x)$$

from $G \times \mathcal{X}$ to $\mathcal{X} \times \mathcal{X}$, the inverse image of each compact set in $\mathcal{X} \times \mathcal{X}$ is compact in $G \times \mathcal{X}$. Under this assumption, the quotient space $\mathcal{X}/G$ is a locally compact Hausdorff space with a countable base for its topology. (A detailed discussion of proper action, its verification in examples, and its comparison to other approaches can be found in Wijsman (1990).)

Let $f \in K(\mathcal{X})$, the space of continuous functions $f : \mathcal{X} \longrightarrow R$ with compact support. Define the function $T(f) : \mathcal{X} \longrightarrow \mathbf{R}$ by

$$T(f)(x) = \int f(gx)\nu_r(dg).$$

The function $T(f)$ is *invariant* in the sense that $T(f)(\hat{g}x) = T(f)(x)$ for all $x \in \mathcal{X}$, $\hat{g} \in G$. Now the natural projection π is a *maximal invariant* function, which means that every invariant function on $\mathcal{X}$ is a function of π. In particular, there exists $f^* : \mathcal{X}/G \longrightarrow \mathbf{R}$ such that

$$T(f)(x) = f^*(\pi(x)).$$

In fact, under our assumption that G acts properly on $\mathcal{X}$, T maps $K(\mathcal{X})$ onto $K(\mathcal{X}/G)$. (See the discussion in Andersson (1982).) Thus the outer integral on the right side of (2.1) will make sense once we specify the measure β.

If the right hand side of (2.1) is well-defined, then, by (2.4),

$$\int_{\mathcal{X}/G}\left[\int_G f(\hat{g}^{-1}gx)\nu_r(dg)\right]d\beta = \Delta^{-1}(\hat{g})\int_{\mathcal{X}/G}\left[\int_G f(gx)\nu_r(dg)\right]d\beta.$$

Thus, in order that (2.1) hold, the measure λ on $\mathcal{X}$ must be *relatively invariant with multiplier* Δ^{-1}; that is,

$$\int f(g^{-1}x)\lambda(dx) = \Delta(g^{-1})\int f(x)\lambda(dx). \tag{2.5}$$

Therefore, we assume that λ is such a measure and define the *quotient measure* β by the formula

$$\int f(x)\lambda(dx) = \int f^*(u)\beta(du) \tag{2.6}$$

for $f \in K(\mathcal{X})$. (For a sketch of the argument that justifies this definition of β, see Andersson(1982).) It follows from (2.6) and the definition of f^* that (2.1) holds for $f \in K(\mathcal{X})$. Hence, it also holds for nonnegative measurable f as well.

Therefore if λ is relatively invariant with multiplier Δ^{-1} and if G acts properly on $\mathcal{X}$, the disintegration formula (2.1) holds. These two conditions are assumed to hold throughout this paper.

As a first illustration of (2.1), we use it to find a formula for the density of a maximal invariant. Let X be an $\mathcal{X}$-valued random variable with density p with respect to the measure λ. Then $Y = \pi(X)$ has density p^* with respect to the quotient measure β. Indeed, for $\phi : \mathcal{X}/G \longrightarrow \mathbf{R}$ bounded measurable, we have by (2.1)

$$\begin{aligned} E\phi(Y) &= \int \phi(\pi(x))p(x)\lambda(dx) \\ &= \int \phi(u)p^*(u)\beta(du). \end{aligned}$$

This is essentially the argument in Andersson (1982).

3. Invariant decision problems

In this section we describe a large class of invariant statistical decision problems. In the next section we will explain a method for finding a best invariant decision rule.

There are six fundamental ingredients for an invariant decision problem: a sample space $\mathcal{X}$, a parameter space Θ, an action space $\mathcal{A}$, a statistical model $\{P_\theta, \theta \in \Theta\}$, a loss function $L : \mathcal{A}\times\Theta\times\mathcal{X} \longrightarrow [0,\infty)$, and a topological group G. The spaces $\mathcal{X}, \Theta, \mathcal{A}$, and the group G are assumed to be locally compact, complete separable metric spaces and each is equipped with its σ-field of Borel sets. We assume that G acts continuously on each of the spaces $\mathcal{X}$, Θ, $\mathcal{A}$ and that its action on $\mathcal{X}$ is proper as in the previous section. An assumption needed in the next section is that G acts *transitively* on Θ; that is, given $\theta_1, \theta_2 \in \Theta$, there exists $g \in G$ such that $g\theta_1 = \theta_2$. Each P_θ specified by the model is, as usual, a probability measure on the sample space $\mathcal{X}$, and we assume that there is a reference measure λ on $\mathcal{X}$ such that

$$P_\theta(dx) = p(x|\theta)\lambda(dx).$$

Here the density $p(x|\theta)$ is jointly measurable in x and θ, and satisfies

$$p\,(x \mid \theta) = \Delta^{-1}(g)\,p\,(gx \mid g\theta). \tag{3.1}$$

The measure λ is relatively invariant with multiplier Δ^{-1} as in (2.5). The loss function $L(a,\theta,x)$ is jointly measurable in a, θ, and x and is *relatively invariant with multiplier* χ in the sense that

$$L(ga, g\theta, gx) = \chi(g)L(a,\theta,x), \tag{3.2}$$

where χ is a given multiplier for the group G. (A *multiplier* χ for G is a homomorphism from G to the multiplicative group of positive reals. For a discussion with many examples, see Section 1.4 of Eaton(1989).) In most treatments of invariant decision problems, the loss function is assumed to be invariant. This amounts to taking the multiplier χ to be identically equal to one.

Here is a simple example.

EXAMPLE 3.1. Let X be a random variable whose distribution depends on a scale parameter $\theta > 0$. Suppose that X has density $\frac{1}{\theta}f(\frac{x}{\theta})$ with respect to Lebesque measure on $\mathbf{R}^1$. Let G be the multiplicative group of positive reals acting on $\mathcal{X} = \mathbf{R}^1$ and $\Theta = (0, \infty)$ in the obvious way. The group G is Abelian and so has modulus $\Delta \equiv 1$; so a trivial calculation shows that (3.1) fails. However, we can replace Lebesque measure by another dominating measure, namely

$$\lambda(dx) = \frac{dx}{|x|},$$

with respect to which X has the density

$$p(x|\theta) = \frac{|x|}{\theta}f(\frac{x}{\theta}).$$

This new density does satisfy (3.1). Another technical problem is that the action of G on $\mathbf{R}^1$ is not proper. (The inverse image of the singleton $\{(0,0)\}$ under the mapping $(g, x) \longrightarrow (gx, x)$ is the set $G \times \{0\}$, which is not compact.) This can be remedied by taking $\mathcal{X}$ to be $\mathbf{R}^1 - \{0\}$. Finally, take the action set to be $\mathcal{A} = (0, \infty)$ and the loss function to be

$$L(a, \theta, x) = L(a, \theta) = (a - \theta)^2,$$

the usual quadratic loss function. Then L is relatively invariant with multiplier $\chi(g) = g^2$.

It is, of course, easy to extend the example to the case of a random sample $(X_1, X_2, \ldots, X_n)$ from a scale family.

A *decision rule* is a measurable mapping $\delta : \mathcal{X} \longrightarrow \mathcal{A}$. (To simplify the exposition, we will not consider randomized decision rules in this paper.) A decision rule is evaluated by its expected loss or *risk function*

$$R(\delta, \theta) = \int L(\delta(x), \theta, x)p(x|\theta)\lambda(dx). \tag{3.3}$$

A decision rule δ is *invariant* if $\delta(gx) = g\delta(x)$ for all $g \in G, x \in \mathcal{X}$. The next section gives conditions that lead to a best invariant rule.

4. Best invariant decision rules

In this section we prove our main result concerning best invariant decision rules. The assumptions of Sections 2 and 3 are all in force. In particular, we assume that G is transitive on Θ, the dominating measure λ on $\mathcal{X}$ is relatively invariant with multiplier Δ^{-1}, and G acts properly on $\mathcal{X}$ so that the basic disintegration formula (2.1) holds.

In order to find a best invariant rule, we will first apply (2.1) to calculate the risk function of an arbitrary invariant rule. Indeed, if δ is an invariant decision rule, then, applying (2.1) with $f(x) = L(\delta(x), \theta, x)\, p\,(x|\theta)$, we have

$$(4.1) \qquad R(\delta,\theta) = \int_{\mathcal{X}/G} \left[\int_G L(\delta(gx), \theta, gx)\, p\,(gx|\theta) \nu_r(dg) \right] d\beta = \int_{\mathcal{X}/G} I_\delta(\theta, x) d\beta,$$

where $I_\delta(\theta, x)$ is, by definition, the inner integral on the right side of the first line above. Now use the properties of the density p in (3.1), the loss function L in (3.2), and the Haar measures ν_r and ν_l in (2.1) to (2.4) to calculate as follows:

$$(4.2) \qquad \begin{aligned} I_\delta(\theta, x) &= \int_G L(g\delta(x), gg^{-1}\theta, gx)\, p\,(x|g^{-1}\theta)\Delta(g)\nu_r(dg) \\ &= \int_G L(g\delta(x), gg^{-1}\theta, gx)\, p\,(x|g^{-1}\theta)\nu_l(dg) \\ &= \int_G \chi^{-1}(g^{-1}) L(\delta(x), g^{-1}\theta, x)\, p\,(x|g^{-1}\theta)\nu_l(dg) \\ &= \int_G \chi^{-1}(g) L(\delta(x), g\theta, x)\, p\,(x|g\theta)\nu_r(dg). \end{aligned}$$

Hence,

$$\begin{aligned} I_\delta(\hat{g}\,\theta, x) &= \int_G \chi(\hat{g})\chi^{-1}(g\hat{g}) L(\delta(x), g\hat{g}\,\theta, x)\, p\,(x|g\hat{g}\,\theta)\nu_r(dg) \\ &= \chi(\hat{g}) I_\delta(\theta, x). \end{aligned}$$

So, by (4.1),

$$(4.3) \qquad R(\delta, g\,\theta) = \chi(g) R(\delta,\, \theta).$$

Now fix a parameter value $\theta_0 \in \Theta$ and define the function $H : \mathcal{A} \times \mathcal{X} \longrightarrow \mathbf{R}^+$ by

$$(4.4) \qquad H(a, x) = \int \chi^{-1}(g) L(a, g\theta_0, x)\, p\,(x|g\theta_0)\nu_r(dg).$$

Then (4.1) can be rewritten as

$$(4.5) \qquad R(\delta, \theta_0) = \int_{\mathcal{X}/G} H(\delta(x), x) d\beta$$

and (4.3) as

$$(4.6) \qquad R(\delta, g\theta_0) = \chi(g) \int_{\mathcal{X}/G} H(\delta(x), x) d\beta.$$

The integrals in (4.5) and (4.6) over $\mathcal{X}/G$ make sense because, as a straightforward calculation shows, H is invariant in the sense that $H(ga, gx) = H(a, x)$ and so $H(\delta(gx), gx) = H(\delta(x), x)$.

THEOREM 4.1. *Suppose that there exists an invariant decision rule δ^* such that*

$$(4.7) \qquad H(\delta^*(x), x) = \inf_a H(a, x)$$

for all $x \in \mathcal{X}$. Then δ^* is a best invariant rule; that is,

$$R(\delta^*, \theta) \le R(\delta, \theta)$$

for all $\theta \in \Theta$ and all invariant decision rules δ.

PROOF. For $\theta \in \Theta$, we can write $\theta = g\theta_0$ for some $g \in G$, because G acts transitively on Θ. By (4.5) and the assumed properties of δ^*, we have

$$\begin{aligned} R(\delta,\theta) &= \chi(g)\int_{\mathcal{X}/G} H(\delta(x),x)d\beta \\ &\geq \chi(g)\int_{\mathcal{X}/G} H(\delta^*(x),x)d\beta \\ &= R(\delta^*,\theta). \end{aligned}$$

□

If there is a decision rule that achieves the infinitum in (4.7), there is typically no difficulty in finding one that is also invariant. However, there are delicate measurability problems in the way of a general proof. In the special case when there is a unique minimizer $\delta^*(x)$ in (4.7), there are no such problems because such a δ^* must be invariant. To see this, use the invariance of H to calculate

$$\begin{aligned} H(g\delta^*(x),gx) &= H(\delta^*(x),x) \\ &\leq H(g^{-1}a,x) \\ &= H(a,gx). \end{aligned}$$

Thus $g\delta^*(x)$ minimizes $H(a,gx)$ over all $a \in \mathcal{A}$ and must equal the unique minimizer $\delta^*(gx)$.

There is a Bayesian interpretation of the best invariant rule δ^* of Theorem 1. First let $H = \{g \in G \mid g\theta_0 = \theta_0\}$ be the *isotropy subgroup* of θ_0 and assume that H is compact. Then, by Weil's Theorem (cf. Nachbin(1965), pp. 125-141), the formula

$$\int f(\theta)\pi(d\theta) = \int f(g\theta_0)\chi^{-1}(g)\nu_r(dg) \tag{4.8}$$

defines a Radon measure $\pi(d\theta)$ on Θ, which we can interpret as an improper prior induced by $\chi^{-1}(g)\nu_r(dg)$. (The prior π will be proper if and only if G is compact.) The formal posterior of π, namely $Q(d\theta|x)$, is given by

$$\int f(\theta)Q(d\theta|x) \propto \int f(g\theta_0)\, p\,(x|g\theta_0)\chi^{-1}(g)\nu_r(dg),$$

where the proportionality symbol means equality up to a function of x. Thus the best invariant rule of δ^* of Theorem 1 is chosen to minimize the posterior risk

$$\int L(a,\theta,x)Q(d\theta|x) \propto H(a,x);$$

that is, δ^* is the formal Bayes rule for the improper prior π.

As a continuation of Example 1, recall that the problem is to estimate the parameter of a scale family using a quadratic loss. It follows from the discussion above that the best invariant rule will be the mean of the formal posterior of the improper prior induced by $\chi^{-1}(g)\nu_r(dg) = g^{-3}dg$ on $G = (0,+\infty)$.

EXAMPLE 4.2. Suppose that, rather than a point estimate, the statistician is required to specify a probability distribution $Q(d\theta|x)$ for the parameter θ given the data x. For a decision theoretic formulation, we can take $\mathcal{A} = M(\Theta)$ to be the set of probability distributions on Θ. The action of G on Θ induces a natural action on $M(\Theta)$ by the rule $(g\gamma)(B) = \gamma(g^{-1}B)$ for $g \in G$, $\gamma \in M(\Theta)$ and all Borel subsets

B of Θ. Thus a decision rule $\delta(x) = Q(\cdot|x)$ is invariant if $Q(gB|gx) = Q(B|x)$ for all g, x, and Borel $B \subseteq \Theta$.

A natural class of loss functions $L(\gamma, \theta, x)$ are the *fair Bayes* loss functions (cf. Eaton (1992)) with the defining property that

$$\inf_{\gamma} \int_{\Theta} L(\gamma, \theta, x)\pi(d\theta) = \int_{\Theta} L(\pi, \theta, x)\pi(d\theta) \tag{4.9}$$

for all x and every $\pi \in M(\Theta)$. Suppose that $\delta_{\pi}(x) = Q(\cdot|x)$ corresponds to the posterior of a proper prior π. Then, by (4.9), $\delta_{\pi}(x)$ minimizes the posterior risk for every x and is therefore a Bayes rule for π.

Suppose that $L(\gamma, \theta, x)$ is a fair Bayes loss function that is *fully invariant* in the sense that $L(g\gamma, g\theta, gx) = L(\gamma, \theta, x)$ for all g, γ, θ, and x. (Thus L satisfies (3.2) with $\chi(g) \equiv 1$.) Let $Q(d\theta|x)$ be the formal posterior for the improper prior π induced by $\nu_r(dg)$. If Q is invariant (and this is true for specific examples that we know), then Q is a best invariant rule by Theorem 1 and the discussion that follows it. Notice that the same Q is a best invariant rule for every fully invariant fair Bayes loss function L. This provides some decision theoretic support to the conclusion in Eaton and Sudderth (2000) – namely, that among all the invariant decision rules, only Q is worthy of serious consideration.

What makes the proof of Theorem 1 so transparent is the formula (2.1) for the quotient measure. Other proofs of special cases of Theorem 1 can be found in Berger (1985, p. 409 ff and the reference there) and Eaton (1989). When $\chi(g) \equiv 1$, other authors have also noted that the best invariant rule is a formal Bayes rule when the improper prior is right Haar measure – for example, see Hora and Buehler (1966) or Zidek (1969). It is the simplicity of our arguments that makes the current version of Theorem 1 attractive.

5. An invariant prediction problem

The extension of Theorem 1 to the prediction case is the subject of this and the next section. Here, notation and assumptions for the prediction problem are stated. As far as possible, the notation established in earlier sections is maintained.

What we mean by a prediction problem is the following. Data $Y \in \mathcal{Y}$ is available for predicting a variable $Z \in \mathcal{Z}$. The joint distribution of (Y, Z) is indexed by a parameter $\theta \in \Theta$ and it is assumed that a group G acts on $\mathcal{Y}, \mathcal{Z}$ and Θ with the action on Θ being transitive. This joint distribution is assumed to have the form

$$P(dy, dz|\theta) = r(y, z|\theta)\lambda(dy)\xi(dz), \tag{5.1}$$

where λ is a relatively invariant measure with multiplier Δ^{-1} and ξ is a relatively invariant measure with multiplier ϕ. The density function $r(y, z|\theta)$ is assumed to satisfy the usual invariance equation

$$r(y, z|\theta) = \Delta^{-1}(g)\phi(g)r(gy, gz|g\theta) \tag{5.2}$$

for all values of y, z, θ and g. Therefore the probability model defined by (5.1) is invariant under the action of the group G.

Next, let $\mathcal{A}$ be an action space on which G acts. Consider a relatively invariant loss function (see Section 3) $L(a, \theta, x, z)$ which satisfies

$$L(ga,\ g\theta,\ gy,\ gz) = \chi(g)\ L(a,\ \theta,\ y,\ z), \tag{5.3}$$

where χ is a multiplier on G. Two particular $\mathcal{A}'s$ of special interest in the prediction problem are:

- $\mathcal{A} = \mathcal{Z}$ so one is interested in a point predictor for the variable Z.
- $\mathcal{A}$ = set of all probability measures on $\mathcal{Z}$. In this case, an action corresponds to stating a probability distribution for Z based on the model and the data $Y = y$.

However, the main result in the next section will be for general action spaces and loss functions satisfying (5.3).

As usual, a decision rule is a measurable mapping $\delta : \mathcal{Y} \longrightarrow \mathcal{A}$ whose risk function is given by

$$R(\delta, \theta) = \int_{\mathcal{Z}} \int_{\mathcal{Y}} L(\delta(y), \theta, y, z)\, r(y, z\,|\theta)\, \lambda(dy)\xi(dz). \tag{5.4}$$

A decision rule is invariant if $\delta(gy) = g\delta(y)$ for $g \in G$ and $y \in \mathcal{Y}$.

6. Best invariant predictors

In this section, we give the prediction version of Theorem 1 that is proved in Section 4. The basic assumptions here are those stated in Section 5 plus those stated at the beginning of Section 4. In particular, it is assumed that G is transitive on Θ, the measure λ on $\mathcal{Y}$ is relatively invariant with multiplier Δ^{-1}, and G acts properly on $\mathcal{Y}$ so that the basic disintegration formula (2.1) holds.

Under the above assumptions, we now want to show how to find a best invariant decision rule. The argument below is quite similar to that given in Section 4 so many of the details are omitted. For any invariant decision rule δ, first apply (2.1) with

$$f(y) = \int_{\mathcal{Z}} L(\delta(y),\, \theta,\, y,\, z)\; r\;(y,\; z|\theta)\; \xi(dz)$$

to obtain

$$\begin{aligned} R(\delta,\, \theta) &= \int_{\mathcal{Y}/G} \left[\int_G \int_{\mathcal{Z}} L(\delta(gy),\, \theta,\, gy,\, z)\; r\;(gy,\; z|\theta)\cdot \right. \\ &\qquad\qquad \left. \xi(dz)\nu_r(dg)\right] d\beta \\ &= \int_{\mathcal{Y}/G} I_\delta(\theta, y) d\beta, \end{aligned} \tag{6.1}$$

where $I_\delta(\theta, y)$ is, by definition, the expression inside the square brackets of the triple integral. Now argue as in Section 4 to show that

$$\left\{ \begin{array}{ccc} I_\delta(g\theta, y) & = & \chi(g) I_\delta(\theta, y) \\ \text{and consequently} & & \\ R(\delta, g\theta) & = & \chi(g) R(\delta, \theta). \end{array} \right. \tag{6.2}$$

Next fix $\theta_0 \in \Theta$ and define $H : \mathcal{A} \times \mathcal{Y} \longrightarrow [0, \infty)$ by

$$H(a, y) = \int_G \int_{\mathcal{Z}} \chi^{-1}(g) L(a, g\theta_0, y, z)\; r\;(y, z|g\theta_0)\xi(dz)\nu_r(dg). \tag{6.3}$$

Then, it follows that

$$\left\{ \begin{array}{ccc} R(\delta, \theta_0) & = & \int_{\mathcal{X}/G} H(\delta(y), y) d\beta \\ \text{and} & & \\ R(\delta, g\theta_0) & = & \chi(g) \int_{\mathcal{X}/G} H(\delta(y), y) d\beta. \end{array} \right. \tag{6.4}$$

Note that H is invariant in the sense that $H(ga, gy) = H(a, y)$. Therefore $H(\delta(y), y)$ is an invariant function of y.

Here is the analog of Theorem 1 for the prediction situation.

THEOREM 6.1. *Suppose that there exists an invariant decision rule δ^* such that*

$$H(\delta^*(y), y = \inf_a H(a, y) \tag{6.5}$$

for all $y \in \mathcal{Y}$. Then δ^* is a best invariant rule; that is,

$$R(\delta^*, \theta) \leq R(\delta, \theta)$$

for all $\theta \in \Theta$ and all invariant rules δ.

PROOF. Just repeat the proof of Theorem 1. □

The discussion given in Section 4 concerning the invariance of minimizers in (6.5) is valid here without modification. However the Bayesian interpretation does require some explanation. Assuming the isotropy subgroup of θ_0 is compact (as in Section 4), let π be the induced Radon measure on Θ as in (4.8). Then the formal posterior distribution of (Z, θ) (given the data $Y = y$), say $Q(dz, d\theta|y)$, is characterized by the equation

$$\int_\Theta \int_\mathcal{Z} f(z, \theta) Q(dz, d\theta|y) \propto \tag{6.6}$$

$$\int_\Theta \int_\mathcal{Z} f(z, \theta) r(y, z|\theta) \xi(dz) \pi(d\theta),$$

where the proportionality sign means equality up to a function of y. From this we see that

$$H(a, y) \propto \int_\Theta \int_\mathcal{Z} L(a, \theta, y, z) Q(dz, d\theta|y).$$

In other words, the best invariant rule δ^* is obtained by minimizing the formal posterior risk based on the conditional distribution of (Z, θ) given the data. This is what we mean by saying that δ^* is the formal Bayes rule for the improper prior $\chi^{-1}(g)\nu_r(dg)$.

A simple application of Theorem 2 follows. Suppose $X_1, \ldots, X_n$ are iid Exponential (θ) with density $f(x|\lambda) = (x/\theta)exp[-x/\theta]$ for $x > 0$ with respect to the measure dx/x on $(0, \infty)$. Hence $Y = (0, \infty)^n$ and $\lambda(dy)$ is the product measure

$$\lambda(dy) = \prod_{i=1}^n \frac{dx_i}{x_i}.$$

Take $Z \in \mathcal{Z} = (0, \infty)$ to also be exponential (θ) and independent of $X_1, \ldots, X_n$, given θ. The group in question is the multiplicative group $G = (0, \infty)$ which is Abelian so $\Delta \equiv 1$. To motivate the decision problem of this application, suppose $X_i = x_i$, $i = 1, \ldots, n$ have been observed, and we now ask "What is the chance that Z will set a record?" That is, what is the chance that Z will exceed $\max\{x_i| \ i = 1, \ldots, n\} = x_{(n)}$? That is, with the x_i's fixed, how might one estimate

$$P_\theta\{Z > x_{(n)}\} \equiv \Psi(x_{(n)}, \theta)?$$

A bit more generally, let x be the vector of observed values and consider $t(x) \in (0, \infty)$ which is equivariant – that is, $t(cx) = ct(x)$, $c \in (0, \infty)$.

Consider

$$P_\theta\{Z > t(x)\} = \Psi(t(x), \theta) \tag{6.7}$$

and suppose we want to estimate (6.7). Let the action space be $(A) = [0, 1]$ and let the loss function be

$$L(a, \theta, x) = (a - \Psi(t(x), \theta))^2.$$

Note that L is invariant when G acts trivially on $\mathcal{A}$ and acts in the obvious way on Θ and $\mathcal{Y}$. A direct application of Theorem 2 shows that the best invariant decision rule is the point estimator:

$$\hat{\Psi}(x) = \int_0^\infty \int_0^\infty \Psi(t(x), \theta)\, Q(dz, d\theta | x),$$

where $Q(dz, d\theta|x)$ is the joint formal posterior distribution of (Z, θ) given the data is x. A routine calculation shows that $\hat{\Psi}(x)$ is

$$\hat{\Psi}(x) = 1 - H(t(x)/\sum_{i=1}^n x_i),$$

where H is the distribution function of the pivotal quantity

$$W = \frac{Z}{\sum_{i=1}^n X_i}.$$

7. Best invariant predictive distributions

In the context of this paper, a predictive distribution for Z given the data $Y = y$ is just a probability distribution for Z which depends on y, say $S(dz|y)$. In other words, $S(dz|y)$ is a Markov Kernel. The general notion of a predictive distribution was introduced by Laplace in 1774 (see Stigler(1986)). More modern discussions can be found in Aitchison and Dunsmore (1975) and Geisser (1993). Under the assumptions (5.1) on the joint model for (Y, Z), we now want to focus on one particular predictive distribution. This predictive distribution will turn out to be invariant in a sense to be made precise below. The primary purpose of this section is to show that this invariant predictive distribution is in fact a best invariant decision rule in a large family of invariant decision problems.

The predictive distribution of interest here can be described as follows. Let θ_0 be a fixed point in Θ and let

$$m(y) = \int_{\mathcal{Z}} \int_G r(y, z | g\theta_0) \nu_r(dg) \xi(dz). \tag{7.1}$$

For simplicity of exposition, it is assumed that

$$0 < m(y) < +\infty \qquad \text{for all } y \in Y. \tag{7.2}$$

Therefore, the function

$$q(z|y) = \frac{\int_G r(y, z|g\theta_0)\nu_r(dg)}{m(y)} \tag{7.3}$$

is a probability density with respect $\xi(dz)$ for each $y \in \mathcal{Y}$. Thus, the Markov Kernel

$$Q^*(dz|y) = q(z|y)\xi(dz) \tag{7.4}$$

is a predictive distribution for Z given $Y = y$ in the sense defined above. Using the assumed invariance (5.2) for the model density and the invariance of the measure ξ, it is not hard to show that Q^* satisfies

$$Q^*(gB|gy) = Q^*(B|y) \tag{7.5}$$

for all $g \in G$, for all Borel sets B in $\mathcal{Z}$, and all $y \in \mathcal{Y}$. In other words, the Markov Kernel Q^* is *invariant.* The Kernel Q^* is often called the formal predictive distribution for Z given $Y = y$ when the improper prior distribution is obtained from right Haar measure on G.

To interpret Q^* as a decision rule, consider a decision problem where the action space $\mathcal{A}$ is the set of all probability measures on $\mathcal{Z}$. The group G acts on $\mathcal{A}$ via

$$(g\gamma)(B) = \gamma(g^{-1}B), \tag{7.6}$$

where $g \in G$, $\gamma \in \mathcal{A}$ and B is a Borel subset of $\mathcal{Z}$. Thus a decision rule δ is a map from $\mathcal{Y}$ to $\mathcal{A}$. Because of the nature of $\mathcal{A}$, the value of a decision rule δ at a point y will be written $\delta(\cdot|y)$. Thus the value of $\delta(\cdot|y) \in \mathcal{A}$ at a point $B \subseteq \mathcal{Z}$ is $\delta(B|y)$ so δ can be thought of as a Markov Kernel. Observe that the invariance of a decision rule δ is expressed as

$$\delta(gB|gy) = \delta(B|y) \tag{7.7}$$

for all $g \in G$, $y \in \mathcal{Y}$ and subsets B. Therefore the predictive distribution Q^* defined above is an invariant decision rule.

Next, let $L(a, \theta, y, z)$ be an invariant loss function when the action space $\mathcal{A}$ is the set of all probability measures on $\mathcal{Z}$. Using assumption (7.2), a conditional distribution of (Z, θ) given the data $Y = y$ is

$$Q(dz, d\theta|y) = \frac{r(y, z|\theta)}{m(y)} \xi(dz)\pi_r(d\theta), \tag{7.8}$$

where π_r is the measure induced on Θ from the right Haar measure on G (see equation (4.8) and set $\chi(g) \equiv 1$). It then follows from Theorem 2 that a best invariant decision rule is found by minimizing (over a) the function

$$H^*(a, y) = \int_\Theta \int_{\mathcal{Z}} L(a, \theta, y, z)\, Q(dz, d\theta|y). \tag{7.9}$$

THEOREM 7.1. *Suppose $L(a, \theta, y, z)$ is an invariant loss function and suppose that for each fixed y, $\tilde{L}_y(a, (\theta, z)) \equiv L(a, \theta, y, z)$ is a Fair Bayes Loss Function as defined in Eaton (1982, see Section 5). Then $Q^*(dz|y)$ as defined in (7.4) is a best invariant decision rule.*

PROOF. Using the discussion in Eaton(1982, Section 5), a minimizer (over a) of $H^*(a, y)$ in (7.9) is given by the z–marginal distribution obtained from Q. That is, let $\tilde{Q}(dz|y)$ be given by $\tilde{Q}(dz|y) = Q(dz, \Theta|y)$. Then $\tilde{Q}$ is a minimizer of (7.9). However, a simple calculation shows that $\tilde{Q}$ and Q^* are the same. □

Many examples of fully invariant Fair Bayes Loss Functions can be constructed using the material in Eaton (1982). The above Theorem shows that the predictive distribution Q^* is a best invariant decision rule for a wide variety of decision problems simultaneously. This observation is in keeping with that in Eaton and Sudderth (1999).

EXAMPLE 7.2. Consider $X_1, \ldots, X_n$ which are iid $N_1(\mu, \sigma^2)$ and Z which is $N_1(\mu, \sigma^2)$. It is assumed $n \geq 2$, Z is independent of the X's, and $\sigma^2 > 0$. For this example $\theta = (\sigma, \mu)$ and the group in question is $G = Al_1-$ the affine group whose elements are (a, b) with $a > 0$ and $b \in R^1$. The group operation is

$$(a_1, b_1)(a_2, b_2) = (a_1 a_2, a_1 b_2 + b_1)$$

and a right Haar measure on G is

$$\nu_r(da, db) = \frac{dadb}{a}.$$

The modular function on G is

$$\Delta((a, b)) = \frac{1}{a}.$$

Set $s^2 = \sum_1^n (X_1 - \bar{X})^2$ and let

$$\mathcal{Y} = \{x | x \in R^n, \ s^2 > 0\}$$

so the complement of $\mathcal{Y}$ has Lebesque measure zero in R^n. The measure

$$\lambda(dy) = \frac{dy}{(\sum_1^n (y_i - \bar{y})^2)^{n/2}}$$

is relatively invariant with multiplier Δ^{-1}. The verification that G acts properly on $\mathcal{Y}$ is not too hard. The predictive distribution for Z given the data $X_1, \ldots, X_n$ can now be calculated using densities and integrating with respect to right Haar measure, but it is much easier to use a Fisherian pivoting argument (The equivalence of the two arguments is established in Eaton and Sudderth (1999, Section 9)). First note that the sampling distribution of the statistic

$$T_{n-1} = (1 + \frac{1}{n})^{-\frac{1}{2}} \frac{Z - \bar{X}}{\sqrt{\frac{s^2}{n-1}}}$$

is a Student's t-distribution with $n - 1$ degrees of freedom. Thus the predictive distribution of Z given the data is that of

$$(1 + \frac{1}{n})^{\frac{1}{2}} (\frac{s^2}{n-1})^{\frac{1}{2}} T_{n-1} + \bar{X},$$

where T_{n-1} is random but s^2 and $\bar{X}$ are fixed. This shows that the predictive distribution of Z obtained from the improper prior equal to right Haar measure is a scaled and translated t_{n-1} distribution. Example 3 is complete.

References

[1] Aitchison, J. and Dunsmore, J.R. (1975). *Prediction Analysis*. Cambridge: Cambridge University Press.

[2] Andersson, S. (1982). Distributions of maximal invariants using quotient measures. *Annals of Statistics*, **10**, 955-961.

[3] Berger, J.O. (1985). *Statistical Decision Theory and Bayesian Analysis*, 2nd Edition. New York: Springer-Verlag.

[4] Eaton, M.L. (1982). A method for evaluating improper prior distributions. In S.S. Gupta and J.O. Berger (eds.), *Statistical Decision Theory and Related Topics III*, **1**, pp. 329 352. New York: Academic Press.

[5] ——— (1989). *Group Invariance Applications in Statistics.* Regional Conference Series in Probability and Statistics **1**. Hayward, CA: Institute of Mathematical Statistics.
[6] Eaton, M.L. and Sudderth, W. (1999). Consistency and Strong Inconsistency of group invariant predictive inferences. *Bernoulli*, **5**, 833–854.
[7] ——— (2000). Group invariant inference and right Haar measure. To appear in *Journal of Statistical Planning and Inference.*
[8] Geisser, S. (1993). *Predictive Inferences: An Introduction.* New York: Chapman & Hall.
[9] Hora, R.B. and Buehler, R.J. (1966). Fiducial theory and invariant estimation. *Annals of Mathematical Statistics* **37**, 643–656.
[10] Nachbin, L. (1965). *The Haar Integral.* Princeton, NJ: Van Nostrand.
[11] Stigler, S. (1986). LaPlace's 1774 memoir on inverse probability. *Statistical Sciences* **1**, 359-363.
[12] Wijsman, R. (1990). *Invariant Measures on Groups and Their Use in Statistics.* Lecture Notes and Monograph Series, Vol. 14. Hayward, CA. Institute of Mathematical Statistics.
[13] Zidek, J. (1969). A representation of Bayes invariant procedures in terms of Haar measure. *Annals of the Institute of Statistical Mathematics* **21**, 291–308.

School of Statistics, University of Minnesota, Minneapolis, Minnesota 55455
E-mail address: eaton@stat.umn.edu

School of Statistics, University of Minnesota, Minneapolis, Minnesota 55455
E-mail address: bill@stat.umn.edu

Contemporary Mathematics
Volume **287**, 2001

A Family Of Probability Densities Related To The Riemann Zeta Function

Werner Ehm

ABSTRACT. The function $\zeta(z)\,(z-1)/z^2$ is shown to generate an exponential family of probability densities $\{p_\sigma,\ \sigma>0\}$ on $[0,\infty)$. Nevanlinna's formula and the connection with Riemann's ξ function are used to obtain information about the cumulants of the densities p_σ, and to derive various "tests" of the Riemann Hypothesis.

1. Introduction

The Riemann zeta function is a complex, meromorphic function with a simple pole at $z=1$ defined as $\zeta(z)=\sum_{n=1}^{\infty} n^{-z}$ in the half plane $\Re z>1$. The intention of this paper is to connect the study of the Riemann zeta function with the theory of Hardy functions, and to explore some of the consequences. This is an old theme dating back (at least) to the work of Nyman [**N**] and Beurling [**Be**]. Some interesting notes about the history of the subject may be found in [**BS**, Introduction]. For recent work we refer, in addition to [**BS**], to [**BSY**] and [**Bu1**] and the references cited therein.

A straightforward way of trying to associate a Hardy function with $\zeta(z)$ consists in multiplying it by $(z-1)/z^2$: the factor $z-1$ removes the pole at $z=1$, and the division by z^2 should make it decay sufficiently fast along vertical lines in the complex plane. Moreover, the zeros remain the same. Thus we are led to study the modified zeta function

$$\eta(z) \;=\; \zeta(z)\,(z-1)/z^2.$$

This recipe of constructing a Hardy function may also be applied to other Dirichlet series having a simple pole at $z=1$, but here we shall only consider the case of the zeta function.

We show that η can be represented as the Laplace transform of a positive function $\phi(x)$ on the positive real half line with linear growth for $x\to\infty$. It follows that η belongs to the Hardy space $\mathcal{H}^2_\sigma$ of every half plane $\Re z>\sigma$, $\sigma>0$, but is not a Hardy function with respect to the entire right half plane $\Re z>0$. (For information about Hardy functions and spaces see the books by Dym and McKean

1991 *Mathematics Subject Classification.* 11M26, 30E20, 60E10.

Key words and phrases. Cumulant, exponential family, Hardy function, Laplace transform, Nevanlinna's formula, Riemann Hypothesis.

[**DMK**] and Hoffman [**H**].) The function η also appears in a recent paper of Burnol [**Bu1**]. Burnol shows that $\eta \in \mathcal{H}^2_{1/2}$ and also states the inversion formula giving ϕ (up to a variable transformation), but he does not mention the positivity aspect.

The non negativity of ϕ makes it possible to associate an exponential family $\mathcal{P} = \{p_\sigma,\ \sigma > 0\}$ of probability densities p_σ on $[0, \infty)$ with ϕ, by taking $p_\sigma(x)$ proportional to $e^{-\sigma x}\,\phi(x)$. Information about $\log \eta$, the cumulant generating function of $\mathcal{P}$, can be gained by exploiting the Hardy function properties of η and its connection with the Riemann ξ function. We thus obtain explicit formulae for the cumulants of the densities p_σ which involve the zeros of the zeta function, as well as various "tests" of the Riemann Hypothesis one of which generalizes a related result of Balazard, Saias and Yor [**BSY**]. However, genuinely probabilistic aspects are touched upon only marginally in this paper, and remain to be explored. Related material emphasizing probabilistic interpretations may be found in the recent work of Biane, Pitman and Yor [**BPY**]. These authors comprehensively review known and new results connecting integral representations of the zeta function to probability laws of certain sums of independent random variables, and of functionals of Brownian motion and related processes.

The statement and proof of the representation theorem is given in Section 2. Section 3 contains the material related to Hardy functions and some tentative remarks concerning probabilistic aspects. The cumulants of the densities p_σ and certain sums built with the zeros of the zeta function are studied in Section 4.

2. Representation of η as a Laplace transform

Our starting point is the following Paley–Wiener type representation of η. See also [**Bu1**].

THEOREM 2.1. *The function η can be represented as*

$$\eta(z) = \int_0^\infty e^{-zx}\,\phi(x)\,dx \qquad (\Re z > 0), \tag{2.1}$$

with a real function ϕ given by

$$\phi(x) = \sum_{1 \le n \le e^x} (1 + \log n - x) \qquad (x \in \mathbb{R}). \tag{2.2}$$

ϕ vanishes for $x < 0$, is strictly positive for $x \ge 0$, and satisfies the growth conditions

$$\begin{aligned} -1 &= \liminf_{x\to\infty} \left(\phi(x) - x/2 - \log\sqrt{2\pi}\right) \\ &\le \limsup_{x\to\infty} \left(\phi(x) - x/2 - \log\sqrt{2\pi}\right) = 0. \end{aligned} \tag{2.3}$$

PROOF. We first prove the statements concerning the function ϕ. $\phi(x) = 0$ for $x < 0$ by definition, so let us consider the second claim, $\phi(x) > 0$ for $x \ge 0$. For every $N \ge 1$ and $x \in J_N = [\log N, \log(N+1))$ one has

$$\begin{aligned} N + \log N! - N\log(N+1) &= \inf_{x\in J_N} \phi(x) \\ \le \phi(x) \le \phi(\log N) &= N + \log N! - N\log N. \end{aligned} \tag{2.4}$$

Therefore, writing $g(N) = N + \log N! - N\log(N+1)$ and using the lower bound in (2.4) one obtains the estimate $\phi(x) \ge \inf\{g(N) : N = 1, 2, \dots\}$ $(x \ge 0)$. The sequence $g(N)$ is increasing, because $g(N) - g(N-1) = 1 - N\log(1+\frac{1}{N}) > 0$ for

$N \geq 2$. Therefore the infimum is attained at $N = 1$, and the positivity assertion follows from $\inf_{x \geq 0} \phi(x) \geq g(1) = 1 - \log 2 > 0$.

The third statement, (2.3), is a consequence of (2.4) and Stirling's formula $\log N! = (N + \frac{1}{2}) \log N - N + \log \sqrt{2\pi} + O(N^{-1})$, which gives

$$N + \log N! - N \log N = \tfrac{1}{2} \log N + \log \sqrt{2\pi} + O(N^{-1}).$$

The values of limsup and liminf then follow by picking $x_N = \log N$ for limsup, and $x_N \in J_N$ close to $\log(N+1)$ for liminf (and noting that the length of J_N tends to zero).

Proceeding to the proof of (2.1) let us introduce truncated versions of the relevant functions,

$$\phi_N(x) = \sum_{1 \leq n \leq \min(N,\, e^x)} (1 - x + \log n),$$

$$\zeta_N(z) = \sum_{n=1}^{N} n^{-z}, \qquad \eta_N(z) = \zeta_N(z)\,(z-1)/z^2,$$

and set $h_{\sigma,N}(x) = e^{-\sigma x}\,\phi_N(x)$, and $h_\sigma(x) = e^{-\sigma x}\,\phi(x)$ $(N \geq 1)$. Below we need the well-known Fourier inversion formula

$$\text{(2.5)} \qquad \frac{1}{2\pi} \int_{-\infty}^{\infty} e^{itx}\,(\sigma + it)^{-q}\,dt = \begin{cases} e^{-\sigma x} x^{q-1}/(q-1)! & (x > 0) \\ 0 & (x < 0) \end{cases}$$

(for every $\sigma > 0$ and integer $q \geq 1$, almost everywhere $[x]$ on the two half lines. Here and throughout the sequel Fourier integrals generally are not to be interpreted pointwise, but in the L^2–sense. This is sufficient for our purposes.) Using (2.5) one gets

$$\begin{aligned}
\text{(2.6)} \qquad & \frac{1}{2\pi} \int_{-\infty}^{\infty} e^{itx}\,\eta_N(\sigma + it)\,dt \\
&= \frac{1}{2\pi} \int_{-\infty}^{\infty} e^{itx} \sum_{n=1}^{N} n^{-\sigma - it}\,(\sigma + it - 1)/(\sigma + it)^2\,dt \\
&= \sum_{n=1}^{N} n^{-\sigma} \frac{1}{2\pi} \int_{-\infty}^{\infty} e^{it\,(x - \log n)} \left(\frac{1}{\sigma + it} - \frac{1}{(\sigma + it)^2} \right) dt \\
&= \sum_{1 \leq n \leq \min(N,\, e^x)} n^{-\sigma}\, e^{-\sigma(x - \log n)}\,(1 - (x - \log n)) \\
&= h_{\sigma,N}(x)
\end{aligned}$$

almost everywhere in x. Now, let $\sigma > 1$. Then the functions $t \mapsto \eta_N(\sigma + it)$ converge to $t \mapsto \eta(\sigma + it)$ in $L^2(dt)$, because we have $|\zeta_N(z) - \zeta(z)| \leq \sum_{n>N} n^{-\sigma} \to 0$ uniformly on $\Re z = \sigma$ and $(z-1)/z^2$ is square integrable along that line. Consequently, $h_{\sigma,N}$ is a Cauchy sequence in $L^2(dx)$, by (2.6) and Parseval's formula. Since the pointwise limit $h_\sigma(x)$ of $h_{\sigma,N}(x)$ is identical a.e. with the $L^2(dx)$ limit, the (L^2) limits $h_\sigma(x)$ and $\eta(\sigma + it)$ represent a Fourier transform pair (for every $\sigma > 1$, by (2.6)). It follows that

$$\text{(2.7)} \qquad \eta(\sigma + it) = \widehat{h_\sigma}(t) = \int_0^{\infty} h_\sigma(x)\,e^{-ixt}\,dx = \int_0^{\infty} e^{-(\sigma + it)x}\,\phi(x)\,dx$$

holds almost everywhere in t ($\sigma > 1$), hence everywhere in $\Re z > 1$, by continuity. Therefore the Laplace transform of ϕ, which by the properties of ϕ established above is well–defined and analytic for $\Re z > 0$, represents the analytic continuation of η to that region. □

Let $\mathcal{H}^2_\sigma$ denote the Hardy space of the half plane $\Re z > \sigma$. It consists of all functions $g(z)$ which are analytic for $\Re z > \sigma$, with $\sup_{s>\sigma} \int_{-\infty}^{\infty} |g(s+it)|^2\, dt < \infty$. Since $h_\sigma \in L^2$ for every $\sigma > 0$, we immediately get from (2.7) and Parseval's formula the following.

COROLLARY 2.2. *η is an element of every Hardy space $\mathcal{H}^2_\sigma$, $\sigma > 0$, but not of $\mathcal{H}^2_0$.*

REMARK 2.3. Although h_σ is square integrable for every $\sigma > 0$, it is *not* true that $h_{\sigma,N} \to h_\sigma$ in L^2 if $\sigma \leq 1$. In fact, one has for all L^p norms

$$\liminf_{N\to\infty} ||h_{\sigma,N} - h_\sigma||_p > 0, \qquad (0 < \sigma \leq 1,\ p > 0). \tag{2.8}$$

PROOF. Let $[y]$ denote the largest integer $\leq y$. Using Stirling's formula and the elementary inequalities $0 \leq x - \log[e^x] \leq 2\, e^{-x}$ ($x > 0$) one obtains for $x > \log N \to \infty$

$$\begin{aligned}
&\phi(x) - \phi_N(x) \\
&= \sum_{N<n\leq e^x} (1 - x + \log n) \;=\; (1-x)([e^x] - N) + \log[e^x]! - \log N! \\
&= (1-x)([e^x] - N) + ([e^x] + \tfrac{1}{2}) \log[e^x] - [e^x] - (N + \tfrac{1}{2}) \log N + N + O(1) \\
&= (N + \tfrac{1}{2})(\log[e^x] - \log N) + ([e^x] - N)(\log[e^x] - x) + O(1) \\
&= (N + \tfrac{1}{2})(x - \log N) + O(1).
\end{aligned}$$

This approximation shows that there exists a finite constant $B > 0$ such that $\phi(x) - \phi_N(x) \geq N(x - \log N)$ for all sufficiently large N and $x > \log N + B$. Therefore

$$\begin{aligned}
||h_{\sigma,N} - h_\sigma||_p^p &\geq \int_{\log N + B}^{\infty} |\phi(x) - \phi_N(x)|^p\, e^{-p\sigma x}\, dx \\
&\geq N^p \int_{\log N + B}^{\infty} (x - \log N)^p\, e^{-p\sigma x}\, dx \\
&= N^{p(1-\sigma)} \int_B^{\infty} y^p\, e^{-p\sigma y}\, dy
\end{aligned}$$

for large N, and the assertion follows. Note that the norm of the difference even tends to infinity if $\sigma < 1$. □

The properties of ϕ established above make it possible to associate with ϕ an exponential family $\mathcal{P} = \{p_\sigma,\ \sigma > 0\}$ of probability densities supported by $[0, \infty)$ by setting

$$p_\sigma(x) = h_\sigma(x)/\eta(\sigma) = \phi(x)\, e^{-\sigma x}/\eta(\sigma) \qquad (x \in \mathbb{R},\ \sigma > 0). \tag{2.9}$$

This has some consequences for the function η. First, η is *completely monotone* on the positive half line, as the Laplace transform of a nonnegative function vanishing on the negative reals. In particular, it is decreasing and convex on $(0, \infty)$, and also *logarithmically convex*, by Hölder's inequality. Secondly, $\eta(z)$ is *positive definite* along every vertical line $\Re z = \sigma$, $\sigma > 0$, since $\eta(\sigma - it)/\eta(\sigma)$, $t \in \mathbb{R}$, is the characteristic function of the density p_σ. Note also that as a consequence of these facts, *$|\eta(z)|$ is uniformly bounded (by $\eta(\sigma)$) in every half plane $\Re z \geq \sigma$, $\sigma > 0$.* Gneiting [**Gn**] has other interesting inequalities for the zeta function on the real line.

3. The logarithmic derivative of η, and "integral tests" of the Riemann Hypothesis

Let us first collect some basic facts about the non–real zeros of the zeta function, which are generically denoted as ρ in the sequel. (a) The ρ's all lie in the "critical strip" $0 < \Re z < 1$ and are symmetric with respect to both the real axis and the "critical line" $\Re z = 1/2$. That is, whenever ρ is a zero then so are $\overline{\rho}$, $1-\rho$, and $1-\overline{\rho}$. (b) Every zero has imaginary part $|\Im\rho| > 14$. (c) There are infinitely many zeros, and the series $\sum(\Im\rho)^{-2}$ is convergent. (d) It is not known whether there exist any zeros not located on the critical line. Riemann conjectured that all zeros have real part equal to 1/2. This is the Riemann Hypothesis, hereafter abbreviated as "RH". Edwards [**E**] provides a wealth of technical and historical information about the ζ function and the RH.

The basic tool for the calculation of the logarithmic derivative of η is the factorization theorem for Hardy functions (see e.g. Dym and McKean [**DMK**, sect. 2.7], or Hoffman [**H**, p. 132, 133]). For our purposes an application of Nevanlinna's formula as given in [**DMK**, sect. 1.9] is sufficient. For the function η, and specialized to real arguments, it reads as follows.

(3.1)
$$\log\,\eta(s) \;=\; \frac{1}{\pi}\int_{-\infty}^{\infty}\log\,|\eta(\sigma+it)|\,\frac{s-\sigma}{(s-\sigma)^2+t^2}\,dt \;+\; \sum_{\Re\rho>\sigma}\log\left|\frac{s-\rho}{s-(2\sigma-\overline{\rho})}\right|$$

for $s > \sigma > 0$, where the sum extends over all zeros whose real part exceeds σ.

Let us indicate how (3.1) follows from the Nevanlinna's formula in [**DMK**, sect. 1.9]. The upper half plane in Dym and McKean's formulation here has to be replaced by $\Re z > \sigma$, and the zeros ρ of η in this latter half plane have to be reflected at the line $\Re z = \sigma$, which gives $2\sigma - \overline{\rho}$ as the mirror images of the ρ's. The assumptions about η are evidently fulfilled. For η is bounded on $\Re z \geq \sigma$, hence certainly of finite exponential type in this region, and it satisfies the condition

$$\int_{-\infty}^{\infty}\log^+|\eta(\sigma+it)|\,\frac{dt}{1+t^2} < \infty.$$

Since $\lim_{s\to\infty}\zeta(s) = 1$ we have $\limsup_{s\to\infty} s^{-1}\log\,\eta(s) = 0$, so that the constant c appearing in the linear term of the general formula is zero. Finally, $\eta(z)$ has an analytic continuation across the line $\Re z = \sigma$ to the entire right half plane, which guarantees continuity and prevents accumulation of zeros. This completes the derivation of (3.1).

As a first consequence of (3.1) one can formulate a kind of "integral test" for the RH. Putting $\sigma = 1/2$ we see that the sum on the right–hand side of (3.1) vanishes if and only if ζ has no zero within the region $\Re z > 1/2$. That is, *the RH holds if and only if for some (and then for all) $s > 1/2$*

(3.2)
$$\frac{1}{\pi}\int_{-\infty}^{\infty}\log\,|\eta(\tfrac{1}{2}+it)|\,\frac{(s-\frac{1}{2})\,dt}{(s-\frac{1}{2})^2+t^2} \;=\; \log\,\eta(s).$$

This criterion amounts to the condition that $\eta(z)$ be an *outer function* for the half plane $\Re z > 1/2$; cf. [**DMK**, sect. 2.7]. For $s = 1$, the eq. (3.1) assumes a particularly neat form. The left–hand side vanishes and the right–hand side can be simplified (the contribution of the factor $(z-1)/z^2$ of $\eta(z)$ to the integral term vanishes), and one gets the following formula

$$\frac{1}{2\pi}\int_{-\infty}^{\infty}\frac{\log|\zeta(\frac{1}{2}+it)|}{\frac{1}{4}+t^2}\,dt \;=\; \sum_{\Re\rho>1/2}\log\left|\frac{\rho}{1-\rho}\right| \tag{3.3}$$

due to Balazard, Saias and Yor [**BSY**]. This paper also provides references to related work and a probabilistic interpretation of the integral in (3.3). Our main application of (3.1) is the following.

PROPOSITION 3.1. *For every* $\sigma > 0$,

$$(\log\eta)'(\sigma) \;=\; \int_{-\infty}^{\infty}\log\left[\,|\eta(\sigma+it)|/\eta(\sigma)\,\right]\frac{dt}{\pi\,t^2} \;-\; 2\sum_{\Re\rho>\sigma}\Re\,(\rho-\sigma)^{-1}. \tag{3.4}$$

The integral and the sum on the right–hand side are both convergent.

PROOF. (3.4) is nothing else than the differentiated version of (3.1). To justify the limiting operations, fix $\sigma > 0$ and let $s > \sigma$. Subtracting $\log\eta(\sigma)$ on both sides of (3.1) and including this constant into the integrand on the right–hand side (which is possible because the Poisson kernel has total mass 1), then dividing by $s-\sigma$, we get

$$\begin{aligned}&\frac{\log\eta(s)-\log\eta(\sigma)}{s-\sigma}\\ &= \frac{1}{\pi}\int_{-\infty}^{\infty}\log\left(\frac{|\eta(\sigma+it)|}{\eta(\sigma)}\right)\frac{dt}{(s-\sigma)^2+t^2} \;+\; \frac{1}{s-\sigma}\sum_{\Re\rho>\sigma}\log\left|\frac{s-\rho}{s-(2\sigma-\overline{\rho})}\right|,\end{aligned}$$

and (3.4) will follow by taking the limit $s\downarrow\sigma$ in each term.

There is no problem with the first term, so let us first consider the sum term. Putting $s-\sigma=\epsilon$, $\rho-\sigma=\delta+i\beta$, one can write

$$\left|\frac{s-\rho}{s-(2\sigma-\overline{\rho})}\right|^2 = 1-\epsilon\,\frac{4\,\delta}{(\epsilon+\delta)^2+\beta^2}.$$

Because all zeros have imaginary part $|\beta| > 14$ (and $\delta,\epsilon > 0$), we have the global estimate

$$\frac{4\,\delta}{(\epsilon+\delta)^2+\beta^2} \le \frac{2}{|\beta|} \le \frac{1}{7}.$$

Therefore the inequalities $x \le -\log(1-x) \le x+x^2$, $0 \le x \le 1/2$, apply to each member of the sum (for small ϵ's). A straightforward argument based on these inequalities and monotone convergence then shows that as $\epsilon\downarrow 0$, the sum term converges to the sum in (3.4). The finiteness of this latter sum follows from the convergence of the series $\sum(\Im\,\rho)^{-2}$. The integrand in the integral term is non–positive because $|\eta(\sigma+it)|/\eta(\sigma) \le 1$ for all $t\in\mathbb{R}$ (by positive definiteness), and it depends monotonically on $s-\sigma$. Therefore the integral term tends to the integral in (3.4), which then also must be finite because of the finiteness of the other two terms. For completeness we remark that the finiteness of the limit integral could also be proved directly by noting that: (a) the convergence of the integral

$$\int_{-\infty}^{\infty} \log\{|\eta(\sigma+it)|/\eta(\sigma)\} \frac{dt}{1+t^2}$$

is a basic fact from the theory of Hardy functions; (b) the singularity induced by the weight $1/t^2$ instead of $1/(1+t^2)$ is compensated by the parabolic behavior of $\log\{|\eta(\sigma+it)|/\eta(\sigma)\}$ at $t=0$, which follows from the fact that $\{|\eta(\sigma+it)|/\eta(\sigma)\}^2$ is the characteristic function of a symmetric probability distribution with a finite second moment. □

The special case $\sigma = 1/2$ of (3.4) yields another, differential version of the above integral tests. *The RH holds if and only if*

$$\int_{-\infty}^{\infty} \log[\,|\eta(\tfrac{1}{2}+it)|/\eta(\tfrac{1}{2})\,] \frac{dt}{\pi t^2} = (\log\eta)'(\tfrac{1}{2}). \tag{3.5}$$

Notice that $-(\log\eta)'(\sigma)$ equals the first moment of the density p_σ. In view of (3.4) and (3.5), this raises the question whether the integral term in these relations has a probabilistic interpretation, too. A partial answer in the case $\sigma>1$ can be obtained from the observation going back to Khintchine that the function $f_\sigma(t)=\zeta(\sigma-it)/\zeta(\sigma)$, $t\in\mathbb{R}$, is the characteristic function of an infinitely divisible distribution; cf. [**GK**, Example 6, p. 75]. Let $F_\sigma = \sum_p \log p \sum_{n=1}^{\infty} p^{-n\sigma}\,\delta_{\log p^n}$ denote the discrete (Lévy) measure on $(0,\infty)$ which puts mass $p^{-n\sigma}\log p$ at the point $\log p^n$ (for every $n\geq 1$, p prime). Moreover, let G_σ denote the signed measure on $(0,\infty)$ with the Lebesgue–density $e^{-(\sigma-1)x} - 2\,e^{-\sigma x}$.

PROPOSITION 3.2. *For $\sigma>1$ one has*

$$-\int_{-\infty}^{\infty} \log[\,|\eta(\sigma+it)|/\eta(\sigma)\,] \frac{dt}{\pi t^2} = \int_0^{\infty} (F_\sigma - G_\sigma)(dx). \tag{3.6}$$

PROOF. Since there are no zeros with $\Re\rho>\sigma>1$, we have by (3.4) and the definition of η

$$\int_{-\infty}^{\infty} \log[\,|\eta(\sigma+it)|/\eta(\sigma)\,] \frac{dt}{\pi t^2} = (\log\eta)'(\sigma) = (\log\zeta)'(\sigma) + (\sigma-1)^{-1} - 2\,\sigma^{-1}.$$

The product representation $\zeta(z)=\prod_p (1-p^{-z})^{-1}$ of the zeta function gives

$$\begin{aligned}\log\frac{\zeta(\sigma+it)}{\zeta(\sigma)} &= \log\prod \frac{1-p^{-\sigma}}{1-p^{-\sigma-it}} = \sum_p\sum_{n=1}^{\infty} \frac{p^{-\sigma n}}{n}\left(e^{-itn\log p}-1\right) \\ &= \int_0^{\infty} (e^{-itx}-1)\,x^{-1}\,F_\sigma(dx),\end{aligned} \tag{3.7}$$

so by differentiating with respect to t at $t=0$ we get $(\log\zeta)'(\sigma) = -\int_0^\infty F_\sigma(dx)$. The proof is completed by noting that $\int_0^\infty G_\sigma(dx) = (\sigma-1)^{-1} - 2\,\sigma^{-1}$. □

More direct calculations are also possible, and instructive. Taking real parts in (3.7) and using $\int_{-\infty}^{\infty}(1-\cos tx)/t^2\,dt = \pi\,|x|$, $x\in\mathbb{R}$, (and Fubini) one obtains

$$\begin{aligned}\int_{-\infty}^{\infty} \log[\,|\zeta(\sigma+it)|/\zeta(\sigma)\,] \frac{dt}{\pi t^2} &= \int_{-\infty}^{\infty}\int_0^{\infty} (\cos tx - 1)\,x^{-1}\,F_\sigma(dx)\,\frac{dt}{\pi t^2} \\ = \int_0^{\infty}\int_{-\infty}^{\infty} \frac{\cos tx-1}{x}\,\frac{dt}{\pi t^2}\,F_\sigma(dx) &= -\int_0^{\infty} F_\sigma(dx).\end{aligned}$$

On the other hand, the total mass of F_σ is

$$\int_0^\infty dF_\sigma = \sum_p \log p \sum_{n=1}^\infty p^{-n\sigma} = \sum_p (p^\sigma - 1)^{-1} \log p = -(\log\zeta)'(\sigma).$$

In any case, (minus) the integral in question equals the mass of the Lévy measure F_σ minus a correction term due to the factor $(z-1)/z^2$, which compensates for the "explosion" of $\int dF_\sigma$ as $\sigma \downarrow 1$. This may lead one to regard G_σ as a smooth "renormalization" of F_σ and to look for alternative renormalizations such as, e.g., $F_\sigma^*(dx) = \theta(e^x)\, e^{-x\sigma}\, dx$, where $\theta(y) = \sum_{p\le y} \log p$. (To see the connection between F_σ^* and F_σ re-write the density of F_σ^* as $\sum_p \log p \,\{e^{-x\sigma}\, I_{[\log p,\,\infty)}(x)\}$ and consider the measure $e^{-x\sigma}\, dx$ on the interval $[\log p, \infty)$ as a continuous version of the measure $\sum_{n=1}^\infty p^{-n\sigma}\, \delta_{\log p^n}$, for every p.) The measures F_σ^*, $\sigma > 1$, give rise to infinitely divisible distributions with characteristic functions $\exp(\int (e^{-itx} - 1)\, x^{-1}\, F_\sigma^*(dx))$, and might be interesting objects for further study.

The infinite divisibility (in fact, compound Poisson character) of f_σ is immediate from (3.7). The first equation in (3.7) shows that f_σ also is the characteristic function of a sum (over all primes) of independent *geometrically distributed* random variables (assuming the value $n \log p$ with probability $(1-p^{-\sigma})\, p^{-n\sigma}$, $n \ge 0$). The probabilistic interpretation of the densities p_σ themselves remains open.

4. Higher–order cumulants

The k–th moment $M_{\sigma,k}$ of the density p_σ can be calculated by k–fold differentiation of the equation $\eta(s)/\eta(\sigma) = \int p_\sigma(x)\, e^{-(s-\sigma)x}\, dx$ at $s = \sigma$, which gives the familiar formula $M_{\sigma,k} = (-1)^k\, \eta^{(k)}(\sigma)/\eta(\sigma)$. Likewise, the k–th *cumulant* of p_σ is defined as the k–th logarithmic derivative of η at the point σ,

$$C_{\sigma,k} = (-1)^k\, (\log \eta)^{(k)}(\sigma).$$

However, higher–order cumulants of p_σ evidently cannot be obtained by straigthforward differentiation of (3.4). In this section information about such cumulants is obtained by passing to an alternative representation of $(\log\eta)'(\sigma)$. Many of the following results are known, however, the perspective taken here is different.

PROPOSITION 4.1. *For every* $\sigma > 0$,

$$(\log\eta)'(\sigma) = -\tfrac{1}{2}(\psi(\sigma/2) - \log \pi) - 3\sigma^{-1} - \sum \Re\, (\rho - \sigma)^{-1}. \tag{4.1}$$

Here $\psi(z) = \Gamma'(z)/\Gamma(z)$ *denotes the logarithmic derivative of the Gamma function, and the summation extends over all (non–trivial) zeros of the zeta function.*

PROOF. This follows by expressing $\eta(z)$ in terms of the Riemann ξ function, here defined as $\xi(z) = z(z-1)\,\zeta(z)\,\Gamma(z/2)\,\pi^{-z/2}$, and using the known formula for $(\log\xi)'(z)$. Since $z^2\,\eta(z) = (z-1)\,\zeta(z)$ we have $\eta(z) = \xi(z)\,\pi^{z/2}/(z^3\,\Gamma(z/2))$, which upon logarithmic differentiation and use of the formula (cf. [**I**, p. 17]) $(\log\xi)'(z) = \sum (\rho^{-1} - (\rho - z)^{-1}) - \sum \Re\,\rho^{-1}$ gives

$$(\log\eta)'(z) = \sum (\rho^{-1} - (\rho - z)^{-1}) - \sum \Re\,\rho^{-1} + \tfrac{1}{2}\log\pi - 3z^{-1} - \tfrac{1}{2}\psi(z/2).$$

Since η and ψ are real–valued on $(0,\infty)$, this formula remains valid for real, positive arguments $z = \sigma$ if one takes the real part on the right–hand side. The latter then simplifies to the right–hand side of (4.1), and the proof is complete. □

All terms in this more standard representation of $(\log \eta)'(\sigma)$ are smooth functions of σ. Therefore the higher–order cumulants of the densities p_σ can be obtained by differentiation of (4.1).

COROLLARY 4.2. *For every $n \geq 2$ and $\sigma > 0$,*

$$(\log \eta)^{(n)}(\sigma) = (-1)^n C_{\sigma,n} \tag{4.2}$$
$$= (-1)^n\, 3\,\Gamma(n)\,\sigma^{-n} - 2^{-n}\,\psi^{(n-1)}(\sigma/2) - \Gamma(n)\sum \Re\,(\rho-\sigma)^{-n}.$$

In the sequel we discuss some interesting special cases of (4.1) and (4.2).

CASE $\sigma = 1$. More elaborate expressions for the cumulants can be obtained in this case by observing the symmetry properties of the zeros and utilizing known properties of the function ψ and its derivatives. For information about these so–called "digamma" and "polygamma" functions see [**AS**, Sections 6.3 and 6.4]. In the sequel, related results from that reference (such as the explicit values of $\psi^{(n-1)}(\frac{1}{2})$, for instance) will be used without citing it in each case anew. Now, since $1-\rho$ runs through all zeros whenever ρ does so, we have $\sum \Re\,(\rho-1)^{-n} = (-1)^n \sum \Re\,\rho^{-n}$. Therefore, denoting Euler's constant by γ, we get

$$-(\log \eta)'(1) = 3 - \gamma/2 - \tfrac{1}{2}\,\log 4\pi - \sum \Re\,\rho^{-1}, \tag{4.3}$$

$$(-1)^n\,(\log \eta)^{(n)}(1) = \Gamma(n)\,\Big(3 - (1-2^{-n})\,\zeta(n) - \sum \Re\,\rho^{-n}\Big) \quad (n \geq 2). \tag{4.4}$$

CASE $\sigma = 1/2$. The symmetry of the ρ's implies that $\sum \Re\,(\rho - 1/2)^{-n}$ vanishes for odd $n \geq 1$. Therefore, observing $\psi(1/4) = -\gamma - \log 8 - \pi/2$ one can evaluate the first cumulant of $p_{1/2}$ explicitly,

$$-(\log \eta)'(\tfrac{1}{2}) = 6 - \pi/4 - \gamma/2 - \tfrac{1}{2}\,\log 8\pi. \tag{4.5}$$

The value of ψ at $z = 1/4$ can be derived from the known value $\psi(\frac{1}{2}) = -\gamma - 2\,\log 2$ by adding the "duplication formula" $\psi(z) + \psi(z+\frac{1}{2}) = 2\psi(2z) - 2\log 2$ and the "reflection formula" $\psi(z) - \psi(1-z) = -\pi \cot \pi z$, and evaluating their sum at $z = 1/4$. By applying the same recipe to the $(n-1)$th derivatives of these formulae one could write down explicit expressions for any cumulant of odd order. For even orders n, useful general formulae for $\psi^{(n-1)}(\frac{1}{4})$ do not seem to be available.

CASE $\sigma \downarrow 0$. The asymptotic behavior of the cumulants of p_σ in this limit is as follows.

$$\lim_{\sigma\downarrow 0}\Big((\log \eta)^{(n)}(\sigma) - (-1)^n\, 2\,\Gamma(n)\,\sigma^{-n}\Big) = -A_n, \qquad \text{where} \tag{4.6}$$

$$A_1 = \sum \Re\,\rho^{-1} - \tfrac{\gamma}{2} - \tfrac{1}{2}\,\log \pi, \quad A_n = \Gamma(n)\,\Big(\sum \Re\,\rho^{-n} + (-\tfrac{1}{2})^n\,\zeta(n)\Big) \quad (n \geq 2).$$

For the proof of this claim one needs the series expansions of the digamma and polygamma functions, which are given by $\psi(z) = -\gamma - z^{-1} + z\,\sum_{k=1}^{\infty}(k(k+z))^{-1}$ and by $\psi^{(n-1)}(z) = (-1)^n\,\Gamma(n)\,\{z^{-n} + \sum_{k=1}^{\infty}(k+z)^{-n}\}$ $(n \geq 2)$, respectively. Using these expansions one finds that as $\sigma \downarrow 0$ one has $\psi(\sigma/2)/2 = -\sigma^{-1} - \gamma/2 + o(1)$ and, for $n \geq 2$,

$$\begin{aligned} 2^{-n}\,\psi^{(n-1)}(\sigma/2) &= (-2)^{-n}\,\Gamma(n)\,\{(\sigma/2)^{-n} + \textstyle\sum_{k=1}^{\infty}(k+\sigma/2)^{-n}\} \\ &= (-1)^n\,\Gamma(n)\,\{\sigma^{-n} + 2^{-n}\zeta(n) + o(1)\}. \end{aligned}$$

The proof is completed by inserting these expressions into (4.1) and (4.2), respectively, and noting that $\lim_{\sigma\downarrow 0}\sum \Re\,(\rho-\sigma)^{-n}=\sum \Re\,\rho^{-n}$.

The previous results yield a connection between the logarithmic derivatives of $\zeta(z)$ at $z=0$ and of $\eta(z)$ at $z=1$. It could be used (along with (4.3), (4.4), and (4.8) to (4.10) below) to derive explicit expressions for the former.

COROLLARY 4.3. *One has* $(\log\zeta)'(0)=\log 2\pi+\gamma-2-(\log\eta)'(1)$, *and*

$$(4.7)\qquad (\log\zeta)^{(n)}(0) = \Gamma(n)\,\{\zeta(n)\,[\,1-(1+(-1)^n)\,2^{-n}\,]-2\} \;+\; (-1)^n\,(\log\eta)^{(n)}(1) \quad (n\geq 2).$$

PROOF. From the definition of η it follows that $(\log\eta)^{(n)}(\sigma)=(\log\zeta)^{(n)}(\sigma)+(-1)^n\,2\,\Gamma(n)\,\sigma^{-n}-\Gamma(n)+o(1)$ as $\sigma\downarrow 0$. Inserting this into (4.6) one gets $(\log\zeta)^{(n)}(\sigma)=\Gamma(n)-A_n+o(1)$, and hence $(\log\zeta)^{(n)}(0)=\Gamma(n)-A_n$, since the zeta function is regular and non–zero at $z=0$. Using (4.3) and (4.4) one can express the quantities $\sum\Re\,\rho^{-n}$ which appear in the definition of the constants A_n in terms of the quantities $(\log\eta)^{(n)}(1)$. After rearrangement the resulting expressions for $(\log\zeta)^{(n)}(0)$ reduce to those stated above. □

COROLLARY 4.4. *As $\sigma\downarrow 0$ the rescaled densities $q_\sigma(x)=p_\sigma(x/\sigma)/\sigma$ tend to the density $q(x)=x\,e^{-x}$, $x\geq 0$, in the sense of weak convergence of probability distributions.*

PROOF. The n–th cumulant of the rescaled densities equals σ^n times the n–th cumulant of p_σ, hence tends to $2\,\Gamma(n)$, by (4.6). The only distribution on $[0,\infty)$ having exactly these cumulants is the one with the density $q(x)$. The proof is completed on noting that the family of probability distributions on $[0,\infty)$ having densities q_σ, $0<\sigma\leq 1$, is tight because the first moments are uniformly bounded. □

An alternative proof, showing pointwise convergence of the densities, and hence convergence in L^1, can be obtained without any cumulant calculations directly from Theorem 2.1. For we have

$$q_\sigma(x) \;=\; \phi(x/\sigma)\,e^{-x}/\{\sigma\,\eta(\sigma)\} \;=\; \sigma\,\phi(x/\sigma)\,e^{-x}/\{\zeta(\sigma)\,(\sigma-1)\},$$

with $\lim_{\sigma\downarrow 0}\zeta(\sigma)\,(\sigma-1)=1/2$, by the regularity of ζ at zero and the known formula $\zeta(0)=-1/2$. The growth behavior of $\phi(x)$ proved in the theorem implies $\lim_{\sigma\downarrow 0}(\sigma\,\phi(x/\sigma)-x/2)=0$, so the pointwise convergence $\lim_{\sigma\downarrow 0}q_\sigma(x)=x\,e^{-x}$, $x\geq 0$, follows by combining these facts.

CASE $\sigma\uparrow\infty$. The pointwise limit of the rescaled densities is easily found using (2.2). The limit density here is the standard exponential density, $\lim_{\sigma\uparrow\infty}q_\sigma(x)=e^{-x}$, $x\geq 0$. Information about the asymptotic behavior of the cumulants of the original densities p_σ can be obtained by means of the approximation

$$(-1)^k\,\eta^{(k)}(\sigma) \;=\; \sigma^{-k-1}\,\{\,k!-\sigma^{-1}\,(k+1)!+O(\sigma^k\,2^{-\sigma})\} \quad (\sigma\uparrow\infty,\ 0\leq k\in\mathbb{Z}).$$

This follows from the representation of $\eta(\sigma)$ as the Laplace transform of $\phi(x)$ on observing that $\phi(x)$ equals $1-x$ for $0\leq x<\log 2$ and is bounded by a constant times x for $x\geq\log 2$. The approximation is then established by writing

$$\begin{aligned}(-1)^k\,\eta^{(k)}(\sigma) \;&=\; \int_0^\infty x^k\,\phi(x)\,e^{-\sigma x}\,dx\\ &=\; \sigma^{-k-1}\Big(\int_0^\infty t^k\,(1-t/\sigma)\,e^{-t}\,dt \;+\; \int_{\sigma\log 2}^\infty t^k\,\{\phi(t/\sigma)-(1-t/\sigma)\}\,e^{-t}\,dt\Big)\end{aligned}$$

and estimating the last integral by $\int_{\sigma \log 2}^{\infty} t^k \{1 + O(t/\sigma)\} e^{-t} \, dt = O\left(\sigma^k e^{-\sigma \log 2}\right)$.

Let us now reverse the perspective and consider the expressions for the cumulants as formulae for series of the form $\sum \Re(\rho - \sigma)^{-n}$. Below we put together a few of the results which can be obtained in this way. The constants γ_n, $n \geq 1$, are defined as the coefficients appearing in the Laurent series expansion of the zeta function around $z = 1$,

$$\zeta(z) = (z-1)^{-1} + \gamma + \sum_{n=1}^{\infty} (-1)^n \frac{\gamma_n}{n!} (z-1)^n.$$

$$\sum \Re \rho^{-1} = 1 + \gamma/2 - \tfrac{1}{2} \log 4\pi . \tag{4.8}$$

$$\sum \Re \rho^{-2} = 1 + \gamma^2 + 2\gamma_1 - \pi^2/8 . \tag{4.9}$$

$$\sum \Re \rho^{-3} = 1 + \gamma^3 + 3\gamma\gamma_1 + \tfrac{3}{2}\gamma_2 - \tfrac{7}{8}\zeta(3) . \tag{4.10}$$

$$\sum \Re (\rho - \tfrac{1}{2})^{-2} = 12 - \tfrac{1}{4}\psi'(\tfrac{1}{4}) - (\log \eta)''(\tfrac{1}{2}) . \tag{4.11}$$

$$\sum \Re (\rho - \tfrac{1}{2} - \sigma)^{-1} = -\tfrac{1}{2} \log \tfrac{\sigma}{2\pi} - \tfrac{7}{4}\sigma^{-1} + O(\sigma^{-2}) \qquad (\sigma \uparrow \infty). \tag{4.12}$$

$$\sum \Re (\rho - \tfrac{1}{2} - \sigma)^{-2} = -\tfrac{1}{2}\sigma^{-1} + \tfrac{7}{4}\sigma^{-2} + O(\sigma^{-3}) \qquad (\sigma \uparrow \infty). \tag{4.13}$$

These relations follow from the previous results by straightforward calculations. They contain indirect information about the distribution of the zeros, which may not be easy to extract, however. Here we will use the above relations to derive yet another type of "tests" of the RH based on series of the form $\sum \Re(\rho - \sigma)^{-n}$. Let us illustrate this by means of the relation (4.11). Denoting the real and the imaginary part of the zero ρ by α and β, respectively, we may write

$$-\sum \Re (\rho - \tfrac{1}{2})^{-2} = \sum (\beta^2 - (\alpha - \tfrac{1}{2})^2) \,/\, (\beta^2 + (\alpha - \tfrac{1}{2})^2)^2 .$$

Since β^2 dominates the second term in the numerator, each member of this series is positive. Therefore

$$-\sum \Re (\rho - \tfrac{1}{2})^{-2} \geq \sum_{\alpha = 1/2} \beta^{-2}, \quad \textit{with equality if and only if the RH is true.}$$

Note that this "test" involves the zeros on the critical line only. In particular, the RH were disproved if it could be shown that $S = \sum_{\alpha=1/2} \beta^{-2}$ is strictly less than minus the number on the right–hand side of (4.11), which can be computed to arbitrary accuracy by means of a computer algebra system. In that case the partial sums of S would saturate somewhere below the upper bound. However, since from extensive computations of the zeta function zeros (for a review see [**O**]) it is known that any zero not located on the critical line must have a very large imaginary part, the "gap" $-\sum \Re (\rho - \tfrac{1}{2})^{-2} - S$, if it were positive at all, would necessarily be very small and thus hard to detect numerically. The same qualification applies to other tests of a similar form, for instance those which could be based on the series $\sum \Re (\rho - \tfrac{1}{2})^{-4}$ or on the linear combination $\sum \Re \rho^{-1} + \tfrac{1}{2} \sum \Re \rho^{-2} - \tfrac{1}{4} \sum \Re \rho^{-4}$.

Let us briefly reconsider the integral test (3.5) from this point of view. Using Proposition 3.1 with $\sigma = 1/2$, (4.5), and the symmetry of the zeros with respect to the critical line one obtains the relation

(4.14)
$$\sum \frac{|\Re\rho - \frac{1}{2}|}{|\rho - \frac{1}{2}|^2} = \int_{-\infty}^{\infty} \log\left[\frac{|\eta(\frac{1}{2}+it)|}{\eta(\frac{1}{2})}\right] \frac{dt}{\pi t^2} - \left(\frac{\gamma}{2} + \frac{1}{2}\log(8\pi) + \frac{\pi}{4} - 6\right),$$

which identifies the (possible) gap between the two sides of (3.5) as a weighted sum of the absolute deviations of the real parts of the zeros from 1/2. Because of the weights, essentially $(\Im\rho)^{-2}$, the gap would be very small in this case, too, complicating possible checks based on a numerical evaluation of the integral in (4.14). Let us conclude by mentioning that a beautiful interpretation of the gap term in the integral test due to Balazard, Saias and Yor [**BSY**] (i.e., of the sum on the right–hand side of (3.3)) was recently given by Burnol [**Bu2**].

Acknowledgments

Very helpful comments by a referee led to substantial improvements especially in Section 2 and in the introduction. Thanks are also due to Harald Atmanspacher, Tilmann Gneiting, and Donald Richards for useful remarks on an earlier version of this paper, and for encouragement.

References

[AS] M. Abramowitz and I.A. Stegun, *Handbook of Mathematical Functions*, Dover Publications, New York, 1965.

[BS] M. Balazard and E. Saias, *Notes sur la fonction ζ de Riemann, 1*, Adv. Math. **139** (1998), 310-321.

[BSY] M. Balazard, E. Saias and M. Yor, *Notes sur la fonction ζ de Riemann, 2*, Adv. Math. **143** (1999), 284-287.

[Be] A. Beurling, *A closure problem related to the Riemann Zeta–function*, Proc. Nat. Acad. Sci. **41** (1955), 312-314.

[BPY] P. Biane, J. Pitman and M. Yor, *Probability laws related to the Jacobi theta and Riemann zeta functions, and Brownian excursions*, <http://www.arXiv.org/abs/math/9912170>

[Bu1] J.–F. Burnol, *An adelic causality problem related to Abelian L–functions*, to appear in J. Number Theory.

[Bu2] J.–F. Burnol, *A note on Nyman's equivalent formulation of the Riemann Hypothesis*, this volume.

[DMK] H. Dym and H.P. McKean, *Gaussian Processes, Function Theory, and the Inverse Spectral Problem*, Academic Press, New York, 1976.

[E] H.M. Edwards, *The Theory of the Riemann Zeta Function*, Academic Press, New York, 1974.

[GK] B.V. Gnedenko and A.N. Kolmogorov, *Limit Distributions for Sums of Independent Random Variables*, Addison–Wesley, Cambridge, 1954.

[Gn] T. Gneiting, *Power–law correlations, related models for long–range dependence, and their simulation*, J. Appl. Probab. **37** (2000), 1104-1109.

[H] K. Hoffman, *Banach Spaces of Analytic Functions*, Dover Publications, New York, 1988.

[I] A. Ivic, *The Riemann Zeta–Function*, Wiley, New York, 1985.

[N] B. Nyman, *On some groups and semigroups of translations*, Thesis, Univ. Uppsala (1950).

[O] A.M. Odlyzko, *Analytic computations in number theory*, Proc. Sympos. Appl. Math., vol. 48, Amer. Math. Soc., Providence, RI, 1994, pp. 451-463.

Institut für Grenzgebiete der Psychologie und Psychohygiene, Wilhelmstr. 3a, D–79098 Freiburg, Germany

E-mail address: ehm@igpp.de

Contemporary Mathematics
Volume **287**, 2001

Local Field U–Statistics

Steven N. Evans

ABSTRACT. Using the classical theory of symmetric functions, a general distributional limit theorem is established for U–statistics constructed from a sequence of independent, identically distributed random variables taking values in a local field with zero characteristic.

1. Introduction

Since the work of Hoeffding and Halmos in the 1940s, U–statistics constructed from sequences of independent, identically distributed real random variables have played a central role in theoretical and applied statistics. They have also attracted considerable attention from probabilists because they exhibit a rich limit theory that parallels that of i.i.d. sequences (for example, strong laws, central limit theorems, large deviation results, and Berry–Esseen–type theorems have been established for them). Surveys with extensive bibliographies may be found in [**Ser80, KB94, Lee90**].

Our aim in this paper is to initiate an investigation into the properties of U–statistics on algebraic structures other than the reals: namely, local fields. A local field $\mathbb{K}$ is any locally compact, non-discrete field other than the field of real numbers or the field of complex numbers. All local fields are totally disconnected, and are either finite algebraic extensions of the field of p-adic numbers – in which case the characteristic is zero – or finite algebraic extensions of the less familiar p-series field (the field of formal Laurent series with coefficients drawn from the finite field with p elements) – in which case the characteristic is non–zero. We give an overview of some of the basic properties of local fields in Section 2.

Probability on local fields has a substantial, if somewhat scattered, literature, and a comprehensive book–length treatment has yet to be written. For the convenience and interest of the reader we have included a representative (but by no means complete) bibliography in the references.

The natural definition of the notion of U–statistic on the local field $\mathbb{K}$ is the following direct translation of the familiar Euclidean definition.

1991 *Mathematics Subject Classification.* 60B15, 60F05, 05E05.
Key words and phrases. local field, p–adic, U–statistic, symmetric function, random walk.
Research supported in part by NSF grants DMS-9703845 and DMS-0071468.

DEFINITION 1.1. Let $\{X_k\}_{k\in\mathbb{N}}$ be an infinite sequence of independent, identically distributed random variables taking values in the local field $\mathbb{K}$. Fix a symmetric Borel function $\Psi : \mathbb{K}^m \to \mathbb{K}$ for some $m \in \mathbb{N}$ (that is, the value of the function Ψ is unchanged by permutations of its arguments). The sequence of *U–statistics* corresponding to $\{X_k\}_{k\in\mathbb{N}}$ and Ψ is the sequence of $\mathbb{K}$–valued random variables $\{Z_k\}_{k\geq m}$ given by

$$Z_k := \sum_{1\leq i_1<i_2<\cdots<i_m\leq k} \Psi(X_{i_1}, X_{i_2}, \ldots, X_{i_m}).$$

The following result is proved in §3. Some remarks on the hypotheses are given after the proof.

THEOREM 1.2. *Suppose that the local field $\mathbb{K}$ has characteristic zero, the support of the common distribution of the random variables X_k is compact, and the function Ψ is continuous. Let $\{k(h)\}_{h\in\mathbb{N}}$ be a sequence of positive integers such that $k(h)$ converges to infinity as $h \to \infty$ and also $k(h)$ thought of as an element of $\mathbb{K}$ converges to some $k^* \in \mathbb{K}$ as $h \to \infty$. Then the sequence $\{Z_{k(h)}\}_{h\in\mathbb{N}}$ of U–statistics converges in distribution as $h \to \infty$.*

2. Local fields

This section is essentially a summary of selected results from [**Tai75, Sch84**]. We refer the reader to these works for a fuller account. Before giving the general definition of a local field, we begin with the prototypical example.

EXAMPLE 2.1. Fix a positive prime p. We can write any non-zero rational number $r \in \mathbb{Q}\backslash\{0\}$ uniquely as $r = p^s(a/b)$, where a and b are not divisible by p. Set $|r| = p^{-s}$. If we set $|0| = 0$, then the map $|\cdot|$ has the properties:

$$\begin{aligned} &|x| = 0 \Leftrightarrow x = 0; \\ &|xy| = |x||y|; \\ &|x+y| \leq |x| \vee |y|. \end{aligned} \tag{2.1}$$

The map $(x, y) \mapsto |x - y|$ defines a metric on $\mathbb{Q}$ and we denote the completion of $\mathbb{Q}$ in this metric by $\mathbb{Q}_p$. The field operations on $\mathbb{Q}$ extend continuously to make $\mathbb{Q}_p$ a topological field called the *p-adic numbers.* The map $|\cdot|$ also extends continuously and the extension continues to have properties (2.1). The closed unit ball around 0, $\mathbb{Z}_p := \{x \in \mathbb{Q}_p : |x| \leq 1\}$, is the closure in $\mathbb{Q}_p$ of the integers $\mathbb{Z}$, and is thus a ring (this is also apparent from the properties (2.1)) called the *p–adic integers.* As $\mathbb{Z}_p = \{x \in \mathbb{Q}_p : |x| < p\}$, the set $\mathbb{Z}_p$ is also open. Any other ball around 0 is of the form $\{x \in \mathbb{Q}_p : |x| \leq p^{-k}\} = p^k\mathbb{Z}_p$ for some integer k. Such a ball is the closure of the rational numbers divisible by p^k, and is thus a $\mathbb{Z}_p$–module (this is again also apparent from the properties (2.1)). In particular, such a ball is an additive subgroup of $\mathbb{Q}_p$. Arbitrary balls are translates (= cosets) of these closed and open subgroups. In particular, the topology of $\mathbb{Q}_p$ has a base of closed and open sets, and hence $\mathbb{Q}_p$ is totally disconnected. Further, each of these balls is compact, and hence $\mathbb{Q}_p$ is also locally compact.

A *local field* is a locally compact, non–discrete, totally disconnected, topological field. (As an aside, a locally compact, non-discrete, topological field that is not totally disconnected is necessarily either the real or the complex numbers. A local field with characteristic zero is a finite algebraic extension of the p–adic number field for some prime p. A local field with non–zero characteristic is a finite algebraic extension of the *p–series field*; that is, the field of formal Laurent series with coefficients drawn from the finite field with p elements for some prime p.

From now on, let $\mathbb{K}$ be a fixed local field. There is a real–valued mapping on $\mathbb{K}$ which we denote by $x \mapsto |x|$. This map has the properties (2.1) and it takes the values $\{q^k : k \in \mathbb{Z}\} \cup \{0\}$, where $q = p^c$ for some prime p and positive integer c (so that for $\mathbb{K} = \mathbb{Q}_p$ we have $c = 1$).

A map with properties (2.1) is called a *non-Archimedean valuation.* The property $|x + y| \leq |x| \vee |y|$ is known as the *ultrametric inequality* or the *strong triangle inequality.* The mapping $(x, y) \mapsto |x - y|$ on $\mathbb{K} \times \mathbb{K}$ is a metric for $\mathbb{K}$ that gives the topology of $\mathbb{K}$. A consequence of (2.1) is that if $|x| \neq |y|$, then $|x + y| = |x| \vee |y|$. This latter result implies that for every "triangle" $\{x, y, z\} \subset \mathbb{K}$ we have that at least two of the lengths $|x - y|$, $|x - z|$, $|y - z|$ must be equal and is therefore called the *isosceles triangle property.*

3. Proof of Theorem 1.2

PROOF. Write E for the support of the common distribution of the X_k. By the ultrametric Stone–Weierstrass theorem (see, for example, §A.4 of [**Sch84**]), polynomials are uniformly dense in the space of continuous functions from the compact set E^m into $\mathbb{K}$. Therefore, for each $\epsilon > 0$ there exists a polynomial Q such that

$$\sup_{(x_1,\dots,x_m)\in E^m} |Q(x_1, \dots , x_m) - \Psi(x_1, \dots , x_m)| < \epsilon.$$

Define a symmetric polynomial $\overline{Q} : E^m \to \mathbb{K}$ by

$$\overline{Q}(x_1, \dots , x_m) = \frac{1}{m!} \sum_{\sigma \in \mathcal{S}_m} Q(x_{\sigma(1)}, \dots , x_{\sigma(m)}),$$

where $\mathcal{S}_m$ denotes the symmetric group on m letters and we have used the assumption that $\mathbb{K}$ has zero characteristic to conclude that $m! \neq 0$. By the strong triangle inequality and the symmetry of Ψ,

$$\begin{aligned}
&\sup_{(x_1,\dots,x_m)\in E^m} \left|\overline{Q}(x_1, \dots , x_m) - \Psi(x_1, \dots , x_m)\right| \\
&= \sup_{(x_1,\dots,x_m)\in E^m} \left| \frac{1}{m!} \sum_{\sigma \in \mathcal{S}_m} \left[Q(x_{\sigma(1)}, \dots , x_{\sigma(m)}) - \Psi(x_{\sigma(1)}, \dots , x_{\sigma(m)})\right]\right| \\
&\leq |m!|^{-1}\epsilon.
\end{aligned}$$

Thus, again by the strong triangle inequality,

$$\left| \sum_{1\leq i_1<i_2<\dots<i_m\leq k} \overline{Q}(X_{i_1}, X_{i_2}, \dots , X_{i_m}) - Z_k \right| \leq |m!|^{-1}\epsilon.$$

It thus clearly suffices to consider the special case of the theorem when Ψ is a symmetric polynomial. By replacing X_k by $X_k - c$ and $\Psi(x_1, \dots , x_m)$ by $\Psi(x_1 + c, \dots , x_m + c)$, we may further suppose that $0 \in E$. Moreover, because

$$\lim_h \sum_{1\leq i_1<i_2<\dots<i_m\leq k(h)} 1 = \lim_h \binom{k(h)}{m} = \frac{k^*(k^* - 1)\dots(k^* - m + 1)}{m!}$$

by assumption, we may suppose that Ψ has no constant term.

Given a positive integer n and integers $\lambda_1 \geq \lambda_2 \geq \ldots \geq 0$ with $0 = \lambda_{n+1} = \lambda_{n+2} = \ldots$, define the corresponding *monomial symmetric function* $\mathbf{M}_{n,\lambda}$ in the variables $(x_1, \ldots, x_n)$ by $\mathbf{M}_{n,\lambda}(x_1, \ldots, x_n) := \sum_\alpha x^\alpha$, where the sum is over all distinct permutations $\alpha = (\alpha_1, \ldots, \alpha_n)$ of $(\lambda_1, \ldots, \lambda_n)$ and $x^\alpha := x_1^{\alpha_1} \ldots x_n^{\alpha_n}$. The symmetric polynomials $\mathbf{M}_{n,\lambda}$ with $\sum_i \lambda_i \leq d$ form a basis for the vector space of symmetric polynomials in $(x_1, \ldots, x_n)$ of total degree at most d (cf. Ch I of [**Mac95**] or Ch 7 of [**Sta99**]). Consequently, we have

$$\Psi(x_1, \ldots, x_m) = \sum_\lambda c_\lambda \mathbf{M}_{m,\lambda}(x_1, \ldots, x_m),$$

for suitable constants c_λ, where the sum is over all λ with $0 = \lambda_{m+1} = \lambda_{m+2} = \ldots$, only finitely many of the c_λ are non–zero, and $c_0 = 0$ (by our added assumption that Ψ has no constant term). Observe for such λ that if $\ell := \max\{r : \lambda_r > 0\}$, then

$$\sum_{1 \leq i_1 < i_2 < \cdots < i_m \leq k} \mathbf{M}_{m,\lambda}(X_{i_1}, \ldots, X_{i_m}) = \binom{k-\ell}{m-\ell} \mathbf{M}_{k,\lambda}(X_1, \ldots, X_k).$$

Note that $\lim_h \binom{k(h)-\ell}{m-\ell}$ exists. It therefore suffices to show that the random vectors

$$(\mathbf{M}_{k,\lambda}(X_1, \ldots, X_k)),$$

where λ ranges over those non–zero partitions such that $0 = \lambda_{m+1} = \lambda_{m+2} = \ldots$ and $\sum_i \lambda_i$ is at most the total degree of Ψ, converge in distribution as $k \to \infty$.

Given a non-negative integer j, define the *power sum symmetric function* $\mathbf{P}_{n,j}$ in the variables $(x_1, \ldots, x_n)$ by

$$\mathbf{P}_{n,j}(x_1, \ldots, x_n) := x_1^j + \cdots + x_n^j,$$

and given integers $\mu_1 \geq \mu_2 \geq \ldots \geq 0$ with $0 = \mu_{n+1} = \mu_{n+2} = \ldots$, set

$$\mathbf{P}_{n,\mu}(x_1, \ldots, x_n) := \prod_i \mathbf{P}_{n,\mu_i}(x_1, \ldots, x_n).$$

We have

$$\mathbf{M}_{n,\lambda}(x_1, \ldots, x_n) = \sum_\mu c_{\lambda\mu} \mathbf{P}_{n,\mu}(x_1, \ldots, x_n),$$

where the sum is over all μ with $\sum_i \mu_i = \sum_i \lambda_i$ and, importantly, the constants $c_{\lambda\mu}$ do not depend on n (cf. Ch I of [**Mac95**] or Ch 7 of [**Sta99**], and note that we are again using the assumption that $\mathbb{K}$ has characteristic 0).

It thus suffices to show for any positive integer J that the random vectors

$$\left(\sum_{i=1}^k X_i, \sum_{i=1}^k X_i^2, \ldots, \sum_{i=1}^k X_i^J\right)$$

converge in distribution as $k \to \infty$. However, this process is just a random walk on the compact subgroup of the additive group of $\mathbb{K}^J$ generated by $\{(x, x^2, \ldots, x^J) : x \in E\}$, and the added assumption that $0 \in E$ ensures that the random walk converges in distribution to Haar measure on this subgroup. □

REMARK 3.1. The hypothesis that $\mathbb{K}$ has zero characteristic was used several times in the above proof. We do not know whether the result has a counterpart for non–zero characteristics.

REMARK 3.2. The role played by the hypothesis that $k(h)$ converges in $\mathbb{K}$ is apparent from the proof. However, because the assumption initially looks rather unusual, we emphasize its importance with the following simple example. Suppose that $\mathbb{K}$ is the field of p-adic numbers $\mathbb{Q}_p$, $m = 2$, and $\Psi(x_1, x_2) = x_1 + x_2$. Then

$$Z_k = (k-1) \sum_{i=1}^{k} X_i.$$

As we observed in the proof, if 0 is in the support of the common distribution of X_k, then $\sum_{i=1}^{k} X_i$ converges in distribution as $k \to \infty$ to Haar measure on the subgroup of $\mathbb{Q}_p$ generated by the support. Note that if k is of the form $p^s + 1$, then $|k-1| = p^{-s}$, whereas if k is of the form $p^s + 2$, $s \geq 1$, then p does not divide $k - 1$ and hence $|k-1| = 1$. Consequently, we must take $k \to \infty$ along a subsequence in order for Z_k to have a limit in distribution.

REMARK 3.3. In principle, the steps in the proof can be reversed to describe the limiting distribution as the push–forward by an appropriate function of Haar measure on the compact additive subgroup of $\mathbb{K}^{\mathbb{N}}$ generated by $(x, x^2, x^3, \dots)$ for x in suitable fixed translate of the support of the distribution of the X_k. It does not appear that one can give a more effective characterization of the limit. In the next section we examine some particularly simple examples where it possible to say something concrete about the limit.

4. Some examples

Suppose that $\mathbb{K}$ is the field of p-adic numbers $\mathbb{Q}_p$ for some prime p and X_k takes only the values 0 and 1 with positive probability. Then $\sum_{i=1}^{k} X_i = \sum_{i=1}^{k} X_i^2 = \dots$, and these sums converge in distribution to Haar measure on the ring of p-adic integers $\mathbb{Z}_p$.

Write $\mathbf{E}_{n,r}$ for the rth *elementary symmetric function* of n variables; that is,

$$\mathbf{E}_{n,r}(x_1, \dots, x_n) := \sum_{1 \leq i_1 < \dots < i_r \leq n} x_{i_1} \dots x_{i_r}, \quad n \geq r.$$

If we put $\Psi = \mathbf{E}_{m,r}$, then $Z_k = \binom{k-r}{m-r} \mathbf{E}_{k,r}(X_1, \dots, X_k)$. Write U for a random variable with Haar distribution on $\mathbb{Z}_p$. Then, by a classical determinantal identity (see Example 8 in §I.2 of [**Mac95**]), $\mathbf{E}_{k,r}(X_1, \dots, X_k)$ converges in distribution as $k \to \infty$ to

$$\frac{1}{r!} \det \begin{pmatrix} U & 1 & 0 & \dots & 0 & 0 \\ U & U & 2 & \dots & 0 & 0 \\ \vdots & \vdots & \vdots & \ddots & \ddots & \vdots \\ U & U & U & \dots & U & r-1 \\ U & U & U & \dots & U & U \end{pmatrix} = \frac{U(U-1)\dots(U-r+1)}{r!} = \binom{U}{r}$$

(this is also clear from elementary considerations: if $\{x_1, \dots, x_n\} \subseteq \{0, 1\}$, then $\mathbf{E}_{n,r}(x_1, \dots, x_n)$ counts how many subsets of size r can be drawn from a set of $(x_1 + \dots + x_n)$ objects).

We note in passing that the random variable U is the simplest example of the natural analogue on $\mathbb{Q}_p$ of a Gaussian random variable. Moreover, the random variables $\binom{U}{r}$ are, in a very natural sense, orthogonal and appear in a theory of stochastic integration and Wiener chaos on $\mathbb{Q}_p$. We refer the reader to [**Eva89a, Eva91, Eva93, Eva95**] for details.

Write $\mathbf{H}_{n,r}$ for the rth *complete symmetric function* of n variables; that is, $\mathbf{H}_{n,r}(x_1, \dots, x_n)$ is the sum of all monomials of total degree r in the variables $x_1, \dots, x_n$. By another classical determinantal identity (see Example 8 in §I.2 of [**Mac95**]), $\mathbf{H}_{k,r}(X_1, \dots, X_k)$ converges in distribution as $k \to \infty$ to

$$\frac{1}{r!} \det \begin{pmatrix} U & -1 & 0 & \dots & 0 & 0 \\ U & U & -2 & \dots & 0 & 0 \\ \vdots & \vdots & \vdots & \ddots & \ddots & \vdots \\ U & U & U & \dots & U & -(r-1) \\ U & U & U & \dots & U & U \end{pmatrix} = (-1)^r \binom{-U}{r}.$$

Note that $\{\mathbf{H}_{k,r}(X_1, \dots, X_k)\}$ is not a sequence of U–statistics for some function Ψ.

Finally, given a partition λ of some integer r, let $\mathbf{S}_{n,\lambda}(x_1, \dots, x_n)$ be the *Schur function* in the variables $x_1, \dots, x_n$ associated with λ (see §I.3 of [**Mac95**]). From a classical determinantal formula expressing Schur functions in terms of the complete symmetric functions (see Equation (3.4) of [**Mac95**])

$$\mathbf{S}_{k,\lambda}(X_1, \dots, X_k) = \det \left(\mathbf{H}_{k,\lambda_i - i + j}(X_1, \dots, X_k)\right)_{1 \le i,j \le N}$$

for any N that is at least the length of the partition λ (that is, the number of non–zero parts in λ). By the above, the right–hand side converges in distribution as $k \to \infty$ to

$$\det \left((-1)^{\lambda_i - i + j} \binom{-U}{\lambda_i - i + j} \right)_{1 \le i,j \le N} = (-1)^r \det \left(\binom{-U}{\lambda_i - i + j} \right)_{1 \le i,j \le N}.$$

From Example 4 in §I.3 of [**Mac95**], we see that the right–hand side is

$$\prod_{x \in \lambda'} \frac{U - c_{\lambda'}(x)}{h_{\lambda'}(x)},$$

where λ' is the partition dual to λ and $c_{\lambda'}(x)$ (resp. $h_{\lambda'}(x)$) denotes the *content* (resp. *hook length*) of λ' at x (that is, if x is the box in the ith row and j^{th} column of the Young diagram of λ', then $c_{\lambda'}(x) := j - i$ and $h_{\lambda'}(x)$ is the number of boxes in the hook with corner at x). There is a natural identification of boxes in the Young diagram of a partition with boxes in the Young diagram of the dual partition. Under this identification, $c_{\lambda'}(x) = -c_\lambda(x)$ and $h_{\lambda'}(x) = h_\lambda(x)$. Consequently, $\mathbf{S}_{k,\lambda}(X_1, \dots, X_k)$ converges in distribution as $k \to \infty$ to

$$\prod_{x \in \lambda} \frac{U + c_\lambda(x)}{h_\lambda(x)}.$$

References

[AK91] S. Albeverio and W. Karwowski, *Diffusion on p-adic numbers*, Gaussian random fields (Nagoya, 1990), Ser. Probab. Statist., no. 1, World Sci. Publishing, River Edge, NJ, 1991, pp. 86–99.

[AK94] ———, *A random walk on p-adics—the generator and its spectrum*, Stochastic Process. Appl. **53** (1994), 1–22.

[AKZ99] S. Albeverio, W. Karwowski, and X. Zhao, *Asymptotics and spectral results for random walks on p-adics*, Stochastic Process. Appl. **83** (1999), 39–59.

[Bik99] A. Kh. Bikulov, *Stochastic equations of mathematical physics over the field of p-adic numbers*, Theoret. and Math. Phys. **119** (1999), 594–604.

[BV97] A. Kh. Bikulov and I.V. Volovich, *p-adic Brownian motion*, Izv. Math. **61** (1997), 537–552.

[Eva89a] S.N. Evans, *Local field Gaussian measures*, Seminar on Stochastic Processes, 1988 (Gainesville, FL, 1988), Progr. Probab., no. 17, Birkhäuser Boston, Boston, MA, 1989, pp. 121–160.

[Eva89b] ———, *Local properties of Lévy processes on a totally disconnected group*, J. Theoret. Probab. **2** (1989), 209–259.

[Eva91] ———, *Equivalence and perpendicularity of local field Gaussian measures*, Seminar on Stochastic Processes, 1990 (Vancouver, BC, 1990), Progr. Probab., no. 24, Birkhäuser Boston, Boston, MA, 1991, pp. 173–181.

[Eva93] ———, *Local field Brownian motion*, J. Theoret. Probab. **6** (1993), 817–850.

[Eva95] ———, *p-adic white noise, chaos expansions, and stochastic integration*, Probability measures on groups and related structures, XI (Oberwolfach, 1994), World Sci. Publishing, River Edge, NJ, 1995, pp. 102–115.

[Gui89] F. Guimier, *Simplicité du spectre de Liapounoff d'un produit de matrices aléatoires sur un corps ultramétrique*, C. R. Acad. Sci. Paris Sr. I Math. **309** (1989), 885–888.

[KB94] V.S. Koroljuk and Yu. V. Borovskich, *Theory of U-statistics*, Mathematics and its Applications, no. 273, Kluwer, Dordrecht, 1994.

[KM94] W. Karwowski and R. Vilela Mendes, *Hierarchical structures and asymmetric stochastic processes on p-adics and adèles*, J. Math. Phys. **35** (1994), 4637–4650.

[Koc97] A.N. Kochubei, *Stochastic integrals and stochastic differential equations over the field of p-adic numbers*, Potential Anal. **6** (1997), 105–125.

[Koc99] ———, *Analysis and probability over infinite extensions of a local field*, Potential Anal. **10** (1999), 305–325.

[Lee90] A.J. Lee, *U-statistics. Theory and practice*, Statistics: Textbooks and Monographs, no. 110, Marcel Dekker, Inc., New York, 1990.

[Mac95] I.G. Macdonald, *Symmetric Functions and Hall Polynomials*, 2nd ed., Oxford Mathematical Monographs, Oxford University Press, Oxford, 1995.

[Mad90] A. Madrecki, *Minlos' theorem in non-Archimedean locally convex spaces*, Comment. Math. Prace Mat. **30** (1990), 101–111.

[Mad91] ———, *Some negative results on existence of Sazonov topology in l-adic Fréchet spaces*, Arch. Math. (Basel) **56** (1991), 601–610.

[Sat94] T. Satoh, *Wiener measures on certain Banach spaces over non-Archimedean local fields*, Compositio Math. **93** (1994), 81–108.

[Sch84] W.H. Schikhof, *Ultrametric Calculus : an Introduction to p-adic Analysis*, Cambridge Studies in Advanced Mathematics, no. 4, Cambridge University Press, Cambridge, 1984.

[Ser80] R.J. Serfling, *Approximation Theorems of Mathematical Statistics*, Wiley Series in Probability and Mathematical Statistics, John Wiley and Sons, Inc., New York, 1980.

[Sta99] R.P. Stanley, *Enumerative Combinatorics*, Cambridge Studies in Advanced Mathematics, no. 62, Cambridge University Press, Cambridge, 1999.

[Tai75] M.H. Taibleson, *Fourier Analysis on Local Fields*, Mathematical Notes, no. 15, Princeton University Press, Princeton, NJ, 1975.

[Yas96] K. Yasuda, *Additive processes on local fields*, J. Math. Sci. Univ. Tokyo **3** (1996), 629–654.

E-mail address: `evans@stat.Berkeley.EDU`

Department of Statistics #3860, University of California at Berkeley, 367 Evans Hall, Berkeley, CA 94720-3860, U.S.A

Contemporary Mathematics
Volume **287**, 2001

Krawtchouk Matrices from Classical and Quantum Random Walks

Philip Feinsilver and Jerzy Kocik

ABSTRACT. Krawtchouk's polynomials occur classically as orthogonal polynomials with respect to the binomial distribution. They may be also expressed in the form of matrices, that emerge as arrays of the values that the polynomials take. The algebraic properties of these matrices provide a very interesting and accessible example in the approach to probability theory known as *quantum probability.* First it is noted how the Krawtchouk matrices are connected to the classical symmetric Bernoulli random walk. And we show how to derive Krawtchouk matrices in the quantum probability context via tensor powers of the elementary Hadamard matrix. Then connections with the classical situation are shown by calculating expected values in the quantum case.

1. Introduction

Some very basic algebraic rules can be expressed using matrices. Take, for example,

$$\begin{array}{lcl} (a+b)^2 & = & a^2 + 2ab + b^2 \\ (a+b)(a-b) & = & a^2 \qquad\; - b^2 \\ (a-b)^2 & = & a^2 - 2ab + b^2 \end{array} \quad \Rightarrow \quad \Phi^{(2)} = \begin{bmatrix} 1 & 1 & 1 \\ 2 & 0 & -2 \\ 1 & -1 & 1 \end{bmatrix}$$

(the expansion coefficients make up the columns of the matrix). In general, we make the definition:

DEFINITION 1.1. The Nth-order Krawtchouk matrix $\Phi^{(N)}$ is an $(N+1)\times(N+1)$ matrix, the entries of which are determined by the expansion:

$$(1.1) \qquad (1+v)^{N-j}\,(1-v)^j = \sum_{i=0}^{N} v^i \Phi_{ij}^{(N)} .$$

The left-hand-side $G(v) = (1+v)^{N-j}\,(1-v)^j$ is thus the *generating function* for the row entries of the jth column of $\Phi^{(N)}$. Expanding gives an explicit expression:

$$\Phi_{ij}^{(N)} = \sum_k (-1)^k \binom{j}{k}\binom{N-j}{i-k} .$$

1991 *Mathematics Subject Classification.* 05B20, 60G50, 47A80, 46L53.
Key words and phrases. Krawtchouk polynomials, Hadamard matrices, symmetric tensors, random walk, quantum computing, quantum probability.

Here are the Krawtchouk matrices of order zero and one:

$$\Phi^{(0)} = \left[\ 1\ \right] \qquad \Phi^{(1)} = \begin{bmatrix} 1 & 1 \\ 1 & -1 \end{bmatrix}. \tag{1.2}$$

More examples can be found in Table 1 of Appendix C. In the remaining of the text, matrix indices run from 0 to N.

One may view the columns of Krawtchouk matrices as *generalized binomial coefficients*. The rows define Krawtchouk *polynomials*: for a fixed order N, the i-th *Krawtchouk polynomial* is the function

$$K_i(j, N) = \Phi_{ij}^{(N)}$$

that takes its corresponding values from the i-th row. One can easily show that $K_i(j, N)$ is indeed a polynomial of degree i in the variable j.

Historically, Krawtchouk's polynomials were introduced and studied by Mikhail Krawtchouk in the late 20's [**12, 13**]. Since then, they have appeared in many areas of mathematics and applications. As orthogonal polynomials, they occur in the classic work by Szëgo [**20**]. They have been studied from the point of view of harmonic analysis and special functions, e.g., in work of Dunkl [**6, 7**]. In statistical considerations, they arose in work of Eagleson [**8**] and later Vere-Jones [**21**]. They play various roles in coding theory and combinatorics, for example, in MacWilliams' theorem on weight enumerators [**17, 14**], and in association schemes [**3, 4, 5**].

A classical probabilistic interpretation has been given in [**10**]. In the context of the classical symmetric random walk, it is recognized that Krawtchouk's polynomials are elementary symmetric functions in variables taking values ± 1. Specifically, if ξ_i are independent Bernoulli random variables taking values ± 1 with probability $\frac{1}{2}$, then if j of the ξ_i are equal to -1, the ith elementary symmetric function in the ξ_i is equal to $\Phi_{ij}^{(N)}$. It turns out that the generating function (1.1) is a martingale in the parameter N. Details are in Section 3 below.

As matrices, they appeared in the 1985 work of N. Bose [**1**] on digital filtering, in the context of the Cayley transform on the complex plane. The symmetric version of the Krawtchouk matrices has been considered in [**9**].

Despite this wide research, the full potential, meaning and significance of Krawtchouk polynomials is far from being complete. In this paper we look at Krawtchouk matrices as *operators* and propose two new ways in which Krawtchouk matrices arise: via classical and quantum random walks. Especially the latter is of current interest. The starting idea is to represent the second Krawtchouk matrix (coinciding with the basic Hadamard matrix) as a sum of two operators

$$\begin{bmatrix} 1 & 1 \\ 1 & -1 \end{bmatrix} = \begin{bmatrix} 0 & 1 \\ 1 & 0 \end{bmatrix} + \begin{bmatrix} 1 & 0 \\ 0 & -1 \end{bmatrix}.$$

Via the technique of tensor products of underlying spaces we obtain a relationship between Krawtchouk matrices and Sylvester-Hadamard matrices.

The reader should consult Parthasarathy's [**18**] for material on quantum probability. It contains the operator theory needed for the subject as well as showing the connections with classical probability theory.

For information on Hadamard matrices as they appear here, we recommend Yarlagadda and Hershey's work [**22**] which provides an overview of the subject of

Sylvester-Hadamard matrices, indicating many interesting applications. For statisticians, they point out that in Yates' factorial analysis, the Hadamard transform provides a useful nonparametric test for association.

Yet another area of significance of this research lies in the quantum computing program [**15, 18**]. Details on this connection will appear in an independent work.

This paper is organized as follows. In the next section, we review basic properties of Krawtchouk matrices. The identities presented, although basic, seem to be new and do not appear in the references cited. Section 3 presents the classical probability interpretation. It may be viewed as a warm-up leading to the *quantum random walk* introduced and studied in Section 4, and to the relationship between Krawtchouk matrices and Sylvester-Hadamard matrices. The generating function techniques used there are original with the present authors. In the last subsection, calculating expected values in the quantum case shows how the quantum model is related to the classical random walk. Appendix C has examples of Krawtchouk and symmetric Krawtchouk matrices so that the reader may see concretely the subject(s) of our discussion. Appendices A (tensor products) and B (symmetric tensor spaces) are included to aid the reader as well as to clarify the notation.

2. Basic properties of Krawtchouk matrices

The square of a Krawtchouk matrix is proportional to the identity matrix, that is, $(\Phi^{(N)})^2 = 2^N I$. This remarkable property allows one to define a Fourier-like *Krawtchouk transform* on integer vectors. The top row is all 1's. The bottom row has ± 1's with alternating signs, starting with $+1$. The leftmost entries are just binomial coefficients, $\Phi^{(N)}_{i0} = \binom{N}{i}$. The rightmost entries are binomial coefficients with alternating signs, $\Phi^{(N)}_{iN} = (-1)^i \binom{N}{i}$. Moreover, there is a four-fold symmetry

$$|\Phi^{(N)}_{i\,j}| = |\Phi^{(N)}_{N-i\,j}| = |\Phi^{(N)}_{i\,N-j}| = |\Phi^{(N)}_{N-i\,N-j}|.$$

Krawtchouk matrices generalize Pascal's triangle in the following sense: Visualize a stack of Krawtchouk matrices, the order N increasing downwards. Pascal's triangle is formed by the leftmost columns. It turns out that Pascal's identity holds for the other columns as well. Less obvious is another identity — call it dual Pascal.

PROPOSITION 2.1. *Set* $a = \Phi^{(N)}_{i-1\,j}$, $b = \Phi^{(N)}_{i\,j}$, $A = \Phi^{(N+1)}_{i\,j}$, $B = \Phi^{(N+1)}_{i\,j+1}$.

1. **(Cross identities)** *The following mutually inverse relations (Pascal and dual Pascal) hold:*

$$\begin{array}{ccc} a+b=A & \quad\textit{and}\quad & A+B=2b \\ b-a=B & & A-B=2a\,. \end{array}$$

2. **(Square identity)** *In a square of any four adjacent entries in a Krawtchouk matrix, the entry in the left-bottom corner is the sum of the other three, i.e.,*

$$\textit{for}\ \ \Phi = \begin{bmatrix} & \vdots & \vdots & \\ \cdots & a & c & \cdots \\ \cdots & b & d & \cdots \\ & \vdots & \vdots & \end{bmatrix} \qquad \textit{one has} \qquad b = a + c + d.$$

PROOF. For $a+b$, consider

$$(1+v)^{N+1-j}(1-v)^j = (1+v)\,(1+v)^{N-j}(1-v)^j\,.$$

For $b-a$, consider

$$(1+v)^{N-j}(1-v)^{j+1} = (1-v)\,(1+v)^{N-j}(1-v)^j\,.$$

The inverse relations are immediate. The square identity follows from the observation $(a+b)+(c+d) = A+B = 2b$, hence $a+c+d = b$. □

The square identity is useful in producing the entries of a Krawtchouk matrix: fill the top row with 1's, the right-most column with sign-alternating binomial coefficients. Then, apply the square identity to reproduce the matrix.

In summary, the identities considered above can be written as follows:

Cross identities:

$$(i)\quad \Phi^{(N)}_{i-1\,j} + \Phi^{(N)}_{i\,j} = \Phi^{(N+1)}_{i\,j} \qquad\qquad (ii)\quad \Phi^{(N)}_{i\,j} + \Phi^{(N)}_{i\,j+1} = 2\Phi^{(N-1)}_{i\,j}$$

$$(iii)\quad \Phi^{(N)}_{i\,j} - \Phi^{(N)}_{i-1\,j} = \Phi^{(N+1)}_{i\,j+1} \qquad\qquad (iv)\quad \Phi^{(N)}_{i\,j} - \Phi^{(N)}_{i\,j+1} = 2\Phi^{(N-1)}_{i-1\,j}\,.$$

Square identity:

$$\Phi^{(N)}_{ij} = \Phi^{(N)}_{i-1\,j} + \Phi^{(N)}_{i-1\,j+1} + \Phi^{(N)}_{i\,j+1}\,.$$

If each column of the matrix is multiplied by the corresponding binomial coefficient, the matrix becomes symmetric. Let $B^{(N)}$ denote the $(N+1)\times(N+1)$ diagonal matrix with binomial coefficients

$$B^{(N)}_{ii} = \binom{N}{i} \tag{2.1}$$

as its non-zero entries. Then, for each $N \geq 0$, one defines the **symmetric Krawtchouk matrix** as

$$S^{(N)} = \Phi^{(N)} B^{(N)}\,.$$

Example: For $N=3$, we have

$$S^{(3)} = \begin{bmatrix} 1 & 1 & 1 & 1 \\ 3 & 1 & -1 & -3 \\ 3 & -1 & -1 & 3 \\ 1 & -1 & 1 & -1 \end{bmatrix} \begin{bmatrix} 1 & 0 & 0 & 0 \\ 0 & 3 & 0 & 0 \\ 0 & 0 & 3 & 0 \\ 0 & 0 & 0 & 1 \end{bmatrix} = \begin{bmatrix} 1 & 3 & 3 & 1 \\ 3 & 3 & -3 & -3 \\ 3 & -3 & -3 & 3 \\ 1 & -3 & 3 & -1 \end{bmatrix}.$$

Some symmetric Krawtchouk matrices are displayed in Table 2 of Appendix C.

3. Krawtchouk matrices and classical random walk

In this section we will give a probabilistic meaning to the Krawtchouk matrices and some of their properties.

Let ξ_i be independent symmetric Bernoulli random variables taking values ± 1. Let $X_N = \xi_1 + \cdots + \xi_N$ be the associated random walk starting from 0. Now observe that the generating function of the elementary symmetric functions in the ξ_i is a martingale, in fact a discrete exponential martingale:

$$M_N = \prod_{i=1}^{N} (1 + v\xi_i) = \sum_k v^k \alpha_k(\xi_1, \ldots, \xi_N),$$

where α_k denotes the kth elementary symmetric function. The martingale property is immediate since each ξ_i has mean 0. Suppose that at time N, the number of the ξ_i that are equal to -1 is j_N, with the rest equal to $+1$. Then $j_N = (N - X_N)/2$ and M_N can be expressed solely in terms of N and X_N, or, equivalently, of N and j_N

$$M_N = (1+v)^{N-j_N}(1-v)^{j_N} = (1+v)^{(N+X_N)/2}(1-v)^{(N-X_N)/2}.$$

From the generating function for the Krawtchouk matrices, (1.1), follows

$$M_N = \sum_i v^i \Phi^{(N)}_{i,j_N},$$

so that as functions on the Bernoulli space, each sequence of random variables $\Phi^{(N)}_{i,j_N}$ is a martingale.

Now we can interpret two basic recurrences of Proposition 2.1. For a fixed column of $\Phi^{(N)}$, the corresponding column in $\Phi^{(N+1)}$ satisfies the Pascal triangle recurrence:

$$\Phi^{(N)}_{i-1\,j} + \Phi^{(N)}_{i\,j} = \Phi^{(N+1)}_{i\,j}.$$

To see this in the probabilistic setting, write $M_{N+1} = (1 + v\xi_N)M_N$. Observe that for j_N to remain constant, ξ_N must take the value $+1$ and expanding $(1+v)M_N$ yields the Pascal recurrence as in the proof of Proposition 2.1. It is interesting how the martingale property comes into play. We have

$$\Phi^{(N)}_{ij_N} = E(\Phi^{(N+1)}_{ij_{N+1}} | \xi_1, \ldots, \xi_N) = \frac{1}{2}\left(\Phi^{(N+1)}_{i\,j_N+1} + \Phi^{(N+1)}_{ij_N}\right),$$

since half the time ξ_{N+1} is -1, increasing j_N by 1, and half the time j_N is unchanged. Thus, writing j for j_N,

$$\Phi^{(N)}_{ij} = \frac{1}{2}\left(\Phi^{(N+1)}_{i\,j+1} + \Phi^{(N+1)}_{ij}\right).$$

Many further properties of Krawtchouk polynomials may be derived from their interpretation as elementary symmetric functions on the Bernoulli space with scope for probabilistic methods as well.

4. Krawtchouk matrices and quantum random walk

In quantum probability, random variables are modeled by self-adjoint operators on Hilbert spaces and independence by tensor products. We can model a symmetric Bernoulli random walk as follows. Consider a 2-dimensional Hilbert space $V = \mathbf{R}^2$ and two special 2×2 operators,

$$F = \begin{bmatrix} 0 & 1 \\ 1 & 0 \end{bmatrix} \quad \text{and} \quad G = \begin{bmatrix} 1 & 0 \\ 0 & -1 \end{bmatrix},$$

satisfying $F^2 = G^2 = I$ (the 2×2 identity). The fundamental Hadamard matrix H coincides with the second Krawtchouk matrix. Now we shall view it as a sum of the above operators

$$H = F + G = \begin{bmatrix} 1 & 1 \\ 1 & -1 \end{bmatrix}.$$

One can readily check that

$$FH = F(F+G) = (F+G)G = HG \tag{4.1}$$

(use $F^2 = G^2 = I$). This, of course, can be viewed as the spectral decomposition of F and we can interpret the Hadamard matrix as a matrix reducing F to diagonal form.

REMARK 4.1. Note that the exponentiated operator

$$\exp(zF) = \begin{bmatrix} \cosh z & \sinh z \\ \sinh z & \cosh z \end{bmatrix}$$

has the expected value in the state e_0 equal to

$$\langle e_0, \exp(zF)e_0 \rangle = \cosh z\,, \tag{4.2}$$

where e_0 denotes the transpose of $[1, 0]$. This coincides with the moment generating function for the symmetric Bernoulli random variable taking values ± 1, showing that indeed we are dealing with the (quantum) generalization of the classical model.

The Hilbert space of states is represented by the N-th tensor product of the original space V, that is, by the 2^N-dimensional Hilbert space $V^{\otimes N}$ (see Appendix A for notation). Define the following linear operators, N in all, in $V^{\otimes N}$

$$\begin{aligned} f_1 &= F \otimes I \otimes \cdots \otimes I \\ f_2 &= I \otimes F \otimes I \otimes \cdots \otimes I \\ \vdots &= \vdots \\ f_N &= I \otimes I \otimes \cdots \otimes F\,, \end{aligned}$$

each f_i describing a flip at the i-th position. These are the quantum equivalents of the random walk variables from Section 3. We shall consider the superposition of these independent actions, setting

$$X_F = f_1 + \cdots + f_N\,.$$

Notation: For notational clarity, since N is fixed throughout the discussion, we drop the index N from the X's.

Analogously, we define:

$$\begin{aligned} g_1 &= G \otimes I \otimes \cdots \otimes I \\ g_2 &= I \otimes G \otimes I \otimes \cdots \otimes I \\ \vdots &= \vdots \\ g_N &= I \otimes I \otimes \cdots \otimes G, \end{aligned}$$

with $X_G = g_1 + \cdots + g_N$. Finally, let us extend H to the N-fold tensor product, setting $H_N = H^{\otimes N}$. These are the well-known Sylvester-Hadamard matrices with the first few given here:

$$H_1 = \begin{bmatrix} \bullet & \bullet \\ \bullet & \circ \end{bmatrix} \quad H_2 = \begin{bmatrix} \bullet & \bullet & \bullet & \bullet \\ \bullet & \circ & \bullet & \circ \\ \bullet & \bullet & \circ & \circ \\ \bullet & \circ & \circ & \bullet \end{bmatrix} \quad H_3 = \begin{bmatrix} \bullet & \bullet & \bullet & \bullet & \bullet & \bullet & \bullet & \bullet \\ \bullet & \circ & \bullet & \circ & \bullet & \circ & \bullet & \circ \\ \bullet & \bullet & \circ & \circ & \bullet & \bullet & \circ & \circ \\ \bullet & \circ & \circ & \bullet & \bullet & \circ & \circ & \bullet \\ \bullet & \bullet & \bullet & \bullet & \circ & \circ & \circ & \circ \\ \bullet & \circ & \bullet & \circ & \circ & \bullet & \circ & \bullet \\ \bullet & \bullet & \circ & \circ & \circ & \circ & \bullet & \bullet \\ \bullet & \circ & \circ & \bullet & \circ & \bullet & \bullet & \circ \end{bmatrix},$$

etc., where, for typographical reasons, we use $\bullet$ for 1 and $\circ$ for -1.

It turns out that our X-operators intertwine the Sylvester-Hadamard matrices. For illustration, consider a calculation for $N = 3$:

$$\begin{aligned} f_1 H_3 &= (F \otimes I \otimes I)(H \otimes H \otimes H) \\ &= (H \otimes H \otimes H)(G \otimes I \otimes I) = H_3 g_1 , \end{aligned}$$

where the relation $FH = HG$ is used. This clearly generalizes to $f_k H_N = H_N g_k$ and, by summing over k, yields an important relation:

$$X_F H_N = H_N X_G .$$

Now, we shall consider the symmetrized versions of the operators (the reader is referred to Appendix B for the theory and methods used here). Since products are preserved in the process of reduction to the symmetric tensor space, we get

$$\overline{X}_F \overline{H}_N = \overline{H}_N \overline{X}_G ,$$

the bars indicating the corresponding induced maps. We know how to calculate $\overline{H}_N$ from the action of H on polynomials in degree N. For symmetric tensors the components in degree N are

$$x_0^{N-k} x_1^k ,$$

where $0 \leq k \leq N$.

Proposition 4.2. *For each $N > 0$, symmetric reduction of the Nth Hadamard matrix results in the transposed N^{th} Krawtchouk matrix:*

$$(\overline{H}_N)_{ij} = \Phi_{ji}^{(N)} .$$

Proof. Writing (x, y) for (x_0, x_1), we have in degree N for the kth component:

$$(x+y)^{N-k}(x-y)^k = \sum_l \overline{H}_{kl}\, x^{N-l} y^l .$$

Scaling out x^N and replacing $v = y/x$ yields the generating function for the Krawtchouk matrices with the coefficient of v^l equal to $\Phi_{lk}^{(N)}$. Thus the result. □

Now consider the generating function for the elementary symmetric functions in the quantum variables f_j. This is the N-fold tensor power

$$\mathcal{F}_N(t) = (I + tF)^{\otimes N} = I^{\otimes N} + t\, X_F + \cdots ,$$

noting that the coefficient of t is X_F. Similarly, define

$$\mathcal{G}_N(t) = (I + tG)^{\otimes N} = I^{\otimes N} + tX_G + \cdots .$$

From $(I + tF)H = H(I + tG)$ we have

$$\mathcal{F}_N H_N = H_N \mathcal{G}_N \qquad \text{and} \qquad \overline{\mathcal{F}}_N \overline{H}_N = \overline{H}_N \overline{\mathcal{G}}_N .$$

The difficulty is to calculate the action on the symmetric tensors for operators, such as X_F, that are not pure tensor powers. However, from $\mathcal{F}_N(t)$ and $\mathcal{G}_N(t)$ we can recover X_F and X_G via

$$X_F = \frac{d}{dt}\bigg|_{t=0} (I + tF)^{\otimes N}, \qquad X_G = \frac{d}{dt}\bigg|_{t=0} (I + tG)^{\otimes N}$$

with corresponding relations for the barred operators. Calculating on polynomials yields the desired results as follows.

$$I + tF = \begin{bmatrix} 1 & t \\ t & 1 \end{bmatrix}, \qquad I + tG = \begin{bmatrix} 1+t & 0 \\ 0 & 1-t \end{bmatrix}.$$

In degree N, using x and y as variables, we get the kth component for $\overline{X}_F$ and $\overline{X}_G$ via

$$\frac{d}{dt}\bigg|_{t=0} (x + ty)^{N-k}(tx + y)^k \quad = \quad (N - k)\, x^{N-(k+1)} y^{k+1} + k\, x^{N-(k-1)} y^{k-1} ,$$

and since $I + tG$ is diagonal,

$$\frac{d}{dt}\bigg|_{t=0} (1 + t)^{N-k}(1 - t)^k\, x^{N-k} y^k \quad = \quad (N - 2k)\, x^{N-k} y^k .$$

For example, calculations for $N = 4$ result in

$$\overline{X}_F = \begin{bmatrix} 0 & 4 & 0 & 0 & 0 \\ 1 & 0 & 3 & 0 & 0 \\ 0 & 2 & 0 & 2 & 0 \\ 0 & 0 & 3 & 0 & 1 \\ 0 & 0 & 0 & 4 & 0 \end{bmatrix}, \qquad \overline{H}_4 = \begin{bmatrix} 1 & 4 & 6 & 4 & 1 \\ 1 & 2 & 0 & -2 & -1 \\ 1 & 0 & -2 & 0 & 1 \\ 1 & -2 & 0 & 2 & -1 \\ 1 & -4 & 6 & -4 & 1 \end{bmatrix},$$

$$\overline{X}_G = \begin{bmatrix} 4 & 0 & 0 & 0 & 0 \\ 0 & 2 & 0 & 0 & 0 \\ 0 & 0 & 0 & 0 & 0 \\ 0 & 0 & 0 & -2 & 0 \\ 0 & 0 & 0 & 0 & -4 \end{bmatrix}.$$

Since $\overline{X}_G$ is the result of diagonalizing $\overline{X}_F$, we observe that

COROLLARY 4.3. *The spectrum of $\overline{X}_F$ is $N, N-2, \ldots, 2-N, -N$, coinciding with the support of the classical random walk.*

4.1. Expected values. To find the probability distributions associated to our X_F operators, we must calculate expected values, cf. Remark 4.1. In the present context, expected values in two particular states are especially interesting. Namely, in the state e_0 and in the normalized trace, which is the uniform distribution on the spectrum. In the N-fold tensor product, we want to consider expected values in the ground state $|\,000\ldots0\,\rangle$ and normalized traces. Then we can go to the symmetric tensors.

The scalar product on the tensor product space factors, corresponding to independence in classical probability. Thus, from (4.2) one obtains the expected value of $\exp(zX_F)$ in the ground state $|\,000\ldots0\,\rangle$ to be $(\cosh z)^N$. Similarly, the trace of the tensor product of operators is the product of their traces. So, for the trace, $\operatorname{tr}\exp(zF)=2\cosh z$ implies $\operatorname{tr}\exp(zX_F)=2^N(\cosh z)^N$ and, after normalizing, this yields $(\cosh z)^N$.

For the barred operators, we consider the symmetric trace. Here we use the **symmetric trace theorem**, detailed in Appendix B. It tells us that the generating function for the symmetric traces of any operator A in the various degrees is $\det(I-tA)^{-1}$. Taking $A=\exp(zF)$, we have

$$\begin{aligned}\det(I-te^{zF})^{-1} &= [(1-te^{z})(1-te^{-z})]^{-1}\\ &= (1-2t\cosh z+t^2)^{-1}\,.\end{aligned}$$

The latter is the generating function for Chebyshev polynomials of the second kind, U_N, so that the normalized symmetric trace is

$$(N+1)^{-1}\operatorname{tr}{}^{N}_{\mathrm{Sym}}\exp(zF)=U_N(\cosh z)/(N+1)\,,$$

which equals as well

$$\frac{e^{z(N+1)}-e^{-z(N+1)}}{(e^z-e^{-z})(N+1)}=\frac{\sinh(N+1)z}{(N+1)\sinh z}\,.$$

This corresponds to a uniform distribution on the support of the random walk at time N, namely, $-N, 2-N, \ldots, N-2, N$.

Appendix A. Tensor products

Fulton and Harris [**11**] is a useful reference for this section and the next. Also Parthasarathy [**18**], Chapter II, is an excellent reference.

Let V be a d-dimensional vector space over $\mathbf{R}$. We fix an orthonormal basis $\{e_0,\ldots,e_\delta\}$ with $d=1+\delta$. Denote tensor powers of V by $V^{\otimes N}$, so that $V^{\otimes 2}=V\otimes V$, etc. A basis for $V^{\otimes N}$ is given by all N-fold tensor products of the basis vectors e_i,

$$|\,n_1n_2\ldots n_N\,\rangle=e_{n_1}\otimes e_{n_2}\otimes\cdots\otimes e_{n_N}\,.$$

Note that we can label these d^N basis elements by all numbers 0 to d^N-1 and recover the tensor products by expressing these numbers in base d, putting leading zeros so that all extended labels are of length N.

Now let $\{A_i:\ 1\le i\le N\}$ be a set of N linear operators on V. On $V^{\otimes N}$, the linear operator $A=A_1\otimes A_2\otimes\cdots\otimes A_N$ acts on a basis vector $|\,n_1n_2\ldots n_N\,\rangle$ by

$$A|\,n_1n_2\ldots n_N\,\rangle=A_1e_{n_1}\otimes\cdots\otimes A_Ne_{n_N}\,.$$

This needs to be expanded and terms regrouped using bilinearity.

If A and B are two $d \times d$ matrices, the matrix corresponding to the operator $A \otimes B$ is the Kronecker product, a $d^2 \times d^2$ matrix having the block form:

$$\begin{bmatrix} a_{00}B & \dots & a_{0\delta}B \\ a_{10}B & \dots & a_{1\delta}B \\ \vdots & \vdots & \vdots \\ a_{\delta 0}B & \dots & a_{\delta\delta}B \end{bmatrix}.$$

This, iteratively, is valid for higher-order tensor products (associating from the left by convention). The rows and columns of the matrix of a linear operator acting on $V^{\otimes N}$ are conveniently labeled by associating to each basic tensor $|\, n_1 n_2 \dots n_N \,\rangle$ the corresponding integer label $\sum_{k=1}^{N} n_k d^{N-k}$, which thus provides a canonical ordering.

Appendix B. Symmetric tensor spaces

Here we review symmetric tensor spaces as spaces of polynomials in commuting variables. This material is presented with a view to the infinite-dimensional case in [**18**], pp. 105ff., however we focus on the finite-dimensional context and include as well an important observation contained in Theorem B.1.

The space $V^{\otimes N}$ can be mapped onto the space of symmetric tensors, $V^{\otimes_S N}$ by identifying basis vectors (in $V^{\otimes N}$) that are equivalent under all permutations. Alternatively, one can identify the basic tensor $|\, n_1 n_2 \dots n_N \,\rangle$ with the monomial $x_{n_1} x_{n_2} \cdots x_{n_N}$ in the commuting variables $x_0, \dots, x_\delta$. Hence we have a linear map from tensor space into the space of polynomials, itself isomorphic to the space of symmetric tensors:

$$\overline{} : \bigcup_{N\ge 0} V^{\otimes N} \longrightarrow \mathbf{R}[x_0, \dots, x_\delta] \cong \bigcup_{N\ge 0} V^{\otimes_S N}.$$

In the symmetric tensor space, tensor labels need to count only occupancy, that is, the number of times a basis vector of V occurs in a given basic tensor of $V^{\otimes N}$. We indicate occupancy by a multi-index which is the exponent of the corresponding monomial. The dimension of $V^{\otimes_S N}$ is thus

$$\dim V^{\otimes_S N} = \binom{N+d-1}{d-1},$$

that is, the number of monomials homogeneous of degree N.

Given an operator A on V, let $A_N = A^{\otimes N}$. Then A_N induces an operator $\overline{A}_N$ on $V^{\otimes_S N}$ from the action of A on polynomials, which we call the **symmetric representation of** A **in degree** N. For convenience we work dually with the tensor components rather with the action on the basis vectors. Denote the matrix elements of the action of $\overline{A}_N$ by $\overline{A}_{mn}$. If A has matrix entries A_{ij}, let

$$y_i = \sum_j A_{ij} x_j\,.$$

Then the matrix elements of the symmetric representation are defined by the relation (expansion):

$$y_0^{m_0} \cdots y_\delta^{m_\delta} = \sum_n \overline{A}_{mn} x_0^{n_0} \cdots x_\delta^{n_\delta}$$

with multi-indices m and n.

Composition of A_1 with A_2 shows that mapping to the symmetric representation is an algebra homomorphism, i.e.,

$$\overline{A_1 A_2} = \overline{A}_1 \overline{A}_2 \,.$$

Explicitly, in basis notation

$$\overline{(A_1 A_2)}_{mn} = \sum_r (\overline{A_1})_{mr} (\overline{A_2})_{rn} \,.$$

Define the **symmetric trace** in degree N of A as the trace of the matrix elements of $\overline{A}_N$, i.e., the sum of the diagonal matrix elements:

$$\operatorname{tr}{}^N_{\mathrm{Sym}} A = \sum_{|m|=N} \overline{A}_{mm},$$

with $|m|$ denoting, as usual, the sum of the components of m. Observe that if A is upper-triangular, with eigenvalues $\lambda_1, \ldots, \lambda_d$, then the trace of this action on the space of polynomials homogeneous of degree N is exactly $h_N(\lambda_1, \ldots, \lambda_d)$, the Nth homogeneous symmetric function in the λ's.

We recall a useful theorem on calculating the symmetric trace. Since the mapping from A to $\overline{A}_N$ is a homomorphism, a similarity transformation on A extends to one on $\overline{A}_N$ thus preserving traces. Now, any matrix is similar to an upper-triangular one with the same eigenvalues, thus follows [**19**]:

THEOREM B.1. **Symmetric trace theorem** *Denoting by* $\operatorname{tr}^N_{\mathrm{Sym}}$ *the trace of the symmetric representation on polynomials homogeneous of degree* N,

$$\frac{1}{\det(I - tA)} = \sum_{N=0}^{\infty} t^N \operatorname{tr}{}^N_{\mathrm{Sym}} A \,.$$

PROOF. With $\{\lambda_i\}$ denoting the eigenvalues of A,

$$\begin{aligned} \frac{1}{\det(I - tA)} &= \prod_i \frac{1}{1 - t\lambda_i} = \sum_{N=0}^{\infty} t^N h_N(\lambda_1, \ldots, \lambda_d) \\ &= \sum_{N=0}^{\infty} t^N \operatorname{tr}{}^N_{\mathrm{Sym}} A \,, \end{aligned}$$

as stated above. □

REMARK B.2. Note that this result is equivalent to *MacMahon's Master Theorem* in combinatorics [**16**].

REMARK B.3. From another point of view, Chen and Louck [**2**], considering powers of the bilinear form $\sum_{i,j} x_i A_{ij} y_j$ rather than just the linear form as done here, study representation functions that are analogs of our symmetric Krawtchouk matrices. Their $L_{\alpha,\beta}$ is our symmetric representation scaled by multinomial factors. In addition they suggest further interesting generalizations beyond symmetric tensors.

Appendix C. Krawtchouk matrices

$$\Phi^{(0)} = \begin{bmatrix} 1 \end{bmatrix}$$

$$\Phi^{(1)} = \begin{bmatrix} 1 & 1 \\ 1 & -1 \end{bmatrix}$$

$$\Phi^{(2)} = \begin{bmatrix} 1 & 1 & 1 \\ 2 & 0 & -2 \\ 1 & -1 & 1 \end{bmatrix}$$

$$\Phi^{(3)} = \begin{bmatrix} 1 & 1 & 1 & 1 \\ 3 & 1 & -1 & -3 \\ 3 & -1 & -1 & 3 \\ 1 & -1 & 1 & -1 \end{bmatrix}$$

$$\Phi^{(4)} = \begin{bmatrix} 1 & 1 & 1 & 1 & 1 \\ 4 & 2 & 0 & -2 & -4 \\ 6 & 0 & -2 & 0 & 6 \\ 4 & -2 & 0 & 2 & -4 \\ 1 & -1 & 1 & -1 & 1 \end{bmatrix}$$

$$\Phi^{(5)} = \begin{bmatrix} 1 & 1 & 1 & 1 & 1 & 1 \\ 5 & 3 & 1 & -1 & -3 & -5 \\ 10 & 2 & -2 & -2 & 2 & 10 \\ 10 & -2 & -2 & 2 & 2 & -10 \\ 5 & -3 & 1 & 1 & -3 & 5 \\ 1 & -1 & 1 & -1 & 1 & -1 \end{bmatrix}$$

$$\Phi^{(6)} = \begin{bmatrix} 1 & 1 & 1 & 1 & 1 & 1 & 1 \\ 6 & 4 & 2 & 0 & -2 & -4 & -6 \\ 15 & 5 & -1 & -3 & -1 & 5 & 15 \\ 20 & 0 & -4 & 0 & 4 & 0 & -20 \\ 15 & -5 & -1 & 3 & -1 & -5 & 15 \\ 6 & -4 & 2 & 0 & -2 & 4 & -6 \\ 1 & -1 & 1 & -1 & 1 & -1 & 1 \end{bmatrix}$$

Table 1: Krawtchouk matrices

$$S^{(0)} = \begin{bmatrix} 1 \end{bmatrix}$$

$$S^{(1)} = \begin{bmatrix} 1 & 1 \\ 1 & -1 \end{bmatrix}$$

$$S^{(2)} = \begin{bmatrix} 1 & 2 & 1 \\ 2 & 0 & -2 \\ 1 & -2 & 1 \end{bmatrix}$$

$$S^{(3)} = \begin{bmatrix} 1 & 3 & 3 & 1 \\ 3 & 3 & -3 & -3 \\ 3 & -3 & -3 & 3 \\ 1 & -3 & 3 & -1 \end{bmatrix}$$

$$S^{(4)} = \begin{bmatrix} 1 & 4 & 6 & 4 & 1 \\ 4 & 8 & 0 & -8 & -4 \\ 6 & 0 & -12 & 0 & 6 \\ 4 & -8 & 0 & 8 & -4 \\ 1 & -4 & 6 & -4 & 1 \end{bmatrix}$$

$$S^{(5)} = \begin{bmatrix} 1 & 5 & 10 & 10 & 5 & 1 \\ 5 & 15 & 10 & -10 & -15 & -5 \\ 10 & 10 & -20 & -20 & 10 & 10 \\ 10 & -10 & -20 & 20 & 10 & -10 \\ 5 & -15 & 10 & 10 & -15 & 5 \\ 1 & -5 & 10 & -10 & 5 & -1 \end{bmatrix}$$

$$S^{(6)} = \begin{bmatrix} 1 & 6 & 15 & 20 & 15 & 6 & 1 \\ 6 & 24 & 30 & 0 & -30 & -24 & -6 \\ 15 & 30 & -15 & -60 & -15 & 30 & 15 \\ 20 & 0 & -60 & 0 & 60 & 0 & -20 \\ 15 & -30 & -15 & 60 & -15 & -30 & 15 \\ 6 & -24 & 30 & 0 & -30 & 24 & -6 \\ 1 & -6 & 15 & -20 & 15 & -6 & 1 \end{bmatrix}$$

Table 2: Symmetric Krawtchouk matrices

Acknowledgments

We would like to thank Marlos Viana for inviting us to participate in the Special Session and we extend our appreciation for all the hard work involved in organizing it, as well as related activities.

References

[1] N. Bose, *Digital filters: theory and applications*, North-Holland, 1985.
[2] W.Y.C. Chen and J.D. Louck, *The combinatorics of a class of representation functions,* Adv. in Math., **140** (1998), 207–236.
[3] P. Delsarte, *Bounds for restricted codes, by linear programming,* Philips Res. Reports, **27** (1972) 272–289.
[4] P. Delsarte, *Four fundamental parameters of a code and their combinatorial significance,* Info. & Control, **23** (1973) 407–438.
[5] P. Delsarte, *An algebraic approach to the association schemes of coding theory,* Philips Research Reports Supplements, No. 10, 1973.
[6] C.F. Dunkl, *A Krawtchouk polynomial addition theorem and wreath products of symmetric groups,* Indiana Univ. Math. J., **25** (1976) 335–358.
[7] C.F. Dunkl and D.F. Ramirez, *Krawtchouk polynomials and the symmetrization of hypergraphs,* SIAM J. Math. Anal., **5** (1974) 351–366.
[8] G.K. Eagelson, *A characterization theorem for positive definite sequences of the Krawtchouk polynomials,* Australian J. Stat, **11** (1969) 29–38.
[9] P. Feinsilver and R. Fitzgerald, *The spectrum of symmetric Krawtchouk matrices,* Lin. Alg. & Appl., **235** (1996) 121–139.
[10] P. Feinsilver and R. Schott, *Krawtchouk polynomials and finite probability theory,* Probability Measures on Groups X, Plenum, 1991, pp. 129–135.
[11] W. Fulton and J. Harris, *Representation theory, a first course,* Graduate texts in mathematics, **129**, Springer-Verlag, 1991.
[12] M. Krawtchouk, *Sur une generalisation des polynomes d'Hermite,* Comptes Rendus, **189** (1929) 620–622.
[13] M. Krawtchouk, *Sur la distribution des racines des polynomes orthogonaux,* Comptes Rendus, **196** (1933) 739–741.
[14] V.I. Levenstein, *Krawtchouk polynomials and universal bounds for codes and design in Hamming spaces,* IEEE Transactions on Information Theory, **41** 5 (1995) 1303–1321.
[15] S.J. Lomonaco, Jr., *A Rosetta Stone for quantum mechanics with an introduction to quantum computation,* http://www.arXiv.org/abs/quant-ph/0007045
[16] P.A. MacMahon, *Combinatory analysis,* Chelsea, New York, 1960.
[17] F.J. MacWilliams and N.J.A. Sloane, *The theory of Error-Correcting Codes,* The Netherlands, North Holland, 1977.
[18] K.R. Parthasarathy, *An introduction to quantum stochastic calculus,* Birkhäuser, 1992.
[19] T.A. Springer, *Invariant theory,* Lecture notes in mathematics, **585**, Springer-Verlag, 1977.
[20] G. Szegö, *Orthogonal Polynomials*, Colloquium Publications, Vol. 23, New York, AMS, revised eddition 1959, 35–37.
[21] D. Vere-Jones, *Finite bivariate distributions and semi-groups of nonnegative matrices,* Q. J. Math. Oxford, **22** 2 (1971) 247–270.
[22] R.K. Rao Yarlagadda and J.E. Hershey, *Hadamard matrix analysis and synthesis: with applications to communications and signal/image processing,* Kluwer Academic Publishers, 1997.

Department of Mathematics, Southern Illinois University, Carbondale, IL 62901
E-mail address: pfeinsil@math.siu.edu

Department of Mathematics, Southern Illinois University, Carbondale, IL 62901
E-mail address: jkocik@siu.edu

Contemporary Mathematics
Volume **287**, 2001

Some Rank-based Hypothesis Tests for Covariance Structure and Conditional Independence

Yonghong Gao and John I. Marden

ABSTRACT. Rank-based procedures are proposed for two classes of multivariate hypothesis testing problems: Invariant models on covariance matrices, and conditional independence models defined by undirected graphical models. Nonparametric tests based on multivariate spatial ranks are presented for the invariant models, which include intraclass correlation models, sphericity, and zero correlation models. The conditional independence testing procedures arise from an extension of a model of Feigin that is placed on coordinatewise rank vectors.

1. Introduction

Rank-based hypothesis testing procedures provide convenient and robust alternatives to the usual normal-theory procedures. This article proposes rank methods for testing hypotheses on certain variance-covariance structures and conditional independence structures.

One popular approach towards robustness is to find a robust estimate of the covariance matrix, then base tests on that estimate. We take a different approach. Rather than estimating the covariance matrix, we use ranks to construct hypothesis testing procedures. There are few methods for assessing covariance parameters using ranks, beyond tests for independence of variables. The work of Visuri, Koivunen, and Oja (2000) is most notable. They show how to estimate a covariance matrix (under some symmetry conditions) using different types of univariate and multivariate ranks.

This article will present rank analogs of two classes of multivariate models which have been defined using simple algebraic structures: invariant normal models (Andersson, 1975) and conditional independence graphical models (Whittaker, 1990). The invariant normal models consist of structural models on covariance matrices, e.g., intraclass correlation structure (the variances are equal, and the correlations are equal), independence, sphericity, and complex and quaternion structure. The

2000 *Mathematics Subject Classification.* Primary 62G10, 62G35, 62H15, 62H20.

Key words and phrases. Hypothesis testing; multivariate analysis; multivariate ranks; rank procedures; invariant normal models; conditional independence graphical models.

Support for this research was provided by National Science Foundation Grant NSF DMS 00-71757.

graphical models describe certain types of independence and conditional independence among variables.

The rank analogs for the two classes are very different. For the structural models on the covariance matrix, we use multivariate spatial ranks, and the assumptions on the distribution are broad. For the graphical models, we use the usual variable-wise ranks, but assume a specific model of Feigin (1993) for the ranks. Recently, Andersson and Madsen (1998) considered models combining structural models with lattice conditional independence models for the multivariate normal. It would be interesting to extend the rank models to these more complicated models, but it is not immediately obvious how to do it.

2. Invariant normal models

Andersson (1975) defines the *invariant normal model* for X being a $p \times 1$ multivariate normal vector with mean 0 and covariance matrix Σ, $X \sim N(0, \Sigma)$. For given $\mathcal{G}$, a subgroup of $\mathcal{O}_p$, the $p \times p$ group of orthogonal matrices, the model is the set of distributions

$$\mathcal{P}(\mathcal{G}) \equiv \{N(0, \Sigma) \mid \Sigma = g\Sigma g' \text{ for all } g \in \mathcal{G}\}. \tag{2.1}$$

Different choices of $\mathcal{G}$ yield different models for Σ. Some examples:

1. Unrestricted: $\mathcal{G} = \{I_p\}$, so that Σ is unrestricted. (I_p is the $p \times p$ identity matrix.)
2. Diagonal: $\mathcal{G} = p \times p$ matrices with ± 1's on diagonals, so that Σ is diagonal.
3. Intraclass correlation model: $\mathcal{G} = p \times p$ permutation matrices, so that

$$\Sigma = \sigma^2 \begin{pmatrix} 1 & \rho & \rho & \cdots & \rho \\ \rho & 1 & \rho & \cdots & \rho \\ \rho & \rho & 1 & \cdots & \rho \\ \vdots & \vdots & \vdots & \ddots & \vdots \\ \rho & \rho & \rho & \cdots & 1 \end{pmatrix}.$$

4. Sphericity: $\mathcal{G} = \{g_1 g_2 \mid g_1$ is a permutation matrix and g_2 is diagonal with ± 1's on the diagonal$\}$, so that $\Sigma = \sigma^2 I_p$. The same model is generated by $\mathcal{G} = \mathcal{O}_p$.
5. Circular stationary model: $\mathcal{G} =$ cyclic group, so that, e.g., when $p = 6$,

$$\Sigma = \sigma^2 \begin{pmatrix} 1 & a & b & c & b & a \\ a & 1 & a & b & c & b \\ b & a & 1 & a & b & c \\ c & b & a & 1 & a & b \\ b & c & b & a & 1 & a \\ a & b & c & b & a & 1 \end{pmatrix}.$$

We are interested in hypothesis testing problems that test two nested invariant models, that is, we test

$$H_0 : \mathcal{G} = \mathcal{G}_0 \text{ versus } H_A : \mathcal{G} = \mathcal{G}_A, \tag{2.2}$$

where $\mathcal{G}_A \subset \mathcal{G}_0 \subset \mathcal{O}_p$. Canonical tests are based on a sample $X_1, \ldots, X_n$ of independent and identically distributed (iid) $N(0, \Sigma)$ random vectors. More specifically, let S_n denote the sample covariance matrix,

$$S_n = \frac{1}{n}\sum_{i=1}^{n} X_i X_i'.$$

The likelihood ratio statistic for testing the hypotheses (2.2) rejects the null hypothesis for large values of

$$LRT \equiv \frac{|\widehat{\Sigma}(\mathcal{G}_0)|}{|\widehat{\Sigma}(\mathcal{G}_A)|}, \tag{2.3}$$

where for group $\mathcal{G}$,

$$\widehat{\Sigma}(\mathcal{G}) = \int_{\mathcal{G}} g S_n g' \nu(dg),$$

ν being Haar probability measure on $\mathcal{G}$. When $\mathcal{G}$ is finite,

$$\widehat{\Sigma}(\mathcal{G}) = \frac{1}{\#\mathcal{G}} \sum_{g\in\mathcal{G}} g S_n g'.$$

Under the null hypothesis, $n\log(LRT)$ is asymptotically distributed as a chi-squared with degrees of freedom being the difference in free parameters in the two hypotheses.

The model (2.1) implies that $gX \stackrel{\mathcal{D}}{=} X$, that is, gX and X have the same distribution, for any $g \in \mathcal{G}$. To generalize the model beyond normality, we start with $\mathcal{G}$ as above and a $p \times 1$ location parameter γ. Then we define the class of distributions to be

$$\mathbf{F}(\mathcal{G},\gamma) \equiv \{F \mid X \sim F \Rightarrow g(X-\gamma) \stackrel{\mathcal{D}}{=} (X-\gamma) \text{ for all } g \in \mathcal{G}\}. \tag{2.4}$$

If (2.4) holds, and $\Sigma \equiv Var(X)$ exists, then

$$g\Sigma g' = \Sigma, \text{ for all } g \in \mathcal{G},$$

just as in the invariant normal model (2.1).

In the next section, we construct multivariate rank tests for hypotheses in this general nonparametric model.

2.1. Multivariate rank tests. We use the definition of multivariate rank given in Möttönen and Oja (1995). See also Chaudhuri (1996), Koltchinskii (1997), and Small (1990). For a multivariate sample $x_1, \ldots, x_n$, the rank of any vector x relative to the data is defined to be the average of the unit vectors pointing from the x_i's to x:

$$R(x; \{x_1, \ldots, x_n\}) = \frac{1}{n}\sum_{i=1}^{n} \frac{x - x_i}{\|x - x_i\|}, \tag{2.5}$$

where we set $z/\|z\| = 0$ if $z = 0$. Similarly, the rank of x relative to a distribution F is

$$R(x; F) = E\left(\frac{x - X}{\|x - X\|}\right), X \sim F.$$

In the univariate ($p = 1$) case, when there are no ties among the x_i's, we have that

$$R(x_k; \{x_1, \ldots, x_n\}) = \frac{1}{n}(\#\{x_i < x_k\} - \#\{x_i > x_k\}) = \frac{1}{n}(2 \cdot rank(x_k) - n - 1),$$

where $rank(x_k)$ is the usual rank of x_k among $x_1, \ldots, x_n$. Thus (2.5) is a direct extension of the usual univariate rank.

The following result (Marden, 1992) allows us to use these ranks to test the hypotheses (2.2).

THEOREM 2.1. *Suppose $X_1, \ldots, X_n$ are iid F for $F \in \mathbf{F}(\mathcal{G}, \gamma)$. Then*

$$\begin{aligned}&(g \cdot R(X_1; \{X_1, \ldots, X_n\}), \ldots, g \cdot R(X_n; \{X_1, \ldots, X_n\})) \\ &\quad \stackrel{\mathcal{D}}{=} (R(X_1; \{X_1, \ldots, X_n\}), \ldots, R(X_n; \{X_1, \ldots, X_n\}))\end{aligned}$$

for any $g \in \mathcal{G}$.

In particular, letting S_{Rn} denote the sample covariance matrix of the ranks,

$$S_{Rn} \equiv \frac{1}{n} \sum_{i=1}^{n} R(X_i; \{X_1, \ldots, X_n\}) R(X_i; \{X_1, \ldots, X_n\})',$$

(the sum of the ranks is 0), we have that

$$g \Sigma_{Rn} g' = \Sigma_{Rn}, \text{ where } \Sigma_{Rn} = E(S_{Rn}).$$

Thus the sample covariance matrix of the ranks can be used to test the hypotheses (2.2).

In analogy with the invariant normal model, we base the tests on the estimators of Σ_{Rn} under the two hypotheses,

$$\widehat{\Sigma}_{Rn}(\mathcal{G}_0) \text{ and } \widehat{\Sigma}_{Rn}(\mathcal{G}_A),$$

where

$$\widehat{\Sigma}_{Rn}(\mathcal{G}) = \int_{\mathcal{G}} g S_{Rn} g' \nu(dg).$$

Work is in progress analyzing statistics based on quadratic forms in the difference $\widehat{\Sigma}_{Rn}(\mathcal{G}_0) - \widehat{\Sigma}_{Rn}(\mathcal{G}_A)$. The statistics are asymptotically chi-squared under the null hypotheses, and simulations and efficiency calculations show the resulting tests to be more robust and nearly as efficient at the normal as the usual normal procedures.

In the following example we use test statistics analogous to (2.3), that is, we reject the null hypothesis for large values of

$$T \equiv \frac{|\widehat{\Sigma}_{Rn}(\mathcal{G}_0)|}{|\widehat{\Sigma}_{Rn}(\mathcal{G}_A)|}. \tag{2.6}$$

We take this approach so that the normal- and rank-based statistics are comparable. We do not expect $n \log(T)$ to have an asymptotic chi-squared distribution, but we suspect that it is asymptotically a linear combination of independent chi-squareds. In this paper, we use an approximate randomization procedure instead of asymptotic considerations to estimate p-values.

2.2. Illustration of the invariant model. To illustrate the above ideas, we will use data from an elementary statistics course with $n = 143$ students. The variables are the three hourly exam scores, X_1, X_2, and X_3, and the final exam score Y. The four models considered are (I) Sphericity; (II) The intraclass correlation model; (III) Restricted intraclass correlation, wherein only the three hourly exams (X_1, X_2, X_3) exhibit the intraclass correlation model, so that $\mathcal{G}_{III}$ consists of matrices of the form

$$\begin{pmatrix} g & 0 \\ 0 & 1 \end{pmatrix},$$

where g is a 3×3 permutation matrix; and (IV) The general model. The form of the covariance under model III is

$$\Sigma = \begin{pmatrix} \sigma^2 & \rho\sigma^2 & \rho\sigma^2 & \rho_*\sigma_*\sigma \\ \rho\sigma^2 & \sigma^2 & \rho\sigma^2 & \rho_*\sigma_*\sigma \\ \rho\sigma^2 & \rho\sigma^2 & \sigma^2 & \rho_*\sigma_*\sigma \\ \rho_*\sigma_*\sigma & \rho_*\sigma_*\sigma & \rho_*\sigma_*\sigma & \sigma_*^2 \end{pmatrix}$$

The others are given earlier in this section. For these data, Model I is a straw man.

There are nine outliers in the data set, consisting of people who received 0 on one or more of the exams. The sample covariance matrices S and S^{-o} are based on the entire data set, and the data set without the nine outliers, respectively:

$$S = \begin{pmatrix} 97.82 & 57.27 & 77.32 & 75.3 \\ 57.27 & 111.23 & 70.29 & 99.71 \\ 77.32 & 70.29 & 434.11 & 155.22 \\ 75.3 & 99.71 & 155.22 & 312.91 \end{pmatrix},$$

$$S^{-o} = \begin{pmatrix} 93.96 & 52.37 & 49.92 & 59.09 \\ 52.37 & 100.49 & 42.12 & 68.39 \\ 49.92 & 42.12 & 126.26 & 66.86 \\ 59.09 & 68.39 & 66.86 & 149.59 \end{pmatrix}.$$

The outliers affect the variances of X_3 and Y quite dramatically, inflating the third variance by a factor of nearly 3.5, and doubling the fourth variance. To compare, the sample covariance matrices of the ranks, S_R for the entire data set, S_R^{-o} without the outliers, are

$$1000 \times S_R = \begin{pmatrix} 95.82 & 36.3 & 20.24 & 38.97 \\ 36.3 & 100.01 & 23.54 & 34.69 \\ 20.24 & 23.54 & 100.65 & 34.96 \\ 38.97 & 34.69 & 34.96 & 133.21 \end{pmatrix},$$

$$1000 \times S_R^{-o} = \begin{pmatrix} 85.4 & 30.02 & 32.46 & 33.2 \\ 30.02 & 95.87 & 24.72 & 40.47 \\ 32.46 & 24.72 & 120.6 & 39.49 \\ 33.2 & 40.47 & 39.49 & 134.61 \end{pmatrix}.$$

Notice how little effect the outliers have on the covariance matrices of the ranks, especially relative to the differences for the regular covariance matrices.

Consider the following three hypothesis testing problems:

1. Sphericity (Model I) versus intraclass correlation (Model II);
2. Intraclass correlation (Model II) versus restricted intraclass correlation (Model III);
3. Restricted intraclass correlation (Model III) versus general (Model IV).

The following table has the test statistics for the normal and rank statistics, $n\log(LRT)$ (2.3) and $n\log(T)$ (2.6), respectively.

	Normal	Normal, w/o outliers	Rank	Rank, w/o outliers
I vs. II	104.82	151.95	64.77	64.57
II vs. III	11.88	8.49	5.84	4.99
III vs. IV	149.64	12.21	21.01	7.20

Again, we see that the rank statistics are much less affected by outliers than are the normal statistics. It is only in the third testing problem that the rank statistics shows appreciable effect of outliers, and in that case the normal statistic's effect is four times as large.

2.3. Estimated p-values. As mentioned in Section 2.1, the rank statistic is not expected to have an approximate chi-squared distribution under the null hypothesis. We will use an approximate randomization procedure to estimate the p-values for the test statistics. Under the model (2.4), $(X_1 - \gamma, \dots, X_n - \gamma)$ and $(g_1(X_1-\gamma), \dots, g_n(X_n-\gamma))$ have the same distribution for any set of $g_1, \dots, g_n \in \mathcal{G}$. Because γ is unknown, we estimate it by $\hat{\gamma}$, the vector of sample medians of the variables. Then for test statistics with the location invariant property $h(x_1, \dots, x_n) = h(x_1 - a, \dots, x_n - a)$ for any vector a, we define the approximate randomization p-value to be

$$\hat{p}_{rand} = P[h(G_1(x_1 - \hat{\gamma}), \dots, G_n(x_n - \hat{\gamma})) \geq h(x_1, \dots, x_n)],$$

where the $G_1, \dots, G_n$ are iid from uniform measure on the group $\mathcal{G}_0$ that defines the null hypothesis. In practice, we estimate $\hat{p}_{rand}$ by simulating the $G_1, \dots, G_n$. For the first and third testing problems, all the estimated p-values were 0 (out of 1000 simulations). The estimated p-values for the second problem are given next:

	Normal	Normal, w/o outliers	Rank	Rank, w/o outliers
II vs. III	0.304 (0.003)	0.035 (0.014)	0.016	0.010

The parenthetical values are the p-values obtained using the usual chi-squared approximation to the likelihood ratio statistic. The main point to notice is the stability of the rank-based p-values compared to the normal ones. Also, note that the rank tests are at least no less powerful than the normal tests in this example.

3. Conditional independence graphical models

Undirected graphs can be used to represent certain types of independence and conditional independence models. Given variables $V_1, \dots, V_p$, an undirected graph consists of a set E of edges between variables, i.e., $E \subset \{(i,j) \mid 1 \leq i < j \leq p\}$, where $(i,j) \in E$ means that there is an edge between V_i and V_j. The following is a graph corresponding to $p = 6$, $E = \{(1,2), (2,3), (2,4), (3,4), (5,6)\}$:

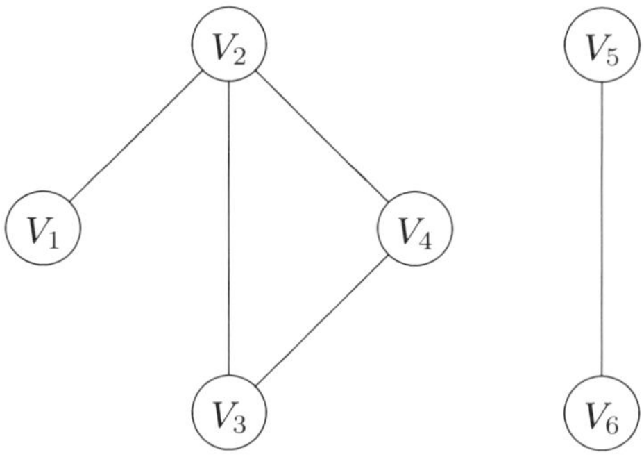

An edge between two variables indicates a direct relationship. Formally, the conditional independence model given by a set E requires that

$$\text{if } (i,j) \notin E, \quad \text{then } V_i \perp V_j \mid \{V_k \mid k \neq i,j\}, \tag{3.1}$$

which means that V_i is conditionally independent of V_j given $\{V_k \mid k \neq i,j\}$, that is, given all the other variables. So, for example, from the above graph, we have that V_1 is conditionally independent of V_3 given $\{V_2, V_4, V_5, V_6\}$, etc. Whittaker (1990) shows that the conditions (3.1) do imply what the graph suggests, such as (V_1, V_2, V_3, V_4) is independent of (V_5, V_6), and $V_1 \perp (V_3, V_4) \mid V_2$.

If the variables are multivariate normal, then the inverse of the covariance matrix has 0 in the components $(i,j) \notin E$. Thus for the example above, if $(V_1, \ldots, V_6) \sim N(0, \Sigma)$,

$$\Sigma^{-1} = \begin{pmatrix} * & * & 0 & 0 & 0 & 0 \\ * & * & * & * & 0 & 0 \\ 0 & * & * & * & 0 & 0 \\ 0 & * & * & * & 0 & 0 \\ 0 & 0 & 0 & 0 & * & * \\ 0 & 0 & 0 & 0 & * & * \end{pmatrix}.$$

3.1. Feigin's multivariate rank model. A rank-based alternative to the normal model above places a model directly on the coordinatewise ranks of the observed variables. That is, suppose the data are $(V_{i1}, \ldots, V_{ip})$, $i = 1, \ldots, n$, where V_{ij} is the ith observation on the jth variable. Then the coordinatewise ranks are the p vectors

$$Y^{(j)} = rank(V_{1j}, \ldots, V_{nj}), j = 1, \ldots, p,$$

the usual vectors of ranks. We extend a model by Feigin (1993), which is itself an extension of Mallows' (1957) ϕ model, to these p rank vectors.

Suppose x and y are n-dimensional rank vectors, that is, $x, y \in \mathcal{S}_n$, the set of permutations of the integers $1, \ldots, n$. Mallows' model is a one-parameter exponential family model with density

$$f_\theta(x,y) = \exp(\theta d(x,y) - \psi(\theta)) \tag{3.2}$$

with respect to uniform probability measure on $\mathcal{S}_n \times \mathcal{S}_n$, where θ is a real-valued parameter, $\psi(\theta)$ is the normalizing constant, and the sufficient statistic $d(x,y)$ is Kendall's distance between the rank vectors x and y given by

$$d(x,y) = \sum_{1 \leq i < j \leq n} I[(x_i - x_j)(y_i - y_j) < 0], \tag{3.3}$$

$I[A]$ being the indicator function of the set A.

Under model (3.2), X and Y are independent if and only if $\theta = 0$.

Because the function d is permutation invariant, $d(\Gamma x, \Gamma y) = d(x, y)$ for any $p \times p$ permutation matrix Γ, the marginal distributions of both X and Y from (3.2) are uniform over $\mathcal{S}_n$, hence the conditional density of Y given $X = x$ is the same as the joint, that is,

$$f_\theta(y|x) = f_\theta(x, y), \tag{3.4}$$

where the conditional density is with respect to the uniform probability measure on $\mathcal{S}_n$. We will write $Y \sim \text{Mallows}(x, \theta)$ to mean that Y has density (3.4).

Feigin's model for three rank vectors $y^{(1)}$, $y^{(2)}$, and $y^{(3)}$ depends on three parameters, θ_{12}, θ_{13}, and θ_{23}. It has density

$$\begin{aligned} f(y^{(1)}, y^{(2)}, y^{(3)}; \theta_{12}, \theta_{13}, \theta_{23}) \;&=\; \exp(\theta_{12} d(y^{(1)}, y^{(2)}) + \theta_{13} d(y^{(1)}, y^{(3)}) \\ &\quad + \theta_{23} d(y^{(2)}, y^{(3)}) - \psi(\theta_{12}, \theta_{13}, \theta_{23})) \end{aligned} \tag{3.5}$$

with respect to uniform probability measure on $\mathcal{S}_n \times \mathcal{S}_n \times \mathcal{S}_n$.

Note that if $\theta_{12} = 0$ in model (3.5), then for fixed $y^{(3)}$, the density factors into functions of $y^{(1)}$ and $y^{(2)}$. That is, $Y^{(1)}$ and $Y^{(2)}$ are conditionally independent given $Y^{(3)}$, which is the graphical model with $E = \{(1,3), (2,3)\}$. Other hypotheses that are easily expressed include testing that $Y^{(1)}$ is independent of $(Y^{(2)}, Y^{(3)})$, which means $\theta_{12} = \theta_{13} = 0$ ($E = \{(2,3)\}$), and the hypothesis that the three variables are mutually independent, which sets $\theta_{12} = \theta_{13} = \theta_{23} = 0$ ($E =$ the empty set).

The model can be extended to any number of variables. If there are p rank vectors $y^{(1)}, \ldots, y^{(p)}$, then we have $\binom{p}{2}$ parameters, $\theta \equiv (\theta_{ij}, 1 \leq i < j \leq p)$, and density

$$f(y^{(1)}, \ldots, y^{(p)}; \theta) = \exp\Big(\sum_{1 \leq i < j \leq p} \theta_{ij} d(y^{(i)}, y^{(j)}) - \psi(\theta)\Big). \tag{3.6}$$

Now setting $\theta_{ij} = 0$ is equivalent to having $Y^{(i)}$ and $Y^{(j)}$ be conditionally independent given all the other variables. Thus any conditional independence graph E corresponds to the model (3.6) with $\theta_{ij} = 0$ for $(i, j) \notin E$.

In principle, maximum likelihood estimates of the θ_{ij}'s can be obtained, as can the likelihood ratio statistic for testing two nested graphical models $E_0 \subset E_A$:

$$H_0 : \theta_{ij} = 0 \text{ for } (i, j) \notin E_0 \quad \text{versus} \quad H_A : \theta_{ij} = 0 \text{ for } (i, j) \notin E_A.$$

Obtaining the exact likelihood quantities is difficult except in special cases because the sample space is very large, $(n!)^p$. The examples below are simple enough to allow easy analysis. Approximate solutions using computationally intensive methods for more complicated models are being explored.

3.2. Illustrations of conditional independence tests.

EXAMPLE 3.1 (Conditional independence given one variable). We will look at scores from an elementary statistics course again, this time with $n = 107$ and $p = 4$. The variables are scores for $H \equiv$ Homework, $L \equiv$ Labs, $E \equiv$ Exams, and $F \equiv$ Final Exam. We wish to test whether the ranks of the Homework, Labs, and Exams are conditionally independent given the ranks of the Final Exam. The corresponding graph is

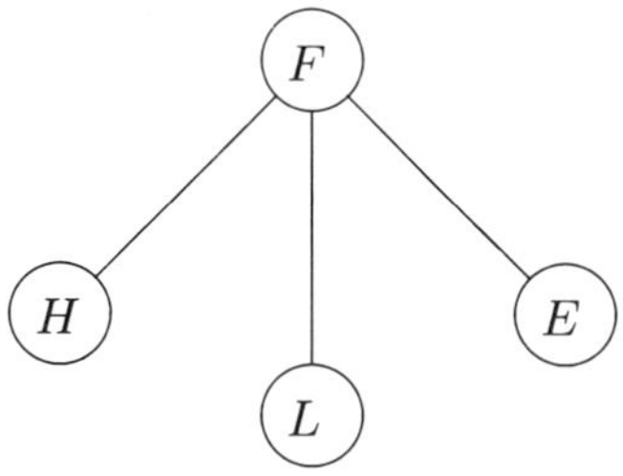

The null hypothesis is the above, i.e., $E_0 = \{(H,F),(L,F),(E,F)\}$. The alternative is the general model, with $E_A = \{(H,F),(L,F),(E,F),(H,L),(H,E),(L,E)\}$.

The sufficient statistic for the general model consists of all the pairwise Kendall distances between variables. (For simplicity, we added small uniform random variables to the data points so that there would be no ties within the variables.) The matrix of these distance follows.

$$\begin{array}{r} \\ Homework \\ Labs \\ Exams \\ Final \end{array} \begin{array}{c} \begin{matrix} HW & Labs & Exams & Final \end{matrix} \\ \begin{pmatrix} 0 & 1595 & 2023 & 1990 \\ 1595 & 0 & 1918 & 2051 \\ 2023 & 1918 & 0 & 1543 \\ 1990 & 2051 & 1543 & 0 \end{pmatrix} \end{array}$$

A more accessible statistic is Kendall's τ, which for x and y is defined to be

$$t(x,y) = 1 - \frac{d(x,y)}{E_{Ind}(d(X,Y))}, \tag{3.7}$$

where E_{Ind} is expected value assuming the variables are independent, so that $E_{Ind}(d(X,Y)) = n(n-1)/4$. See Kendall and Gibbons (1990). If the variables are indeed independent, then the expected value of $t(X,Y)$ is 0, and if the variables have observations in exactly the same order, then $d(x,y) = 0$, so that $t(x,y) = 1$. The matrix of Kendall's τ's for the four variables is next:

$$\begin{array}{r} \\ Homework \\ Labs \\ Exams \\ Final \end{array} \begin{array}{c} \begin{matrix} HW & Labs & Exams & Final \end{matrix} \\ \begin{pmatrix} 1 & 0.44 & 0.29 & 0.30 \\ 0.44 & 1 & 0.32 & 0.28 \\ 0.29 & 0.32 & 1 & 0.46 \\ 0.30 & 0.28 & 0.46 & 1 \end{pmatrix} \end{array}$$

All the τ's are statistically significant with p-value essentially 0, that is, no pair of variables shows independence.

Now consider the null hypothesis that the ranks of the Homework, Labs, and Exams are conditionally independent given the ranks of the Final Exam, which is equivalent to

$$H_0 : \theta_{12} = \theta_{13} = \theta_{23} = 0 \tag{3.8}$$

in (3.6). It follows that

$$f_{\theta_{i4}}(y^{(i)} \mid y^{(4)}) = \exp(\theta_{i4} d(y^{(i)}, y^{(4)}) - \psi(\theta_{i4})), i = 1, 2, 3. \tag{3.9}$$

We test this null hypothesis based on the vector

$$U = (d(Y^{(1)}, Y^{(2)}), d(Y^{(1)}, Y^{(3)}), d(Y^{(2)}, Y^{(3)})),$$

which has value $u = (1595, 2023, 1918)$ for these data. We do not wish to compare this to what is expected under complete independence, but under conditional independence given $Y^{(4)}$, that is, we desire the distribution of U under (3.9), where we replace the θ_{ij}'s with their estimates under the null. Although the exact distribution of U is difficult, it is easy to find its mean, and also easy to simulate from the distribution.

LEMMA 3.2. *If $X \sim$ Mallows(z, θ_x) and $Y \sim$ Mallows(z, θ_y), where X and Y are independent, then*

$$E(d(X,Y) \mid z) = \frac{1}{2} \sum_{1 \le i < j \le n} (h(|i-j|, \theta_x) h(|i-j|, \theta_y) - 1), \tag{3.10}$$

where

$$h(k, \theta) = \frac{2(k+1)}{1 - exp(-(k+1)\theta)} - \frac{2k}{1 - exp(-k\theta)} - 1. \tag{3.11}$$

The proof is in Section 4.

The maximum likelihood estimates of the parameters under the null hypothesis are $\hat{\theta}_{14} = -0.02638, \hat{\theta}_{24} = -0.02421$, and $\hat{\theta}_{34} = -0.04517$. Thus $\widehat{E_0(U)} = (2505.2, 2309.8, 2346.6)$. From these values, we can find the *conditional* τ's, defined as

$$t(x, y \mid z) = 1 - \frac{d(x,y)}{E_0(d(X,Y)|Z = z)},$$

where E_0 denotes expected value under the null hypothesis (3.8). Analogous to the regular Kendall's τ (3.7), this coefficient has mean 0 when X and Y are conditionally independent given $Z = z$, and is 1 if $d(x, y) = 0$.

REMARK 3.3. This conditional τ is not the same as the *partial* τ of Kendall. See Chapter 8 in Kendall and Gibbons (1990) for the partial τ.

Below we have estimates of the conditional τ's between the first three variables given the fourth:

	HW	*Labs*	*Exams*
Homework	1	0.36	0.12
Labs	0.36	1	0.18
Exams	0.12	0.18	1

These values are somewhat less than their marginal counterparts, which suggests that the Final Exam score does explain some of the correlation between the other three variables, but probably not all the correlation.

To test the null hypothesis (3.8), we use the statistic

$$W = \sum_{1 \le i < j \le 3} (d(y^{(i)}, y^{(j)}) - E(d(Y^{(i)}, Y^{(j)}) | y^{(4)}))^2$$

The maximum W among 10,000 simulations was about 650,000, which is far less than the observed statistic $W = 1,038,000$. Thus we have to reject the null hypothesis.

Looking more closely, we can test whether each of the individual pairs among the first three rank vectors exhibits conditional independence give the fourth. The two-sided p-values for these individual tests are given below.

Variables	Conditional τ	p-value
Homework and Labs	0.36	0.000
Homework and Exams	0.12	0.110
Labs and Exams	0.18	0.012

Thus the conditional relationship between Homework and Exams given Final is not quite strong enough to reject conditional independence, but the other two are statistically significant.

EXAMPLE 3.4 (Analysis of covariance). Chapter 14 of Snedecor and Cochran (1989) contains an analysis of covariance problem on leprosy patients. Thirty subjects were divided into three groups of ten. One group received drug "A", one received drug "D", and the third was the control group, receiving a placebo. The variable of interest is a bacterial count **after** receiving the drug. A covariate was also measured, that being the bacterial count **before** the treatments were administered. The null hypothesis can be represented by the following graph:

For the drug variable, we have two contrasts: one comparing the two active drugs to the placebo, and the other comparing the Drug A to Drug D. Let y be the rank vector for the bacterial counts after treatment, where the observations are indexed with the first 10 being the drug A subjects, the next ten being the drug D subjects, and the final ten being the control group. Define $x^{(1)}$ to be the vector for the first contrast: it has 1's as its first 20 elements, and 2's as the last ten. Then $d(y, x^{(1)})$ turns out to be equivalent to the Mann-Whitney/Wilcoxon statistic for comparing the drug groups to the control group. The second vector $x^{(2)}$ represents the second contrast. It has only twenty elements, the first ten being 1 and the rest being 2. The statistic will be $d(y^*, x^{(2)})$, where y^* consists of the first twenty elements of y. The statistic is then the Mann-Whitney/Wilcoxon statistic for comparing drug A to drug D. (See Marden, 1992, and Chung and Marden, 1993, for more discussion of such contrasts.) Letting z be the rank vector of the before-treatment bacterial counts, so that $d(y, z)$ represents the relationship of the before and after bacterial counts, we have the model

$$\begin{aligned} f_{\theta_1,\theta_2,\theta_z}(y, z \mid x^{(1)}, x^{(2)}) &= \exp(\theta_1 d(y, x^{(1)}) + \theta_2 d(y^*, x^{(2)}) + \theta_z d(y, z) \\ &\quad -\psi(\theta_1, \theta_2, \theta_z)). \end{aligned}$$

This model is a little different than the one in (3.6) in that one of the variables (the drug variable) is fixed, and has two associated vectors, each with only two distinct values.

The null hypothesis is that the after-treatment ranks do not depend on the drug, that is, on $x^{(1)}$ and $x^{(2)}$, given the before-treatment ranks. This hypothesis is expressed as

$$H_0 : \theta_1 = \theta_2 = 0.$$

Under this null hypothesis, the estimate of θ_z is $\hat{\theta}_z = -0.2406$. The statistics of interest are $(d(y, x^{(1)}), d(y^*, x^{(2)})) = (45, 47)$. Simulating 1000 y's from the null hypothesis, the number of times the simulated $d(y, x^{(1)})$'s were less than or equal to 45 was 107, leading to an estimated one-sided p-value of 0.107 for that statistic. Thus there may be some evidence of the drugs' effectiveness, but it is not statistically significant. The p-value for the other contrast was 0.477, showing no significant difference between the two drugs.

Redoing the analysis but without using the covariate z (so that under the null the simulated y's are uniform over $\mathcal{S}_n$), the p-values are 0.011 and 0.426. Now the difference between the drugs and the control is quite significant. Looking at the data, it is clear than the control group had, by chance, higher bacterial counts before treatment than the drug groups. Thus the more modest results given by the covariate analysis are warranted.

4. Proof of Lemma 3.2

Start by noting that when $a \neq 0$ and $b \neq 0$,

$$I[ab < 0] = \frac{1 - (1 - 2I[a < 0])(1 - 2I[b < 0])}{2}.$$

Setting $a = X_i - X_j$ and $b = Y_i - Y_j$, and taking expectations, we have that for $i \neq j$,

(4.1)

$$E(I[(X_i - X_j)(Y_i - Y_j) < 0] \mid z) = \frac{1 - (1 - 2P(X_i < X_j))(1 - 2P(Y_i < Y_j))}{2},$$

where we use the assumption that X and Y are independent given z. Now from Cohen and Mallows (1983), for $j > i$,

$$P(X_i < X_j) = 1 - \frac{j - i + 1}{1 - exp(-\theta_x(j - i + 1))} + \frac{j - i}{1 - exp(-\theta_x(j - i))},$$

so that from (3.11),

$$1 - 2P(X_i < X_j) = h(j - i, \theta_x). \tag{4.2}$$

Then (4.1), (4.2), and (3.3) prove (3.10).

References

[1] S. Andersson, *Invariant normal models*, Annals of Statistics **3** (1975), 132 – 154.

[2] S. Andersson and J. Madsen, *Symmetry and lattice conditional independence in a multivariate normal distribution*, The Annals of Statistics **26** (1998), 525–572.

[3] P. Chaudhuri, *On a geometric notion of quantiles for multivariate data*, Journal of the American Statistical Association **91** (1996), 862–872.

[4] L. Chung and J. I. Marden, *Extensions of Mallows' ϕ model*, Probability Models and Statistical Analyses for Ranking Data (New York) (M. A. Fligner and J. S. Verducci, eds.), Springer-Verlag, 1993, pp. 108 – 139.

[5] P. D. Feigin, *Modelling and analysing paired ranking data*, Probability Models and Statistical Analyses for Ranking Data (New York) (M. A. Fligner and J. S. Verducci, eds.), Springer-Verlag, 1993, pp. 75 – 91.

[6] M. Kendall and J. D. Gibbons, *Rank correlation methods*, Edward Arnold, London, 1990.
[7] V. I. Koltchinskii, *M-estimation, convexity and quantiles*, Annals of Statistics **25** (1997), 435 – 477.
[8] C. L. Mallows, *Non-null ranking models: I*, Biometrika (1957), 114 – 130.
[9] J. I. Marden, *Use of nested orthogonal contrasts in analyzing rank data*, Journal of the American Statistical Association **87** (1992), 307 – 318.
[10] J. Möttönen and H. Oja, *Multivariate spatial sign and rank methods*, Nonparametric Statistics **5** (1995), 201 – 213.
[11] C. G. Small, *A survey of multidimensional medians*, International Statistical Review **58** (1990), 263–277.
[12] G. W. Snedecor and W. G. Cochran, *Statistical methods*, Iowa State University Press, Ames, Iowa, 1989.
[13] S. Visuri, V. Koivunen, and H. Oja, *Sign and rank covariance matrices*, Journal of Statistical Planning and Inference **91** (2000), 557 – 575.
[14] J. Whittaker, *Graphical models in applied multivariate statistics*, John Wiley, Chichester, 1990.

Yonghong Gao, Department of Mathematics and Statistics, University of Missouri-Kansas City, 206 Haag Hall, 5100 Rockhill Rd., Kansas City, MO 64110
E-mail address: GaoY@umkc.edu

John I. Marden, Department of Statistics, University of Illinois at Urbana-Champaign, 725 South Wright Street #101, Champaign, IL 61820
E-mail address: marden@stat.uiuc.edu

Contemporary Mathematics
Volume **287**, 2001

Gaussian Measures as Limits on Irreducible Symmetric Spaces and Cones

Piotr Graczyk

ABSTRACT. We prove central limit theorems of Lindeberg-Lévy and Lindeberg-Feller type for any K-invariant random variables on all irreducible symmetric spaces and irreducible symmetric cones, completing in this way the numerous partial results known before. In all cases the limit measures turn out to be Gaussian and being such a limit characterizes these measures. On the other hand we show that other classical characterizations of Gaussian measures on $\mathbb{R}^n$, like 2-stability and the Bernstein theorem, are not true on symmetric spaces.

1. Introduction

The important role of Gaussian measures on symmetric spaces of non-compact type is justified by the fact that they can be characterized as limit measures in central limit theorems on these spaces.

Symmetric cones are reducible symmetric spaces playing an important role in multivariate statistics(see e.g., [L],[LM], [M],[MN] and the references in [L]). Studying central limit theorems on symmetric cones requires Jordan algebra techniques as well as methods of classical harmonic analysis and probability on symmetric spaces.

Different central limit theorems on symmetric spaces have been studied intensely and proved in recent years ([G1],[G2],[R],[S],[T],[Z]).

The central limit theorems of Lindeberg-Lévy type give the asymptotic behavior of normalized products of independent identically distributed random variables. Such theorems were proved for K-invariant random variables with absolutely continuous distributions on symmetric cones ([R], [T],[Z]) and on the space $SO_o(p,q)/SO(p) \times SO(q)$([S]), in the first case without identifying limit measures as Gaussian ones. In this paper we prove them for any K-invariant random variables on any irreducible symmetric space or irreducible symmetric cone. Note that

1991 *Mathematics Subject Classification.* Primary 60F05, 62H05, 43A05, Secondary 62E10.

Key words and phrases. probability measures on non-commutative groups, Gaussian measures on symmetric spaces, central limit theorems, stable measures .

The author is partially supported by the *European Commission* (TMR 1998-2001 Network *Harmonic Analysis*).

the central limit theorems on symmetric cones do not follow from central limit theorems on irreducible symmetric spaces. However the main ideas of their proofs are similar.

The central limit theorems of Lindeberg-Feller type deal with infinitesimal arrays of K-invariant random variables and are based on the knowledge of dispersions of these random variables, defined as expected values of some generalized quadratic forms on symmetric spaces. Such theorems were proved on irreducible symmetric spaces([G2]) and on the space $GL(n,\mathbb{R})/O(n)$([G1]). In this article we extend these results to all irreducible symmetric cones.

The basic tool for all our work is a Taylor expansion of spherical functions obtained using invariant differential operators.

We present our results on irreducible symmetric spaces in Section 2. In Section 3 we show that classical characterizations of Gaussian measures on $\mathbb{R}^n$ as 2-stable measures or as measures verifying the Bernstein property (independence of $X+Y$ and $X-Y$ when X, Y are independent random variables with the considered law) both fail on symmetric spaces. We discuss both a stability property related to the Cartan decomposition of the symmetric space and the stability in the sense of Tortrat. The symmetric cone case is studied in Section 4 which we wanted to make self-contained and accessible for readers familiar with symmetric cones but not with the general theory of symmetric spaces.

2. Lindeberg-Lévy central limit theorem on irreducible symmetric spaces

Let G be a semi simple non-compact Lie group with finite center and let K be a maximal compact subgroup of G. Let $\mathfrak{g} = \mathfrak{k} + \mathfrak{p}$ be a Cartan decomposition of the Lie algebra $\mathfrak{g}$ of the group G. The Riemannian symmetric space $X = G/K$ is *irreducible* if Ad_K acts irreducibly on $\mathfrak{p}$. Let $\mathfrak{a} \cong \mathbb{R}^r$ be a Cartan subspace of $\mathfrak{p}$. r is the rank of the symmetric space G/K. The dimension of $X = G/K$ is equal to $n = \dim\mathfrak{p}$.

Recall the Cartan decomposition $G = K(\exp\overline{\mathfrak{a}^+})K$ where $\mathfrak{a}^+$ is the positive Weyl chamber. We will denote by $x^+ \in \overline{\mathfrak{a}^+}$ the middle component of $x \in G$ in this decomposition

$$x = k_1 \exp(x^+) k_2.$$

On the level of the symmetric space $X = G/K$ we write $y = k \exp(y^+).x_o$, $y \in X$, $x_o = eK$.

In the sequel we identify the spaces $M^\natural(G)$ of K-biinvariant measures on G and $M^\natural(X)$ of K-invariant measures on $X = G/K$ (see e.g. [Ga]). In particular the convolution of K-invariant bounded measures on X is defined as the canonical projection on G/K of the convolution of their K-biinvariant counterparts on G.

The spherical functions on G are defined as K-biinvariant functions which are eigenfunctions of all G-invariant differential operators on G and are normalized by $\phi(e) = 1$. They are all given by the Harish-Chandra formula

$$\phi_\lambda(x) = \int_K e^{\langle i\lambda - \rho, H(xk)\rangle} dk, \quad \lambda \in \mathfrak{a}^*_\mathbb{C}, \tag{2.1}$$

where $g = k \exp H(g) n \in KAN$ is the Iwasawa decomposition of $g \in G$ and $\rho = 12 \sum_{\alpha\in\Sigma^+} m_\alpha \alpha$.

The *spherical Fourier transform* of a bounded measure $\mu \in M^{\natural}(G)$ is defined by

$$\hat{\mu}(\lambda) = \int_G \phi_\lambda(x^{-1}) d\mu(x),$$

for $\lambda \in \mathfrak{a}^*_{\mathbb{R}} + iC(\rho)$ where $C(\rho)$ is the convex hull of the orbit $W\rho$. The knowledge of $\hat{\mu}(\lambda)$ for $H \in \mathfrak{a}^*_{\mathbb{R}}$ is sufficient for the probabilistic applications of this paper. In particular, the spherical Fourier transforms of two different K-biinvariant measures on G differ on $\mathfrak{a}^*_{\mathbb{R}}$.

As X is irreducible, the only K-invariant Gaussian semigroup on X is given by the heat semigroup $(\kappa_t)_{t>0}$ which is generated by the Laplace-Beltrami operator L on X(see [G2]). We have

$$\hat{\kappa}_t(\lambda) = e^{-t(\|\lambda\|^2 + \|\rho\|^2)} = e^{t\gamma_L(\lambda)},$$

where $\gamma_L(\lambda) = -(\|\lambda\|^2 + \|\rho\|^2)$ is the eigenvalue of L on ϕ_λ: $L\phi_\lambda = \gamma_L(\lambda)\phi_\lambda$.

In this section we systematically use the notation, definitions and details concerning spherical harmonic analysis on symmetric spaces found in the monograph [H].

THEOREM 2.1. *The Taylor expansion of order 2 of the spherical function $\phi_\lambda(\exp H)$ at $H = 0$ is*

$$\phi_\lambda(\exp H) = 1 + \gamma_L(\lambda) 2n \|H\|^2 + \sum f_\beta(\lambda) P_\beta(H), \tag{2.2}$$

where n is the dimension of X and P_β are W-invariant polynomials on $\mathfrak{a}$, homogeneous of order greater or equal to 3.

PROOF. In [G2] it was shown that

$$\phi_\lambda(\exp H) = 1 + b(\lambda)\|H\|^2 + \sum f_\beta(\lambda) P_\beta(H),$$

with $b(\lambda) = \gamma_L(\lambda) L(\|H\|^2)|_{H=0}$. Using the well known formula for the radial part of the Laplace-Beltrami operator acting on a K-invariant function f,([H, p. 267]),

$$Lf = L_{\mathfrak{a}}\tilde{f} + \sum_{\alpha \in \Sigma^+} m_\alpha \coth(\alpha(H)) \langle \alpha, \operatorname{grad} \tilde{f} \rangle,$$

where $\tilde{f}(H) = f(\exp H)$, we get $L(\|H\|^2)|_{H=0} = 2r + 2n_o$. Here $n_o = \sum_{\alpha \in \Sigma^+} m_\alpha$ is the dimension of the nilpotent group N in the Iwasawa decomposition $G = KAN$. As $r + n_o = n$, the proof is finished. □

Let Y and Z be two independent K-invariant random variables with values in X and probability distributions μ_Y and μ_Z. Their composition $Y \circ Z$ is defined as any K-invariant random variable on X having the distribution $\mu_Y * \mu_Z$.

THEOREM 2.2. *Let $(Y_m)_{m \geq 1}$ be a sequence of independent identically distributed K-invariant random variables on an irreducible symmetric space of non-compact type X. Suppose that*

$$\mathbb{E}(\|Y_1^+\|^2) = t > 0.$$

Let $S_m = Y_1 \circ Y_2 \circ \ldots \circ Y_m$ and $Y_j = k_j \exp(Y_j^+).x_o$. We normalize S_m by setting

$$S_m^\sharp = (k_1 \exp(Y_1^+ \sqrt{m}).x_o) \circ (k_2 \exp(Y_2^+ \sqrt{m}).x_o) \circ \ldots \circ (k_m \exp(Y_m^+ \sqrt{m}).x_o).$$

Then

$$S_m^\sharp \Rightarrow \kappa_{t2n}, \qquad m \to \infty.$$

PROOF. We apply (2.2) in the study of the asymptotic behavior of the spherical Fourier transform of $S^\sharp_m$, which is equal to

$$\Big(\int_G \phi_\lambda(\exp(-x^+\sqrt{m}))d\mu(x)\Big)^m,$$

where μ is the K-biinvariant probability measure on G corresponding to the probability distribution of the random variable Y_1. We easily see that this expression tends to $\exp(t\gamma_L(\lambda)2n)$ when $m \to \infty$. The proof is finished after an application of the Lévy continuity theorem on symmetric spaces([Ga]). $\square$

EXAMPLE 2.3. Let $X = SO_o(p,q)/SO(p) \times SO(q)$ with $p \le q$. The dimension of X is equal to pq. Theorem 2.2 implies that if $\mathbb{E}(\|Y_1^+\|^2) = t > 0$, then

$$S^\sharp_m \Rightarrow \kappa_{t2pq}.$$

In particular, when $t = p$, which is equivalent to $\mathbb{E}(Y_1^+)^2_j = 1$ for all $j = 1, \dots, p$, then the limit measure is κ_{12q}.

This is the result of [S] obtained for random variables with densities (our normalization $\mathbb{E}(\|Y_1^+\|^2) = p$ leads to that of [S] for absolutely continuous measures).

EXAMPLE 2.4. If $X = SL(k,\mathbb{R})/SO(k)$ then the dimension n of X is equal to the dimension of the space $\mathfrak{p}$ of anti-symmetric $k \times k$ matrices with zero trace, so that $2n = k^2 + k - 2$. It follows from Theorem 2.2 that if $\mathbb{E}(\|Y_1^+\|^2) = t > 0$ then

$$S^\sharp_m \Rightarrow \kappa_{tk^2+k-2},$$

and in particular, if $t = r = k - 1$, then $S^\sharp_m \Rightarrow \kappa_{1k+2}$.

REMARK 2.5. Theorem 2.2 can be expressed equivalently for K-biinvariant random variables on G. The composition of two such random variables ξ and ζ is simply given by their group product $\xi\zeta$.

For any random variable $Z = k\exp(Z^+).x_o$ and $c > 0$ let us denote by $Z^\flat_c$ the random variable $k\exp(Z^+c).x_o$. Then

$$S^\sharp_m = (Y_1)^\flat_{\sqrt{m}} \circ (Y_2)^\flat_{\sqrt{m}} \circ \dots \circ (Y_m)^\flat_{\sqrt{m}}.$$

REMARK 2.6. Observe that $(S_m)^\flat_{\sqrt{m}} \neq S^\sharp_m$, i.e., in order to obtain the normalized product $S^\sharp_m$ one must first normalize the random variables Y_j and then compose them and not the other way round. This may be deduced from the fact that $\mathbb{E}f(Y \circ Z) = \mathbb{E}f(Y)\mathbb{E}f(Z)$ for all independent K-invariant random variables Y and Z if and only if f is a bounded spherical function. On the other hand, if $c \neq 1$, then the function $H \to \phi_\lambda(Hc)$ is not spherical when ϕ_λ is not constant.

REMARK 2.7. Let D_j be a dispersion as defined in [G2]. We are not able to compute $D_j(Y^\flat_{\sqrt{m}})$ knowing $D_j(Y)$. This is related to the impossibility of expressing of $\phi_\lambda(\exp(Hc))$ in terms of $\phi_{\lambda'}(\exp H)$. That is why Theorem 2.2 cannot be deduced from the central limit theorem for infinitesimal arrays with the dispersion D_j equal to t, as proved in [G2].

Conversely, it follows from both theorems that, if $\mathbb{E}(\|Y_1^+\|^2) = t$ then

$$D_j(S^\sharp_m) = mD_j((Y_1)^\flat_{\sqrt{m}}) \to t2n, \quad m \to \infty.$$

REMARK 2.8. On any irreducible symmetric space $D_j\kappa_t = t$. However, the computation of $\mathbb{E}(\|W_t^+\|^2)$ when the random variable W_t has the distribution κ_t does not seem to be feasible in general.

If $X = SL(2,\mathbb{C})/SU(2)$, the density of the heat kernel in the Cartan coordinates is $\kappa_t(dh) = (4\pi t)^{-3/2} h \mathrm{sh} h \exp(-t - h^2 4t) J(h)dh$, $h \in \mathbb{R}$, $J(h) = 2\pi\mathrm{sh}^2 h$, and we can compute

$$\mathbb{E}(\|W_t^+\|^2) = \int_{\mathbb{R}} h^2\kappa_t(dh) = 6t + 4t^2.$$

3. Gaussian measures and stability on symmetric spaces

The Lindeberg-Lévy central limit theorem on $\mathbb{R}^n$ trivializes for independent random variables $Y_1, Y_2, \ldots$ having all the same Gaussian law $N(0,\Sigma)$. Their normalized sum

$$Y_1 + \ldots + Y_m\sqrt{m}$$

has then the same law $N(0,\Sigma)$. In this section we show that this is not the case on symmetric spaces of non-compact type. The above-mentioned property of Gaussian measures on $\mathbb{R}^n$ is called 2-stability. It is natural to define an analogue of this property for K-invariant measures on symmetric spaces *via* the Cartan decomposition, used in the previous section in the normalization of sums of independent random variables.

We will say that a measure $\mu \in M^\natural(X)$ is *stable in a Cartan decomposition sense* if, for any $m \in \mathbb{N}$ and m independent random variables $Y_1, \ldots, Y_m$ with the distribution μ, there exists a constant $c > 0$ such that

$$(Y_1)_c^\flat \circ (Y_2)_c^\flat \circ \ldots \circ (Y_m)_c^\flat \stackrel{\mathrm{d}}{=} Y_1,$$

where $\stackrel{\mathrm{d}}{=}$ denotes equality in law.

PROPOSITION 3.1. *If X is an irreducible symmetric space of non-compact type, then the K-invariant Gaussian measures $(\kappa_t)_{t>0}$ are not stable in the above defined Cartan sense.*

PROOF. Suppose that $Y_1, Y_2, \ldots$ are independent random variables with the same law κ_t. By the Cramér theorem on symmetric spaces([G3]), if the random variable product $(Y_1)_c^\flat \circ (Y_2)_c^\flat \circ \ldots \circ (Y_m)_c^\flat$ had a Gaussian law then each random variable $(Y_1)_c^\flat$ would be Gaussian, too. It is then sufficient to show that if Y has the law κ_t, then $Y_c^\flat$ is not Gaussian when $c \neq 1$.

Recall that the integration of a K-biinvariant function f on G can be written in the Cartan decomposition

$$\int_G f(x)dx = \int_{\mathfrak{a}^+} f(\exp H)J(H)dH,$$

where $J(H) = \mathrm{vol}(K/M)\prod_{\alpha\in\Sigma^+} \mathrm{sh}^{m_\alpha}\langle\alpha, H\rangle$. As usual M denotes here the centralizer of $A = \exp\mathfrak{a}$ in K and $\alpha \in \Sigma^+$ are positive roots with multiplicities m_α.

Let us call $h_t(x), x \in G$, the density of the measure κ_t and $h_t^{(c)}$ the density of the probability distribution of $Y_c^\flat$. We have then

$$h_t^{(c)}(\exp H) = c^r J(cH)J(H)h_t(\exp(cH)), \quad H \in \mathfrak{a}^+.$$

Let us suppose that $Y_c^\flat$ is Gaussian with distribution κ_s. Then $h_t^{(c)} = h_s$. By [AJ, Th. 3.7(i)], if we fix $H \in \mathfrak{a}^+$ and consider uH, $u \to \infty$, we have the estimates

$$\begin{aligned} h_t^{(c)}(\exp(uH)) &\asymp P_t(ucH)e^{u(c-2)\langle\rho,H\rangle}e^{-u^2c^2\|H\|^2/4t}, \\ h_s(\exp(uH)) &\asymp P_s(uH)e^{-u\langle\rho,H\rangle}e^{-u^2\|H\|^2/4s}, \end{aligned}$$

where $\asymp$ means that the quotient of two expressions on both sides of $\asymp$ has a lower bound $C_1 > 0$ and an upper bound $C_2 < +\infty$ (for u sufficiently large), and

$$P_t(H) = \prod_{\alpha\in\Sigma^{++}} \langle\alpha, H\rangle(t + \langle\alpha, H\rangle)^{[(m_\alpha+m_{2\alpha})/2]-1},$$

with Σ^{++} denoting the set of positive indivisible roots. If $h_t^{(c)} = h_s$, than comparing the dominating factors of these estimates we get $s = t/c^2$. If this is the case, one must have $c - 2 = -1$ and $c = 1$. □

COROLLARY 3.2. *If $Y_1, \dots, Y_m$ are independent K-invariant random variables with a Gaussian law on X, then their normalized product $S_m^\sharp$ has not a Gaussian law.*

REMARK 3.3. One could also define the Cartan stability of a measure $\mu \in M^\natural(X)$ by the existence of a constant $c > 0$ such that

$$(Y_1 \circ \ldots \circ Y_m)_c^\flat \stackrel{\mathrm{d}}{=} Y_1 \tag{3.1}$$

(cf. [F, p.170]). In the case of $\mathbb{R}^n$, both definitions are equivalent, but not on symmetric spaces (cf. Remark 2.2). Using the same argument as in the second part of the proof of Theorem 3.1, we see that the Gaussian measures on X do not verify the condition (3.1), either. The asymptotic behavior of different components of the product $Y_1 \circ \ldots \circ Y_m$ normalized like in (3.1),

$$S_m^\flat = (Y_1 \circ Y_2 \circ \ldots \circ Y_m)_{\sqrt{m}}^\flat,$$

is studied in [LP] and [GR]. It does not seem possible to study it by methods of the spherical harmonic analysis.

On any topological group G one may consider probability measures which are stable in the sense of Tortrat([To]). A probability measure μ on G is strictly 2-stable in the sense of Tortrat if for any positive integer n and a random variable ξ with probability distribution μ, the law of ξ^n is equal to μ^{*n^2}. In the case of $G = \mathbb{R}^n$ this definition is equivalent to the standard one (cf.[F]). Note that μ^{*n^2} is the law of the product $\xi_1 ... \xi_{n^2}$ of n^2 independent random variables with the law μ. The 2-stability of μ means that

$$\xi_1^n \stackrel{\mathrm{d}}{=} \xi_1 \ \dots \ \xi_{n^2}.$$

PROPOSITION 3.4. *Let G/K be a Riemannian symmetric space of non-compact type. There exists $t_o > 0$ such that, if $\xi_1, \xi_2, \xi_3, \xi_4$ are independent K-invariant random variables with Gaussian distribution κ_t, with $t > t_o$, then*

$$\mathbb{E}\phi_o(\xi_1^2) \neq \mathbb{E}\phi_o(\xi_1\xi_2\xi_3\xi_4).$$

PROOF. We have $\mathbb{E}\phi_o(\xi_1\xi_2\xi_3\xi_4) = \hat{\kappa}_{4t}(0) = e^{-4\|\rho\|^2 t}$.
On the other hand, using the K-invariance of ϕ_o and κ_t we observe that

$$\mathbb{E}\phi_o(\xi_1^2) = \int_G \phi_o(x^2)d\kappa_t(x) = \int_K \int_G \phi_o(xkx)d\kappa_t(x) = \int_G \phi_o^2(x)d\kappa_t(x).$$

For $H \in \mathfrak{a}^+$ with $|H| < 1$ and $t > 0$ we deduce from the lower estimate of the heat kernel h_t [AJ, Th. 3.7(ii)] that

$$h_t(\exp H) \geq ct^{-n/2}(1+t)^{(m/2)-|\Sigma^{++}|}e^{-\|\rho\|^2 t}e^{-\|H\|^2/4t},$$

where $m = \dim N$ and the constant c is independent of $t > 0$. It follows that for $t \geq 1$,

$$\begin{aligned} \mathbb{E}\phi_o(\xi_1^2) &= \int_{\mathfrak{a}^+} \phi_o^2(\exp H)h_t(\exp H)J(H)dH \\ &\geq c_1 t^{-n/2}(1+t)^{(m/2)-|\Sigma^{++}|}e^{-\|\rho\|^2 t}\int_{H\in\mathfrak{a}^+,|H|<1} e^{-\|H\|^2/4}J(H)dH \\ &= c_2 t^{-n/2}(1+t)^{(m/2)-|\Sigma^{++}|}e^{-\|\rho\|^2 t}, \end{aligned}$$

where $c_1, c_2 > 0$. Hence, for t large enough, $\mathbb{E}\phi_o(\xi_1^2) \neq e^{-4\|\rho\|^2 t}$. □

COROLLARY 3.5. *There exists $t_o > 0$ such that for $t > t_o$ the Gaussian measures κ_t are not 2-stable in the sense of Tortrat.*

REMARK 3.6. Similarly, one proves that for sufficiently large t the Gaussian measures κ_t are not p-stable in the sense of Tortrat for any positive integer p.

We say that a probability measure μ on a group G has the Bernstein property (B) if, given two independent random variables ξ and η with values in G and having the same probability distribution μ, the random variables

$$(B) \qquad \xi\eta \text{ and } \xi\eta^{-1} \text{ are independent.}$$

COROLLARY 3.7. *The Gaussian measures κ_t on a Riemannian symmetric space of non-compact type do not have the Bernstein property (B) for sufficiently large values of t.*

PROOF. If $\xi \overset{d}{=} \xi^{-1}$, what is true when the distribution of ξ is equal to κ_t, then the property (B) implies 2-stability,

$$\eta^2 = (\eta\xi^{-1})(\xi\eta),$$

and the law of $(\eta\xi^{-1})(\xi\eta) = (\xi\eta^{-1})^{-1}(\xi\eta)$ is equal to κ_{4t}. □

REMARK 3.8. The Gaussian measures κ_t on $X = G/K$ with K non-commutative have not the Bernstein property (B) also for sufficiently small $t > 0$. In fact, if they verified the condition (B), then by the Proposition 11[GL] and the remark after it, then the Haar measure on K would also verify the condition (B). This is equivalent to

$$\int f(k^2)dk = \int f(k)dk$$

for any bounded Borel function f on K, which may be shown to be false if K is a non-commutative connected compact Lie group.

REMARK 3.9. In [NRS] one can find a result contrary to the Corollary 3.7 and the Remark 3.8. The proof in [NRS] is not correct because a pair (ξ, η) of K-invariant random variables is in general not $K \times K$-invariant without the hypothesis of independence of ξ and η.

4. A reducible symmetric space case: symmetric cones

In this section we show that our methods make it possible to obtain, in a simple way, central limit theorems on Riemannian symmetric spaces which are not irreducible, provided that one has some information about their structure. An important class of such spaces is given by symmetric cones associated to simple Euclidean Jordan algebras (which is equivalent to say that the considered cones are irreducible, see [FK, III.4]).

Central limit theorems for normalized products of independent random variables on symmetric cones were studied by Terras ([T]) and Richards ([R]) in the case of $\mathbf{P}_r \cong GL(r,\mathbb{R})/O(r)$, the cone of positive definite real symmetric $r \times r$ matrices, and by Zhang ([Z]) for all irreducible symmetric cones. Their results concern random variables whose distributions have densities.

In this section we prove this central limit theorem on irreducible symmetric cones without the assumption of absolute continuity of considered measures and with a different description of the limit measure as a convolution of two Gaussian measures. Our method of finding coefficients of the Taylor expansion of spherical functions, based on an application of invariant differential operators, is much simpler than the very computational methods of before-mentioned authors. We also prove a central limit theorem for an infinitesimal array of K-invariant measures on an irreducible symmetric cone, generalizing [G1]. For any needed information about analysis on symmetric cones we refer the reader to the monograph [FK]. Throughout this section we use the notations of this book.

We assume that V is a simple Euclidean Jordan algebra of dimension n as a real vector space. Let Ω be the associated symmetric cone:

$$\Omega = \{x^2 \mid \ x \in V \text{ invertible}\}.$$

The cone Ω is a Riemannian symmetric space G/K where G is the connected component of the identity in the automorphism group $G(\Omega) \subset GL(V)$ and $K = G \cap O(V)$.

Let $c_1, \ldots, c_r$ be a Jordan frame in V with $r = \operatorname{rank} V = \operatorname{rank} G/K$ as a symmetric space. The polar decomposition of any point $x \in \Omega$ writes

$$x = k.\sum_{j=1}^{r} e^{t_j} c_j, \qquad k \in K, t_j \in \mathbb{R},$$

which we abbreviate to

$$x = k.e^{\mathbf{t}} \ \text{ where } \ \mathbf{t} \in R = \{\mathbf{t} = \sum_{j=1}^{r} t_j c_j \mid \ t_j \in \mathbb{R}\} \cong \mathbb{R}^r.$$

Let us denote

$$R^+ = \{\mathbf{t} = \sum_{j=1}^{r} t_j c_j \mid \ t_1 < t_2 < \ldots < t_r\}.$$

Then every element $x \in \Omega$ may be represented as $x = k.e^{\mathbf{t}}$ with $\mathbf{t} \in R^+$ and such a $\mathbf{t}$ is unique. We define in this way the map

$$x \in \Omega \to \mathbf{t}(x),$$

which may be understood as the projection of Ω on R^+ via the polar decomposition; $\mathbf{t}(x)$ is the vector of logarithms of eigenvalues of x, ordered increasingly. Let L be the Laplace-Beltrami operator on Ω and E the Euler operator on Ω:

$$Ef(x) = ddtf(tx)\big|_{t=1}.$$

All the elliptic second order G-invariant differential operators on Ω are of the form

$$D = a(L - 1rE^2) + bE^2 + cE + d, \quad a, b > 0.$$

A K-invariant probability measure is called *Gaussian* if it belongs to the continuous semigroup of K-invariant measures $(\mu_t)_{t>0}$ generated by such an operator. It is called *Gaussian degenerate* if $a = 0$ or $b = 0$. The semigroup $(\kappa_t)_{t>0}$ generated by L is called the *heat semigroup*. The semigroups $(\eta_t)_{t>0}$ generated by $12E^2$ and $(\nu_t)_{t>0}$ generated by $L_1 = L - 1rE^2$ are Gaussian degenerate. They may be interpreted in the following way: Let us call

$$\Omega_1 = \{x \in \Omega | \det(x) = 1\}.$$

Then Ω_1 is an irreducible symmetric space. For example, if $\Omega = \mathbf{P}_r$ is the cone of positive definite symmetric matrices of dimension r, we have $\Omega_1 \cong SL(r, \mathbb{R})/SO(r)$.

We can identify in an evident way Ω with $\Omega_1 \times \mathbb{R}^+$. The operators E and E^2 restrained to Ω_1 are equal to zero. The operator $L - 1rE^2$ restrained to Ω_1 is the Laplace-Beltrami operator on Ω_1. Hence, the semigroups $(\eta_t)_{t>0}$ and $(\nu_t)_{t>0}$ are equal to the heat semigroups $(\eta_t^o)_{t>0}$ and $(\nu_t^o)_{t>0}$ on, respectively, $\mathbb{R}^+$ and Ω_1 extended to Ω by

$$\eta_t(B) = \eta_t^o(B \cap \mathbb{R}^+ e) \quad \text{and} \quad \nu_t(B) = \nu_t^o(B \cap \Omega_1), \quad B \in B_\Omega.$$

The spherical functions φ_λ on Ω ($\lambda \in \mathbb{C}^r$) which are traditionally used on symmetric cones ([FK, XIV.3]) differ slightly from spherical functions ϕ_λ given by the Harish-Chandra formula (2.1) on the symmetric space G/K:

$$\varphi_\lambda = \phi_{i\lambda}.$$

We still use systematically the approach and the notation of [FK]. If $\mu \in M^\natural(\Omega)$ then its spherical Fourier transform

$$\hat{\mu}(\lambda) = \int_\Omega \varphi_\lambda(x^{-1}) d\mu(x), \quad \lambda \in i\mathbb{R}^r$$

is always well defined and uniquely determines the measure μ. Recall that $L\varphi_\lambda = \gamma_L(\lambda)\varphi_\lambda$ and $E\varphi_\lambda = \gamma_E(\lambda)\varphi_\lambda$ with

$$\gamma_L(\lambda) = \|\lambda\|^2 - \|\rho\|^2, \quad \gamma_E(\lambda) = \sum_{j=1}^r \lambda_j,$$

where $\rho = (\rho_j)_{j=1\ldots r}$, $\rho_j = d4(2j - r - 1)$ with d equal to the degree of the Jordan algebra V. It follows that the spherical Fourier transforms of κ_t, η_t and ν_t are:

$$\hat{\kappa}_t(\lambda) = e^{t\gamma_L(\lambda)}, \quad \hat{\eta}_t(\lambda) = e^{t2\gamma_E^2(\lambda)}, \quad \hat{\nu}_t(\lambda) = e^{t(\gamma_L(\lambda) - 1r\gamma_E^2(\lambda))}.$$

THEOREM 4.1. *The Taylor expansion of order 2 of the spherical function* $\varphi_\lambda(e^{\mathbf{t}})$ *in* $\mathbf{t}=0$ *is*

$$\begin{aligned}\varphi_\lambda(e^{\mathbf{t}}) = 1 \quad &+\quad a(\lambda)\sum_{j=1}^r t_j + b(\lambda)\sum_{j=1}^r t_j^2 + c(\lambda)(\sum_{j=1}^r t_j)^2 + R_\lambda(\mathbf{t}),\\ \text{where} \qquad & a(\lambda) = 1r\gamma_E(\lambda)\\ & b(\lambda) = r\gamma_L(\lambda) - \gamma_E^2(\lambda)dr(r-1)(r+2d^{-1})\\ & c(\lambda) = (dr-d+2)\gamma_E^2(\lambda) - 2\gamma_L(\lambda)2dr(r-1)(r+2d^{-1}),\end{aligned}$$

and $R_\lambda(\mathbf{t}) = \sum f_\beta(\lambda)P_\beta(\mathbf{t})$, *with* $P_\beta(\mathbf{t})$ *symmetric homogeneous polynomials in* $t_1,\dots,t_r$ *of order greater or equal to 3.*

PROOF. The proof is similar to the proof of the Theorem 3 in [G1]. We apply the operators E, E^2 and L to the Taylor expansion of $\varphi_\lambda(e^{\mathbf{t}})$ in order to determine its coefficients up to the order 2. □

Let Y and Z be two independent K-invariant random variables with values in Ω. Their composition $Y\circ Z$ is defined, like on any symmetric space, as a random variable with the law $\mu_Y * \mu_Z$, the convolution of K-invariant measures on Ω being defined as the projection on Ω of the convolution of the corresponding K-biinvariant measures on G.

In the symmetric cone case one may give an equivalent definition of the composition $Y\circ Z$, based on the following fact. We denote by $P(x)$, $x\in V$ the quadratic representation of V and we recall that if $x\in\Omega$ then there exists a unique $y\in\Omega$ such that $y^2=x$. We then write $y=x^{1/2}$.

PROPOSITION 4.2. *Let* Y *and* Z *be two independent* K*-invariant random variables on* Ω *with probability distributions* μ *and* ν. *Then the convolution* $\mu*\nu$ *is the probability distribution of the random variable* $P(Y^{1/2})Z$ *and of the random variable* $P(Z^{1/2})Y$.

PROOF. Let $e\in\Omega$ be fixed by the action of K. If $g\in G$ then $ge=x\in\Omega$ and there exists $k_o\in K$ such that $g=P(x^{1/2})k_o$ ([FK, III.5.1]). Let $f(x)$ be a bounded Borel function on Ω and F a bounded right K-invariant Borel function on G. The correspondence between a measure m on Ω and its right K-invariant analogue $\tilde m$ on G is given by

$$\int_\Omega f(x)dm(x) = \int_G f(ge)d\tilde m(g); \qquad \int_G F(g)d\tilde m(g) = \int_\Omega F(P(x^{1/2}))dm(x).$$

We get, using these equalities and the K-invariance of Y,

$$\begin{aligned}\langle \mu*\nu, f\rangle &= \int_G\int_G f(ghe)d\tilde\mu(g)d\tilde\nu(h) = \int_G\Big(\int_\Omega f(gy)d\nu(y)\Big)d\tilde\mu(g) =\\ &= \int_G\int_K\Big(\int_\Omega f(gky)d\nu(y)\Big)dk\,d\tilde\mu(g)\\ &= \int_\Omega\int_K\Big(\int_\Omega f(P(x^{1/2})ky)d\nu(y)\Big)dk\,d\mu(x)\\ &= \int_\Omega\int_\Omega f(P(x^{1/2})y)d\mu(x)d\nu(y) = \mathbb{E}f(P(Y^{1/2})Z).\end{aligned}$$

□

Hence, on the symmetric cones we may define

$$Y \circ Z = P(Y^{1/2})Z.$$

In particular, on the matrix symmetric cones $\Pi_r(\mathbb{R}) = \mathbf{P}_r$, $\Pi_r(\mathbb{C}), \Pi_r(\mathbb{H})$ and $\Pi_3(\mathbb{O})$ we may put

$$Y \circ Z = Y^{12} Z Y^{12}.$$

Note that we have $P(Y^{1/2})Z \overset{\mathrm{d}}{=} P(Z^{1/2})Y$, but in general $P(Y^{1/2})Z \neq P(Z^{1/2})Y$.

THEOREM 4.3. *Let $(Y_m)_{m\geq 1}$ be a sequence of independent identically distributed K-invariant random variables with values in Ω such that* $\mathbb{R}$

$$\begin{aligned} \mathbb{E}(\mathbf{t}(Y_1)) &= \mathbf{0} \\ \mathbf{Cov}(\mathbf{t}(Y_1)) &= \mathbb{E}(\mathbf{t}(Y_1)^T \mathbf{t}(Y_1)) = \mathbf{Id}. \end{aligned}$$

Let $S_m = Y_1 \circ Y_2 \circ \ldots \circ Y_m$. Let us normalize S_m by putting

$$S_m^\sharp = (k_1.e^{\mathbf{t}(Y_1)\sqrt{m}}) \circ \ldots \circ (k_m.e^{\mathbf{t}(Y_m)\sqrt{m}})$$

when $Y_j = k_j.e^{\mathbf{t}(Y_j)}$.

Then $\quad S_m^\sharp \Rightarrow \kappa_{1r+2d^{-1}} * \eta_{1r+2d^{-1}} = \nu_{1r+2d^{-1}} * \eta_{r+2r(r+2d^{-1})}, \qquad m \to \infty.$

PROOF. Let μ be the law of Y_1 and μ_m the law of $S_m^\sharp$. We study the behavior of the spherical Fourier transform

$$\hat{\mu}_m(-\lambda) = \Big(\int_\Omega \phi_\lambda(e^{\mathbf{t}(x)\sqrt{m}}) d\mu(x) \Big)^m,$$

when $m \to \infty$ and $\lambda \in i\mathbb{R}^r$. By Theorem 4.1 we see that

$$\hat{\mu}_m(\lambda) \to e^{r(b(\lambda)+c(\lambda))} = \exp(12(r+2d^{-1})\gamma_E^2(\lambda) + 1r + 2d^{-1}\gamma_L(\lambda)).$$

An application of the Levy-Gangolli continuity theorem([Ga]) ends the proof. □

REMARK 4.4. When the distribution μ of Y_1 has a density with respect to the G-invariant measure $d\dot{x} = (\det x)^{-n/r} dx$ on Ω then the normalization hypothesis of the Theorem 4.3 coincide with those of [R],[T] and [Z].

Let us now present briefly the Lindeberg-Feller type central limit theorem for irreducible symmetric cones, generalizing the main result of [G1].

The *dispersions* D_j, $j = 1, \ldots r-1$ having the property

$$D_j(Y) \geq 0, \qquad \text{and} \quad D_j(Y \circ Z) = D_j(Y) + D_j(Z)$$

for independent K-invariant random variables Y, Z and such that

$$D_j(\kappa_t) = D_j(\nu_t) = t, \quad t > 0$$

are defined by

$$D_j(Y) = \mathbb{E}Q_j(Y),$$

where the generalized quadratic forms Q_j of the symmetric pair (G, K) (see e.g. [FH]) are defined by

$$Q_j(x) = 1d\big(\partial\partial\lambda_{j+1} - \partial\partial\lambda_j\big)\big|_{\lambda=\rho}\varphi_\lambda(x) = 1d\big(\partial\partial\lambda_j - \partial\partial\lambda_{j+1}\big)\big|_{\lambda=-\rho}\varphi_\lambda(x).$$

If we denote $\partial_j = 1d(\partial\partial\lambda_{j+1} - \partial\partial\lambda_j)|_{\lambda=\rho}$ and $\tilde{\partial}_j = 1d(\partial\partial\lambda_j - \partial\partial\lambda_{j+1})|_{\lambda=-\rho}$ then

$$D_j(Y) = \partial_j(\check{\widehat{\mu_Y}}) = \tilde{\partial}_j(\check{\widehat{\mu_Y}}),$$

where $\check{f}(\lambda) = f(-\lambda)$.

For example, if for $p > d2(r-1)$ and $c > 0$

$$\gamma_{p,ce}(dx) = c^{-rp}\Gamma_\Omega(p)\, e^{-1c\mathrm{tr}(x)} \det(x)^{p-nr}\, dx$$

is a K-invariant Wishart distribution on Ω, then

$$\hat{\gamma}_{p,ce}(\lambda) = c^{-\sum_{j=1}^{r}\lambda_j}\Gamma_\Omega(p)\Gamma_\Omega(\rho - \lambda + p),$$

for $\mathrm{Re}\lambda_j < p - d4(r-1)$ and

$$D_j(\gamma_{p,ce}) = 1d\big[\psi(p-(j-1)d2) - \psi(p-jd2)\big],$$

where ψ is the digamma function $\psi = \Gamma'/\Gamma$.

THEOREM 4.5. *Let (Y_{mj}), $m \in \mathbb{N}, 1 \le j \le k_m$ be a triangular array of K-invariant random variables such that for each $m \in \mathbb{N}$ the variables $Y_{m1}, \ldots, Y_{mk_m}$ in the m-th row of the array are independent. Define*

$$Z_m = Y_{m1} \circ Y_{m2} \circ \ldots \circ Y_{mk_m}.$$

Suppose that the random variables (Y_{mj}) satisfy

$$\mathbb{E}(\log(\det(Y_{mj})) = 0, \tag{4.1}$$

$$1r^2\mathbf{Var}(\log(\det(Z_m)) \to u, \qquad m \to \infty \tag{4.2}$$

$$D_1(Z_m) \to t, \qquad m \to \infty. \tag{4.3}$$

If the technical condition
(4.4)

$$\lim_{m\to\infty} \sum_{j=1}^{k_m} \big[\mathbb{E}(\|\mathbf{t}(Y_{mj})\|^3 1 + \|\mathbf{t}(Y_{mj})\|^2) + \mathbb{E}(\log^2(\det(Y_{mj})), \|\mathbf{t}(Y_{mj})\| > 1)\big] = 0$$

is verified, then

$$Z_m \Rightarrow \nu_t * \eta_u, \qquad m \to \infty.$$

PROOF. Using Theorem 4.1 we obtain the following expansion of the spherical function $\varphi_\lambda(e^{\mathbf{t}})$ in $\mathbf{t} = 0$:

$$\begin{aligned} \varphi_\lambda(e^{\mathbf{t}}) &= 1 + \gamma_E(\lambda) r \sum_{j=1}^{r} t_j + (\gamma_L(\lambda) - 1r\gamma_E^2(\lambda))Q_1(e^{\mathbf{t}}) \\ &+ 12r^2\gamma_E^2(\lambda)(\sum_{j=1}^{r} t_j)^2 + \tilde{R}_\lambda(\mathbf{t}) \end{aligned}$$

with the remainder $\tilde{R}_\lambda(\mathbf{t})$ of the same form as the remainder $R_\lambda(\mathbf{t})$ in the Theorem 4.1. Observe that for $x \in \Omega$,

$$\log(\det x) = \sum_{j=1}^{r} t_j(x).$$

The conclusion of the proof follows along the lines of the proof of Theorem 4 in [G1]. □

If we consider a triangular array (Y_{mj}) as in the Theorem 4.5 with normalization conditions analogous to those of the Theorem 4.3, then one proves, similarly to Theorem 4.5, the following central limit theorem.

THEOREM 4.6. *Suppose that*

$$\begin{aligned} \mathbb{E}(\mathbf{t}(Y_{mj})) &= \mathbf{0} \\ \sum_{j=1}^{k_m} \mathbf{Cov}(\mathbf{t}(Y_{mj})) &= \mathbf{Id}. \end{aligned}$$

If the following technical condition

$$\lim_{m\to\infty} \sum_{j=1}^{k_m} \mathbb{E}\big(\|\mathbf{t}(Y_{mj})\|^4 1 + \|\mathbf{t}(Y_{mj})\|^2\big) = 0$$

is verified then

$$Z_m \Rightarrow \nu_{1r+2d^{-1}} * \eta_{r+2r(r+2d^{-1})}, \qquad m \to \infty.$$

REMARK 4.7. Note that Theorem 4.3 is a special case of the Theorem 4.6. In fact, if $Y_1, Y_2, \ldots$ is a sequence of independent identically distributed K-invariant random variables as in the Theorem 4.3, set

$$Y_{mj} = k.e^{\mathbf{t}(Y_j)\sqrt{m}} \quad \text{when} \quad Y_j = k.e^{\mathbf{t}(Y_j)}.$$

Then the random variables Z_m and $S_m^\sharp$ are equal.

REMARK 4.8. The technical condition of the Theorem 4.6 is stronger than the condition (4.4) of the Theorem 4.5.

REMARK 4.9. If we replace the condition (4.4) by

$$\forall \epsilon > 0 \quad \lim_{m\to\infty} \sum_{j=1}^{k_m} \mathbb{E}\big[Q_1(Y_{mj}) + \log^2(\det(Y_{mj})), \|\mathbf{t}(Y_{mj})\| > \epsilon\big] = 0, \tag{4.5}$$

then the normalization hypotheses (4.1),(4.2),(4.3) together with the condition (4.5) are sufficient and necessary for the weak convergence of Z_m to $\nu_t * \eta_u$. The condition (4.4) implies (4.5) and in the proof of the Theorem 4.5 we prove in fact the sufficiency of (4.1),(4.2),(4.3) and (4.5). The proof of necessity is similar to the classical proof of necessity in the Lindeberg-Feller Central Limit Theorem on $\mathbb{R}^n$([F]).

REMARK 4.10. The Remarks 4.7, 4.8 and 4.9 also apply to irreducible symmetric spaces. For the irreducible symmetric spaces $SL(2,\mathbb{R})/SO(2)$ and $SL(2,\mathbb{C})/SU(2)$ the conditions (4.4) and (4.5) (simplified by the disappearance of terms with the determinant) are equivalent (see the explicit formulae for Q in these cases in [G1] and [G2]).

Acknowledgments

We thank Jacques Faraut, Hélène Massam and Donald St.P. Richards for discussions and comments on probability and statistics problems on symmetric cones. We thank Jean-Jacques Loeb for discussions on the Bernstein property of measures on groups.

References

[AJ] J.-Ph. Anker, L. Ji, Heat kernel and Green function estimates on noncompact symmetric spaces, Geom. Funct. Anal. (GAFA) **9**, no. 6 (1999), 1035-1091.

[FH] J. Faraut, Kh. Harzallah, Distances Hilbertiennes invariantes sur un espeace homogène, Ann. Inst. Fourier, **24** (1974), 171-217.

[FK] J. Faraut, A. Koranyi, Analysis on Symmetric Cones, Clarendon Press, Oxford, 1994.

[F] W. Feller, An Introduction to Probability Theory, Vol.2, Wiley, New York, 1966.

[Ga] R. Gangolli, Isotropic infinitely divisible measures on symmetric spaces, *Acta Math.* **111**(1964), 213-246.

[G1] P. Graczyk, A central limit theorem on the space of positive definite symmetric matrices, *Ann.Inst.Fourier* **42**(1992), 857-874.

[G2] P. Graczyk, Dispersions and a central limit theorem on symmetric spaces, *Bull.Sciences Math.* **118**(1994), 105-116.

[G3] ——— Cramer Theorem on Symmetric Spaces of Noncompact Type, *J.Theor.Prob.* **7**(1994),609-613

[GL] P. Graczyk, J.J. Loeb, A Bernstein property of measures on groups, to appear in *Prob. Math. Stat.*(2000).

[GR] Y. Guivarc'h, A. Raugi, Frontière de Furstenberg, propriétés de contraction et théorèmes de convergence, *Z.Wahrscheinlichkeitstheorie verw. Gebiete* **69**(1985), 187-242.

[H] S. Helgason, Groups and Geometric Analysis, Academic Press, New York, 1984.

[L] G. Letac, Symmetric cones as Gelfand pairs: probabilistic applications, *Contemporary Mathematics*, **261**(2000), 109-119.

[LM] G. Letac, H. Massam, Quadratic and inverse regression for Wishart distributions, *The Annals of Statistics*, **26**(1998).

[LP] E. Le Page, Théorèmes limites pour les produits de matrices, *Séminaire de probabilités de Rennes*(1981); CRAS **290** (1980), 559-562.

[M] H. Massam, An exact decomposition for a class of exponential dispersion models: Application to sequential testing, *Canadian Journal of Statistics*, **23** (1995), 67–84.

[MN] H. Massam, E. Neher, On transformations and determinants of Wishart variables on symmetric cones, *J. Theor. Prob.*, **10**, (1997), 867-902 .

[NRS] D. Neuenschwander, B. Roynette, and R. Schott, Characterization of Gauss measures on nilpotent Lie groups and symmetric spaces , C.R.Acad. Sci. Paris,**324**(1997), 87-92.

[R] D.St.P. Richards, The central limit theorem on spaces of positive definite matrices, *J.Multivariate Anal.* **29**(1989), 326-332.

[S] P. Sawyer, The central limit theorem on $\mathbf{SO}_0(p,q)/\mathbf{SO}(p)\times\mathbf{SO}(q)$ to appear in *Journal of Theor. Prob.*(2000).

[T] A. Terras, Noneuclidean harmonic analysis, the central limit theorem and long transmission lines with random inhomogeneities, *J.Multivariate Anal.* **15**(1984), 261-276.

[To] A. Tortrat, Lois stables dans un groupe, *Ann. Inst. Henri Poincaré*, Prob.Stat.**17**(1981), 51-61.

[Z] G. Zhang, The Asymptotics of Spherical Functions and the Central Limit Theorem on Symmetric Cones, *Ann. Inst. Fourier*, **45**(1995), 565-575.

DÉPARTEMENT DE MATHÉMATIQUES, UNIVERSITÉ D'ANGERS, 2 BLV.LAVOISIER, 49045 ANGERS, FRANCE

E-mail address: `graczyk@univ-angers.fr`

Contemporary Mathematics
Volume **287**, 2001

The Covariance Structure of the Multivariate Liouville Distributions

Rameshwar D. Gupta and Donald St. P. Richards

ABSTRACT. We study the covariance structure of the multivariate Liouville distributions and their power-scale transformations. For any Liouville distribution, we show that the off-diagonal entries of the corresponding correlation matrix all have the same sign; and we show that this result holds also for power-scale transformations of the Dirichlet distributions. We prove that the correlations between log-ratios of Liouville-distributed variables all are non-negative. Noting that these results are counter to some of the properties of models satisfactory for compositional data analysis (Aitchison, 1986), we urge caution in the use of Liouville distributions, or their power-scale transformations, for the statistical analysis of compositional data. In the case of the matrix Dirichlet distributions, we derive formulas for all moments of degree one and two, leading to the discovery of some remarkable relationships among the underlying covariances; in particular, we deduce that the resulting covariance matrices are of block-diagonal form. In the case of the matrix Liouville distributions we also derive formulas for all moments of degree one and two, thereby deducing completely their covariance structure.

1. Introduction

Although a wealth of information about the multivariate Liouville distributions is now available (cf. Gupta and Richards, 1987, 1991, 1992, 1995, 1997), little appears in the literature about the *covariance structure* (Aitchison, 1986) of these distributions. In this paper, we analyze the covariance matrices of the Liouville distributions, and their power-scale transformations. As a special case of our results we recover classical results on the covariance structure of the Dirichlet distributions, and we obtain new results on the covariance structure of power-scale transformations of the Dirichlet distributions.

1991 *Mathematics Subject Classification.* Primary 62H05; Secondary 62H10.

Key words and phrases. Compositional data analysis, confluent hypergeometric function of matrix argument, Dirichlet distributions, Kronecker product, Liouville distributions, power-scale transformations, size and shape analysis, zonal polynomials.

Research supported in part by the Natural Sciences and Engineering Research Council, grant OGP0004850 (R. D. Gupta) and in part by a grant to the Institute for Advanced Study from The Bell Fund, and by the National Science Foundation under grant DMS-9703705 (D. St. P. Richards).

We will also consider the extent to which power-scale transformations of the Liouville distributions are suitable for the modeling of compositional data. By application of our results about the covariance structure of the Liouville distributions, and basic principles for the proper analysis of compositional data sets (cf. Aitchison, 1986), we shall delineate those Liouville distributions, or their power-scale transformations, which may be appropriate for the analysis of compositional data. In the case of the Dirichlet distributions, our conclusions agree with those of Aitchison (1986).

Additionally, we consider the matrix analogs of the Liouville distributions, a family of distributions which contains the matrix Dirichlet distributions. Here, it was also an open problem to determine the covariance structure of these distributions. Using some basic results from the theory of zonal polynomials, we derive all moments of degree one and two for the matrix Dirichlet distributions, and then we apply those formulas to derive the corresponding results for the matrix Liouville distributions.

An outline of this article is as follows. In Section 2 we review basic ideas and notation. In Section 3 we develop a complete description of the covariance structure of the multivariate Liouville distributions, and we obtain a simple condition which ensures that the cross-covariances of a Liouville distribution are negative. For the Liouville random vectors we also study the covariances between log-ratios of components, a device shown by Aitchison (1986) to be of fundamental importance in the study of compositional data sets. We also study the covariance structure of random vectors obtained from multivariate Liouville vectors by power-scale transformations; in the case of a power-scale transformation of the Dirichlet distribution in which the powers are positive, we obtain the surprising result that the cross-covariances all are nonnegative.

In Section 4 we provide a review of our conclusions. We note that the Liouville distributions and their power transformations, due to the special nature of their correlation structure, generally are inappropriate for the modeling of compositional data. These conclusions refute some recent claims in the literature, and we provide an explanation of the flaws intrinsic to those claims. In Section 5 we derive the covariance structure of the matrix Dirichlet distributions; in the process, we discover some remarkable relationships in the underlying covariance matrices. In particular, we deduce that the covariance matrices are of block-diagonal form. In the case of the matrix Liouville distributions we also derive formulas for all moments of degree one and two, thereby describing completely their covariance structure.

2. Preliminaries

A random vector $(X_1, \ldots, X_n)$, assumed for simplicity to be absolutely continuous with a continuous density, has a *multivariate Liouville distribution* (Gupta and Richards, 1987) if it is supported on the n-dimensional orthant $\mathbb{R}_+^n = \{(x_1, \ldots, x_n) : x_i > 0, i = 1, \ldots, n\}$ and has a density function proportional to

$$x_1^{a_1-1} \cdots x_n^{a_n-1} f(x_1 + \cdots + x_n), \tag{2.1}$$

where $f : \mathbb{R}_+ \to \mathbb{R}_+$ is a continuous function and $a_i > 0$ for all $i = 1, \ldots, n$. We write $(X_1, \ldots, X_n) \sim L_n[f(\cdot); a_1, \ldots, a_n]$ whenever the density function of $(X_1, \ldots, X_n)$ is of the form (2.1). Moreover, throughout the paper, we let $a := a_1 + \cdots + a_n$.

If the function f in (2.1) has support $(0,\infty)$ then $(X_1,\dots,X_n)$ is said to have a Liouville distribution *of the first type*, denoted $L_n^{(1)}[f(\cdot);a_1,\dots,a_n]$. If the support of f is a finite interval which, after scaling, may be taken to be the unit interval $(0,1)$, then $(X_1,\dots,X_n)$ is said to have a Liouville distribution *of the second type*, denoted $L_n^{(2)}[f(\cdot);a_1,\dots,a_n]$; in this case $(X_1,\dots,X_n)$ is supported on the n-dimensional simplex

$$\mathcal{S}_n = \{(x_1,\dots,x_n) : x_i > 0, i = 1,\dots,n; x_1 + \cdots + x_n < 1\}.$$

By Proposition 3.6 of Gupta and Richards (1987, p. 238), there is a one-to-one transformation between the Liouville distributions of the first and second types.

The following *stochastic representation* for the Liouville distributions will be needed in the sequel: If $(X_1,\dots,X_n) \sim L_n[f(\cdot);a_1,\dots,a_n]$ then

$$(X_1,\dots,X_n) =^{\mathrm{st}} (Y_1,\dots,Y_n)R, \tag{2.2}$$

where $R =^{\mathrm{st}} \sum_{i=1}^n X_i \sim L_1[f(\cdot);a]$ and $(Y_1,\dots,Y_n)$ is independent of R. Furthermore $Y_n = 1 - \sum_{i=1}^{n-1} Y_i$; and $(Y_1,\dots,Y_{n-1}) \sim D(a_1,\dots,a_{n-1};a_n)$, a Dirichlet distribution, with density function

$$\frac{\Gamma(a)}{\Gamma(a_1)\cdots\Gamma(a_n)} y_1^{a_1-1}\cdots y_{n-1}^{a_{n-1}-1}\Big(1-\sum_{i=1}^{n-1} y_i\Big)^{a_n-1}, \tag{2.3}$$

$(y_1,\dots,y_{n-1}) \in \mathcal{S}_{n-1}$. It can be shown, using (2.2), that several properties of the Dirichlet distributions, arising in compositional data analysis, remain valid for all Liouville distributions of the second type. These properties include the amalgamation, sub-compositional and partition properties (cf. Aitchison, 1986, §2.5-2.7). For instance, to establish the sub-compositional property for the Liouville distributions of the second type, suppose that $(X_1,\dots,X_n) \sim L_n^{(2)}[f(\cdot);a_1,\dots,a_n]$. For $k < n$, let $\{X_{i_1},\dots,X_{i_k}\}$ be a subset of $\{X_1,\dots,X_n\}$, and define

$$Y_{i_j} = X_{i_j}/(X_{i_1} + \cdots + X_{i_k}),$$

$j = 1,\dots,k$. Since $(X_{i_1},\dots,X_{i_k})$ has a marginal Liouville distribution (cf. Gupta and Richards, 1987), then a stochastic representation of the type described in (2.2) also holds for $(X_{i_1},\dots,X_{i_k})$. Then we find that $(Y_{i_1},\dots,Y_{i_k}) \sim SD(a_{i_1},\dots,a_{i_k})$, a singular Dirichlet distribution. In the terminology of Aitchison (1986, §2.5), we have shown that every sub-composition of a Liouville-distributed composition vector is a Dirichlet sub-composition.

3. Covariance structure of the Liouville distributions

From the stochastic representation (2.2), and the property that each Y_i has a marginal beta distribution, we obtain

$$\begin{aligned}\operatorname{Var}(X_i) &= \operatorname{Var}(RY_i) = E(R^2)E(Y_i^2) - [E(R)E(Y_i)]^2\\ &= \frac{a_i}{a^2(a+1)}(a(a_i+1)E(R^2) - a_i(a+1)[E(R)]^2)\\ &= \frac{a_i}{a^2(a+1)}(a(a_i+1)\operatorname{Var}(R) + (a-a_i)[E(R)]^2),\end{aligned} \tag{3.1}$$

for all $i = 1, \ldots, n$. Similarly, for all $i < j$,

$$\begin{aligned} \operatorname{Cov}(X_i, X_j) &= E(R^2)E(Y_iY_j) - [E(R)]^2 E(Y_i)E(Y_j) \\ &= \frac{a_i a_j}{a^2(a+1)}(aE(R^2) - (a+1)[E(R)]^2) \\ &= \frac{a_i a_j}{a^2(a+1)}(a\operatorname{Var}(R) - [E(R)]^2). \end{aligned} \tag{3.2}$$

Note that for each $i < j$, the terms in (3.2) all have the same sign. Furthermore (3.2) is negative if and only if $\mathrm{CV}(R) := \sqrt{\operatorname{Var}(R)}/E(R)$, the *coefficient of variation* of R, satisfies the condition

$$\mathrm{CV}(R) < a^{-1/2}. \tag{3.3}$$

We now derive a sufficient condition under which (3.3) holds.

PROPOSITION 3.1. *Suppose that* $(X_1, \ldots, X_n) \sim L_n[f(\cdot); a_1, \ldots, a_n]$, *where* f *is differentiable, strictly log-concave and*

$$\lim_{u \to 0+} u^a f(u) = \lim_{u \to \infty} u^{a+1} f(u) = 0.$$

Then (3.3) holds.

PROOF. Since $R \sim L_1[f(\cdot); a]$ then the density function of R is

$$h(t) = c^{-1} t^{a-1} f(t), \qquad t > 0, \tag{3.4}$$

where

$$c = \int_0^\infty u^{a-1} f(u) du. \tag{3.5}$$

It follows from (3.4) and (3.5) that

$$\begin{aligned} &(aE(R^2) - (a+1)[E(R)]^2)c^2 \\ &\quad = \int_0^\infty \int_0^\infty \big(at^{a+1}u^{a-1} - (a+1)t^a u^a\big) f(t) f(u) dt\, du \\ &\quad = \int_0^\infty \int_0^\infty \big(t^{a+1}(u^a)' - t^a (u^{a+1})'\big) f(t) f(u) dt\, du \\ &\quad = \int_0^\infty \int_0^\infty \big(t^a u^{a+1} - t^{a+1} u^a\big) f(t) f'(u) dt\, du, \end{aligned} \tag{3.6}$$

where the last equality is obtained by integrating-by-parts with respect to the u-variable in (3.6), bearing in mind the stated regularity conditions. Next we partition the domain of integration into the regions $\{0 < u \le t < \infty\}$ and $\{0 < t < u < \infty\}$, apply a symmetry argument and simplify the outcome; then we obtain the result

$$\begin{aligned} &(aE(R^2) - (a+1)[E(R)]^2)c^2 \\ &\quad = \int_0^\infty \int_0^u t^a u^a (u - t)\big(f(t) f'(u) - f'(t) f(u)\big) dt\, du. \end{aligned} \tag{3.7}$$

Since f is log-concave; equivalently, the function $(\log f(t))' = f'(t)/f(t)$ is strictly decreasing; then we have $f'(t)/f(t) > f'(u)/f(u)$, or $f(t)f'(u) - f'(t)f(u) < 0$, for all $t < u$. Therefore (3.7) is negative under the stated conditions. □

As an example, consider the case in which $f(t) = (1-t)^{a_{n+1}-1}$, $0 < t < 1$, and $f(t) = 0$ elsewhere, where $a_{n+1} > 1$. Then $(X_1, \ldots, X_n) \sim D(a_1, \ldots, a_n; a_{n+1})$, a Dirichlet distribution. It can be checked that the regularity conditions in the preceding result are valid in this case, and then we deduce that $\text{Cov}(X_i, X_j) < 0$ for all $i < j$, a result which is well-known for the Dirichlet distributions.

Notice also that $\text{CV}(R) = a^{-1/2}$ if and only if $\text{Cov}(X_i, X_j) = 0$ for some, and hence for all, $i < j$. From this result, we can develop new examples of multivariate Liouville random variables which are pairwise uncorrelated but no two of which are mutually independent.

EXAMPLES 3.2. (i) Consider the case in which $f(t) = t^{\alpha}(1-t)^{\beta-1}$, $0 < t < 1$, where α and β are constants; the resulting distributions are the beta-Liouville distributions (Fang, *et al.*, 1990, p. 147), and no two of the X_i are mutually independent. Moreover, R has a beta distribution with parameters $(\alpha + a, \beta)$ and it is straightforward to calculate that

$$\text{CV}(R) = \left[\frac{\beta}{(a+\alpha)(a+\alpha+\beta+1)}\right]^{1/2}.$$

Hence, $\text{CV}(R) = a^{-1/2}$ if and only if $(a+\alpha)(a+\alpha+1) + \alpha\beta = 0$; thus by choosing $\alpha \in (-a, 0)$ and $\beta = -(a+\alpha)(a+\alpha+1)/\alpha$, we obtain an example of a Liouville distribution in which each pair (X_i, X_j) are pairwise uncorrelated but are not mutually independent.

(ii) Let $Z \sim G(\alpha, \beta)$, a gamma distribution with index and scale parameters α and β, respectively. Further, let $R = e^{-Z}$ (cf. Fang, *et al.*, 1990, p. 149). Then

$$\text{CV}(R) = \left[\frac{(1+\beta)^{2\alpha}}{(1+2\beta)^{\alpha}} - 1\right]^{1/2}.$$

Here we can choose the parameters α and β so that any pair of X_i's exhibit positive, negative or zero correlation; however, mutual independence cannot hold for this model.

(iii) Consider the case in which $R \sim IG(\lambda, \alpha)$, an inverse Gaussian distribution with parameters λ and α. Then $\text{CV}(R) = (\lambda/\alpha)^{1/2}$ and hence $\text{Cov}(X_i, X_j) = 0$ if and only if $\alpha = a\lambda$. Therefore, we can choose the parameters λ and α so that any pair of X_i can exhibit positive, negative or zero correlation.

In Examples 3.2(i) and (ii) given above, it is clear that neither model is appropriate for compositional data modeling if only because the corresponding correlation structures are not sufficiently general.

We can also obtain an explicit formula for the characteristic polynomial of $\text{Cov}(X)$, the covariance matrix of X. Let us introduce the notation $c_1 = aE(R^2)$; $c_2 = aE(R^2) - (a+1)[E(R)]^2$; $\Delta = \text{diag}(a_1, \ldots, a_n)$, a diagonal matrix; and $v = (a_1, \ldots, a_n)'$, a column vector. By (3.1) and (3.2), it follows that

$$a^2(a+1)\text{Cov}(X) = \left(c_1\Delta + c_2 vv'\right). \tag{3.8}$$

Then the characteristic polynomial of the matrix (3.8) is

$$\begin{aligned}
\det(\lambda I - c_1\Delta - c_2vv') &= \det(I - c_2vv'(\lambda I - c_1\Delta)^{-1}) \cdot \det(\lambda I - c_1\Delta) \\
&= (1 - c_2v'(\lambda I - c_1\Delta)^{-1}v) \det(\lambda I - c_1\Delta) \\
&= \Bigl(1 - c_2 \sum_{i=1}^{n} \frac{a_i^2}{\lambda - c_1 a_i}\Bigr) \prod_{i=1}^{n} (\lambda - c_1 a_i) \qquad (3.9) \\
&= \prod_{i=1}^{n} (\lambda - c_1 a_i) - c_2 \sum_{i=1}^{n} a_i^2 \prod_{j=1, j \neq i}^{n} (\lambda - c_1 a_j).
\end{aligned}$$

For the special case in which $a_1 = \cdots = a_n$, it follows directly from (3.1) and (3.2) that $\operatorname{Cov}(X)$ has an intraclass correlation structure (Rao, 1973). Then the characteristic roots are c_1a_1 (with multiplicity $n-1$) and $(c_1 + nc_2a_1)a_1$; this can also be deduced from (3.9) which, in this special case, reduces to

$$\det(\lambda I - c_1\Delta - c_2vv') = \bigl(\lambda - c_1a_1\bigr)^{n-1}\bigl(\lambda - (c_1 + nc_2a_1)a_1\bigr).$$

Next, we consider the covariance structure of the log-ratios $\log(X_i/X_j)$, $i, j = 1, \ldots, n$.

PROPOSITION 3.3. *If* $(X_1, \ldots, X_n) \sim L_n[f(\cdot); a_1, \ldots, a_n]$ *then the covariance between any two distinct log-ratios* $\log(X_i/X_j)$ *and* $\log(X_k/X_l)$ *is nonnegative.*

PROOF. It follows from (2.2), and a well-known relationship between the Dirichlet and gamma distributions, that

$$\frac{X_i}{X_j} = \frac{Y_i}{Y_j} =^{\text{st}} \frac{T_i}{T_j}, \qquad (3.10)$$

where $T_1, \ldots, T_n$ are mutually independent gamma variables, $T_i \sim G(a_i, \lambda)$, $i = 1, \ldots, n$. Therefore

$$\operatorname{Cov}\Bigl(\log\frac{X_i}{X_j}, \log\frac{X_k}{X_l}\Bigr) = \operatorname{Cov}\Bigl(\log\frac{T_i}{T_j}, \log\frac{T_k}{T_l}\Bigr) = 0 \qquad (3.11)$$

whenever the indices i, j, k and l are distinct. If i, k and $j \equiv l$ are distinct then, by (3.10) and the mutual independence of $T_1, \ldots, T_n$, we have

$$\begin{aligned}
\operatorname{Cov}\Bigl(\log\frac{X_i}{X_j}, \log\frac{X_k}{X_j}\Bigr) &= \operatorname{Cov}\Bigl(\log\frac{T_i}{T_j}, \log\frac{T_k}{T_j}\Bigr) \\
&= \operatorname{Cov}(\log T_i - \log T_j, \log T_k - \log T_j) \\
&= \operatorname{Var}(\log T_j) > 0.
\end{aligned}$$

Similarly, if j, l and $i \equiv k$ are distinct then (3.11) reduces to $\operatorname{Var}(\log T_i) > 0$. The remaining cases in (3.11) are those in which $i \equiv k$ and $j \equiv l$; but then (3.11) is the variance of $\log(X_i/X_j)$ which, clearly, is positive. □

Next we consider the covariance structure of the random vector $(U_1, \ldots, U_n)$ where, for $i = 1, \ldots, n$, $U_i = (X_i/\alpha_i)^{\beta_i}$ is a power-scale transformation of X_i, with $(X_1, \ldots, X_n) \sim L_n[f(\cdot); a_1, \ldots, a_n]$, and the constants α_i and β_i, $i = 1, \ldots, n$, are positive. These distributions arise in reliability theory as mixtures of independent Weibull random variables (cf. Gupta and Richards, 1992, p. 41).

When $i < j$ it follows from the stochastic representation (2.2) that

$$\begin{aligned}\operatorname{Cov}(U_i, U_j) &= \operatorname{Cov}\big((X_i/\alpha_i)^{\beta_i}, (X_j/\alpha_j)^{\beta_j}\big) \\ &= \alpha_i^{-\beta_i}\alpha_j^{-\beta_j} \\ &\quad \times \big[E(Y_i^{\beta_i}Y_j^{\beta_j}E(R^{\beta_i+\beta_j}) - E(Y_i^{\beta_i})E(Y_j^{\beta_j})E(R^{\beta_i})E(R^{\beta_j})\big] \\ &= \frac{\Gamma(a)\Gamma(a_i+\beta_i)\Gamma(a_j+\beta_j)}{\Gamma(a_i)\Gamma(a_j)\Gamma(a+\beta_i)\alpha_i^{\beta_i}\alpha_j^{\beta_j}} \\ &\quad \times \Big[\frac{\Gamma(a+\beta_i)E(R^{\beta_i+\beta_j})}{\Gamma(a+\beta_i+\beta_j)} - \frac{\Gamma(a)E(R^{\beta_i})E(R^{\beta_j})}{\Gamma(a+\beta_j)}\Big].\end{aligned} \tag{3.12}$$

For special choices for the distribution of R, and particular values of β_i and β_j, it is straightforward to verify that $\operatorname{Cov}(U_i, U_j)$ can attain positive or negative values. However, in the case in which $(X_1, \ldots, X_n)$ follows a Dirichlet distribution, we have the following result.

PROPOSITION 3.4. *Suppose that $(X_1, \ldots, X_n)$ follows a Dirichlet distribution $D(a_1, \ldots, a_n; a_{n+1})$. Then $Cov(U_i, U_j) < 0$ for all $i < j$.*

PROOF. By applying (2.2) and (3.12) we deduce that, for $i < j$, $\operatorname{Cov}(U_i, U_j)$ is a positive multiple of

$$\frac{\Gamma(a+\beta_i)}{\Gamma(a+\beta_i+\beta_j)} - \frac{\Gamma(a)}{\Gamma(a+\beta_j)}. \tag{3.13}$$

Recall (cf. Artin, 1964) that the gamma function is strictly log-convex; equivalently, the function $\Gamma(t)/\Gamma(t+\beta)$ is strictly decreasing in $t > 0$ for any $\beta > 0$. Therefore (3.13), and hence $\operatorname{Cov}(U_i, U_j)$, is negative. □

4. Some cautionary remarks

Compositional data are observations of a random vector $(W_1, \ldots, W_n, W_{n+1})$, where $(W_1, \ldots, W_n)$ is supported on the simplex $\mathcal{S}_n$ and $W_{n+1} = 1-(W_1+\cdots+W_n)$ is the *fill-up value*. Aitchison (1986) provided a comprehensive and systematic treatment of the theory of compositional data analysis, with extensive information about parametric families of distributions appropriate for modeling compositional data.

To model compositional data using a given parametric family, it is a basic requirement that the off-diagonal entries of the corresponding correlation matrix be able to exhibit arbitrary sign patterns. As noted by Aitchison (1986), since the off-diagonal entries of the covariance matrix of the Dirichlet distributions (2.3) are all negative; and because of the invariance property under sub-compositions; then those distributions may not be suitable for modeling compositional data. It follows from the results in Section 3 that the same conclusion holds for any Liouville distribution or for any power-scale transformation of the Dirichlet distributions in which the powers are positive. These conclusions are reinforced further by the non-negativity of the covariances between the log-ratios $\log(X_i/X_j)$ (cf. Aitchison, 1986, p. 65).

A related point is that if the support of $(X_1, \ldots, X_n)$ is the simplex $\mathcal{S}_n$, then the support of the power-scale transformed variables $(U_1, \ldots, U_n)$ is the region

$$\Big\{u_i > 0, i = 1, \ldots, n; \sum_{i=1}^{n} \alpha_i u_i^{1/\beta_i} < 1\Big\},$$

which is generally different from the simplex support set, viz. $\mathcal{S}_n$, for compositional data. Some researchers (cf. Rayens and Srinivasan, 1994) have tried to circumvent this obstacle by the usage of the so-called "generalized Liouville distribution," which pertain to a random vector $(U_1, \ldots, U_n)$ having a joint density function which is proportional to

$$f\Big(\sum_{i=1}^{n} (u_i/q_i)^{\beta_i}\Big) \prod_{i=1}^{n} u_i^{a_i - 1}, \tag{4.1}$$

$(u_1, \ldots, u_n) \in \mathcal{S}_n$, where the function $f : \mathbb{R}_+ \to \mathbb{R}_+$.

Since the simplex $\mathcal{S}_n$ is required to be the support set for the density function (4.1), then the evaluation of the normalizing constant is a recondite problem (cf. Rayens and Srinivasan, 1994, p. 1467). If, in calculating the normalizing constant, an attempt is made to simplify the problem by means of the power-scale transformation $X_i = (U_i/q_i)^{\beta_i}$, $i = 1, \ldots, n$, this will transform the density function (4.1) into a simpler form, proportional to

$$f(\sum_{i=1}^{n} x_i) \prod_{i=1}^{n} x_i^{(a_i/\beta_i) - 1},$$

but the transformation also complicates the support set, changing it into

$$\Big\{(x_1, \ldots, x_n) : x_i > 0, i = 1, \ldots, n; \sum_{i=1}^{n} q_i x_i^{1/\beta_i} \le 1\Big\}.$$

It is easy to see that the evaluation of the normalizing constant has been made no simpler by this transformation. These observations make it clear that the classical Liouville densities on the simplex $\mathcal{S}_n$ cannot be transformed into the densities given in (4.1) by means of a power-scale transformation. Consequently, and unfortunately, the algorithm proposed by Rayens and Srinivasan (1994, p. 1467), to simulate the distributions corresponding to (4.1), is not valid. Specifically, steps A3, A4 and A5 of the algorithm proposed by Rayens and Srinivasan (*loc. cit.*, p. 1468) are incorrect. For these reasons, we urge caution in the use of Liouville-type distributions as models for the unconditional distributions of compositional data sets.

We note that there have appeared parametric models which have been shown to be appropriate for modeling data on the simplex (cf. Aitchison, 1986; Connor and Mosimann, 1969; Johnson and Young, 1960). We also refer to Mosimann (1987) for bibliography on applications of the Dirichlet distributions to the analysis of data arising in studies of size and shape; and to Bandeen-Roche (1994) for Dirichlet-based covariate models of data on the simplex. Bandeen-Roche (1994), p. 1457, observes too that choosing "the associated Dirichlet parameters to be functions of measured covariates ... may increase flexibility of model covariance structure ... [and] may also allow a more precise interpretation of the mixing distribution."

5. Covariance structure of the matrix Liouville distributions

In the literature, the covariance structure of the matrix Liouville distributions has received little attention, this no doubt being due to their extremely complicated nature. Even in the simplest cases, the problem of calculating the covariance matrix has remained unsolved. To set the stage for the results to follow, we shall denote by I_m the $m \times m$ identity matrix, and we write $Y > 0$ whenever Y is a real symmetric positive-definite $m \times m$ matrix. We also let

$$\mathcal{S}_{n,m} := \{(Y_1, \dots, Y_n) : Y_j > 0 \text{ for all } 1 \le j \le n, \text{ and } I_m - Y_1 - \cdots - Y_n > 0\}$$

denote the matrix simplex. For $\alpha > (m-1)/2$, define the multivariate gamma function

$$\Gamma_m(\alpha) = \pi^{m(m-1)/4} \prod_{k=1}^{m} \Gamma\Big(\alpha - \frac{1}{2}(k-1)\Big);$$

cf., Muirhead (1982), p. 62.

Then the $m \times m$ random matrices $Y_1, \dots, Y_{n-1}$ are said to follow a multivariate Dirichlet distribution with parameters $a_1, \dots, a_n$, denoted $(Y_1, \dots, Y_{n-1}) \sim D_m(a_1, \dots, a_{n-1}; a_n)$, if, relative to Lebesgue measure on the simplex $\mathcal{S}_{n,m}$, the joint probability density function of $(Y_1, \dots, Y_{n-1})$ is

$$(5.1) \qquad \frac{\Gamma_m(a)}{\Gamma_m(a_1) \cdots \Gamma_m(a_n)} |I_m - Y_1 - \cdots - Y_{n-1}|^{a_n - (m+1)/2} \prod_{j=1}^{n-1} |Y_j|^{a_j - (m+1)/2},$$

where $(Y_1, \dots, Y_{n-1}) \in \mathcal{S}_{n,m}$ and, as before, we use the notation $a := a_1 + \cdots + a_n$.

For the case in which $n = 2$, (5.1) reduces to the well-known multivariate beta distribution, denoted by $Y_1 \sim \boldsymbol{\beta}_m(a_1, a_2)$. Even in this case, the problem of calculating the covariance matrix has remained unsolved; we now derive a solution using results from the theory of zonal polynomials.

Returning to the general case in (5.1), we need to calculate the first moment of each Y_i in (5.1). Recall that the marginal distribution of Y_i is $\boldsymbol{\beta}_m(a_i, a - a_i)$; hence we lose no generality in evaluating $E(Y_i)$ for $i = 1$ only. To this end, denote the (i,j)th entry of Y_1 by $y_{1,ij}$, and let $T = (t_{ij})$ be an $m \times m$ real symmetric matrix. If a is not a half-integer, it follows from Muirhead (1982), Theorem 7.4.2, p. 264, that the moment-generating function of Y_1, is

$$(5.2) \qquad E(e^{\operatorname{tr} TY_1}) = {}_1F_1(a_1; a; T),$$

where ${}_1F_1$ denotes the confluent hypergeometric function of matrix argument. To derive a series expansion of (5.2) leading to the covariance matrix of the entries of Y_1, we need to recall some results from the theory of zonal polynomials.

A *partition* $\kappa = (k_1, \dots, k_n)$ is an n-tuple of integers $k_1, \dots, k_n$ such that $k_1 \ge \cdots \ge k_n \ge 0$. The *length* of the partition κ is given by $|\kappa| := k_1 + \cdots + k_n$. For any $\alpha \in \mathbb{R}$, the *partitional rising factorial* is defined by

$$(\alpha)_\kappa := \prod_{j=1}^{n} \Big(\alpha - \frac{1}{2}(j-1)\Big)_{k_j},$$

where $(\alpha)_k = \alpha(\alpha+1)\cdots(\alpha+k-1)$, $k = 0, 1, 2, \dots$, is the classical rising factorial.

Corresponding to each partition κ is a *zonal polynomial*, $C_\kappa(T)$ (cf. Muirhead, 1982, Chapter 7). The polynomial $C_\kappa(T)$ is homogeneous of degree $|\kappa|$ in T; moreover, in the case of partitions of length no more than two (cf. Muirhead, 1982, p.

232 ff.), the polynomial $C_\kappa(T)$ is given explicitly by the formulas

$$C_\kappa(T) = \begin{cases} 1, & \kappa = (0) \\ \operatorname{tr}(T), & \kappa = (1) \\ \frac{1}{3}\big[(\operatorname{tr} T)^2 + 2\operatorname{tr}(T^2)\big], & \kappa = (2) \\ \frac{2}{3}\big[(\operatorname{tr} T)^2 - \operatorname{tr}(T^2)\big], & \kappa = (1,1) \end{cases} \tag{5.3}$$

Next, we recall that the confluent hypergeometric function in (5.2) can be expanded in a series of zonal polynomials,

$${}_1F_1(a_1; a; T) = \sum_{k=0}^{\infty} \frac{1}{k!} \sum_{|\kappa|=k} \frac{(a_1)_\kappa}{(a)_\kappa} C_\kappa(T), \tag{5.4}$$

where the inner sum is over all partitions κ of length k. To derive the covariance matrix of the elements of Y_1 we apply the standard procedure of differentiating the generating-function (5.2) with respect to the t_{ij} and evaluating the outcome at $T = 0$. Since the zonal polynomials are homogeneous then it suffices to apply the expansion in (5.4) up to terms of degree two.

Applying (5.3) and (5.4) to (5.2), we obtain

$$E(e^{\operatorname{tr} TY_1}) = 1 + \frac{a_1}{a}\operatorname{tr}(T) + b_1\operatorname{tr}(T^2) + b_2(\operatorname{tr} T)^2 + O(T^3), \tag{5.5}$$

where

$$b_1 = \frac{1}{3}\Big[\frac{(a_1)_{(2)}}{(a)_{(2)}} - \frac{(a_1)_{(1,1)}}{(a)_{(1,1)}}\Big] = \frac{a_1(a - a_1)}{a(a+1)(2a-1)}$$

and

$$b_2 = \frac{1}{3}\Big[\frac{(a_1)_{(2)}}{2(a)_{(2)}} + \frac{(a_1)_{(1,1)}}{(a)_{(1,1)}}\Big] = \frac{a_1(2a_1 a + a_1 - 1)}{2a(a+1)(2a-1)}.$$

For future reference, we note that

$$b_1 + b_2 = \frac{a_1(a_1+1)}{2a(a+1)}.$$

Differentiating (5.5) once with respect to each t_{ij} and evaluating the outcome at $T = 0$ we obtain

$$E(Y_1) = \frac{a_1}{a} I_m, \tag{5.6}$$

and this evaluates the first moments.

To evaluate the second moments, we differentiate the terms of degree two in (5.5), thereby obtaining

$$E(y_{1,i_1 j_1} y_{1,i_2 j_2}) = \begin{cases} 2(b_1 + b_2), & \text{if } i_1 = j_1 = i_2 = j_2, \\ b_1, & \text{if } i_1 = i_2 \neq j_1 = j_2, \\ 2b_2, & \text{if } i_1 = j_1 \neq i_2 = j_2, \\ 0, & \text{otherwise.} \end{cases}$$

Combining this result with (5.6) we deduce that the variances of, and covariances between, elements of Y_1 are given by

$$\text{(5.7)} \qquad \operatorname{Cov}(y_{1,i_1j_1}, y_{1,i_2j_2}) = \begin{cases} (2a-1)b_1/a, & \text{if } i_1 = j_1 = i_2 = j_2, \\ b_1, & \text{if } i_1 = i_2 \neq j_1 = j_2, \\ -b_1/a, & \text{if } i_1 = j_1 \neq i_2 = j_2, \\ 0, & \text{otherwise.} \end{cases}$$

To describe the structure of the covariance matrix of the elements of Y_1, we define the $\frac{1}{2}m(m+1)$-dimensional vector

$$\text{(5.8)} \qquad V_1 := \big(\operatorname{Vec}(\operatorname{diag}(Y_1)), \operatorname{Vec}(\operatorname{odiag}(Y_1))\big),$$

where $\operatorname{diag}(Y_1) := (y_{1,11}, \dots, y_{1,mm})'$ is a column vector containing the diagonal entries of Y_1 and $\operatorname{odiag}(Y_1) := (y_{1,12}, \dots, y_{1,1m}, y_{1,23}, \dots, y_{1,2m}, \dots, y_{1,m-1,m})'$ is a column vector containing all off-diagonal entries of Y_1 listed by descending row. Therefore $\operatorname{Cov}(V_1)$, the covariance matrix of V_1, is the matrix

$$\begin{bmatrix} \operatorname{Cov}\big(\operatorname{Vec}(\operatorname{diag}(Y_1))\big) & \operatorname{Cov}\big(\operatorname{Vec}(\operatorname{diag}(Y_1)), \operatorname{Vec}(\operatorname{odiag}(Y_1))\big) \\ \operatorname{Cov}\big(\operatorname{Vec}(\operatorname{diag}(Y_1)), \operatorname{Vec}(\operatorname{odiag}(Y_1))\big)' & \operatorname{Cov}\big(\operatorname{Vec}(\operatorname{odiag}(Y_1))\big) \end{bmatrix}.$$

Denote by E_m the $m \times m$ matrix with all entries equal to 1 and by $\mathbf{0}$ the zero matrix of order $m \times m(m-1)/2$. Then it follows from (5.7) that

$$\text{(5.9)} \qquad \operatorname{Cov}(V_1) = \frac{a_1(a-a_1)}{a^2(a+1)(2a-1)}\Omega,$$

where

$$\text{(5.10)} \qquad \Omega = \begin{bmatrix} 2aI_m - E_m & \mathbf{0} \\ \mathbf{0}' & aI_{m(m-1)/2} \end{bmatrix}.$$

We can also deduce the covariance structure of the general Dirichlet distributions (5.1). Given that the first moments already have been calculated, the evaluation of the covariance structure requires that we calculate the expectation of the Kronecker (or tensor) product $Y_i \otimes Y_j$, for all $i < j$. Here again, we lose no generality by assuming that $i = 1$ and $j = 2$. Starting from (5.1), we calculate that the marginal density function of (Y_1, Y_2) is

$$\frac{\Gamma_m(a)}{\Gamma_m(a_1)\Gamma_m(a_2)\Gamma_m(a-a_1-a_2)} \times |Y_1|^{a_1-(m+1)/2}|Y_2|^{a_2-(m+1)/2}|I_m - Y_1 - Y_2|^{a-a_1-a_2-(m+1)/2},$$

where $Y_1 > 0$, $Y_2 > 0$ and $I_n - Y_1 - Y_2 > 0$. Next, we introduce the transformation $Y_1 = W_1$, $Y_2 = (I_m - W_1)^{1/2}W_2(I_m - W_1)^{1/2}$; then W_1 and W_2 are mutually independent beta matrices, with $W_1 \sim \boldsymbol{\beta}_m(a_1, a-a_1)$ and $W_2 \sim \boldsymbol{\beta}_m(a_2, a-a_1-a_2)$. By applying the mutual independence of W_1 and W_2, we have

$$\text{(5.11)} \qquad \begin{aligned} E(Y_1 \otimes Y_2) &= E\big(W_1 \otimes (I_m - W_1)^{1/2}W_2(I_m - W_1)^{1/2}\big) \\ &= E\big(W_1 \otimes (I_m - W_1)^{1/2}E(W_2)(I_m - W_1)^{1/2}\big). \end{aligned}$$

By the same argument which led to (5.6), we have

$$\text{(5.12)} \qquad E(W_j) = \frac{a_j}{\sum_{i=j}^n a_i} I_m$$

for $j = 1, 2$. Therefore (5.11) reduces to

$$\begin{aligned} E(Y_1 \otimes Y_2) &= \frac{a_2}{a - a_1} E\big(W_1 \otimes (I_m - W_1)\big) \\ &= \frac{a_2}{a - a_1} \big(E(W_1 \otimes I_m) - E(W_1 \otimes W_1)\big). \end{aligned}$$

However, by (5.12),

$$E(W_1 \otimes I_m) = E(W_1) \otimes I_m = \frac{a_1}{a} I_m \otimes I_m = \frac{a_1}{a} I_{m^2};$$

therefore we obtain

$$E(Y_1 \otimes Y_2) = \frac{a_2}{a - a_1} \Big(\frac{a_1}{a} I_{m^2} - E(W_1 \otimes W_1)\Big),$$

and this provides a solution to the problem once we recall that all entries of the matrix $E(W_1 \otimes W_1)$ are provided by (5.7). Denoting the (i, j)th entry of Y_2 by $y_{2,ij}$ it follows, *mutatis mutandis*, from (5.7) that

$$E(y_{1,i_1j_1} y_{2,i_2j_2}) = \begin{cases} a_1a_2/a(a+1), & \text{if } i_1 = j_1 = i_2 = j_2, \\ -a_1a_2/a(a+1)(2a-1), & \text{if } i_1 = i_2 \neq j_1 = j_2, \\ a_1a_2(2a+1)/a(a+1)(2a-1), & \text{if } i_1 = j_1 \neq i_2 = j_2, \\ 0, & \text{otherwise,} \end{cases}$$

and hence

$$\text{(5.13)} \quad \operatorname{Cov}(y_{1,i_1j_1}, y_{2,i_2j_2}) = \begin{cases} -a_1a_2/a^2(a+1), & \text{if } i_1 = j_1 = i_2 = j_2, \\ -a_1a_2/a(a+1)(2a-1), & \text{if } i_1 = i_2 \neq j_1 = j_2, \\ a_1a_2/a^2(a+1)(2a-1), & \text{if } i_1 = j_1 \neq i_2 = j_2, \\ 0, & \text{otherwise.} \end{cases}$$

Defining V_2 as in (5.8), i.e.,

$$V_2 := \big(\operatorname{Vec}(\operatorname{diag}(Y_2)), \operatorname{Vec}(\operatorname{odiag}(Y_2))\big),$$

we apply (5.13) to deduce that $\operatorname{Crov}(V_1, V_2)$, the matrix of *cross-covariances* between V_1 and V_2, is

$$\text{(5.14)} \quad \begin{aligned} \operatorname{Crov}(V_1, V_2) &= -\frac{a_1a_2}{a^2(a+1)(2a-1)} \begin{bmatrix} 2aI_m - E_m & \mathbf{0} \\ \mathbf{0}' & aI_{m(m-1)/2} \end{bmatrix} \\ &= -\frac{a_1a_2}{a^2(a+1)(2a-1)} \Omega, \end{aligned}$$

where Ω is given in (5.10). It is remarkable that $\operatorname{Crov}(V_1, V_2)$ is a constant multiple of $\operatorname{Cov}(V_1)$.

Now we collect together the results (5.10) and (5.14). With $(Y_1, \ldots, Y_{n-1}) \sim D_m(a_1, \ldots, a_{n-1}; a_n)$ and $Y_n := I_m - Y_1 - \cdots - Y_{n-1}$, let

$$\operatorname{Cov}(V_1, \ldots, V_n) := \begin{bmatrix} \operatorname{Cov}(V_1) & \operatorname{Crov}(V_1, V_2) & \cdots & \operatorname{Crov}(V_1, V_n) \\ \operatorname{Crov}(V_2, V_1) & \operatorname{Cov}(V_2) & \cdots & \operatorname{Crov}(V_2, V_n) \\ \cdots & \cdots & \cdots & \cdots \\ \operatorname{Crov}(V_n, V_1) & \cdots & \cdots & \operatorname{Cov}(V_n) \end{bmatrix},$$

denote the joint covariance matrix of all elements of $Y_1, \dots, Y_n$. Then we deduce from (5.10) and (5.14), *mutatis mutandis*, that

$$\operatorname{Cov}(V_1, \dots, V_n) = \frac{1}{a^2(a+1)(2a-1)} \begin{bmatrix} a_1(a-a_1)\Omega & -a_1a_2\Omega & \cdots & -a_1a_n\Omega \\ -a_2a_1\Omega & a_2(a-a_2)\Omega & \cdots & -a_2a_n\Omega \\ \cdots & \cdots & \cdots & \cdots \\ -a_na_1\Omega & \cdots & -a_na_{n-1}\Omega & a_n(a-a_n)\Omega \end{bmatrix},$$

which is a multiple of the block matrix with (i,j)th block $a_j(a\delta_{ij} - a_i)\Omega$, $i,j = 1, \dots, n$, where δ_{ij} denotes the familiar Kronecker's delta. Equivalently, using the notation $\Delta = \operatorname{diag}(a_1, \dots, a_n)$ and $v = (a_1, \dots, a_n)'$ from (3.8), we obtain

$$\operatorname{Cov}(V_1, \dots, V_n) = \frac{1}{a^2(a+1)(2a-1)}(a\Delta - vv') \otimes \Omega.$$

We record here an interesting observation about the matrix Dirichlet distributions. Since $Y_1 + \cdots + Y_n = I_m$, almost surely, then $y_{1,ij} + y_{2,ij} + \cdots + y_{n,ij} = \delta_{ij}$, almost surely, for all $i,j = 1, \dots, m$. By Gupta and Richards (2000), Theorem 3.2, we know that $(y_{1,ii}, \dots, y_{n-1,ii}) \sim D(a_1, \dots, a_{n-1}; a_n)$, $i = 1, \dots, m$; therefore, the covariance structure of $(y_{1,ii}, \dots, y_{n,ii})$ is given by the foregoing results. For $i \neq j$, the joint distribution of $(y_{1,ij}, \dots, y_{n,ij})$ is extremely complicated; however, its covariance structure is obtained directly from the results presented before; indeed it follows from the foregoing results that its covariance matrix is $(a\Delta - vv')/a(a+1)(2a-1)$, and this is also a constant multiple of (3.8).

In the case of the matrix Liouville distributions, an approach analogous to the one above also applies to derive the covariance structure. To see this, suppose that the $m \times m$ matrices $X_1, \dots, X_n$ follow a Liouville distribution, denoted $(X_1, \dots, X_n) \sim L_n[f(\cdot); a_1, \dots, a_n]$, where f is a real-valued function on the space of positive-definite $m \times m$ symmetric matrices (cf. Gupta and Richards, 1987). Generalizing (2.2), there holds the stochastic representation

$$(X_1, \dots, X_n) =^{\text{st}} R(Y_1, \dots, Y_n)R', \tag{5.15}$$

where R is an upper-triangular matrix with $RR' \sim L_1[f(\cdot); a]$, $(Y_1, \dots, Y_n) \sim D_m(a_1, \dots, a_n)$, and R and $(Y_1, \dots, Y_n)$ are mutually independent (cf. Gupta and Richards, 1987). Using the stochastic representation (5.15) we can determine the covariance structure of $X_1, \dots, X_n$ in terms of the covariance structure of R.

As before, we lose no generality by considering the means and covariances of X_1 and X_2 only. As a consequence of (5.15), we have $X_i =^{\text{st}} RY_iR'$, $i = 1, \dots, n$. Applying the mutual independence of R and $(Y_1, \dots, Y_n)$, and formula (5.6), we obtain

$$E(X_1) = E\big(RY_1R'\big) = E\big(RE(Y_1)R'\big) = \frac{a_1}{a}E(RR').$$

Again applying the stochastic representation (5.15) we find that, for $j = 1, 2$,

$$\begin{aligned} E(X_1 \otimes X_j) &= E\big(RY_1R' \otimes RY_jR'\big) \\ &= E\big((R \otimes R)(Y_1 \otimes Y_j)(R' \otimes R')\big) \\ &= E\big((R \otimes R)E(Y_1 \otimes Y_j)(R \otimes R)'\big). \end{aligned} \tag{5.16}$$

Since $E(Y_1 \otimes Y_1)$ and $E(Y_1 \otimes Y_2)$ already have been determined, we see that both $E(X_1 \otimes X_1)$ and $E(X_1 \otimes X_2)$ have been reduced to expectations in terms of R. Thus,

(5.16) provides a general approach to the evaluation of the covariance structure of many Liouville-distributed random matrices.

An illuminating special case is that in which RR' is the Bartlett decomposition of a Wishart-distributed matrix. Then the non-zero entries of R are mutually independent, the diagonal entries are chi-square distributed up to a constant multiple, and the off-diagonal entries are normally distributed. Then the expectations in (5.16) can all be evaluated explicitly; we leave this exercise to the reader.

Acknowledgments

We are grateful to the referees for their comments on the initial version of this article.

References

[1] Aitchison, J. (1986). *The Statistical Analysis of Compositional Data.* Chapman & Hall, New York.

[2] Artin, E. (1964). *The Gamma Function.* Holt-Rinehart-Winston, New York.

[3] Bandeen-Roche, K. (1994). Resolution of additive mixtures into source components and contributions: A compositional approach. *J. Amer. Statist. Assoc.*, **89**, 1450 – 1458.

[4] Connor, R. J., and Mosimann, J. E. (1969). Concepts of independence for proportions with a generalization of the Dirichlet distribution. *J. Amer. Statist. Assoc.*, **64**, 194 – 206.

[5] Fang, K.-T., Kotz, S., and Ng, K. W. (1990). *Symmetric Multivariate and Related Distributions.* Chapman & Hall, New York.

[6] Gupta, R. D., and Richards, D. St. P. (1987). Multivariate Liouville distributions. *J. Multivariate Analysis*, **23**, 233 – 256.

[7] Gupta, R. D., and Richards, D. St. P. (1991). Multivariate Liouville distributions, II. *Probab. and Math. Statist.*, **12**, 291 – 309.

[8] ______ (1992). Multivariate Liouville distributions, III. *J. Multivariate Analysis*, **43**, 29 – 57.

[9] ______ (1995). Multivariate Liouville distributions, IV. *J. Multivariate Analysis*, **54**, 1 – 17.

[10] ______ (1997). Multivariate Liouville distributions, V. In *Advances in the Theory and Practice of Statistics: A Volume in Honor of Samuel Kotz*, (N. L. Johnson and N. Balakrishnan, eds.), pp. 377 – 396; Wiley, New York.

[11] ______ Moment properties of the multivariate Dirichlet distributions. *J. Multivariate Analysis*, to appear.

[12] Johnson, N. L., and Young, D. H. (1960). Some applications of two approximations to the multinomial distribution. *Biometrika* **47**, 463 – 469.

[13] Mosimann, J. E. (1987). Size and shape analysis. *Encyclopedia of Statistical Sciences*, Vol. 8, (S. Kotz and N. L. Johnson, eds.), pp. 497 – 506. Wiley, New York.

[14] Muirhead, R. J. (1982). *Aspects of Multivariate Statistical Theory.* Wiley, New York.

[15] Rao, C. R. (1973). *Linear Statistical Inference and its Applications*, 2nd. ed. Wiley, New York.

[16] Rayens, W. S., and Srinivasan, C. (1994). Dependence properties of generalized Liouville distributions on the simplex. *J. Amer. Statist. Assoc.* **89**, 1465 – 1470.

Department of Mathematics, Statistics and Computer Science, University of New Brunswick, St. John, New Brunswick, Canada E2L 4L5.

E-mail address: gupta@unbsj.edu

School of Mathematics, Institute for Advanced Study, Princeton, NJ 08540; and Department of Statistics, University of Virginia, Charlottesville, VA 22904, U.S.A.

E-mail address: dsr2j@unix.mail.virginia.edu

Contemporary Mathematics
Volume **287**, 2001

Reduction of Regression Models under Symmetry

Inge S. Helland

ABSTRACT. When regression methods for collinear data are equivariant under the group of rotations in the x-space, it is argued that the regression parameters along the orbits of the group can be estimated in an optimal way by a Pitman-type estimator. When the space is high-dimensional, any reduction in the regression model must therefore take place via the group's orbit index. The solution which emerges from this, is closely related to the population version of the chemometricians' partial least squares regression. Estimation under the reduced model is briefly discussed, as is model reduction in the corresponding classification problem.

1. Introduction

Collinear data in areas such as linear regression or classification, are a major practical problem, with a large number of seemingly unrelated methods, many of them derived by ad hoc arguments. The present paper is an attempt to describe a theory for such problems, a theory whose essence is model reduction in the presence of rotational symmetry invariance. Our point of departure is a rather heuristic statement, which will be made more concrete in the next sections: When the number of parameters in a statistical model is large relative to the number of observations, it may be of interest to switch attention from the original model to a simplified sub-model. Collinearity is indeed very often related to cases in which the number of model parameters is large compared to the number of data points.

Assume that there is a natural symmetry group attached to the model. Then the model parameter can be divided into two components: (i) an orbit index (or cross-section), which is a maximal invariant under the group action; and (ii) a parameter along the orbit. The second component can be estimated in a rather satisfying optimal way, at least in principle, by finding the minimum risk equivariant estimator along the orbit. Hence, the potential for gaining anything via model reduction is related to the orbit index. The reduced model should also be invariant under the chosen group. Again, this can be achieved by making the model reduction through reducing the dimension of the orbit index parameter.

1991 *Mathematics Subject Classification.* Primary 62J99; Secondary 20G99, 62H30, 62H99.
Key words and phrases. Collinearity, group orbits, model reduction, multiple regression, partial least squares regression, Pitman estimator, regression methods, rotational symmetry.

The models that we will consider in this paper are random regression models, e.g., Breiman and Spector (1992). Furthermore, we assume that all the explanatory variables have the same scale, a situation which is not uncommon in chemometry, where the explanatory variables often are light absorptions at different wave lengths (Brown, 1993). Other data with such a structure can be found in situations with space location, psychometric scales, imaging data or topographic models, for example. In particular, then, in these situations the space of the explanatory variables can in a natural way be viewed as an inner product space. Furthermore, it is not unnatural for such cases to focus upon the group of rotations in the space of explanatory variables. An extra argument for the rotation group is that several well-known regression methods for collinear data are equivariant (see, e.g., Lehmann and Casella, 1998) under this group, but not under the full linear group. The use of such methods is very common in chemometry (Næs and Martens, 1985; Martens and Næs, 1989). For the remaining of the paper, the group of matrix rotations will be denoted by G.

After a brief example to motivate the use of model reduction in regression as a general procedure, we argue in Section 3 that common regression methods for collinear data are equivariant under G. In Section 4 the orbits of G are characterized, and in Section 5 the Pitman estimator under G for multiple regression is found. In Section 6 the preceding results are used to reduce the multiple regression model under G-symmetry. A main conclusion of this paper is that this reduced model is identical to the population model of the chemometricians' partial least squares method. This population model is more closely described in Section 7, where Theorem 7.1 gives a precise characterization of the connection. Finally, in Section 8 various approaches to estimation are discussed, and classification is briefly discussed in Section 9.

2. Model reduction in simple regression

Consider first the elementary linear regression model $y_i = \alpha + \beta x_i + \epsilon_i$, $i = 1, \ldots, n$, where the error terms ϵ_i are i.i.d. $N(0, \sigma^2)$, and where we for simplicity assume that the values $x_1, \ldots, x_n$ of the explanatory variable satisfy $\sum x_i = 0$, so that the mean squared prediction error at the fixed point x_0 under the nonrandom explanatory variable model equals

$$P = \frac{\sigma^2}{n} + x_0^2 \frac{\sigma^2}{\sum x_i^2} + \sigma^2. \tag{2.1}$$

The present model has a small number of parameters, but even here model reduction can be considered. The strength of the example is that an explicit solution is easy to find. A natural reduced model here will be one with null slope, leading to the prediction error implied by using the reduced model when data are generated by the full model:

$$P^R = \frac{\sigma^2}{n} + \beta^2 x_0^2 + \sigma^2. \tag{2.2}$$

Comparison with (2.1) shows that in terms of prediction error, model reduction is worthwhile (for all non-zero x_0) if and only if

$$\beta^2 < \frac{\sigma^2}{\sum x_i^2}. \tag{2.3}$$

This condition can also be written as $|t| < 1$, where t is the 'theoretical t-value' $\beta/\text{std}(\hat{\beta})$ with $\text{std}(\hat{\beta}) = \sigma/\sqrt{\sum x_i^2}$. In this form, it can be shown that the condition also can be generalized to the question of deleting a single variable from a multiple regression model, e.g., Snedecor and Cochran (1989).

In this paper we shall concentrate mainly on the random x regression, where the assumption on the error terms is that ϵ_i, given all the x-variables, are i.i.d. $N(0, \sigma^2)$, and we typically may assume that $x_1, \dots, x_n$ are i.i.d. $N(0, \sigma_x^2)$. Thus the assumption $\sum x_i = 0$ is replaced by $\text{E}(x_i) = 0$. Then the new prediction error is found by taking the expectation over the x-variables in (2.1) and (2.2), leading to the following criterion for model reduction:

$$\beta^2(n-2) < \frac{\sigma^2}{\sigma_x^2}. \tag{2.4}$$

This illustrates explicitly the statement made earlier that model reduction may be beneficial when the number of data points is small or moderate. Unfortunately, the criterion depends upon unknown parameters - a general problem in this area.

In Bickel (1984) a general discussion can be found on the problem of estimating under a simple model when wanting to guard against deviations towards a larger model. A general large-sample discussion of model reduction using likelihood theory can be found in Hjort (1991). A number of specific examples of model reduction can also be found in these two papers and in references given there, whereas other specific cases are discussed elsewhere. For instance, a treatment of model reduction in Kalman filter models is given in Desai et al. (1985).

3. Rotational symmetry in multiple random regression

Consider now in general a p-dimensional x-distribution and the corresponding $(p+1)$-dimensional joint (x, y)-distribution with expectation $(\mu_x^t, \mu_y)^t$ and covariance matrix

$$\begin{pmatrix} \Sigma_{xx} & \sigma_{xy} \\ \sigma_{xy}^t & \sigma_y^2 \end{pmatrix}. \tag{3.1}$$

From now on we will usually replace these covariance parameters by the equivalent parameter set $\theta = (\Sigma, \beta, \sigma_y^2)$, where $\Sigma = \Sigma_{xx}$ and $\beta = \Sigma_{xx}^{-1}\sigma_{xy}$. The parameter of interest is β, and we will be interested in prediction problems, assuming a linear conditional expectation $\text{E}(y|x) = \alpha + \beta^t x$ (this includes the multinormal case, and, slightly more generally, the case of elliptical distributions; see references in Helland, 2000a). We also assume a linear predictor $\hat{y}_x = \hat{\alpha} + \hat{\beta}^t x$. From this, considering the expectation of $(\hat{y}_x - y)^2$ over future x's and using $\hat{\alpha} = \bar{y} - \hat{\beta}^t \bar{x}$, a natural loss function (cf. also Helland and Almøy, 1994, Theorem 1) is

$$L(\theta; \hat{\beta}) = (\hat{\beta} - \beta)^t \Sigma (\hat{\beta} - \beta). \tag{3.2}$$

The data from which estimation shall be made consist of n independent observations from this model, summarized in the usual way as (X, y). In this paper we will consider a random regression model where the x-variables have the same scale. For such cases, rotations in the x-space are meaningful transformations of the data. In fact, most of the solutions that have been proposed for prediction in regression models, such as principal component regression, latent root regression, partial least squares regression or continuum regression are equivariant under rotation in the x-space of the above model, meaning that the estimator transforms under the group in the same way as the parameter. Therefore, a natural task might be to try to find the best or nearly best estimator among these equivariant ones. For completeness we review the following definitions, discussed in detail, for example, in Lehmann and Casella (1998).

The rotation group G in question has group elements g which can be linearly represented by the orthogonal $p \times p$ matrices Q with determinant $+1$. The more general linear group acts on the data by $(\mathcal{X}, y) \to (\mathcal{X}Q, y)$, which induces the action $\bar{G}$

$$(\Sigma, \beta, \sigma_y^2) \to (Q^t \Sigma Q, Q^{-1}\beta, \sigma_y^2). \tag{3.3}$$

on the parameter space. For the rotation group, the matrix Q is orthogonal, and one can replace $Q^{-1}\beta$ by $Q^t\beta$ here.

The parametric function $\beta = \beta(\theta)$ is trivially seen to be invariantly estimable (see also Helland, 2000b), $\beta_1 = \beta_2$ implies $Q^{-1}\beta_1 = Q^{-1}\beta_2$ for all Q. An estimator $\hat{\beta}$ is equivariant if we also have $\hat{\beta} \to Q^{-1}\hat{\beta}$ when $(\mathcal{X}, y) \to (\mathcal{X}Q, y)$. For equivariant estimators the loss (3.2) will be invariant. Since the parameter space is closed under the transformations in $\bar{G}$, we have what Lehmann and Casella (1998) call an invariant estimation problem.

To illustrate the concept of equivariance, consider the ridge regression estimator $\hat{\beta}_{ridge} = (\mathcal{X}^t\mathcal{X} + \nu I)^{-1}\mathcal{X}^t y$. Under the transformation $(\mathcal{X}, y) \to (\mathcal{X}Q, y)$ we have $\hat{\beta}_{ridge} \to Q^{-1}\hat{\beta}_{ridge}$ if and only if Q is orthogonal. In a similar way one can show that principal component regression is equivariant only under orthogonal transformations. Thus, restricting ourselves to the rotation group G, i.e., demanding $|Q| = 1$, it is of definite interest if one can find the best regression estimator equivariant under rotation.

It should be noted, however, that G does not act transitively on the parameter space. For transitive actions, the risk (expected loss) will be a constant function of the parameter, which is a strong indication that the problem of finding an equivariant estimator that minimizes the risk uniformly, has a unique solution. Such best equivariant estimators are indeed found quite generally as the Pitman estimator and its generalizations to non-location groups. In the present case, such uniqueness can only be expected when estimating parameters on the orbits of $\bar{G}$. When estimating orbit indices (maximally invariant parameters under $\bar{G}$), other methods must be used. It is only with respect to this last part of the estimation that we can expect any gain from trying to reduce the parameter space when p is large.

4. Non-exactness, orbits and maximal invariants in the parameter space

First we describe those transformations in G which conserve the real, positive definite, matrix Σ. Write

$$\Sigma = \sum_{k=1}^{q} \lambda_k P_k, \tag{4.1}$$

where the P_k are projection matrices upon orthogonal eigenvector spaces V_k (which we will call *strata* in analogy with the use of this term in Nelder, 1965). Here all the λ_k are positive and different, otherwise strata could have been combined. Without loss of generality we can take the λ_k's in decreasing order. Obviously, $q \leq p$. We will assume that Σ has full rank p, which is equivalent to

$$\sum_{k=1}^{q} \dim(V_k) = \sum_{k=1}^{q} \operatorname{tr}(P_k) = p,$$

or again to the requirement that the direct sum of the spaces V_k equals the full p-dimensional Euclidean space R^p. We have:

PROPOSITION 4.1. *For $Q \in G$, the following are equivalent:*

(1) $Q^t \Sigma Q = \Sigma$;
(2) $QP_j = P_j Q$ *for* $j = 1, \dots, q$;
(3) Q *can be written as the direct sum of V_k-stable rotations.*

PROOF. The equivalence of (1) and (2) follows either from a simplified version of the proof of Theorem 4.3 below or directly from the spectral theorem for self-adjoint linear transformations (e.g., Halmos, 1958). Applying (2) to $v \in V_j$ shows that Qv satisfies $P_j Qv = Qv$, so $Qv \in V_j$. Thus the transformation given by Q must conserve all the spaces V_j; hence it is a direct sum of rotations within each single V_j. The opposite implication is trivial. □

Because the rotations satisfying $Q^t \beta = \beta$ are rotations around the axis determined by β, we get:

COROLLARY 4.2. *A transformation $Q \in G$ conserves (Σ, β) if and only if Q is the direct sum of V_k-stable rotations, Q_k, such that for each stratum V_k upon which β has a non-zero component $P_k \beta$, Q_k is a rotation around the axis given by $P_k \beta$.*

Since σ_y^2 is unaffected by the transformations, this corollary characterizes the transformations that conserve $\theta = (\Sigma, \beta, \sigma_y^2)$, i.e., it shows the degree of non-exactness (or rotation symmetry) in the parameter space.

The following theorem gives a characterization of the G-orbits in the parameter space. The proof is given in the Appendix A.

THEOREM 4.3. *Two parameters $\theta_1 = (\Sigma_1, \beta_1, \sigma_{y,1}^2)$ and $\theta_2 = (\Sigma_2, \beta_2, \sigma_{y,2}^2)$ with $\Sigma_r = \sum_{k=1}^{q_r} \lambda_k^{(r)} P_k^{(r)}$, $r = 1, 2$ are on the same orbit of G if and only if:*

(1) $q_1 = q_2 = q$;
(2) $\lambda_k^{(1)} = \lambda_k^{(2)}$, $(k = 1, 2, \dots, q)$;
(3) *There is a rotation matrix Q such that $P_k^{(2)} = Q^t P_k^{(1)} Q$, $(k = 1, \dots, q)$. In particular, the set of dimensions of $\{V_k^{(2)}\}$ must match the dimensions of $\{V_k^{(1)}\}$;*
(4) *The same Q gives $\beta_2 = Q^t \beta_1$, or equivalently, $P_k^{(2)} \beta_2 = Q^t (P_k^{(1)} \beta_1)$, $(k = 1, \dots, q)$;*
(5) $\sigma_{y,2}^2 = \sigma_{y,1}^2$.

From this result we also obtain the maximal invariant of the parameter group, since this always equals the index of the orbits.

COROLLARY 4.4. *The orbit index of the parameter group is given by:*

(1) *The relative orientation of the ordered set of strata, and their corresponding dimensions and eigenvalues λ_k;*

(2) *the norms of the projected regression coefficients* $\gamma_k = ||P_k\beta||$;
(3) σ_y^2.

PROOF. From Theorem 4.3 it is only left to prove that it is enough to characterize the β-dependence of the orbits by the norms γ_i of each stratum component. This can be seen from Proposition 4.1. By that result, the matrix Q of Theorem 4.3(3) has a non-uniqueness corresponding to any set of rotations within some or all of the strata. Using such a rotation within stratum i, we see that the direction of the stratum component of β within each stratum is arbitrary, and only its norm is fixed. □

5. Optimizing on orbits

In this section we will assume multinormality, at least as a useful approximation, for the data (X, y) from n independent units. The estimation of the expectations is then trivial, and by sufficiency, the estimation of the covariance structure can be assumed to depend only on

$$S = S_{xx} = n^{-1}\sum_{i=1}^{n}(x_i - \bar{x})(x_i - \bar{x})^t, \quad s = s_{xy} = n^{-1}\sum_{i=1}^{n}(x_i - \bar{x})(y_i - \bar{y}),$$

and $s_y^2 = n^{-1}\sum_{i=1}^{n}(y_i - \bar{y})^2$. Now fix an orbit in the rotation group in the parameter space. This group is compact, and thus has a unique Haar measure, which can be taken as a prior probability. Then using well-known results generalizing the Pitman estimator (see, for instance Helland, 2000b; the result can also be derived from Berger, 1985, Result 3, p. 410, or from Eaton, 1989, Theorem 6.4) and using the loss function (3.2), the minimum risk estimator, given the orbit, will be

$$\hat{\beta}_O = \mathrm{E}_{post}(\Sigma_{xx})^{-1}\mathrm{E}_{post}(\sigma_{xy}) = \mathrm{E}_{post}(\Sigma)^{-1}\mathrm{E}_{post}(\Sigma\beta), \tag{5.1}$$

where E_{post} is the corresponding posterior expectation. Because $\beta^t\Sigma\beta = \sum_k \lambda_k\gamma_k^2$ is a constant of the orbit, then $\mathrm{E}_{post}(\beta^t\Sigma\beta) = \beta^t\Sigma\beta$, and, consequently, the minimal risk (given orbit) will then be

$$R_{min} = \beta^t\Sigma\beta - \mathrm{E}_{post}(\Sigma\beta)^t\mathrm{E}_{post}(\Sigma)^{-1}\mathrm{E}_{post}(\Sigma\beta). \tag{5.2}$$

As usual, the posterior density is proportional to the product of the likelihood and the prior. Using a known form for the multinormal likelihood, found by first taking the likelihood of X, and then multiplying by the conditional likelihood of y, given X, we find:

THEOREM 5.1. *Fix an orbit of the rotation group in the above situation. Then the minimum risk equivariant estimator* $\hat{\beta}_O$ *for* β*, given the orbit, has the form*

$$\hat{\beta}_O = D^{-1}N \tag{5.3}$$

with

$$N = \int Q^t\Sigma\beta\exp[\frac{n}{2\sigma^2}(2\beta^tQs - \beta^tQSQ^t\beta) - \frac{n}{2}\mathrm{tr}(Q^t\Sigma^{-1}QS)]d\gamma(Q) \tag{5.4}$$

and

$$D = \int Q^t\Sigma Q\exp[\frac{n}{2\sigma^2}(2\beta^tQs - \beta^tQSQ^t\beta) - \frac{n}{2}\mathrm{tr}(Q^t\Sigma^{-1}QS)]d\gamma(Q), \tag{5.5}$$

where (Σ, β) *correspond to a fixed point on the orbit and* $d\gamma(Q)$ *is Haar measure for* G. *Also,* $\sigma^2 = \sigma_y^2 - \beta^t \Sigma \beta = \sigma_y^2 - \sum_k \lambda_k \gamma_k^2$ *is the residual variance, a constant of the orbit.*

Note that the orbit-estimator of β from Theorem 5.1 does not depend on the whole orbit index as given in Corollary 2. The following Proposition characterizes the minimal set of parameters upon which $\hat{\beta}_O$ depends, more precisely, the largest such set as the data varies.

PROPOSITION 5.2. *The parameter dependence of (5.3)-(5.5) is given by*

(1) σ^2;
(2) *The number* m *of strata such that* $\gamma_k \neq 0$;
(3) *For given* σ^2 *and* m, *a vector function* $\hat{\beta}_O((\lambda_1, \gamma_1), \dots, (\lambda_m, \gamma_m); \text{data})$ *which for all data-values is symmetric under any permutation of the* m *arguments.*

PROOF. For fixed σ^2, the denominator D in $\hat{\beta}_O$ depends upon

$$Q^t \Sigma Q = \sum_i \lambda_i (Q^t e_i)(Q^t e_i)^t, \quad Q^t \beta = \sum_i \beta_i (Q^t e_i),$$

and on

$$\text{tr}(Q^t \Sigma^{-1} Q S) = \sum_i \frac{1}{\lambda_i} (Q^t e_i)^t S (Q^t e_i),$$

where $\{e_i\}$ is a set of eigenvectors for Σ and $\beta_i = \beta^t e_i$. When the set $\{Q^t e_i, i = 1, \dots, p\}$ is given a uniform distribution under rotation, the expectation of any such function will depend upon $\{(\lambda_i, \beta_i), i = 1, \dots, p\}$ in a symmetric way. One particular such symmetric function is the number q of strata. So assume, say, that $V_1 = \text{span}(e_1, \dots, e_t)$ is a stratum with constant λ_1. Then, in particular, the function D, expression 5.4, should be symmetric under any rotation within this stratum. This means that it can be regarded as a function of $((\lambda_1, \gamma_1), (\lambda_1, 0), \dots, (\lambda_1, 0))$ together with the rest of the parameters, or equivalently a function of (λ_1, γ_1) and these parameters. Hence an alternative, and more convenient way to state the result above, is that, given σ^2 and q, we have that D is a symmetric function of $\{(\lambda_k, \gamma_k), k = 1, \dots, q\}$.

A similar analysis can be made for the numerator N in expression 5.5. In addition to the stratum number q, the function $m =$ (number of strata such that $\gamma_k \neq 0$) is also a symmetric function of $\{(\lambda_i, \beta_i), i = 1, \dots, p\}$ (obviously, $m \leq q$). A closer look at (5.3)-(5.5) shows that, without loss of generality, we can let Σ in this formula be diagonal with eigenvalues on the diagonal. Hence any contribution from the strata with $\gamma_k = 0$ will vanish, and consequently we can take $\hat{\beta}_O$ as the symmetric function of the m bivariate orbit parameters as stated. □

COROLLARY 5.3. *The minimal risk given the orbit,* R_{min}, *depends upon* σ^2 *and* m, *and given these parameters it is a function of* $((\lambda_1, \gamma_1), \dots, (\lambda_m, \gamma_m))$ *which is symmetric under any permutation of the* m *arguments.*

The fact remains that the estimator $\hat{\beta}_O$ depends upon orbit indices, which are unknown parameters. The estimation of these parameters will be discussed in Section 8.

6. Model reduction

Fix $\theta_0 \in \Theta$. We are interested in the effect of reducing the statistical model by restricting θ to some set $\Gamma \subset \Theta$ such that $\theta_0 \in \Gamma$. Because an optimal equivariant estimator can be found quite generally on the orbit of G containing θ_0. Thus we can without loss of generality let Γ contain this orbit. This gives the minimal useful choice of Γ. In practice, Γ should be chosen as a set Φ in the orbit index space of G taken together with the corresponding orbits. Our task is to say something about the choice of Φ.

PROPOSITION 6.1. *Assume the multiple regression model, and let G be the group of rotations in the parameter space. Let β be estimated as in Theorem 5.1 for each fixed orbit. Then the only model reductions that can affect the estimate are given by combinations of some or all of the following possibilities:*

(1) *fix the residual variance σ^2,*
(2) *fix the number m of relevant strata (those with $\gamma_k \neq 0$),*
(3) *fix some specific symmetric function of $((\lambda_1, \gamma_1), \dots, (\lambda_m, \gamma_m))$.*

PROOF. This follows immediately from Proposition 5.2. □

Looking at these three possibilities, the first one seems not very attractive from an applied point of view; it will hardly lead to any decrease in the prediction error. Also, the last type of model reduction may be an option when one has particular model information, but it seems very difficult to specify any such function which can serve as an input to a general regression method. Furthermore, this reduction method only makes sense in combination with the second method. This leaves one in general with the option of fixing the number m at some value $m < p$ as definitely the most natural approach to model reduction from the present point of view.

It is interesting, then, that a practical regression method closely related to this particular choice has been used by chemometricians by decades now in a large spectrum of applications, namely partial least squares regression. The method is claimed to have great success, empirically, and it has received a large amount of attention. However, a precise motivation for the method via statistical modelling has up to now been difficult to find. A special issue of the journal Chemometrics and Intelligent Laboratory Systems devoted to partial least squares regression is planned to appear shortly, and two international conferences solely devoted to this method have been arranged.

7. Partial least squares regression in population form

Partial least squares regression (PLS) was proposed as a method to solve collinearity problems in multiple regression and calibration by Wold et al. (1983), and has gained an increasing popularity, especially among chemometricians, see Martens and Næs (1989), and also the forthcoming special journal issue mentioned above. The method was proposed as an algorithm, in fact, several algorithms were available; two of them are shown to be equivalent in Helland (1988). In the present context the corresponding population algorithm, treated in Helland (1990), is of special interest. It can be formulated as follows, taking the $p+1$ dimensional variable (x, y) as the point of departure. Replacing variables by data vectors and (co)variances by their sample counterparts here, leads to the ordinary PLS algorithm.

(1) Initialization: $e_0 = x - \mu_x$, $f_0 = y - \mu_y$.
(2) Next, do for $a = 1, 2, \ldots$:
   (a) Weights and scores: $w_a = \text{Cov}(e_{a-1}, f_{a-1})$, $t_a = e_{a-1}^t w_a$;
   (b) Loadings by least squares:

$$p_a = \text{Cov}(e_{a-1}, t_a)/\text{Var}(t_a), \quad q_a = \text{Cov}(f_{a-1}, t_a)/\text{Var}(t_a).$$

   (c) New residuals: $e_a = e_{a-1} - p_a t_a$, $f_a = f_{a-1} - q_a t_a$.

Stopping at step $a = k$, this leads to the bilinear representation

$$x = \mu_x + p_1 t_1 + \ldots + p_k t_k + e_k, \quad y = \mu_y + q_1 t_1 + \ldots + q_k t_k + f_k, \tag{7.1}$$

where loadings and scores are determined by the algorithm. The corresponding sample equation is much used in the chemometric literature (see, e.g., Martens and Næs, 1989) in the same way as statisticians use factor analysis, and it is often claimed that the corresponding plots are useful for interpreting complex data sets.

Rather than interpretation, our main theme in the present paper is prediction at some set of x-values x^0. Then the natural procedure is to use the partial least squares algorithm to construct new scores t_a^0 from x^0 and, in analogy with the last equation in (7.1), take

$$\hat{y}_{k,PLS} = \mu_y + q_1 t_1^0 + \ldots + q_k t_k^0. \tag{7.2}$$

One can then show that the above predictor can be written as

$$\hat{y}_{k,PLS} = \mu_y + \beta_{k,PLS}^t (x^0 - \mu_x), \tag{7.3}$$

with

$$\beta_{k,PLS} = W_k (W_k^t \Sigma W_k)^{-1} W_k^t \sigma, \tag{7.4}$$

where σ here is σ_{xy} and $W_k = [w_1, \ldots, w_k]$ is defined iteratively by $w_1 = \sigma$ and $w_{k+1} = \sigma - \Sigma W_k (W_k^t \Sigma W_k)^{-1} W_k^t \sigma$. (The corresponding sample case is proved in Helland (1988), Theorem 3.1.) The formula (7.4) indicates that the space $\mathcal{S}_k$ spanned by the columns of W_k is of some interest.

Further progress can be made if we note that this space can also be spanned by a so called Krylov sequence, namely as long as w_k is nonzero, an alternative basis for $\mathcal{S}_k$ is given by the vectors

$$\sigma, \Sigma\sigma, \ldots, \Sigma^{k-1}\sigma.$$

The corresponding sample statement is proved in Helland (1988), Proposition 3.1.

Here are some algebraic conclusions from these results (for more details, see Helland, 1990): First, if there should be some value of k such that the corresponding weight vanishes, i.e., $w_k = 0$, then the PLS algorithm will be terminated at this value (and only at such values). Secondly, the first value $k = m$ such that $w_k = 0$ for $k > m$ is characterized by the property that the k vectors of the Krylov sequence will span the same m-dimensional space $\mathcal{S}_m$ also when $k > m$. If this happens, we will say that the population PLS algorithm stops at step m. Of course, the algorithm will always stop at $m = p$; the interesting case is when it stops before that. Note that by the results cited above, the regression vector $\beta_{k,PLS}$ will always belong to the space $\mathcal{S}_m$.

From these observations there is in fact a short step to the main result of this section, which gives a fairly precise interpretation of the result of the population PLS algorithm in the light of model reduction and rotational symmetry.

THEOREM 7.1. *The population PLS algorithm stops at step m if and only if the number of strata in the population model with non-zero regression coefficient is equal to m. The resulting PLS regression vector $\beta_{m,PLS}$ will then equal $\beta = \Sigma^{-1}\sigma$.*

OUTLINE OF PROOF. From (4.1) a linear dependency in the Krylov sequence is given by

$$\sum_{j=1}^{k} c_j \Sigma^{j-1}\sigma = \sum_{h=1}^{q} \{\sum_{j=1}^{k} c_j \lambda_h^{j-1}\} P_h \sigma = 0.$$

This can happen if and only if $k > m$, where now m is the number of strata with $P_h\sigma$ nonzero. Note that for $k = m$ there is a Vandermonde determinant among the different λ_h involved in this argument. □

Thus with this natural specification of the population PLS algorithm, there is a close connection to the optimality results discussed in Section 5 and Section 6 above. Briefly, the termination of this algorithm at $m(< p)$ steps is equivalent to a reduced model of the type which results in a natural way from the orbit behavior of the optimal equivariant regressor under rotation invariance. Moreover, the regression vector that results from the corresponding population PLS algorithm is what it should be.

In practice, chemometricians use the sample PLS algorithm (cf. Martens and Næs, 1989 and Helland, 2000c), and nothing can be said about this from the above discussion, except a vague indication that it will be sensible if n is not too small and a good stopping rule is used. The stopping rule used in practice for PLS regression is based on cross validation using prediction performance, and has only an indirect link to the population stopping criterion mentioned above: If $w_k = 0$, then one should expect that the prediction at step $k - 1$ is good enough to affect the cross-validation.

The assessment of the sample PLS algorithm is still rather varied. Many practitioners are satisfied with the fact that it gives some automatic procedure which empirically seems to give good results in many cases, and where they also have a possibility to combine prediction will some latent variable analysis. Chemometricians have generalized the algorithm to cases with several y-variables and even to cases with many blocks of variables. There are also attempts to provide a theoretical framework for these methods, see, e.g. Lorber et al (1987). On the other hand, sceptical remarks about the procedure have been made, for instance, in Frank and Friedman (1993). Butler and Denham (2000) have recently shown that the sample PLS procedure, while shrinking globally, has minimal shrinking properties at each factor that it seems impossible that it can satisfy any reasonable optimality property.

To summarize, we have a population procedure which is equivalent to a model reduction with reasonably well-defined good properties related to rotation invariance. We have integral formulae for the optimal estimated regression vector given the orbit indices. In addition to finding practical ways of calculating that integral - where MCMC and related techniques probably may help - the remaining challenge is to find good estimates of the remaining orbit index parameters. A closer discussion of the sample and population PLS algorithms can be found in Helland (2000c).

8. Estimation of orbit parameters

In the formulae (5.3)-(5.5), the point $(\Sigma, \beta, \sigma_y^2)$ is a fixed point on some orbit of the rotation group in the parameter space. But by Proposition 5.2 we may, without loss of generality, let Σ be diagonal with first diagonal elements $\lambda_1, \ldots, \lambda_m$ here, and then take $\beta = (\gamma_1, \ldots, \gamma_m, 0, 0, \ldots, 0)^t$. In addition to its data dependence, the solution will then depend upon m, these parameters and σ^2. This means that we are left with $2m+1$ parameters to estimate. This is in general a very small number compared to the $(p+1)(p+2)/2$ parameters in the original covariance structure. Note, however, that any value of the parameter $(\Sigma, \beta, \sigma_y^2)$ which belongs to the same orbit (or more generally, to orbits which can be linked to that one through the parameter dependence corresponding to the one described in Proposition 5.2) will give the same estimate (5.3)-(5.5).

Taking maximum likelihood under multinormality as a natural approach to the estimation of these parameters, the last remark becomes useful when it is noted that, if $\hat{\theta}$ is a unique maximum likelihood estimator for θ, then, for any f which is constant on a subset Γ of the parameter space Θ and one-to-one on $\Theta\backslash\Gamma$, we have that $f(\hat{\theta})$ is the unique maximum likelihood estimator for $f(\theta)$. Thus, to find maximum likelihood estimators of the $2m+1$ parameters described above, it is enough to give maximum likelihood estimators of the covariance parameters under the hypothesis that the number of strata in Σ with non-zero regression component ('number of relevant components', cf. Næs and Helland, 1993) is equal to m. This latter problem was discussed in Helland (1992). The maximum likelihood solution found there, can be formulated as follows:

(1) With $A = S_{xx} - s_y^{-2} s_{xy} s_{xy}^t$ and $B = S_{xx}^{-1}$, find a $p \times m$ matrix $R = \hat{R}$ which minimizes

$$K(R) = (|R^t AR|/|R^t R|) \cdot (|R^t BR|/|R^t R|);$$

(2) With $\hat{R}$ as in (a), and taking $\hat{P}_R = \hat{R}(\hat{R}^t\hat{R})^{-1}\hat{R}^t$, the maximum likelihood estimates under the hypothesis of m relevant components are given by

$$\hat{\Sigma} = \hat{P}_R S_{xx} \hat{P}_R + (I - \hat{P}_R) S_{xx} (I - \hat{P}_R), \quad \hat{\beta} = \hat{R}(\hat{R}^t S_{xx} \hat{R})^{-1} \hat{R}^t s_{xy},$$

and $\hat{\sigma}_y^2 = s_y^2$. Note that $K(R)$ will be the same here if R is replaced by another matrix spanning the same space, similarly for the estimators. Hence what in effect is estimated, is an $m-$dimensional space, the space of relevant components.

In Helland (1992), an approximate algorithm for doing the minimization in (a) was proposed, but the prediction method based upon this algorithm performed rather poorly in simulations done by Almøy (1996). A reasonable conjecture is that performance will be better when combined with the integral (5.3), since the effective number of parameters estimated by the maximum likelihood part then will be reduced considerably, as discussed above.

There are many questions concerning estimation and prediction under the reduced regression model which need further investigations. In addition to questions related to performance under the reduced model itself, the robustness questions in the spirit of the paper by Bickel (1984) will need to be addressed. In addition to addressing optimality problems of the kind discussed above, it is also of interest to further investigate the performance of the ordinary sample partial least squares algorithm, since this is the algorithm used by chemometricians today. Asymptotic expansions, as in Helland and Almøy (1994) is one of several ways of approaching such questions.

9. Classification

The sample partial least squares algorithm is also used by chemometricians outside the regression context, for instance when doing classification. A large number of applications of this can be found in the chemometrical literature, and the technique has also spread outside the field chemometrics, for instance to psychiatry (Gottfries et al, 1995). It is obvious that techniques of this nature meet a latent need among researchers with near collinear data, both in regression problems and in classification problems.

In this Section we will discuss model reduction under rotational symmetry in the classification problem using ordinary discriminant analysis as the classification method, and point at similarities with and differences from the regression situation. We will concentrate on the simplest possible case: Two-class discriminant analysis with equal prior for the two classes, equal covariance matrix for the observations, equal cost of misclassification and equal number n of observations in the two classes. We have in mind a situation where the number p of classification variables is rather large, also compared to n. We also assume a multinormal model, the standard point of departure in discriminant analysis: In sample i we have n independent observations which are $N(\mu_i, \Sigma)$, and having mean $\bar{x}_i$, $i = 1, 2$. The standard way of classifying a new observation x is to allocate it to class 1 if and only if

$$x^t S^{-1}(\bar{x}_1 - \bar{x}_2) - \frac{1}{2}(\bar{x}_1 + \bar{x}_2)^t S^{-1}(\bar{x}_1 - \bar{x}_2) > 0,$$

where S is the pooled sample covariance matrix from the $2n$ observations.

When p is large, or if we have other sources of collinearity, we may want to replace $S^{-1}(\bar{x}_1 - \bar{x}_2)$ by another observation vector b. Thus x is classified to be of class 1 if and only if

$$[x - \frac{1}{2}(\bar{x}_1 + \bar{x}_2)]^t b > 0,$$

and the probability of misclassification, given the samples, is given by

$$\frac{1}{2}\Phi\left(\frac{(\mu_2 - \frac{1}{2}(\bar{x}_1 + \bar{x}_2))^t b}{\sqrt{b^t \Sigma b}}\right) + \frac{1}{2}\Phi\left(\frac{(\frac{1}{2}(\bar{x}_1 + \bar{x}_2) - \mu_1)^t b}{\sqrt{b^t \Sigma b}}\right),$$

where Φ is the standard normal cumulative distribution function.

For simplicity of notation assume that $\mu_1 = -\mu_2 = \mu$. Also, it is natural to assume that $\frac{1}{2}(\bar{x}_1 + \bar{x}_2)$ is independent of b, which is true for the original choice of b. Integrating over the probability distribution of this variable, using the fact that $\mathrm{E}(\Phi(\nu - u)) = \Phi(\nu/\sqrt{1+\tau^2})$ when $u \sim N(0, \tau)$, we arrive at a natural loss function

$$L(\theta, b) = \Phi\left(-\frac{\mu^t b}{\sqrt{b^t \Sigma b}}\sqrt{\frac{2n}{2n+1}}\right), \tag{9.1}$$

for the classification problem, where $\theta = (\Sigma, \mu)$. With methodology similar to that of Section 4, nearly everything can now be repeated here. In particular, the decomposition (4.1) of Σ will hold and will have the same interpretation. Under a rotation determined by the orthogonal matrix Q we have

$$(\Sigma, \mu) \to (Q^t \Sigma Q, Q^t \mu), \quad (x, b) \to (Q^t x, Q^t b).$$

The loss function L is invariant under the rotation group, and the group acts non-transitively on the parameter space $\{(\Sigma, \mu)\}$ with orbit index given by the ordered set of strata (their relative orientation, their dimensions and the corresponding eigenvalues λ_k) and by the norm of the projected components $\gamma_k = \|P_k\mu\|$. This is proved in the same way as Corollary 4.2.

As a function of b, the loss function (9.1) takes its minimal value at all vectors of the form $\kappa\Sigma^{-1}\mu$ where $\kappa > 0$. A general feature is that $L(\theta, b)$ is independent of the norm of b. Similarly, there is an invariance of L as a function of θ under scale transformations $(\Sigma, \mu) \to (a^2\Sigma, a\mu)$. This implies that the estimation problem as such - regarding b as some estimator - is invariant under a group larger than the rotation group. A scale change in b does not change the classification rule, however.

In general, consider a loss function L of the form $f(b^t\mu\mu^t b/b^t\Sigma b)$, where f is an increasing function. With b as an estimator, this formally gives an estimation problem which is invariant under the rotation group. In this general problem we cannot obtain an explicit formula for the Pitman type estimator, in the way we did in Theorem 5.1. The estimator can be found, in principle, by minimizing

$$\int f(b^tQ^t\mu\mu^tQb/b^tQ^t\Sigma Qb)l(Q, \Sigma, \mu, \{x_{1i}\}, \{x_{2i}\})d\gamma(Q), \tag{9.2}$$

where

$$l = \exp\{-\frac{1}{2}\mathrm{tr}(Q^t\Sigma Q)^{-1}[\sum_i(x_{1i} - Q^t\mu)(x_{1i} - Q^t\mu)^t + \sum_i(x_{2i} + Q\mu)(x_{2i} + Q\mu)^t]\},$$

and $d\gamma(\cdot)$ is Haar measure for the rotation group.

Assuming that f is differentiable (which it is in the concrete application we have in mind), we find that the equation from which (the direction of) b can be determined takes the form:

$$\int f'(r(\cdot))\frac{l(\cdot)}{b^tQ^t\Sigma Qb}Q^t[\mu\mu^t - r(\cdot)\Sigma]Qb\, d\gamma(Q) = 0, \tag{9.3}$$

with

$$r(\cdot) = \frac{b^tQ^t\mu\mu^tQb}{b^tQ^t\Sigma Qb}.$$

Similarly to the proof of Proposition 5.2, we have that every parameter dependence is through $Q^t\Sigma Q = \sum_i \lambda_i(Q^te_i)(Q^te_i)^t$ and $Q^t\mu = \sum_i \mu_i(Q^te_i)$, where e_i are the eigenvectors of Σ and μ_i are the corresponding components of μ. Also, when μ has a vanishing component for some stratum, it follows from equation (9.3) that the corresponding component of b must vanish, and by further inspection: The solution can not depend upon the eigenvalue λ_i of that stratum. Hence the argument of the proof of Proposition 5.2 and the model reduction argument of Section 6 can be repeated in verbatim for the present situation, and we conclude:

The only rotational equivariant model reduction that can affect a standard discrimination rule, is given by:

(1) Fix the number m of relevant component, i.e., the number of strata of Σ with nonzero component γ_i of μ;
(2) Possibly also: Fix some symmetric function of $((\lambda_1, \gamma_1, \ldots, (\lambda_m, \gamma_m))$, where λ_j now is the eigenvalue corresponding to stratum j.

The most natural general procedure, having no particular information about the problem, is to skip step (2) here. Similarly to Section 7 above, this leads to the population version of the partial least squares algorithm for the discriminant analysis problem. Again, of course, the sample algorithm implied by the above argument is different than the sample PLS-algorithm. A necessary part of this procedure will be to solve equation (9.3) for each dimension. It is an open question if such a procedure can be made practical.

Appendix A. Proof of Proposition 4.3

Assume first that $\theta_1 = (\Sigma_1, \beta_1, \sigma^2_{y,1})$ and $\theta_2 = (\Sigma_2, \beta_2, \sigma^2_{y,2})$ are on the same orbit. Then there is a rotation Q such that $Q^t\beta_1 = \beta_2$ and $Q^t\Sigma_1 Q = \Sigma_2$, that is, $\Sigma_1 Q = Q\Sigma_2$, or

$$\sum_{i=1}^{q_1} \lambda_i^{(1)} P_i^{(1)} Q = \sum_{i=1}^{q_2} \lambda_i^{(2)} Q P_i^{2)}.$$

This implies

$$\Sigma_1 Q P_j^{(2)} = \lambda_j^{(2)} Q P_j^{(2)} \text{ and } P_k^{(1)} Q \Sigma_2 = \lambda_k^{(1)} P_k^{(1)} Q.$$

Multiplying the first equation here from the left by $P_k^{(1)}$ and the second equation from the right by $P_j^{(2)}$ and using again $\Sigma_1 Q = Q\Sigma_2$, we find

$$\lambda_j^{(2)} P_k^{(1)} Q P_j^{(2)} = \lambda_j^{(1)} P_k^{(1)} Q P_j^{(2)}.$$

It follows from this that either $\lambda_j^{(2)} = \lambda_k^{(1)}$ or $P_k^{(1)} Q P_j^{(2)} = 0$. Now fix j. If $P_k^{(1)} Q P_j = 0$ for all k, it follows from $I = \sum_{k=1}^{q_2} P_k^1$ that $QP_j^{(2)} = 0$, hence $P_j^{(2)} = 0$, which is impossible. Therefore there must be at least one k such that $\lambda_k^{(1)} = \lambda_j^{(2)}$. Furthermore, there can not be more than one such k, since the $\lambda_k^{(1)}$ are assumed to be different. Hence, if the λ's are ordered according to size, it follows that $\lambda_1^{(1)} = \lambda_1^{(2)}, \ldots, \lambda_q^{(1)} = \lambda_q^{(2)}$ and $q_1 = q_2 = q$.

Finally, $P_k^{(1)} Q P_j^{(2)} = 0$ $(j \neq k)$ implies:

$$P_k^{(1)} Q \sum_{j \neq k} P_j^{(2)} = P_k^{(1)} Q(I - P_k^{(2)}) = 0, \text{ hence } P_k^{(1)} Q = P_k^{(1)} Q P_k^{(2)},$$

$$\sum_{k \neq j} P_k^{(1)} Q P_j^{(2)} = (I - P_j^{(1)}) Q P_j^{(2)}) = 0, \text{ hence } Q P_j^{(2)} = P_j^{(1)} Q P_j^{(2)}.$$

Putting $j = k$ one gets $P_k^{(1)} Q = Q P_k^{(2)}$, that is $P_k^{(2)} = Q^t P_k^{(1)} Q$ $(k = 1, \ldots, q)$. The opposite implication is trivial.

Acknowledgments

First of all I want to thank the editors for the opportunity of writing this paper. Also, I am grateful to three referees for correcting my formulations not only when they were directly wrong, but also when they were unclear and inadequate.

References

[1] Almøy, T. (1996) A simulation study on comparison of prediction methods when only a few components are relevant. *Computational Statistics & Data Analysis* **21**, 87-107.

[2] Berger, J.O. (1985) *Statistical Decision Theory and Bayesian Analysis*, Springer-Verlag, New York. Bickel, P.J. (1984) Parametric robustness: Small biases can be worthwhile. *Ann. Statistics* **12**, 864-879.

[3] Breiman, L. and P. Spector (1992) Submodel selection and evaluation in regression. The x-random case. *Intern. Statist. Rev.* **60**, 291-319.

[4] Brown, P.J. (1993) *Measurement, Regression, and Calibration.* Claredon Press, Oxford.

[5] Butler, N.A. and M.C. Denham (2000) The peculiar shrinkage properties of partial least squares regression. *J. Royal Statist. Soc.* B **62**, 585-593.

[6] Desai, U.B., P. Debajyoti and R.D. Kirkpatrick (1985) *Int. J. Control* **42**, 821-838.

[7] Eaton, M.L. (1989) *Group Invariance Applications in Statistics* IMS-ASA Regional Conference Series in Probability and Statistics Volume 1.

[8] Frank, I.E. and J.H. Friedman (1993) A statistical view of some chemometric tools. *Technometrics* **35**, 109-135.

[9] Gottfries, J., K. Blennow, A. Wallin and C.G. Gottfries (1995) Diagnosis of dementia using partial least squares discriminant analysis. *Dementia* **6**, 83-88.

[10] Halmos, P.S. (1958) *Finite-Dimensional Vector Spaces* D. Van Nostrand, Princeton.

[11] Helland, I.S. (1988) On the structure of partial least squares regression. *Commun. Statist.-Simula.* **17**, 581-607.

[12] ——— (1990) Partial least squares regression and statistical models. *Scand. J. Statist.* **17**, 97-114.

[13] ——— (1992) Maximum likelihood regression on relevant components. *J. R. Statist. Soc.* B **54**, 637-647.

[14] ——— (2000a) Model reduction for prediction in regression models. *Scand. J. Statist.* **27**, 1-20.

[15] ——— (2000b) Statistical inference under a fixed symmetry group. Submitted.

[16] ——— (2000c) Some theoretical aspects of partial least squares regression. To appear in Chemometrics and Intelligent Laboratory Systems.

[17] Helland, I.S. and T. Almøy (1994) Comparison of prediction methods when only a few components are relevant. *J. Amer. Statist. Ass.* **89**, 583-591.

[18] Hjort, N.L. (1991) Estimation in moderately misspecified models. *Statistical Research Report* No. 8/1991, Department of Mathematics, University of Oslo.

[19] Lehmann, E.L. and G. Casella (1998) *Theory of Point Estimation.* Springer, New York.

[20] Lorber, A., L.E. Wangen and B.R. Kowalski (1987) A theoretical foundation for the PLS algorithm. *J. Chemometrics* **1**, 19-31.

[21] Martens, H. and T. Næs (1989) *Multivariate Calibration.* Wiley, New York.

[22] Nelder, J.A. (1965) The analysis of randomized experiments with orthogonal block structure. I. Block structure and null analysis of variance. *Proc. Royal Soc.* A **283**, 147-162.

[23] Næs, T. and I.S. Helland (1993) Relevant components in regression. *Scand. J. Statist.* **20**, 239-250.

[24] Næs, T. and H. Martens (1985) Comparison of prediction methods for collinear data. *Commun. Statist. Simulat. Comput.* **14**, 545-576.

[25] Snedecor, G.W. and W.G. Cochran (1989) *Statistical Methods.* 8th ed. Iowa State University Press, Ames.

[26] Wold, S., H. Martens and H. Wold (1983). The multivariate calibration problem in chemistry solved by the PLS method. *Proc. Conf. Matrix Pencils* (A. Ruhe, B. Kågstrøm, eds.) Lecture Notes in Mathematics, Springer Verlag, Heidelberg, 286-293.

Department of Mathematics, University of Oslo, P.O.Box 1053 Blindern, N-0316 Oslo, Norway.

E-mail address: `ingeh@math.uio.no`

Contemporary Mathematics
Volume **287**, 2001

Deconvolution Density Estimation on Compact Lie Groups

Peter T. Kim and Donald St. P. Richards

ABSTRACT. Let Y be a random variable taking values in G, a compact connected Lie group. Suppose that Y is of the form $Y = X\epsilon$ where $X \in G$ denotes a true random measurement and $\epsilon \in G$ is an independent random error (or noise). We consider the statistical inverse problem of deconvolution density estimation, viz., estimation of the density function of X from a random sample $Y_1, \dots, Y_n$ from Y. By defining the smoothness of the (distribution of the) error variable ϵ through conditions on the operator norm of the density function of ϵ, we obtain optimality results for the deconvolution problem under certain smoothness assumptions on ϵ. For "smooth" error distributions, we prove that the optimal rate of convergence is polynomial, and for "super-smooth" error distributions we show that the optimal rate of convergence is logarithmic. Motivated by the von Mises-Fisher matrix distribution on SO(N), the group of $N \times N$ proper orthogonal matrices, we define a new smoothness class of error distributions on G called "log-super-smooth." We show that this new class is smoother than the class of super-smooth error distributions, and we also derive the corresponding optimal rate of convergence.

1. Introduction

In this paper, we consider the deconvolution problem for probability density functions on a compact connected Lie group G. This problem may be described as follows. Let Y be a random observation, taking values in G, which is composed of a true random measurement X and an independent random noise (or error) ϵ. We consider the model

$$Y = X\epsilon, \tag{1.1}$$

where $X, \epsilon \in G$, and the multiplication in (1.1) is the natural product on G. We assume that the random variables Y, X and ϵ are absolutely continuous with respect to the normalized Haar measure on G; and we denote the corresponding density

---

1991 *Mathematics Subject Classification.* Primary 62G05, 62G07; Secondary 43A75, 58G25.

Key words and phrases. Clebsch-Gordan formula, Fano's lemma, Fourier analysis on groups, irreducible characters, irreducible representations, Koo's lemma, positive roots, Sobolev class, weights, Weyl's formula.

Research supported in part by NSERC grant OGP46204 (P. T. Kim), and by a grant from the Bell Fund to the Institute for Advanced Study, and by NSF grant DMS-9703705 (D. St. P. Richards).

functions by f_Y, f_X and f_ϵ, respectively. By (1.1) it then follows that f_Y, the density of Y, is given by the convolution $f_Y = f_X * f_\epsilon$ where

$$(f_X * f_\epsilon)(u) = \int_G f_X(g^{-1}u) f_\epsilon(g)\, dg, \tag{1.2}$$

$u \in G$, and where dg denotes the Haar measure on G, normalized to have total volume 1.

Given $Y_1, \ldots, Y_n$, a random sample of n mutually independent observations on the variable Y, each having the density function f_Y, the deconvolution problem is to estimate the density function f_X, based on the random sample $Y_1, \ldots, Y_n$. For identifiability reasons, we assume that f_ϵ is known.

The deconvolution problem is an example of an inverse estimation problem in which we attempt to estimate a "signal" f_X indirectly in the presence of a random "noise" ϵ. Here, we are unable to observe the true signal f_X, and instead we can observe only the transformed signal f_Y, which is related to f_X and f_ϵ by the convolution integral (1.2). On the basis of the random observations $Y_1, \ldots, Y_n$, the inverse problem is to estimate the signal f_X.

Recently, there has developed much interest in deconvolution and related density estimation problems on spaces possessing a Riemannian structure. Hendriks (1990), van Rooij and Ruymgaart (1991), Healy and Kim (1996), Healy, Hendriks and Kim (1998), and Kim (1998) considered these problems, studying deconvolution problems on spheres, orthogonal groups and Stiefel manifolds. The work of those authors constitute extensions, to a non-Euclidean setting, of results developed in the Euclidean case (cf., Zhang, 1990; Fan, 1991; Diggle and Hall, 1993; and the references contained in those articles). In both the Euclidean and Riemannian settings, all of the references cited here apply Fourier analytic methods to estimate f_X from the data $Y_1, \ldots, Y_n$.

In this paper we also apply Fourier analysis, and the related representation theory, for compact Lie groups, to solve the deconvolution problem for general G. Given a random sample $Y_1, \ldots, Y_n$, we construct a nonparametric deconvolution density estimator, f_X^n, of f_X. In particular, we obtain optimal, or minimax, rates of convergence of f_X^n to f_X as $n \to \infty$, under varying forms of ill-posedness.

We now describe briefly the form of the estimator f_X^n. Let $\widehat{G}$ denote the dual object of G, viz., the category of (equivalence classes of) inequivalent, irreducible representations of G. Let $\{U_\nu : \nu \in \widehat{G}\}$ be a complete enumeration of the elements of $\widehat{G}$. Denote by d_ν the dimension of the representation space of U_ν. It is well-known that the components of the irreducible representations U_ν are eigenfunctions of Δ, the Laplace-Beltrami operator on G; for each ν, we denote by λ_ν the eigenvalue corresponding to U_ν. Let $\widehat{f}$ denote the Fourier transform, as given below in (2.2), of a function $f \in L^2(G)$. For each ν in a certain finite subset of $\widehat{G}$, called the *Weyl polyhedra*, we assume that $\widehat{f_\epsilon}(\nu)$, the Fourier transform of f_ϵ evaluated at ν, is invertible; and we denote the inverse of $\widehat{f_\epsilon}(\nu)$ by $\widehat{f}_{\epsilon^{-1}}(\nu)$. Then our nonparametric deconvolution density estimator of f_X, based on the random sample $Y_1, \ldots, Y_n$, is

$$f_X^n(g) = \sum_{\nu \in \widehat{G} : \lambda_\nu < T} d_\nu \operatorname{tr}\Big(U_\nu(g) \widehat{f}_Y^n(\nu) \widehat{f}_{\epsilon^{-1}}(\nu)\Big),$$

$g \in G$, where for $\nu \in \widehat{G}$,

$$\widehat{f_Y^n}(\nu) = \frac{1}{n}\sum_{l=1}^{n} U_\nu(Y_l^{-1}),$$

is the empirical Fourier transform (or characteristic function) of f_Y, and the *smoothing parameter* $T = T(n) \to \infty$ as $n \to \infty$. Although we do not discuss here the problem of estimation of the smoothing parameter T, we note that this can be carried out using cross-validation methods for inverse problems along the lines developed by Dey, et al. (1996).

We also want to make special mention of the seminal ideas of Fan (1991) and Fan (1991a) which have guided us throughout the development of our results. Indeed, our paper is part of a program to extend results of Fan and others from the commutative, Euclidean setting of $\mathbb{R}^n$ to the non-commutative, non-Euclidean setting of a compact Lie group G. Clearly, the technical details behind the present portion of this program are formidable, and the ideas of Fan (1991) were crucial in guiding us to concrete realizations of the various smoothness classes of error densities arising in the deconvolution problem.

We now provide a summary of the results given here. In Section 2 we provide some preliminaries on Fourier analysis and representation theory on compact Lie groups. This material can be found in the standard references (cf. Helgason, 1984; Bröcker and tom Dieck, 1985; Fegan, 1991), and we will present the necessary results in a concise manner accessible to readers unfamiliar with the area.

In Section 3 we state the deconvolution problem on G. Motivated by the results of Fan (1991) and Kim and Koo (2001), we define the classes of *smooth* and *super-smooth* error distributions on G. In particular, as done by Kim and Koo (2001), polynomial and exponential bounds are invoked through the operator norm on the Fourier transforms. We also define in Section 3 a new smoothness class of error distributions, called the *log-super-smooth* class, which is smoother, in a sense to be made precise, than the super-smooth class of distributions.

In Section 4 we state the main results. Optimal-rate consistent deconvolution density estimators are established for all three smoothness classes over certain function spaces of Sobolev type. Further we discuss two extreme examples in which the distribution of the error variable ϵ follows the Haar measure or a Dirac measure.

In Section 5 we study an example of each of the smooth, super-smooth and log-super-smooth error distributions. These are the Laplace, Gaussian and von Mises-Fisher distributions, respectively. The Laplace distribution is an example of a smooth error distribution first studied for the case of SO(3) by Healy, Hendriks and Kim (1998). The Gaussian distribution is of course well-known, and the von Mises-Fisher matrix distribution was introduced by Khatri and Mardia (1977).

The proofs of the main results are provided in Section 6. The basic idea behind the proofs is, first, to apply a famous formula of H. Weyl to establish upper bounds for the rates of convergence. Second, a restricted subclass is developed and, through applications of Koo's lemma and Fano's lemma, lower bound rates of convergence are determined. Once we observe that the upper and lower bounds for the rates are the same, optimality of the rate of convergence is confirmed.

2. Notation and preliminaries

Let G be a (nontrivial) compact connected Lie group. With no loss of generality, we may view G as a closed, connected and bounded subgroup of $\mathrm{GL}(r, \mathbb{C})$, the general linear group of nonsingular complex $r \times r$ matrices, for some value of r. Two classic examples of compact connected Lie groups are the *unitary group* $\mathrm{U}(r)$, containing all $r \times r$ unitary matrices; and $\mathrm{SU}(r)$, the *special unitary group* consisting of all unitary matrices with determinant one.

Let us denote by $\mathbb{C}^{r\times r}$ the space of complex $r \times r$ matrices. Then a Lie algebra may be viewed as a subspace of $\mathbb{C}^{r\times r}$ which is closed under the *commutation* $[x, y] := xy - yx$, $x, y \in \mathbb{C}^{r\times r}$. In particular, by identifying G with a subgroup of $\mathrm{GL}(r, \mathbb{C})$, the Lie algebra $\mathfrak{g}$ of G is obtained as the vector space of all $r \times r$ complex matrices x such that

$$e^{tx} := \sum_{j=0}^{\infty} \frac{t^j}{j!} x^j \tag{2.1}$$

is an element of G for all $t \in \mathbb{R}$. The *exponential map* $x \mapsto e^x$, although not injective, is surjective from $\mathfrak{g}$ to G.

As an example, in the case of $\mathrm{SU}(r)$, the special unitary group, it is not difficult to verify that the corresponding Lie algebra is $\mathfrak{su}(r)$, the space of $r \times r$ complex skew-Hermitian matrices with trace zero.

For each $g \in G$, define a linear transformation $\mathrm{Ad}(g)$ on the Lie algebra $\mathfrak{u}$ by

$$\mathrm{Ad}(g) \cdot x = gxg^{-1}$$

for $x \in \mathfrak{g}$; it is simple to verify that $\mathrm{Ad}(g_1 g_2) = \mathrm{Ad}(g_1)\mathrm{Ad}(g_2)$ for all $g_1, g_2 \in G$. The mapping $g \mapsto \mathrm{Ad}(g)$ is called the *adjoint representation* of G. Similarly, we define a mapping

$$\mathrm{ad}(x)y := [x, y]$$

for $x, y \in g$; this defines a linear transformation $x \mapsto \mathrm{ad}(x)$ on g, known as the *adjoint representation* of g. These two adjoint representations, ad on g and Ad on G, satisfy the relationship

$$\mathrm{Ad}(e^{tx}) = e^{t\,\mathrm{ad}(x)}$$

for all $x \in g$ and $t \in \mathbb{R}$.

A *torus* in G is a connected Abelian subgroup of G. From the general structure theory of compact Lie groups it is well-known that, up to isomorphism, there exists a unique *maximal torus*. Any maximal torus is isomorphic to a product of one-dimensional unitary groups $\mathrm{U}(1)$ and is also a maximal Abelian subgroup of G.

Let us fix, once and for all, a maximal torus $\mathbb{T} \subset G$. Denote by t the Lie algebra of $\mathbb{T}$. Then t is a maximal Abelian subalgebra of g. Let t_0 denote the *semisimple* part of t, i.e., $t = z \oplus t_0$ where z is the center of g and $t_0 = t \cap [g, g]$. Let $h = t + it$ be the *complexification* of t. Then h is a maximal Abelian subalgebra, called the *Cartan subalgebra*, of g such that the linear transformations $\mathrm{ad}(H)$, $H \in h$, are simultaneously diagonalizable. Let h^* denote the dual space of h. For $\alpha \in h^*$ we set

$$g^\alpha = \{X \in g : [H, X] = \alpha(H)X \text{ for all } H \in h\}.$$

If g^α is nonzero then α is called a *root* of g with respect to h and g^α is called the corresponding *root space*. We denote the set of roots by $\widetilde{\Phi}$. Then each root space is one-dimensional; $g^0 = h$; and g decomposes as a direct sum,

$$g = h \oplus \sum_{\alpha \in \widetilde{\Phi}} \oplus\, g^{\alpha},$$

called the *root space decomposition* of g.

There exists on the space g a non-degenerate complex bilinear form $\langle \cdot | \cdot \rangle$ which is g-invariant, i.e. $\langle [x, y] | z \rangle = -\langle y | [x, z] \rangle$ for all $x, y, z \in g$. From the invariance and the root space decomposition, it follows that the restriction of this form to h is non-degenerate. The restriction of this form to the real subspace it of h also defines an inner product on it. In particular, for each $\alpha \in \widetilde{\Phi}$ there exists a unique element $H_\alpha \in it_0$ such that $\alpha(H) = \langle H | H_\alpha \rangle$ for all $H \in it$.

The *Killing form* ψ of g is the symmetric bilinear form $\psi(x, y) = \mathrm{tr}(\mathrm{ad}(x) \circ \mathrm{ad}(y))$, $x, y \in g$. We shall henceforth assume that G is semisimple. Then ψ is negative definite; and the negative Killing form, $-\psi$, is the canonical inner product on $\widetilde{\Phi}$, the set of roots.

Let $N(\mathbb{T}) = \{g \in G : g\mathbb{T}g^{-1} = \mathbb{T}\}$ denote the *normalizer* of $\mathbb{T}$ in G; then the quotient group $W = N(\mathbb{T})/\mathbb{T}$ is a finite group called the *Weyl group* of G. Indeed, W can be identified with a subgroup of the group of permutations of the finite set $\widetilde{\Phi}$. We also denote by t^* the dual space of t possessing the Weyl group invariant inner product. To each root $\alpha \in \widetilde{\Phi}$ we associate the hyperplane $\mathcal{H}_\alpha \subset h^*$ which is orthogonal to α. The collection of all hyperplanes $\{\mathcal{H}_\alpha : \alpha \in \widetilde{\Phi}\}$ partitions h^* into a finite number of open convex regions called the *Weyl chambers* of h^*. Then the Weyl group W is the group of permutations of the set of Weyl chambers. We will choose and fix a *fundamental Weyl chamber*, denoted by K.

Denote by Φ the set of real roots, and let $\Phi_+ = \{\alpha \in \Phi : \langle \alpha, \beta \rangle > 0 \text{ for all } \beta \in K\}$ denote the set of *positive roots*. Denote by I the kernel of the exponential map in (2.1). The *integral portion* of t^* is the set $I^* = \{\alpha \in t^* : \alpha I \subset \mathbb{Z}\}$. Since $I^* \subset t^*$ then the set $\bar{K} \cap I^*$ is well-defined, and we shall use it as an index set.

A *representation* U of the group G is a mapping from G into some $\mathrm{Aut}(V)$, the space of linear transformations of a vector space V, such that (i) at the unit element $e \in G$ we have $U(e) = 1$, the identity transformation on V; (ii) $U(g_1 g_2) = U(g_1)U(g_2)$ for all $g_1, g_2 \in G$; and (iii) for each $g \in G$, $U(g)$ is continuous on V.

A *G-invariant* subspace for a representation U of G is a closed subspace $V_1 \subseteq V$ such that $U(g)v \in V_1$ for all $g \in G$ and $v \in V_1$. A nonzero representation U is called *irreducible* if its only G-invariant subspaces are $\{0\}$ and V. It is well-known that all irreducible representations of G are finite-dimensional. By the Peter-Weyl theorem we may assume, without loss of generality, that the irreducible representations are also unitary; cf. Bröcker and tom Dieck (1985, pp. 133-136).

In the case of the compact connected Lie group G, the collection of inequivalent irreducible representations of G can be enumerated using the elements of the set $\bar{K} \cap I^*$; i.e., for each $\nu \in \bar{K} \cap I^*$ there exists an inequivalent irreducible representation, which we denote by U_ν. Further, the correspondence between ν and U_ν is one-to-one up to conjugation, cf. Bröcker and tom Dieck (1985, p. 242). For $\nu \in \bar{K} \cap I^*$ the *irreducible character* is defined by $\chi_\nu = \mathrm{tr}\, U_\nu$, and the *dimension* of the irreducible representation U_ν is $d_\nu = \chi_\nu(e)$. Further, we denote by

$$\rho = \frac{1}{2} \sum_{\alpha \in \Phi_+} \alpha,$$

one-half the sum of the positive roots.

Let dg denote the normalized Haar measure on G and $L^2(G)$ denote the space of complex-valued functions on G which are square-integrable with respect to the Haar measure; we set

$$||f||_2 := \left(\int_G |f(g)|^2 dg \right)^{1/2}$$

for $f \in L^2(G)$. The Fourier transform of $f \in L^2(G)$ is

$$\widehat{f}(\nu) = \int_G U_\nu(g^{-1}) f(g) dg, \tag{2.2}$$

$\nu \in \bar{K} \cap I^*$. Further, the inverse Fourier transform is given by

$$f(g) = \sum_{\nu \in \bar{K} \cap I^*} d_\nu \operatorname{tr} U_\nu(g) \widehat{f}(\nu), \tag{2.3}$$

$g \in G$.

A noteworthy special case is that in which f is a *class function*, i.e., $f(g_1 g g_1^{-1}) = f(g)$ for all $g, g_1 \in G$. By an application of Schur's lemma (cf. Bröcker and tom Dieck, (1985), p. 69), we then deduce that (2.2) reduces to

$$\widehat{f}(\nu) = \gamma_\nu I_\nu \tag{2.4}$$

where I_ν denotes the $d_\nu \times d_\nu$ identity matrix and

$$\gamma_\nu = d_\nu{}^{-1} \int_G f(g) \overline{\chi_\nu(g)} dg, \tag{2.5}$$

with *overbar* denoting complex conjugation. By inverse Fourier transforms, we then obtain

$$f(g) = \sum_{\nu \in \bar{K} \cap I^*} \gamma_\nu \chi_\nu(g), \tag{2.6}$$

$g \in G$.

Returning to the general case, if $f \in L^2(G)$ is a probability density function, we will call (2.2) the *characteristic function* of f. This coincides with the usual notion of characteristic function for probability distributions defined on Euclidean spaces in that, among other things, the set (2.9) below is a complete orthonormal basis for $L^2(G)$. As in the Euclidean case, it can be shown that the mapping $f \mapsto \widehat{f}$ is one-to-one; this is a consequence of the Peter-Weyl theorem; see Bröcker and tom Dieck (1985, pp. 133-136).

Given two functions $f, h \in L^2(G)$, define the convolution $f * h$ by

$$(f * h)(u) = \int_G f(g^{-1}u) h(g) dg,$$

$u \in G$. It is not difficult to verify the convolution property,

$$(\widehat{f * h})(\nu) = \widehat{f}(\nu)\,\widehat{h}(\nu), \tag{2.7}$$

where $f, h \in L^2(G)$ and $\nu \in \bar{K} \cap I^*$.

Let Δ denote the Laplace-Beltrami operator on G; then the components of the irreducible representations are eigenfunctions of Δ and the eigenvalue of U_ν is

$$\lambda_\nu = \|\nu + \rho\|^2 - \|\rho\|^2, \tag{2.8}$$

$\nu \in \bar{K} \cap I^*$, where the norm is that induced by the Killing form. Since the components of the irreducible representations also form a complete orthogonal class in $L^2(G)$ then the set

$$\{d_\nu^{1/2} U_\nu : \nu \in \bar{K} \cap I^*\} \tag{2.9}$$

is a complete orthonormal basis of $L^2(G)$. Moreover, the multiplicity of the eigenfunctions with respect to a fixed eigenvalue is therefore ${d_\nu}^2$ for $\nu \in \bar{K} \cap I^*$.

A relationship between d_ν and λ_ν for $\nu \in \bar{K} \cap I^*$ may be obtained from a famous formula of H. Weyl, viz., if $\dim G$ and $\operatorname{vol} G$ denote the *dimension* and the *volume* of G, respectively; and $\Gamma(\cdot)$ denotes the classical gamma function; then

$$\lim_{T\to\infty} T^{-(\dim G)/2} \sum_{\nu \in \bar{K}\cap I^* : \lambda_\nu < T} {d_\nu}^2 = \frac{\operatorname{vol} G}{(2\sqrt{\pi})^{\dim G}\,\Gamma(1 + \frac{1}{2}\dim G)}; \tag{2.10}$$

cf. Minakshisundaram and Pleijel (1949), Giné (1975), or Hendriks (1990).

We note that our work is also related to the theory of summability on compact Lie groups, cf. Colzani, et al. (1990) and the references contained therein. Our results are concerned not only with summability, but also with rates of convergence. Unlike summability theory, in which a finite subset of $\widehat{G}$ is chosen and then expanded to cover the set $\bar{K} \cap I^*$, our results require an approach which utilizes expansions of eigenvalues of the Laplace operator. This approach allows us to utilize (2.10) from which rates of convergence can be estimated.

3. The deconvolution problem and the smoothness classes

Recall that in the deconvolution problem we consider the statistical model $Y = X\epsilon$ where X is a true random measurement and ϵ is a random noise. We assume that f_ϵ, the density function of ϵ, is known, while the density functions f_X and f_Y of X and Y, respectively, both are unknown. Given a random sample $Y_1, \ldots, Y_n$ from Y, we shall estimate the density function f_X. By (2.7),

$$\widehat{f_Y}(\nu) = \widehat{f_X}(\nu)\widehat{f_\epsilon}(\nu), \tag{3.1}$$

$\nu \in \widehat{G}$. Provided $\widehat{f_\epsilon}(\nu)$ is invertible we denote $[\widehat{f_\epsilon}(\nu)]^{-1}$ by $\widehat{f_{\epsilon^{-1}}}(\nu)$, and then (3.1) becomes

$$\widehat{f_X}(\nu) = \widehat{f_Y}(\nu)\widehat{f_{\epsilon^{-1}}}(\nu)$$

for $\nu \in \widehat{G}$. Next we form the *empirical characteristic function*

$$\widehat{f_Y^n}(\nu) = \frac{1}{n}\sum_{l=1}^{n} U_\nu(Y_l^{-1}). \tag{3.2}$$

By substitution of (3.2) in the Fourier inversion formula (2.3) we are led to define, as an estimator of f_X, the statistic

$$f_X^n(g) = \sum_{\nu \in \bar{K}\cap I^* : \lambda_\nu < T} d_\nu \operatorname{tr}\Big(U_\nu(g)\widehat{f_Y^n}(\nu)\widehat{f_{\epsilon^{-1}}}(\nu)\Big), \tag{3.3}$$

where the smoothing parameter $T = T(n) \to \infty$ as $n \to \infty$. Thus, (3.3) represents our nonparametric deconvolution density estimator for f_X. Alternatively, by defining

$$K_n^\epsilon(g) = \sum_{\nu \in \bar{K}\cap I^* : \lambda_\nu < T} d_\nu \operatorname{tr}\Big\{U_\nu(g)\widehat{f_{\epsilon^{-1}}}(\nu)\Big\} \tag{3.4}$$

then (3.3) can be rewritten as

$$f_X^n(g) = \frac{1}{n}\sum_{l=1}^{n} K_n^{\epsilon}(gY_l^{-1}), \tag{3.5}$$

for $g \in G$.

In classical deconvolution density estimation, it is well-known that the rate of recovery of the density function f_X can be characterized by the smoothness of f_ϵ (cf. Stefanski and Carroll, 1990). Following the seminal ideas of Fan (1991), combined with suitable modifications to the approach of Kim and Koo (2001) as developed for the case of the Lie group SO(3), we now define the smoothness of the characteristic function of f_ϵ. The necessary modification arises from the fact that Fourier transforms on G are matrices which grow in dimension. Consequently, the quality of smoothness needs to be adapted for this change, and this can be done most naturally through the use of the matrix (or operator) norm $\|\cdot\|_{\rm op}$ where $||A||_{\rm op}$ is the largest of the absolute values of all eigenvalues of the matrix A. In the following definitions we recall that λ_ν, given in (2.8), is the eigenvalue of the irreducible representation U_ν under the Laplace-Beltrami operator Δ.

DEFINITION 3.1. The distribution of ϵ is called *smooth of order* $\beta \geq 0$, if there exist positive constants c_0, c_1, such that $\widehat{f_\epsilon}$, the characteristic function of f_ϵ, satisfies

$$\|\widehat{f}_{\epsilon^{-1}}(\nu)\|_{\rm op} \leq c_0^{-1}\lambda_\nu^{\beta} \ \text{ and } \ \|\widehat{f}_\epsilon(\nu)\|_{\rm op} \leq c_1\lambda_\nu^{-\beta}\ , \tag{3.6}$$

as $\lambda_\nu \to \infty$, where $\nu \in \bar{K} \cap I^*$.

DEFINITION 3.2. The distribution of ϵ is called *super-smooth of order* $\beta > 0$ if there exist positive constants c_0, c_1, γ, and $\beta_0, \beta_1 \in \mathbb{R}$, such that $\widehat{f_\epsilon}$ satisfies

$$\|\widehat{f}_{\epsilon^{-1}}(\nu)\|_{\rm op} \leq c_0^{-1}\lambda_\nu^{-\beta_0}\exp(\gamma^{-1}\lambda_\nu^{\beta}) \ \text{ and } \ \|\widehat{f}_\epsilon(\nu)\|_{\rm op} \leq c_1\lambda_\nu^{\beta_1}\exp(-\gamma^{-1}\lambda_\nu^{\beta}), \tag{3.7}$$

as $\lambda_\nu \to \infty$, where $\nu \in \bar{K} \cap I^*$.

These two definitions are natural adaptations to the general group-theoretic setting of definitions provided by Fan (1991) in the Euclidean case and by Kim and Koo (2001) in the case of SO(3). The constant β in Definition 3.2 characterizes the degree of super-smoothness for this class of error distributions; thus the larger the value of β, the smoother the error distribution. There is also a third smoothness class which at first glance may resemble the super-smooth class but is quite distinct from the latter.

DEFINITION 3.3. The distribution of ϵ is called *log-super-smooth of order* $\beta > 0$ if there exist positive constants c_0, c_1, γ, and $\beta_0, \beta_1, \gamma_0, \gamma_1 \in \mathbb{R}$ such that $\widehat{f_\epsilon}$ satisfies

$$\begin{aligned} \|\widehat{f}_{\epsilon^{-1}}(\nu)\|_{\rm op} &\leq c_0^{-1}\lambda_\nu^{-\beta_0}\exp\big(-\lambda_\nu^{\beta}(\gamma_0^{-1}-\gamma^{-1}\log\lambda_\nu)\big) \ \text{ and} \\ \|\widehat{f}_\epsilon(\nu)\|_{\rm op} &\leq c_1\lambda_\nu^{\beta_1}\exp\big(\lambda_\nu^{\beta}(\gamma_1^{-1}-\gamma^{-1}\log\lambda_\nu)\big), \end{aligned} \tag{3.8}$$

as $\lambda_\nu \to \infty$, where $\nu \in \bar{K} \cap I^*$.

For any positive constants C, η we have $C \leq \log\lambda_\nu \leq \lambda_\nu^{\eta}$ as $\lambda_\nu \to \infty$. Consequently if f_ϵ is log-super-smooth then

$$\begin{aligned} \|\widehat{f}_{\epsilon^{-1}}(\nu)\|_{\rm op} &\leq c_0^{-1}\lambda_\nu^{-\beta_0}\exp\big(\gamma^{-1}\lambda_\nu^{\beta(1+\eta)}\big) \ \text{ and} \\ \|\widehat{f}_\epsilon(\nu)\|_{\rm op} &\leq c_1\lambda_\nu^{\beta_1}\exp\big(-\gamma^{-1}C\lambda_\nu^{\beta}\big), \end{aligned} \tag{3.9}$$

as $\lambda_\nu \to \infty$. By comparing (3.7) with (3.8), it follows that $\|\widehat{f}_{\epsilon^{-1}}(\nu)\|_{\rm op}$ has an upper bound which is similar to that of the super-smooth class of order $\beta(1+\eta) > \beta > 0$. This bound differs from that of $\|\widehat{f}_\epsilon(\nu)\|_{\rm op}$ which has a bound similar to that of the super-smooth class of order $\beta > 0$. In this sense, the log-super-smooth class of distributions is distinct from the super-smooth class.

4. Optimal estimation and the main Results

Let $C^\infty(G)$ be the space of infinitely differentiable functions on G. Each function $h \in C^\infty(G)$ may be represented by Fourier inversion,

$$h(g) = \sum d_\nu \mathrm{tr}\, U_\nu(g)\widehat{h}(\nu),$$

$g \in G$. Then we define the *Sobolev norm* $\|\cdot\|_{H_s}$ *of order* s on $C^\infty(G)$ by

$$\|h\|_{H_s} := \Big(\sum_{\nu\in\bar{K}\cap I^*} (1+\lambda_\nu)^s\, d_\nu\, \mathrm{tr}\, |\widehat{h}(\nu)|^2 \Big)^{1/2}, \tag{4.1}$$

$h \in C^\infty(G)$. It is straightforward to verify that (4.1) is a norm, and we denote by $H_s(G)$ the completion of $C^\infty(G)$ with respect to (4.1). For a constant $M > 0$, we define the smoothness class of functions

$$H_s(G, M) := \{h \in H_s(G) : \|h\|_{H_s} \le 1 + M\}.$$

DEFINITION 4.1. (Stone, 1980) For $f \in H_s(G, M)$ let P_f be a probability distribution. A sequence $\{b_n : n = 1, 2, \dots\}$ of positive numbers is called:
(i) A *lower bound* for the function f if

$$\lim_{c\to 0+} \liminf_n \inf_{f^n} \sup_{f\in H_s(G,M)} P_f\left(\|f^n - f\|_2 \ge cb_n\right) = 1, \tag{4.2}$$

where the infimum is over all possible estimators f^n based on a random sample $Y_1, \dots, Y_n$.
(ii) An *upper bound* for f if there exists a sequence of estimators $\{f^n : n = 1, 2, \dots\}$ of f such that

$$\lim_{c\to\infty} \limsup_n \sup_{f\in H_s(G,M)} P_f\left(\|f^n - f\|_2 \ge cb_n\right) = 0. \tag{4.3}$$

(iii) The *optimal rate of convergence* for f if it is both a lower bound and an upper bound; in this case the associated estimators $\{f^n : n = 1, 2, \dots\}$ are called *asymptotically optimal.*

We are now in place to state our main results. The following theorems establish that the deconvolution density estimator (3.5) is asymptotically optimal, and the optimal rates of convergence depend on the smoothness of the characteristic function of the error distribution.

In the sequel, we use the following notation. For two sequences $\{a_n : n = 1, 2, \dots\}$ and $\{c_n : n = 1, 2, \dots\}$ of positive numbers, we write $a_n \ll c_n$ to denote that $a_n/c_n \le C$ as $n \to \infty$. Further we write $a_n \asymp c_n$ if both $a_n \ll c_n$ and $c_n \ll a_n$ are valid.

THEOREM 4.2. *Suppose that f_ϵ is smooth and $f_X \in H_s(G, M)$, $s > (\dim G)/2$. Then the optimal rate of convergence is $n^{-s/(2s+4\beta+\dim G)}$.*

THEOREM 4.3. *Suppose that f_ϵ is super-smooth and $f_X \in H_s(G, M)$, $s > (\dim G)/2$. Then the optimal rate of convergence is $(\log n)^{-s/(2\beta)}$.*

THEOREM 4.4. *Suppose that f_ϵ is log-super-smooth and $f_X \in H_s(G,M)$, $s > (\dim G)/2$. Then the optimal rate of convergence is*

$$\left(\frac{\log n}{\log\log n}\right)^{-s/(2\beta)}.$$

Based on (3.9), we see that the appearance of the logarithm term in the exponent allows us to bound the optimal rate of convergence between those of the super-smooth bound (with different rates).

At this point we will discuss two extreme cases with respect to distribution of the errors ϵ. At one extreme is the Haar measure (or uniform distribution) on G for which $f_\epsilon(g) \equiv 1$, $g \in G$. By applying the Weyl orthogonality relations for characters (cf. Bröcker and tom Dieck, 1985, p. 81), it follows from (2.2) that the corresponding Fourier transform is

$$\widehat{f_\epsilon}(\nu) = \int_G U_\nu(g^{-1})\,dg = \begin{cases} 1, & \text{if } \nu = 0, \\ \mathbf{0}_\nu, & \text{if } \nu \neq 0, \end{cases}$$

where $\mathbf{0}_\nu$ denotes the $d_\nu \times d_\nu$ zero matrix. Therefore, the Fourier transform $\widehat{f_\epsilon}(\nu)$ is not invertible for any $\nu \neq 0$ so that deconvolution is not possible. In short, because the true measurements are uniformly corrupted, there is no possibility of recovering the density f_X.

At the other extreme is the case in which the error ϵ has a probability distribution which is the Dirac measure (or point mass), δ_{g_0}, at a point $g_0 \in G$. Without loss of generality we assume that g_0 is e, the identity element in G. Then the Fourier transform of f_ϵ is

$$\widehat{f_\epsilon}(\nu) = \int_G U_\nu(g^{-1})\,\delta_e(g)\,dg = U_\nu(e) = I_\nu,$$

$\nu \in \bar{K} \cap I^*$. Then $\|\widehat{f_\epsilon}(\nu)\|_{\text{op}} = 1$ for all ν and, by Definition 3.1, the distribution of ϵ is smooth of order $\beta = 0$. By (3.4) we obtain

$$K_n^\epsilon(g) = \sum_{\nu \in \bar{K} \cap I^* : \lambda_\nu < T} d_\nu \chi_\nu(g),$$

$g \in G$, and (3.5) reduces to

$$f_X^n(g) = \frac{1}{n}\sum_{l=1}^{n} \sum_{\nu \in \bar{K} \cap I^* : \lambda_\nu < T} d_\nu \chi_\nu(gY_l^{-1}), \tag{4.4}$$

$g \in G$. Since the observations are made without error, (4.4) represents the case of ordinary nonparametric density estimation on G. Together with Theorem 4.2 this observation leads to the following corollary stating that, for nonparametric density estimation on G, the L^2–upper bound rate of convergence obtained by Hendriks (1990, Theorem 2.1, p. 834) is optimal.

COROLLARY 4.5. *Suppose the distribution of ϵ is concentrated at the identity $e \in G$. If $f_X \in H_s(G,M)$, $s > (\dim G)/2$, then the optimal rate of convergence is $n^{-s/(2s+\dim G)}$.*

5. Examples of smoothness classes

In this section we discuss three error distributions, one each from the smooth, super-smooth and log-super-smooth classes. The first two distributions are characterized spectrally and are class functions, while the third example is a model well-known in directional statistics.

5.1. The Laplace Distribution on G. This distribution is the analogue on G of the classical Laplace (or double exponential) distribution on the real line. For the case in which $G = \mathrm{SO}(3)$, Healy, Hendriks and Kim (1998, Theorem 3.5) provide a detailed discussion of this distribution, including a closed-form expression for the density function. In the case of a general compact Lie group G, we define the Laplace density function by the Fourier transform

$$\widehat{f_\epsilon}(\nu) = (1 + \sigma^2 \lambda_\nu)^{-1} I_\nu, \tag{5.1}$$

$\nu \in \bar{K} \cap I^*$, where σ is a positive parameter. Since $\widehat{f_\epsilon}(0) = 1$ then f_ϵ integrates to 1. Further, because $\widehat{f_\epsilon}(\nu)$ is a positive-definite matrix then, by Bochner's theorem for compact Lie groups (cf. Hewitt and Ross, 1970, Theorem 30.2, p. 157; Lo and Ng, 1998), it follows that $f_\epsilon(g) \geq 0$, $g \in G$. Therefore f_ϵ is a valid probability density function on G. It also follows from Definition 3.1 that the Laplace distribution is a smooth distribution of order $\beta = 1$.

On applying the inverse Fourier transform to (5.1) we obtain

$$f_\epsilon(g) = \sum_{\nu \in \bar{K} \cap I^*} (1 + \sigma^2 \lambda_\nu)^{-1} \, d_\nu \, \chi_\nu(g), \tag{5.2}$$

$g \in G$. By (2.8) $\Delta \chi_\nu = \lambda_\nu \chi_\nu$, and this leads to

$$(1 + \sigma^2 \Delta) f_\epsilon(g) = \sum_{\nu \in \bar{K} \cap I^*} d_\nu \, \chi_\nu(g), \tag{5.3}$$

$g \in G$, which is a differential equation for f_ϵ. In the case of SO(3), this differential equation was solved by Healy, Kim and Hendriks (1998), Theorem 3.5, to obtain an explicit, closed form, expression for f_ϵ. For the higher-rank Lie groups, however, it seems that only series expansions can be obtained.

5.2. The Gaussian Distribution on G. The Gaussian distribution on G arises from the case in which

$$\widehat{f_\epsilon}(\nu) = \exp(-\sigma^2 \lambda_\nu) I_\nu,$$

$\nu \in \bar{K} \cap I^*$, where σ^2 is a positive parameter; this is an example of a super-smooth distribution with $\gamma = \sigma^{-2}$, $\beta = 1$ and $\beta_0 = \beta_1 = 0$. Applying the inverse Fourier transform we obtain

$$f_\epsilon(g) = \sum_{\nu \in \bar{K} \cap I^*} \exp(-\sigma^2 \lambda_\nu) d_\nu \chi_\nu(g), \tag{5.4}$$

$g \in G$. It is also well-known (cf. Fegan, 1991, p. 105, eq. (12.3)) that the Gaussian distribution arises as the heat kernel in connection with the heat equation on G. In the case of SO(2) it is simple to show that (5.4) reduces to the well-known wrapped normal distribution on the circle, with density function

$$\frac{1}{2\pi}\Big\{1 + \sum_{j=1}^{\infty} \exp(-j^2\sigma^2/2)\cos(j\theta)\Big\},$$

$\theta \in [0, 2\pi)$. In the case of SO(3), we can apply the methods used by Healy, Kim and Hendriks (1998) to deduce a multiple series expansion for the density function (5.4).

5.3. The Matrix von Mises-Fisher Distribution on SO(N). Within the context of directional data, the von Mises-Fisher distribution has been widely studied. Khatri and Mardia (1977) extended this distribution to matrix spaces; for the case in which $G = \mathrm{SO}(N)$, the corresponding probability density function, relative to normalized Haar measure on SO(N) is

$$f_{\kappa,\mu}(g) = c(\kappa) \exp(\kappa \operatorname{tr} \mu^{-1} g), \tag{5.5}$$

$g \in \mathrm{SO}(N)$, where $\kappa > 0$ is the *concentration parameter*, $\mu \in \mathrm{SO}(N)$ is the *mean direction*, and $c(\kappa)$ is the normalizing constant. To determine the degree of smoothness of (5.5) we need to calculate the Fourier transform of this distribution.

The dimension of the factor space $\mathrm{SO}(N)/\mathbb{T}$ is;

$$\dim(\mathrm{SO}(N)/\mathbb{T}) = \dim(\mathrm{SO}(N)) - \dim(\mathbb{T}) = \begin{cases} k(2k-1)-k, & \text{if } N = 2k, \\ k(2k+1)-k, & \text{if } N = 2k+1. \end{cases}$$

By an application of the generalized Clebsch-Gordan formula (Bröcker and tom Dieck, 1985, p. 259, Proposition 3.4) it may be shown that, for some $\kappa > 0$,

$$\|\widehat{f}_{\kappa,g}(\ell)\|_{\mathrm{op}} \ll c(\kappa)\lambda_\nu^{-\dim(\mathrm{SO}(N)/\mathbb{T})} \times \exp\left\{-\frac{1}{2}\log\lambda_\nu^{1/2} + (1+\log(\kappa N))\lambda_\nu^{1/2} - \lambda_\nu^{1/2}\log\lambda_\nu^{1/2}\right\}$$

and

$$\left\|\widehat{f}_{\kappa,g}(\ell)^{-1}\right\|_{\mathrm{op}} \ll c(\kappa)^{-1}\lambda_\nu^{\dim(\mathrm{SO}(N)/\mathbb{T})} \times \exp\left\{\frac{1}{2}\log\lambda_\nu^{1/2} - (1+\log\kappa)\lambda_\nu^{1/2} + \lambda_\nu^{1/2}\log\lambda_\nu^{1/2}\right\}$$

as $\lambda_\nu \to \infty$, $\nu \in \bar{K} \cap I^*$ (cf. Kim, 2000). Consequently the von Mises-Fisher distribution on SO(N) is of the log-super-smooth class with parameters $c_0 = c_1 = c(\kappa)$, $\beta_0 = \beta_1 = -\dim(\mathrm{SO}(N)/\mathbb{T})$, $\gamma_0^{-1} = 1 + \log\kappa$, $\gamma_1^{-1} = 1 + \log(\kappa N)$, $\gamma = 2$ and $\beta = 1/2$. Therefore, the optimal rate of convergence pertains to the case treated by Theorem 4.4.

6. Proofs

In this section we first deduce the upper bounds, and then we deduce the lower bounds. Next we show that the two bounds coincide, thereby establishing optimality.

6.1. Upper Bounds.

LEMMA 6.1. *There holds the upper bound*

$$n\int_G \mathrm{Var}(f_X^n) \ll \begin{cases} T^{2\beta+(\dim G)/2}, & f_\epsilon \textit{ smooth}, \\ T^{-2\beta_0+(\dim G)/2}\exp(2\gamma^{-1}T^\beta), & f_\epsilon \textit{ super-smooth}, \\ T^{-2\beta_0+(\dim G)/2}\exp(-2T^\beta(\gamma_0^{-1} - \gamma^{-1}\log T)), & f_\epsilon \textit{ log-super-smooth}, \end{cases}$$

as $n \to \infty$.

PROOF. Since f_Y is bounded then

$$\begin{aligned}\int_G \mathrm{Var}(f_X^n(g))\,dg &= n^{-1}\int_G \mathrm{Var}(K_n^\epsilon(gY^{-1}))\,dg \\ &\ll n^{-1}\int_G E\,K_n^\epsilon(gY^{-1})\overline{K_n^\epsilon(gY^{-1})}\,dg \\ &= n^{-1}\int_G K_n^\epsilon(gy^{-1})\overline{K_n^\epsilon(gy^{-1})}f_Y(y)\,dy \\ &\ll n^{-1}\int_G |K_n^\epsilon(g)|^2\,dg. \end{aligned}\tag{6.1}$$

By (3.4) we have

$$\begin{aligned}\int_G |K_n^\epsilon(g)|^2\,dg = \int_G K_n^\epsilon(g)\,\overline{K_n^\epsilon(g)}\,dg &= \sum_{\nu\in\bar{K}\cap I^*:\lambda_\nu<T} d_\nu \operatorname{tr}|\widehat{f}_{\epsilon^{-1}}(\nu)|^2 \\ &\le \sum_{\nu\in\bar{K}\cap I^*:\lambda_\nu<T} d_\nu{}^2\,\|\widehat{f}_{\epsilon^{-1}}(\nu)\|_{\mathrm{op}}^2. \end{aligned}\tag{6.2}$$

By applying Weyl's formula (2.10) to (3.6), (3.7) and (3.8), followed by a substitution into (6.2), we obtain

$$\|K_n^\epsilon\|_2^2 \ll \begin{cases} T^{2\beta+(\dim G)/2}, & f_\epsilon \text{ smooth} \\ T^{-2\beta_0+(\dim G)/2}\exp\left(2\gamma^{-1}T^\beta\right), & f_\epsilon \text{ super-smooth} \\ T^{-2\beta_0+(\dim G)/2}\exp(-2T^\beta(\gamma_0^{-1}-\gamma^{-1}\log T)), & f_\epsilon \text{ log-super-smooth.}\end{cases}\tag{6.3}$$

Applying (6.3) to (6.1) we obtain the desired result. □

It is shown in Kim (1998) that, for $f_X \in H_s(G,M)$ with $s > (\dim G)/2$,

$$\|Ef_X^n - f_X\|^2 \le \|\Delta^{s/2}f\|_2^2\,T^{-s}.$$

This implies that the integrated variance bias decomposition satisfies

$$\begin{aligned} E\|f_X^n - f_X\|_2^2 &= \int_G \mathrm{Var}(f_X^n(g))dg + \|Ef_X^n - f_X\|^2 \\ &\ll \int_G \mathrm{Var}(f_X^n(g))dg + T^{-s} \end{aligned}\tag{6.4}$$

as $n \to \infty$.

By Lemma 6.1, we now see that the rate in (6.4) is optimized if T, the smoothing parameter, satisfies

$$T \asymp \begin{cases} n^{1/[s+2\beta+\dim G/2]}, & f_\epsilon \text{ smooth,} \\ (\log n)^{1/\beta}, & f_\epsilon \text{ super-smooth,} \\ (\log n/\log\log n)^{1/\beta}, & f_\epsilon \text{ log-super-smooth,}\end{cases}$$

as $n \to \infty$. Consequently, (6.4) leads to the upper bound

$$E\|f_X^n - f_X\|_2^2 \ll \begin{cases} n^{-2s/[2s+4\beta+\dim G]}, & f_\epsilon \text{ smooth,} \\ (\log n)^{-s/\beta}, & f_\epsilon \text{ super-smooth} \\ (\log n/\log\log n)^{-s/\beta}, & f_\epsilon \text{ log-super-smooth,}\end{cases}$$

as $n \to \infty$.

6.2. Lower Bounds. In calculating the lower bounds we follow the standard Euclidean approach of specifying a subproblem, followed by applications of Fano's lemma and Koo's lemma to calculate the difficulty of the subproblem; cf. Koo (1993) and Koo and Chung (1998) for details.

Let $\{h_n : n = 1, 2, \dots\}$ be a sequence of positive integers such that $h_n \to \infty$ as $n \to \infty$, and define

$$\mathcal{V}_n = \{\nu \in \bar{K} \cap I^* : h_n^2 \le \lambda_\nu \le 2h_n^2\}. \tag{6.5}$$

The set $\mathcal{V}_n$ is a finite set, and we denote its cardinality by $|\mathcal{V}_n|$.

Let $\tau_\nu \in \{0, 1\}$ for $\nu \in \mathcal{V}_n$ where τ_ν is the same for χ_ν and $\overline{\chi_\nu}$; thus,

$$\tau_\nu(\chi_\nu + \overline{\chi_\nu}) = 2\tau_\nu \operatorname{Re} \chi_\nu \tag{6.6}$$

for $\nu \in \mathcal{V}_n$. Let $\tau = \{\tau_\nu : \nu \in \mathcal{V}_n\}$ and define the function

$$f_\tau(g) = 1 + M_1 h_n^{-s-(\dim G)/2} \sum_{\mathcal{V}_n} \tau_\nu d_\nu \chi_\nu(g), \tag{6.7}$$

$g \in G$, where M_1 is a sufficiently large positive constant to be made precise below; by (6.6), the function f_τ in (6.7) is real-valued. Finally, define

$$\mathcal{H}_n = \left\{ f_\tau : \tau \in \{0, 1\}^{|\mathcal{V}_n|} \right\}. \tag{6.8}$$

Now we have the following result.

LEMMA 6.2. *For sufficiently large n there exists $M_1 > 0$ such that $\mathcal{H}_n \subseteq H_s(G, M)$, where $s > (\dim G)/2$.*

PROOF. By (2.10) and (6.5),

$$\sum_{\mathcal{V}_n} {d_\nu}^2 (1 + \lambda_\nu)^s \le C h_n^{2s + \dim G}$$

for some constant $C > 0$, as $n \to \infty$. Applying the Sobolev norm to (6.7), we obtain

$$\begin{aligned} \|f_\tau\|_{H_s}^2 &= 1 + M_1^2 h_n^{-2s-\dim G} \sum_{\mathcal{V}_n} \tau_\nu {d_\nu}^2 (1 + \lambda_\nu)^s \\ &\le 1 + M_1^2 h_n^{-2s-\dim G} \sum_{\mathcal{V}_n} {d_\nu}^2 (1 + \lambda_\nu)^s \\ &\ll 1 + C M_1^2, \end{aligned}$$

as $n \to \infty$. Consequently, by choosing $M_1^2 \le M/C$, we have $f_\tau \in H_s(G, M)$. Since $s > (\dim G)/2$ then we have the desired result. □

Now let $f, g \in \mathcal{H}_n$ with $f \ne g$. By the orthonormality of χ_ν, it follows from (6.7) that

$$\|f - g\|_2 \ge M_1 h_n^{-s-(\dim G)/2}. \tag{6.9}$$

We have the following version of Lemma 3.1 of Koo (1993).

LEMMA 6.3. (Koo) *For a given $\mathcal{H}_n$, there exists a set $\mathcal{H}_n^0 \subseteq \mathcal{H}_n$ such that, for all $u, v \in \mathcal{H}_n^0$ with $u \ne v$,*

$$\|u - v\|_2 \ge M_2 h_n^{-s}$$

and

$$\log(|\mathcal{H}_n^0| - 1) \geq M_3 h_n^{\dim G}.$$

where M_2 *and* M_3 *are positive constants independent of* n.

Next we choose u and v in $\mathcal{H}_n$ such that

$$u - v = M_1\, h_n^{-s-(\dim G)/2} \sum_{\mathcal{V}_n} \tau_\nu d_\nu \chi_\nu.$$

Then

$$\begin{aligned} \|f_\epsilon * (u-v)\|_2^2 &= \sum_{\mathcal{V}_n} |(\widehat{f_\epsilon * (u-v)})(\nu)|^2 \\ &= M_1^2 h_n^{-2s-\dim G} \sum_{\mathcal{V}_n} |\tau_\nu d_\nu f_\epsilon * \chi_\nu|^2 \\ &\leq M_1^2 h_n^{-2s-\dim G} \sum_{\mathcal{V}_n} {d_\nu}^2 \|\widehat{f}_\epsilon(\nu)\|_{\text{op}}^2. \end{aligned}$$

Consequently,

$$\|f_\epsilon * (u-v)\|_2^2 \tag{6.10}$$

$$\ll \begin{cases} h_n^{-2s-4\beta}, & f_\epsilon \text{ smooth}, \\ h_n^{-2s+4\beta_1} \exp(-2\gamma^{-1} h_n^{2\beta}), & f_\epsilon \text{ super-smooth}, \\ h_n^{-2s+4\beta_1} \exp(2h_n^{2\beta}[\gamma_1^{-1} - 2\gamma^{-1}\log h_n]), & f_\epsilon \text{ log-super-smooth}, \end{cases}$$

as $n \to \infty$.

Recall that the Kullback-Leibler information divergence, $D(f_1\|f_2)$, between two densities f_1 and f_2 is defined by $D(f_1\|f_2) = \int f \log(f_1/f_2)$ and satisfies

$$D(f_1\|f_2) \leq \int \frac{(f_1 - f_2)^2}{f_2}. \tag{6.11}$$

By (6.11) with f_1 and f_2 replaced by $f_\epsilon * u$ and $f_\epsilon * v$, respectively, and an application of Jensen's inequality, we obtain

$$D(f_\epsilon * u\|f_\epsilon * v) \ll \|f_\epsilon * (u-v)\|_2^2 \tag{6.12}$$

for all $u, v \in \mathcal{H}_n^0$, as $n \to \infty$.

The following version of Fano's lemma is noted by Birgé (1983), Yatracos (1988) and Koo (1993).

LEMMA 6.4. (Fano) *If* f^n *is an estimator of a density function* f *then, for some constant* $c > 0$,

$$\begin{aligned} \sup_{f \in H_s(G,M)} P_f\left(\|f^n - f\|_2 > c h_n^{-s}\right) &\geq \sup_{f \in \mathcal{H}_n^0} P_f\left(\|f^n - f\|_2 > c h^{-s}\right) \\ &\geq 1 - \frac{nD(f_\epsilon * u\|f_\epsilon * v) + \log 2}{\log(|\mathcal{H}_n^0| - 1)}. \end{aligned} \tag{6.13}$$

Now we let

$$h_n \asymp \begin{cases} n^{1/[2s+4\beta+\dim G]}, & f_\epsilon \text{ smooth}, \\ (\log n)^{1/(2\beta)}, & f_\epsilon \text{ super-smooth}, \\ (\log n/\log\log n)^{1/(2\beta)}, & f_\epsilon \text{ log-super-smooth}, \end{cases}$$

as $n \to \infty$. Applying Lemma 6.3, (6.10) and (6.12) to (6.13) we deduce that, for all three smoothness classes,

$$\lim_{c \to 0} \liminf_{n} \inf_{f^n} \sup_{f \in H_s(G,M)} P_f \left(\|f^n - f\|_2 > ch^{-s} \right) = 1,$$

thus establishing the lower bound.

Having established the lower bound rates, our final task is to combine these rates with the upper bound results of Section 6.1. The completion of the proofs to Theorems 4.2, 4.3 and 4.4 arises from the observation that the upper and lower rates of convergence are identical. Therefore the upper bounds are achievable, hence they are optimal.

References

[1] Birgé, L. (1983). Approximation dans les espaces metriques et theorie de l'estimation. *Z. Wahrsch. verw Gebiete* **65**, 181-237.

[2] Bröcker, T., and tom Dieck, T. (1985). *Representations of Compact Lie Groups*. Springer, New York.

[3] Colzani, L., Giulini, S., Travaglini, G., and Vignati, M. (1990). Pointwise convergence of Fourier series on compact Lie groups. *Colloq. Math.* **58**, 379-386.

[4] Dey, A.K., Ruymgaart, F. H., and Mair, B.A. (1996). Cross-validation for parameter selection in inverse estimation problems. *Scand. J. Statist.* **23**, 609-620.

[5] Diggle, P.J., and Hall, P. (1993). A Fourier approach to nonparametric deconvolution of a density estimate. *J. Roy. Statist. Soc.*, Ser. B, **55**, 523-531.

[6] Fan, J. (1991). On the optimal rates of convergence for nonparametric deconvolution problems. *Ann. Statist.* **19**, 1257-1272.

[7] ——— (1991a). Global behavior of deconvolution kernel estimates. *Statist. Sinica* **1**, 541–551.

[8] Fegan, H.D. (1991). *Introduction to Compact Lie Groups*. World Scientific Publ., Singapore.

[9] Giné, E.M. (1975). The addition formula for the eigenfunctions of the Laplacian. *Adv. Math.* **18**, 102-107.

[10] Healy, D.M., and Kim, P.T. (1996). An empirical Bayes approach to directional data and efficient computation on the sphere. *Ann. Statist.* **24**, 232-254.

[11] Healy, D.M., Hendriks, H., and Kim, P.T. (1998). Spherical deconvolution. *J. Multivariate Anal.* **67**, 1-22.

[12] Helgason, S. (1984). *Groups and Geometric Analysis*. Academic Press, Orlando, FL.

[13] Hendriks, H. (1990). Nonparametric estimation of a probability density on a Riemannian manifold using Fourier expansions. *Ann. Statist.* **18**, 832-849.

[14] Hewitt, E., and Ross, K.A. (1970). *Abstract Harmonic Analysis*, Vol. II. Springer, New York.

[15] Khatri, C.G., and Mardia, K.V. (1977). The von Mises-Fisher matrix distribution in orientation statistics. *J. Roy. Statist. Soc.*, Ser. B, **39**, 95-106.

[16] Kim, P.T. (1998). Deconvolution density estimation on SO(N). *Ann. Statist.* **26**, 1083-1102.

[17] Kim, P.T. (2000). On the characteristic function of von Mises-Fisher matrix distribution. In *High Dimensional Probability II*, E. Giné, D.M. Mason and J.A. Wellner, editors. Birkhäuser: Boston, 475-490.

[18] Kim, P.T., and Koo, J.Y. (2001). Optimal spherical deconvolution. *J. Multivariate Anal.*, in press.

[19] Koo, J.Y. (1993). Optimal rates of convergence for nonparametric statistical inverse problems. *Ann. Statist.* **21**, 590-599.

[20] Koo, J.Y., and Chung, H.Y. (1998). Log-density estimation in linear inverse problems. *Ann. Statist.* **26**, 335-362.

[21] Lo, J.T., and Ng, S.K. (1988). Characterizing Fourier series representation of probability distributions on compact Lie groups. *SIAM J. Appl. Math.* **48**, 222-228.

[22] Minakshisundaram, S., and Pleijel, Å. (1949). Some properties of the eigenfunctions on the Laplace operator on Riemannian manifolds. *Canad. J. Math.* **1**, 242-256.

[23] Rooij, A.C.M. van, and Ruymgaart, F.H. (1991). Regularized deconvolution on the circle and the sphere. In: *Nonparametric Functional Estimation and Related Topics* (G. Roussas, ed.)., pp. 679–690, Kluwer Acad. Publ., Dordrecht, 1991.

[24] Stefanski, L.A., and Carroll, R.J. (1990). Deconvoluting kernel density estimators. *Statistics* **21**, 169-184.

[25] Stone, C.J. (1980). Optimal rates of convergence for nonparametric estimators. *Ann. Statist.* **8**, 1348-1360.

[26] Yatracos, Y.G. (1988). A lower bound on the error in nonparametric regression type problems. *Ann. Statist.* **16**, 1180-1187.

[27] Zhang, C.H. (1990). Fourier methods for estimating mixing densities and distributions. *Ann. Statist.* **18**, 806-830.

Department of Mathematics and Statistics, University of Guelph, Guelph, Ontario N1G 2W1, Canada.

E-mail address: pkim@uoguelph.ca

School of Mathematics, Institute for Advanced Study, Princeton, NJ 08540, and Department of Statistics, University of Virginia, Charlottesville, VA 22904, USA.

E-mail address: drichards@virginia.edu

Contemporary Mathematics
Volume **287**, 2001

On Efficiency of Indirect Estimation of Nonparametric Regression Functions

Chris A. J. Klaassen, Eun-Joo Lee, and Frits H. Ruymgaart

ABSTRACT. The usual criterion for asymptotic optimality of curve estimators, both in direct and indirect models, is based on the mean integrated squared error. In this paper we focus on asymptotic efficiency in the sense of Hájek (1970). In van Rooij *et al.* (1999) it was shown that the usual estimators of indirectly observed densities are efficient in this sense. As a complement to this result we will show here, however, that in the indirect regression model the usual estimators are not in general efficient, except possibly when the error distribution is normal.

1. Introduction

In order to assess the quality of a curve estimator in a direct or indirect curve estimation model, as a criterion one usually employs the rate at which its mean integrated squared error tends to zero as the sample size tends to infinity. Estimators with highest possible rate are then considered as optimal. Very precise information about optimal rates for mean integrated squared error can be found in Pinsker (1980).

One may, however, also approach the question of optimality from the perspective of the Hájek-LeCam convolution theorem (Hájek (1970), Bickel *et al.* (1993), van der Vaart (1988)), and try to find an asymptotically Gaussian curve estimator with smallest possible covariance operator. This objective, however, is too pretentious since it is well-known that such curve estimators cannot converge weakly to a Gaussian limit as random elements in some function space, like for instance a Hilbert space (Parzen (1962), Schuster (1972), Ruymgaart (1998)). Since certain marginal distributions may still converge weakly, restriction to weak convergence of finite dimensional or even one-dimensional distributions appears to be natural in this context. In Hilbert spaces this boils down to weak convergence of linear functionals of the curve estimator. The problem now reduces to showing that such

1991 *Mathematics Subject Classification.* Primary 65J22, 62G08; Secondary 45R05.

Key words and phrases. Noisy integral equation, indirect nonparametric regression, Hájek-LeCam efficiency.

The last two authors have been partially supported by National Security Agency grant MDA 904-99-1-0029.

random linear functionals, as estimators of the same linear functional of the unknown true curve, are asymptotically normal with smallest possible variance.

Estimation of linear functionals is also of independent interest. For direct models they have been extensively studied in Ibragimov & Hasminskii (1984) and Ibragimov & Khas'minskii (1991). In this note we want to address the issue whether optimal estimators (in the sense of Hájek-LeCam) of linear functionals of the indirectly observed regression function can be obtained from a global estimator of that regression function. In order to prepare for a more detailed discussion let us first present a generic example of an indirect regression model. Such models are usually formulated as noisy integral equations (Wahba (1977)). These contain deconvolution, unfolding as in Wicksell's stereology problem, and tomography, based on noisy data, as special cases.

Example 1. Let $K : L^2([0,1]) \to L^2([0,1])$ be the integral operator with kernel

$$K(s,t) = s \wedge t, \ (s,t) \in [0,1] \times [0,1], \tag{1.1}$$

without confusion denoted by the same symbol. This kernel is the covariance kernel of Brownian motion on [0,1] and related to the kernel

$$\tilde{K}(s,t) = \mathbf{1}_{[s,1]}(t) \frac{1}{\sqrt{t-s}}, \ (s,t) \in [0,1] \times [0,1], \tag{1.2}$$

of the Abel equation with index $\frac{1}{2}$ that plays a role in Wicksell's problem (Groeneboom & Jongbloed (1995), Mair *et al.* (2000)). The operator is injective and compact. (The general theory will be derived for bounded injective operators.) Observations are given on the image

$$(K\theta)(s) = \int_0^s t \cdot \theta(t)dt + s \int_s^1 \theta(t)dt, \ s \in [0,1], \tag{1.3}$$

where $\theta \in \Theta \subset L^2([0,1])$, Θ a given smoothness class, and θ unknown. The problem is to recover θ from the observations and assess the quality of the estimator.

To describe the observations let $\varepsilon_1, ..., \varepsilon_n$ be independent and identically distributed (IID) error variables with mean zero, finite variance, and density f with respect to Lebesgue measure. In the case of a random design we are given IID random variables $X_1, ..., X_n$ with known density concentrating mass 1 on [0,1], for instance the Uniform (0,1) density. The observations are the pairs (X_i, Y_i), where

$$Y_i = (K\theta)(X_i) + \varepsilon_i, \ i = 1, ..., n. \tag{1.4}$$

In the case of a deterministic design we are given nonrandom quantities $x_i \in [0,1]$, for instance $x_i = i/n$, and the observations are the pairs (x_i, Y_i), with

$$Y_i = (K\theta)(x_i) + \varepsilon_i, \ i = 1, ..., n. \tag{1.5}$$

The indirect regression model has a number of features that distinguishes it from indirect density estimation, viz. the presence of the density of the errors that plays the role of a nuisance parameter when unknown, and two possible sampling schemes rather than one. The indirect regression model we will be dealing with is in fact a semi parametric model, where both the parameter of interest and the nuisance parameter are infinite dimensional. Both for indirect density and random design indirect regression models asymptotic normality of estimators of linear functionals, based on the curve estimator, was established in van Rooij *et al.* (1999). In the notation of Example 1, let be given a suitable $h \in L^2([0,1])$ and an estimator $\hat{\theta}$ of θ. Then $\langle \hat{\theta}, h \rangle$ will be an asymptotically normal estimator of $\langle \theta, h \rangle$ in the random design case. The question that we want to address here is whether this estimator is efficient in the sense of Hájek-LeCam (HL-efficient).

In van Rooij *et al.* (1999) this question was only considered for indirect density estimation, where it could be answered in the affirmative. In order to restrict the technicalities, in this paper we will deal with HL- efficiency for a fixed but arbitrary error density, although in practice this density will not usually be known. As we will see in Section 3, the variance of the limiting normal distribution of the estimator $\langle \hat{\theta}, h \rangle$ will not in general be minimal, except possibly when the errors have a normal density. Consequently the estimators will not be HL -efficient, unless the errors are normally distributed. In Section 4 we will briefly comment on these results. The general model, the basic assumptions, as well as the construction of the estimator of the curve θ will be reviewed in Section 2.

The problem of estimating the infinite dimensional nuisance parameter is beyond the purpose and scope of this paper. In the ordinary linear regression model the error distribution has been efficiently estimated in Klaassen & Putter (1999). The random and deterministic designs are instances of a structural and a functional model respectively. Van den Heuvel (1996) deals with the problem of efficiently estimating a finite dimensional parameter in the functional model and how to relate this to efficient estimation in the structural model. See also Bickel & Klaassen (1986), Pfanzagl (1993), and Strasser (1993, 1995). The random design, considered here, leads to IID observations. When a deterministic design is used the observations are no longer IID. This is the cause of additional technical difficulties and therefore the deterministic design will not be dealt with here.

2. The model, the basic assumptions, and the estimator

Let $(\mathbf{X}, \mathcal{X}, \mu)$ and $(\mathbf{Y}, \mathcal{Y}, \nu)$ be measurable spaces with $L^2(\mu)$ and $L^2(\nu)$ real separable Hilbert spaces. We are given a bounded, injective linear operator $K : L^2(\nu) \to L^2(\mu)$. In the random design case we observe a random sample $(X_1, Y_1), ...,$ (X_n, Y_n) of independent copies of a random element (X, Y), where

$$Y = (K\theta)(X) + \varepsilon. \tag{2.1}$$

The indirectly observed regression function $\theta \in \Theta \subset L^2(\mu)$ is unknown, X is the random design variable with density g with respect to μ, and ε the real valued error variable with density f with respect to Lebesgue measure. The error variable ε is independent of the design X and has zero mean and finite variance. Without losing too much generality we will simply assume that

$$\mu(\mathbf{X}) = 1, \; g(x) = 1, \; \text{for } x \in \mathbf{X}. \tag{2.2}$$

To compute the joint density of (X, Y) choose $A \in \mathcal{X}$ and $B \in \mathcal{B}$ ($\mathcal{B}$ is the σ-field of Borel sets in $\mathbb{R}$), and note that

$$\begin{aligned} \mathbf{P}\{X \in A, Y \in B\} &= \mathbf{EP}\{X \in A, (K\theta)(X) + \varepsilon \in B | X\} \\ &= \int \mathbf{1}_A(x) \{ \int_B f(y - (K\theta)(x)) dy \} d\mu(x) \\ &= \int_A \int_B f(y - (K\theta)(x)) d\mu(x) dy. \end{aligned} \tag{2.3}$$

It follows that the joint density is given by

$$p_{\theta,f}(x,y) = f(y-(K\theta)(x)), (x,y) \in \mathbf{X}\times\mathbb{R}. \tag{2.4}$$

If f is unknown this is clearly a semi parametric model with $\theta \in \Theta$ the actual parameter of interest and f a nuisance parameter.

Adopting the approach in van Rooij *et al.* (1999) let us first observe that the polar decomposition (Riesz & Nagy (1990)) claims the existence of a partial isometry $V : L^2(\nu) \to L^2(\mu)$ such that $K = VR$ and $K^* = RV^*$ (a "star" indicates the adjoint), where $R = (K^*K)^{\frac{1}{2}}$. Since K is injective, R is strictly positive Hermitian. Application of Halmos's (1963) version of the spectral theorem entails the existence of a σ-finite measure space $(\mathbf{S},\mathcal{S},\Sigma)$, a unitary operator $U : L^2(\nu) \to L^2(\Sigma)$, and a function $\rho \in L^\infty(\Sigma)$ with $\rho > 0$ in this case, such that

$$Rh = U^{-1}(\rho\cdot Uh),\ \ h \in L^2(\nu). \tag{2.5}$$

Because V^*V is the identity on the range of R we may write $K\theta = VU^{-1}(\rho\cdot U\theta)$, or

$$\chi = WK\theta = \rho\cdot U\theta,\ \text{ where } W = UV^*. \tag{2.6}$$

Let us refer to χ as the generalized characteristic function.

We will now assume that the operators U and W are integral operators with bounded kernels

$$U \in L^\infty(\Sigma\times\nu)\ ,\ W \in L^\infty(\Sigma\times\mu). \tag{2.7}$$

Such an assumption is not needed for V, so that the simple case where V equals the identity may be included. Furthermore we will assume

$$\Sigma\{\rho\geq\alpha\} < \infty\ ,\ \text{for all } \alpha > 0, \tag{2.8}$$

$$\mathbf{E}|Y|^{2+\delta} < \infty,\ \text{for some } \delta > 0. \tag{2.9}$$

To construct a class of estimators for θ let us first estimate the generalized characteristic function χ in (2.6).

Theorem 1. *An unbiased estimator of χ is given by*

$$\hat{\chi}(s) = \frac{1}{n}\sum_{i=1}^{n} Y_i W(s,X_i),\ \ s \in \mathbf{S}. \tag{2.10}$$

Proof. Just note that

$$\begin{aligned}\mathbf{E}YW(s,X) &= \mathbf{E}\{(K\theta)(X)+\varepsilon\}W(s,X) = \\ &= \mathbf{E}(K\theta)(X)W(s,X) + \mathbf{E}\varepsilon W(s,X) = \\ &= \int_{\mathbf{X}} W(s,x)(K\theta)(x)d\mu(x) = \chi(s).\end{aligned} \tag{2.11}$$

At the third step we use the stochastic independence of X and ε, assumed in the beginning of this section.

We see from (2.6) that $\theta = U^{-1}(\chi/\rho)$. Division by ρ will in general lead to ill-posedness since ρ will assume values arbitrarily close to zero. For estimation we will therefore apply truncation in order to regularize and consider the estimators

$$\hat{\theta}_\alpha = U^{-1}(\mathbf{1}_{\{\rho\geq\alpha\}}\frac{\hat{\chi}}{\rho}),\ \ \alpha > 0. \tag{2.12}$$

The unbiasedness of $\hat{\chi}$ entails at once that

$$\theta_\alpha = \mathbf{E}\hat{\theta}_\alpha = U^{-1}(\mathbf{1}_{\{\rho\geq\alpha\}}\frac{\chi}{\rho}),\ \alpha > 0. \tag{2.13}$$

Since K has a real valued kernel it can be shown that $\hat{\theta}_\alpha$ is real valued as well (van Rooij & Ruymgaart (1999)).

It appears convenient to restrict the linear functionals $\langle\theta, h\rangle$ to those with

$$h \in \mathcal{H} = \{g \in L^2(\nu) : Ug = \rho\cdot\gamma, \gamma \in L^1(\Sigma)\cap L^2(\Sigma)\}. \tag{2.14}$$

Without a condition of this type the estimators of the functional may not be weakly convergent at the $\sqrt{n}$ - rate.

Let us conclude this section with some notation. Since $\hat{\chi}$ is an average of IID elements its covariance function equals

$$\mathbf{E}\{\hat{\chi}(s) - \chi(s)\}\{\overline{\hat{\chi}(t)} - \overline{\chi(t)}\} = \frac{1}{n}D(s,t), s, t \in \mathbf{S}, \tag{2.15}$$

where

$$\begin{aligned} D(s,t) &= \mathbf{E}\{YW(s,X) - \chi(s)\}\{Y\ \overline{W(t,X)} - \overline{\chi(t)}\} = \\ &= \mathbf{E}Y^2W(s,X)\ \overline{W(t,X)} - \chi(s)\overline{\chi(t)},\ \ s,t\in\mathbf{S}. \end{aligned} \tag{2.16}$$

It should be noted that

$$\sup_{s,t\in\mathbf{S}} |D(s,t)| < \infty, \tag{2.17}$$

by assumption (2.7).

Example 2. In Example 1 we have $L^2(\mu) = L^2(\nu) = L^2([0,1])$. The operator K itself is strictly positive Hermitian so that $K = R$ and V reduces to the identity. There exists an orthonormal basis of eigenfunctions $e_1, e_2, ...$ with corresponding eigenvalues $\rho_1, \rho_2, ...,$ where

$$e_k(x) = \sqrt{2}\sin(k-\frac{1}{2})\pi x,\ x\in[0,1],\ \rho_k = \frac{1}{\{(k-\frac{1}{2})\pi\}^2},\ k\in\mathbb{N}. \tag{2.18}$$

For each $h \in L^2([0,1])$ we have the expansion

$$Kh = Rh = \sum_{k=1}^{\infty} \rho_k \langle h, e_k\rangle e_k. \tag{2.19}$$

This is precisely what is claimed in (2.5) if we choose $L^2(\Sigma) = L^2(\#) = \ell^2$, where $\#$ denotes the counting measure on $\mathbb{N}$, and define $U : L^2([0,1]) \to \ell^2$ by $U(k,x) = e_k(x), k \in \mathbb{N}, x \in [0,1]$. Note that $W = U, \rho = (\rho_1, \rho_2, ...) \in L^\infty(\#) = \ell^\infty$, and that

$$Uh = (h_1, h_2, ...) \in \ell^2,\ h_k = \langle h, e_k\rangle, \tag{2.20}$$

$$\rho\cdot Uh = (\rho_1h_1, \rho_2h_2, ...) \in \ell^2, \tag{2.21}$$

and finally, indeed,

$$U^{-1}(\rho\cdot Uh) = \sum_{k=1}^{\infty}\rho_kh_ke_k = Rh. \tag{2.22}$$

Furthermore, from (2.6) we have $\chi = \rho \cdot U\theta = (\rho_1\theta_1, \rho_2\theta_2, ...)$, with $\theta_k = \langle \theta, e_k \rangle$, and from (2.10) since $W = U$

$$\hat{\chi} = (\hat{\chi}_1, \hat{\chi}_2, ...), \ \hat{\chi}_k = \frac{1}{n}\sum_{i=1}^{n} Y_i e_k(X_i). \tag{2.23}$$

This leads to the curve estimator (cf. (2.12))

$$\hat{\theta}_\alpha = \sum_{k:\rho_k \geq \alpha} \frac{1}{\rho_k} \cdot \hat{\chi}_k \cdot e_k(x), \ x \in [0,1]. \tag{2.24}$$

Clearly (2.7) and (2.8) are satisfied (see (2.18)). The covariance in (2.16) equals

$$D(j,k) = \langle K\theta \cdot e_j, K\theta \cdot e_k \rangle + (\mathbf{E}\varepsilon^2)\delta_{j,k} - \langle K\theta, e_j \rangle \langle K\theta, e_k \rangle, \tag{2.25}$$

for k, $j \in \mathbb{N}$. The class $\mathcal{H}$ in (2.14) consists of all

$$g(x) = \sum_{k=1}^{\infty} \rho_k \gamma_k e_k(x), \ x \in [0,1], \ \sum_{k=1}^{\infty} |\gamma_k| < \infty. \tag{2.26}$$

3. On asymptotic efficiency

It has been shown in van Rooij *et al.* (1999) that

$$\sqrt{n}\langle \hat{\theta}_{\alpha(n)} - \theta, h \rangle \to_d \text{Normal } (0, \sigma^2(h)), \ \text{as } n \to \infty, \tag{3.1}$$

where (henceforth $\gamma = \frac{1}{\rho} U h$)

$$\sigma^2(h) = \int_{\mathbf{S}} \int_{\mathbf{S}} \overline{\gamma(s)} \ D(s,t)\gamma(t) d\Sigma(s) d\Sigma(t), \tag{3.2}$$

provided that $\alpha(n) \to 0$ in such a way that

$$\sqrt{n}\langle \theta_{\alpha(n)} - \theta, h \rangle \to 0, \ \text{as } n \to \infty. \tag{3.3}$$

Example 3. In the situation of Examples 1 and 2 let us take $h = \rho_m e_m$ so that $Uh = \rho \cdot \gamma$ with $\gamma = (0, ..., 0, 1, 0, ...) \in \ell^1 \cap \ell^2$, where the 1 is at the m-th position. We see from (2.23) and (2.24) that

$$\langle \hat{\theta}_\alpha, h \rangle = \hat{\chi}_m = \frac{1}{n}\sum_{i=1}^{n} Y_i e_m(X_i), \ \text{for all } 0 < \alpha \leq \rho_m. \tag{3.4}$$

In this simple case the estimator is unbiased for such α, since $\mathbf{E}\hat{\chi}_m = \chi_m = \rho_m\theta_m = \langle \theta, h \rangle$. The central limit theorem applies at once and yields

$$\sqrt{n}\langle \hat{\theta}_\alpha - \theta, h \rangle = \sqrt{n}(\hat{\chi}_m - \chi_m) \to_d \ \text{Normal } (0, \sigma_m^2), \tag{3.5}$$

where

$$\begin{aligned} \sigma_m^2 &= \mathbf{E}(Y e_m(X))^2 - \chi_m^2 = \\ &= \langle K\theta \cdot e_m, K\theta \cdot e_m \rangle + (\ \mathbf{E}\varepsilon^2) - \langle K\theta, e_m \rangle^2. \end{aligned} \tag{3.6}$$

On the other hand (3.1) applies since (3.3) is trivially fulfilled with $\theta_\alpha = \theta$. It is immediate from (2.25) that $\sigma_m^2 = \sigma^2(h)$ for the current h.

For some $\varepsilon > 0$ let $t \mapsto \theta_t$, $|t| < \varepsilon$, be a curve in $L^2(\nu)$, write $\theta_{t,K} = K\theta_t$, and assume that $\overset{\bullet}{\theta}_{t,K} = \partial K\theta_t/\partial t$ exists on $\mathbf{X}$. Let us keep the error density f fixed, and assume that

$$f > 0, \text{ bounded and continuously differentiable on } \mathbb{R}, \tag{3.7}$$

$$\frac{f'}{\sqrt{f}} \text{ bounded}, \quad \lim_{|y|\to\infty} yf(y) = 0, \quad \int_{-\infty}^{\infty} \left| \frac{yf'(y)}{\sqrt{f(y)}} \right| dy < \infty, \tag{3.8}$$

$$\theta_{t,K}(x), \overset{\bullet}{\theta}_{t,K}(x) \text{ continuous in } t \text{ for all } x \in \mathbf{X}, \tag{3.9}$$

$$\sup_{t,x} |\theta_{t,K}(x)|, \quad \sup_{t,x} |\overset{\bullet}{\theta}_{t,K}(x)| < \infty. \tag{3.10}$$

An alternative set of conditions will be needed to deal with the special case where the errors have the normal density

$$f(y) = \frac{1}{\sqrt{2\pi}} \exp(-\frac{1}{2}\, y^2), \; y \in \mathbb{R}. \tag{3.11}$$

In this case we will require that

$$\exists \text{ a curve}: \theta_{0,K}(x) \cdot \overset{\bullet}{\theta}_{0,K}(x) = 1, \; \overset{\bullet}{\theta}_{0,K}(x) = (W^*\gamma)(x), \; x \in \mathbf{X}, \tag{3.12}$$

in addition to (3.9) and (3.10).

We will now summarize the conditions under which the general theory and the special case of normal errors will be dealt with. Unless stated otherwise Assumption 1 will be in force throughout.

Assumption 1. Arbitrary but fixed error density. Conditions (2.2), (2.7) - (2.9), (3.3), and (3.7) - (3.10) are satisfied.

Assumption 2. Normal errors. Conditions (2.2), (2.7) - (2.9), (3.3), and (3.9) - (3.12) are satisfied.

Normal distributions obviously satisfy (3.7) and (3.8). The standard normal density f in (3.11) has, moreover, the well-known property

$$f'(y) = -yf(y), \; y \in \mathbb{R}, \tag{3.13}$$

that will have further consequences. A starting point for constructing curves satisfying (3.12) is to try

$$\theta_{t,K} = \frac{1}{W^*\gamma} + tW^*\gamma, \; |t| < \varepsilon. \tag{3.14}$$

Before proceeding with the main theorem let us investigate the differentiability in quadratic mean of the functional

$$T(\sqrt{p_t}) = \langle \theta_t, h \rangle, \tag{3.15}$$

as a functional of $\sqrt{p_t}$, where we write briefly p_t rather than $p_{\theta_t,f}$; see (2.4). Because $\int_{-\infty}^{\infty} yp_t(x,y)dy = \theta_{t,K}(x)$, $x \in \mathbf{X}$, and $(K^{-1})^*h = W^*\gamma$, it follows that we may write

$$T(\sqrt{p_t}) = \langle K^{-1}K\theta_t, h \rangle = \langle \int_{-\infty}^{\infty} yp_t(\bullet, y)dy, \; W^*\gamma \rangle. \tag{3.16}$$

For the differentiability of T we first need to verify the differentiability of $\sqrt{p_t}$. By the mean value theorem we have

$$\frac{1}{t}\left\{\sqrt{f(y-\theta_{t,K}(x))}-\sqrt{f(y-\theta_{0,K}(x))}\right\}= \tag{3.17}$$
$$=\frac{f'(y-\theta_{\tilde{t},K}(x))}{2\sqrt{f(y-\theta_{\tilde{t},K}(x))}}\cdot\dot{\theta}_{\tilde{t},K}(x),\ x\in\mathbf{X},$$

where $\tilde{t}=\tilde{t}(x)$ is between 0 and t. Under the present conditions there is pointwise convergence to

$$\dot{p}_0(x,y)=\frac{f'(y-\theta_{0,K}(x))}{2\sqrt{f(y-\theta_{0,K}(x))}}\cdot\dot{\theta}_{0,K}(x), \tag{3.18}$$

and by applying the dominated convergence theorem termwise in the expression for the norm of the difference it follows that

$$\frac{1}{t}\{\sqrt{p_t}-\sqrt{p_0}\}\to_{qm}\dot{p}_0,\ \text{as } t\to 0,\ \text{in } L^2(\mu\times\lambda), \tag{3.19}$$

where λ denotes Lebesgue measure on $\mathbb{R}$.

Let us now return to T and note that

$$\frac{1}{t}\{T(\sqrt{p_t})-T(\sqrt{p_0})\}=\langle\psi_t,W^*\gamma\rangle, \tag{3.20}$$

where, once more by the mean value theorem,

$$\psi_t(x)=\int_{-\infty}^{\infty}y\frac{1}{t}\{\sqrt{p_t(x,y)}-\sqrt{p_0(x,y)}\}\times\{\sqrt{p_t(x,y)}+\sqrt{p_0(x,y)}\}dy= \tag{3.21}$$
$$=\left\{\int_{-\infty}^{\infty}\frac{yf'(y)}{2\sqrt{f(y)}}dy+\int_{-\infty}^{\infty}\frac{f'(y)}{2\sqrt{f(y)}}dy\cdot\theta_{\tilde{t},K}(x)\right\}$$
$$\times\ \dot{\theta}_{\tilde{t},K}(x)\cdot\{\sqrt{p_t(x,y)}+\sqrt{p_0(x,y)}\},\ x\in\mathbf{X},$$

with $\tilde{t}=\tilde{t}(x)$ between 0 and t. Because $\int_{-\infty}^{\infty}yf'(y)dy=-1$ and $\int_{-\infty}^{\infty}f'(y)dy=0$ these functions clearly converge pointwise to

$$\psi_0(x)=-\dot{\theta}_{0,K}(x)=2\int_{-\infty}^{\infty}\dot{p}_0(x,y)y\sqrt{p_0(x,y)}\,\dot{\theta}_{0,K}(x)dy,\ x\in\mathbf{X}. \tag{3.22}$$

Because the convergence turns out to be dominated in the required sense we arrive at

$$\psi_t\to_{qm}\psi_0,\ \text{as } t\to 0,\ \text{in } L^2(\mu). \tag{3.23}$$

By continuity of the inner product it follows that

$$\frac{1}{t}\{T(\sqrt{p_t})-T(\sqrt{p_0})\}\to\langle\psi_0,W^*\gamma\rangle= \tag{3.24}$$
$$=2\int_{-\infty}^{\infty}\int_{\mathbf{X}}\dot{p}_0(x,y)y\sqrt{p_0(x,y)}(W^*\gamma)(x)d\mu(x)dy.$$

This means that the derivative or gradient of T at $\sqrt{p_0}$ is the linear functional defined by the function

$$\dot{T}_0(x,y)=2y\sqrt{p_0(x,y)}(W^*\gamma)(x),\ x\in\mathbf{X},\ y\in\mathbb{R}, \tag{3.25}$$

where $\dot{T}_0\in L^2(\mu\times\lambda)$.

Theorem 2. *Let Assumption 1 be satisfied, i.e., we have a fixed, arbitrary error density. Then the estimators $\langle \hat{\theta}_{\alpha(n)}, h\rangle$ will not in general be HL-efficient for estimating $\langle \theta, h\rangle$. If the error density is normal and Assumption 2 is fulfilled, these estimators are HL -efficient.*

Proof. It is immediate from (2.16) that $\sigma^2(h) = \sigma_1^2(h) - \sigma_2^2(h)$, where

$$
\begin{aligned}
\sigma_1^2(h) &= \int_{\mathbf{S}} \int_{\mathbf{S}} \overline{\gamma(s)} \{ \int_{-\infty}^{\infty} \int_{\mathbf{X}} y^2 W(s,x) \overline{W(t,x)} \\
&\qquad \times p_0(x,y) d\mu(x) dy \} \gamma(t) d\Sigma(s) d\Sigma(t) \\
&= \int_{-\infty}^{\infty} \int_{\mathbf{X}} y^2 \{ \int_{\mathbf{S}} W(s,x) \overline{\gamma(s)} d\Sigma(s) \} \\
&\qquad \times \{ \int_{\mathbf{S}} \overline{W(t,x)} \gamma(t) d\Sigma(t) \} p_0(x,y) d\mu(x) dy \\
&= \int_{-\infty}^{\infty} \int_{\mathbf{X}} y^2 \overline{(W^*\gamma)(x)} (W^*\gamma)(x) p_0(x,y) d\mu(x) dy \\
&= \int_{-\infty}^{\infty} \int_{\mathbf{X}} y^2 |(W^*\gamma)(x)|^2 p_0(x,y) d\mu(x) dy,
\end{aligned} \tag{3.26}
$$

$$
\begin{aligned}
\sigma_2^2(h) &= \int_{\mathbf{S}} \int_{\mathbf{S}} \overline{\gamma(s)} \{ \int_{-\infty}^{\infty} \int_{\mathbf{X}} y W(s,x) p_0(x,y) d\mu(x) dy \} \\
&\qquad \times \{ \int_{-\infty}^{\infty} \int_{\mathbf{X}} y \overline{W(t,x)} p_0(x,y) d\mu(x) dy \} \gamma(t) d\Sigma(s) d\Sigma(t) \\
&= | \int_{\mathbf{X}} \int_{-\infty}^{\infty} y (W^*\gamma)(x) p_0(x,y) d\mu(x) dy |^2.
\end{aligned} \tag{3.27}
$$

Let $\dot{\mathcal{P}}_0$ be the tangent space at $\sqrt{p_0}$ to the model $\mathcal{P}$ in $L^2(\mu \times \lambda)$ of all $\sqrt{p_{\theta,f}}$. The optimal variance equals (van der Vaart (1988))

$$
\frac{1}{4} ||\text{projection of } \dot{T}_0 \text{ onto } \dot{\mathcal{P}}_0||^2. \tag{3.28}
$$

Write $[\sqrt{p_0}]$ for the one-dimensional subspace generated by $\sqrt{p_0}$ and $[\sqrt{p_0}]^\perp$ for its orthogonal complement. We always have

$$
\dot{\mathcal{P}}_0 \subset [\sqrt{p_0}]^\perp. \tag{3.29}
$$

Now observe that

$$
\begin{aligned}
&\frac{1}{4} ||\text{projection of } \dot{T}_0 \text{ onto } [\sqrt{p_0}]^\perp||^2 = \\
&= \frac{1}{4} (||\dot{T}_0||^2 - |\langle \dot{T}_0, \sqrt{p_0} \rangle|^2) = \\
&= \frac{1}{4} \{\sigma_1^2(h) - \sigma_2^2(h)\} = \frac{1}{4} \sigma^2(h).
\end{aligned} \tag{3.30}
$$

Hence the estimators $\langle \hat{\theta}_{\alpha(n)}, h\rangle$ will indeed be in general sub efficient if $\dot{\mathcal{P}}_0{}^\perp$ is strictly larger than $[\sqrt{p_0}]$. To see this, first note that the assumptions entail $f' \in L^2(\lambda)$. Choose

$$
u \in L^2(\lambda) : u \perp f', \ v \in L^2(\mu). \tag{3.31}
$$

It is immediate that every function of type

$$q_0(x,y) = \sqrt{p_0(x,y)}u(y-\theta_{0,K}(x))v(x), \ (x,y)\in \mathbf{X}\times\mathbb{R}, \tag{3.32}$$

is in $L^2(\mu\times\lambda)$ and orthogonal to $\overset{\bullet}{p}_0$:

$$\langle q_0, \sqrt{p_0}\rangle = 0, \ \text{for all } \overset{\bullet}{p}_0 \text{ in (3.18)}. \tag{3.33}$$

Since the collection of all $\overset{\bullet}{p}_0$ should be dense in $\overset{\bullet}{\mathcal{P}}_0$, the codimension of $\overset{\bullet}{\mathcal{P}}_0$ must be strictly larger than 1 which proves the general part of the theorem.

Let us observe that

$$\text{projection } \overset{\bullet}{T}_0 \text{ onto } \overset{\bullet}{\mathcal{P}}_0 = \text{ projection } \overset{\bullet}{T}_0 \text{ onto } [\sqrt{p_0}]^{\perp} \tag{3.34}$$

holds true if and only if

$$\beta\sqrt{p_0} + \overset{\bullet}{T}_0 \in \overset{\bullet}{\mathcal{P}}_0, \ \text{for some } \beta\in\mathbb{R}. \tag{3.35}$$

It is not hard to see, however, that there are distributions for which (3.34) cannot be satisfied. We may take the student(3) distribution, for instance. This proves the claim that there exist error distributions for which the estimators are not HL-efficient.

Let us finally consider the statement regarding the normal error densities. To see that now (3.35) does holds true, first note that in the normal case the generic element $\overset{\bullet}{p}_0$ of $\overset{\bullet}{\mathcal{P}}_0$ can be written as (cf. (3.13)).

$$\begin{aligned} \overset{\bullet}{p}_0(x,y) &= -(y-\theta_{0,K}(x))\sqrt{f(y-\theta_{0,K}(x))}\overset{\bullet}{\theta}_{0,K}(x) = \\ &= \sqrt{p_0(x,y)}\theta_{0,K}(x)\overset{\bullet}{\theta}_{0,K}(x) - y\sqrt{p_0(x,y)}\overset{\bullet}{\theta}_{0,K}(x), \ x\in\mathbf{X},\ y\in\mathbb{R}. \end{aligned} \tag{3.36}$$

On the other hand

$$\begin{aligned} &\beta\sqrt{p_0(x,y)} + \overset{\bullet}{T}_0(x,y) = \\ &= \beta\sqrt{p_0(x,y)} + 2y\sqrt{p_0(x,y)}(W^*\gamma)(x), \ x\in\mathbf{X},\ y\in\mathbb{R}. \end{aligned} \tag{3.37}$$

It is immediate from the technical assumption (3.12) that (3.35) is fulfilled for $\beta = -2$.

4. Concluding remarks

The quality of the curve estimators in (2.12) has been investigated by considering the quality of the induced estimators of inner products. If estimation of a particular inner product were a problem in its own right we might as well employ a simpler estimator.

Recall that $Uh = \rho\cdot\gamma$, so that $\langle\theta,h\rangle = \langle U\theta, Uh\rangle = \langle\chi/\rho, \rho\cdot\gamma\rangle = \langle\chi,\gamma\rangle$. Therefore a natural estimator seems to be $\langle\hat{\chi},\gamma\rangle$. It is immediate from (2.10) that

$$\mathbf{E}\langle\hat{\chi},\gamma\rangle = \langle\chi,\gamma\rangle = \langle\theta,h\rangle, \tag{4.1}$$

so that the estimator is unbiased. Since it is an average of IID components the central limit theorem applies and yields asymptotic normality. The asymptotic variance equals the variance of the generic component $\langle YW(\bullet,X),\gamma\rangle$ and is easily seen to be equal to $\sigma^2(h)$. Hence, like the estimators $\langle\hat{f}_\alpha,h\rangle$, this estimator cannot be HL-efficient. In other words, there exist error densities for which $\sigma^2(h)$ is strictly larger than the number in (3.28).

Both estimators are $\sqrt{n}$-consistent and could therefore be used in the construction of an HL-efficient estimator of the inner product. Such a construction would also involve estimation of the error density (Klaassen & Putter (1999)) and is beyond the scope of the paper.

Returning to the curve estimator itself, it is well-known from the literature that its mean integrated squared error converges at the optimal rate, for suitably chosen sequences $\alpha(n)$ depending on the smoothness of θ. Here we have seen that (for possibly different sequences $\alpha(n)$) the asymptotic normality (3.1) holds true. In the case of normal errors the asymptotic variance is as small as possible for a subclass of functions h, and hence the estimator will be HL-efficient. For non normal errors HL-efficiency will no longer in general be satisfied.

For deterministic designs we expect similar results. The convolution theorem which is behind the HL-efficiency concept was established in Hájek (1970) for any underlying joint density satisfying a local asymptotic normality property. Hence in principle it should be possible to consider the deterministic design case that leads to sample elements that are still independent but no longer identically distributed.

Acknowledgments

The authors are grateful to a referee for some useful comments.

References

[1] Bickel, P.J. & Klaassen, C.A.J. (1986). Empirical Bayes estimation in functional and structural models, and uniformly adaptive estimation of location. *Adv. Appl. Math.* **7**, 55-69.

[2] Bickel, P.J., Klaassen, C.A.J., Ritov, Y. & Wellner, J.A. (1993). *Efficient and Adaptive Estimation for Semiparametric Models.* Johns Hopkins Univ. Press, Baltimore.

[3] Groeneboom, P. & Jongbloed, G. (1995). Isotonic estimation and rates of convergence in Wicksell's problem. *Ann. Statist.* **23**, 1518-1542.

[4] Hájek, J. (1970). A characterization of limiting disbtributions of regular estimates. *Z. Wahrsch. Verw. Gebiete* **14**, 323-330.

[5] Halmos, P.R. (1963). What does the spectral theorem say? *Amer. Math. Monthly* **70**, 241-247.

[6] Heuvel, E.R. van den (1996). *Bounds for Statistical Estimation.* Dissertation, Univ. of Amsterdam.

[7] Ibragimov, I.A. & Hasminskii, R.Z. (1984). On nonparametric estimation of the value of a linear functional in Gaussian white noise. *Th. Probab. Appl.* **29**, 18-32.

[8] Ibragimov, I.A. & Khas' minskii, R.Z. (1991). Asymptotically normal families of distributions and efficient estimation. *Ann. Statist.* **19**, 1681-1724.

[9] Klaassen, C.A.J. & Putter, H. (1999). Efficient estimation of Banach parameters in semiparametric models. Tech. Report, Dept. of Mathematics, Univ. of Amsterdam.

[10] Mair, B.A., Ruymgaart, F.H. & Urrabazo, T. (2000). Some comments on Wicksell's problem. *J. Statist. Pl. Inf.* **87**, 31-42.

[11] Parzen, E. (1962). On estimation of a probability density function and mode *Ann. Math. Statist.* **33**, 520-531.

[12] Pfanzagl, J. (1993). Incidental versus random nuisance parameters. *Ann. Statist.* **21**, 1663-1691.

[13] Pinsker, M.S. (1980). Optimal filtering of square integrable signals in Gaussian white noise. *Problems Inform. Transmission* **16**, 52-68.

[14] Riesz, F.& Sz.-Nagy, B. (1990). *Functional Analysis.* Dover, New York.

[15] Rooij, A.C.M. van & Ruymgaart, F.H. (1999). On inverse estimation. In: *Asymptotics, Nonparametrics, and Time Series,* S. Ghosh, ed. (pp 579-613). Dekker, New York.

[16] Rooij, A.C.M. van, Ruymgaart, F.H. & Zwet, W.R. van (1999). Asymptotic efficiency of inverse estimators.. *Th. Probab. Appl.* **44,** 826-844.

[17] Ruymgaart, F.H. (1998). A note on weak convergence of density estimators in Hilbert spaces. *Statist.* **30,** 331-343.

[18] Schuster, E.F. (1972). Joint asymptotic distribution of the estimated regression function at a finite number of distinct points. *Ann. Math. Statist.* **43,** 84-88.

[19] Strasser, H. (1993). Asymptotic efficiency of estimators for models with incidental nuisance parameters. Tech. Report, Univ. of Economics and Business Administration, Vienna.

[20] Strasser, H. (1995). Perturbation invariant estimates and incidental nuisance parameters. Tech. Report, Univ. of Economics and Business Administration, Vienna.

[21] Vaart, A. van der (1988). Statistical Estimation in Large Parameter Spaces. CWI Tract #44. Centre for Mathematics and Computer Science, Amsterdam.

[22] Wahba, G. (1977). Practical approximate solutions to linear operator equations when the data are noisy. *SIAM J. Numer. Anal.* **14,** 651-667.

DEPARTMENT OF MATHEMATICS, UNIVERSITY OF AMSTERDAM, AMSTERDAM, THE NETHERLANDS

E-mail address: `chrisk@science.uva.nl`

DEPARTMENT OF MATHEMATICS AND STATISTICS, TEXAS TECH UNIVERSITY, LUBBOCK, TX 79409, USA

E-mail address: `elee@math.ttu.edu`

DEPARTMENT OF MATHEMATICS AND STATISTICS, TEXAS TECH UNIVERSITY, LUBBOCK, TX 79409, USA

E-mail address: `ruymg@math.ttu.edu`

Contemporary Mathematics
Volume **287**, 2001

Patterned Matrices Treated Via Linear Spaces

Tõnu Kollo and Dietrich von Rosen

ABSTRACT. Patterned matrices, including symmetric, skew-symmetric, diagonal, triangular and Toeplitz are considered as generators of linear spaces, where a basis can be utilized in their explicit representation. The argument can be of use when, for example, Jacobians are calculated.

1. Introduction

The notion *patterned matrix* has been used with different meanings. Graybill [**G**] defines it "... by recognizing a particular structure, or pattern, in certain matrices." Nel [**DN**] uses the notion in a more restricted meaning: A matrix is said to be *patterned*, if a certain number of its elements are constants or repeated according to a given rule. In statistics, matrices with a patterned structure are often used. Symmetric structures, for example, make some problems more difficult than is the case with unstructured matrices; indeed, the basic the reason why the derivation of the inverse moments of the Wishart matrix is such a challenging problem is that we are dealing with a symmetric matrix. Furthermore, there exist problems for which we need only some part of a matrix and it is not of any importance at all what kind of relation this part has with the rest of the matrix. For example, when examining the asymptotic distribution of the multiple correlation coefficients (cf. [**P**]), as well as when we have a pattern, only portions of the matrix need to be identified.

The present paper will treat pattern matrices via a linear space approach. The basic idea is that to each pattern there exists a basis which will be utilized in its representation. In order to work with the basis, we introduce a pattern identifier which connects the pattern and the basis by a simple construction. The basis is then used to construct a transformation matrix and its Moore-Penrose inverse. The point is that we can carry out this explicitly for any arbitrary pattern. Special cases of these matrices have been considered previously, such as the duplication matrix (cf. [**MN**]), skew-symmetric and diagonal matrices (cf. [**N**]), and matrices with correlation-type structures.

1991 *Mathematics Subject Classification.* Primary 15A99; Secondary 15A03.

Key words and phrases. Toeplitz matrices, linear structures, linear spaces, Moore-Penrose inverse, matrix derivatives, Jacobians, transition matrices.

We emphasize that the aim of the paper is to present a new approach for dealing with structures and linear relationships among elements in a matrix. It would have been possible to derive new results for special matrix structures, but we believe that it is better to illustrate our approach via known results. The reason for considering Jacobians is the opportunity to show that matrix derivatives of patterned matrices via linear spaces leads to a systematic way of obtaining Jacobians. Other potential applications of our results include solving linear equations where matrices include linear structures, deriving statistical moments, in particular multivariate moments of higher order and deriving distribution approximations.

The paper is organized as follows: In Section 2 the material necessary to read the article is presented, including a derivation of the transformation matrix and its Moore-Penrose inverse; Section 3 derives explicit expressions for the basis for any pattern, which in turn leads to an explicit expression for the transformation matrix and its Moore-Penrose inverse; in Section 4 the results are applied to various patterns as well as some relations between the spaces obtained from the patterns are exploited. Among other results, it is observed that symmetric and skew-symmetric matrices generate mutually orthogonal spaces; finally Section 5 demonstrates how these results can be applied when deriving Jacobians.

2. Background

First we consider patterned matrices as subsets of matrix elements, without specifying a relation among the elements of the matrix. Specifically, given a $p \times q$ matrix $\mathbf{A}$, non-empty sets $I \subset \{1,\dots,p\}$ and $J \subset \{1,\dots,q\}$ and a non-empty subset K of $I \times J$, let

$$\mathbf{A}(K) = \{a_{ij} \in \mathbf{A};\ (i,j) \in K\},$$

and denote by $\mathrm{vec}\mathbf{A}(K)$ the vector of dimension $|K| = r$ obtained by lexicographically listing the elements of $\mathbf{A}(K)$ according to their indices. We say that the elements of $\mathbf{A}$ are selected according to the *pattern* described by K.

If all the information about the matrices $\mathbf{A}$ and $\mathbf{A}(K)$ is given by an index-set K, we can find a transformation matrix $\mathbf{T}(K)$, with 0-1 entries, satisfying

$$\mathrm{vec}\mathbf{A}(K) = \mathbf{T}(K)\mathrm{vec}(\mathbf{A}).$$

Formally, the transformation matrix $\mathbf{T}(K)$ is a $r \times pq$ partitioned matrix of columns consisting of $r \times p$ blocks. If $(\mathrm{vec}\mathbf{A}(K))_f = a_{ij}$, where $f \in \{1,\dots,r\}$, $i \in \{1,\dots,p\}$, $j \in \{1,\dots,q\}$, the general element of the matrix $\mathbf{T}(K)$ is given by

$$(\mathbf{T}(K))_{f(g,h)} = \begin{cases} 1, & (i,j) = (g,h) \quad g = 1,\dots,p;\ h = 1,\dots,q; \\ 0, & \text{elsewhere}, \end{cases} \tag{2.1}$$

where, following Anderson [**A**], the numeration of elements follows the relation

$$a_{(k,l)(g,h)} = a_{\sum_{i=1}^{k-1} p_i + l, \sum_{i=1}^{g-1} q_j + h} \tag{2.2}$$

and $a_{(k,l)(g,h)}$ is identified to be the l-th row of the k-th row block and the h-th column in the g-th column block of the block matrix

$$\mathbf{A} = \begin{pmatrix} \mathbf{A}_{11} & \mathbf{A}_{12} & \dots & \mathbf{A}_{1v} \\ \vdots & \vdots & \ddots & \vdots \\ \mathbf{A}_{u1} & \mathbf{A}_{u2} & \dots & \mathbf{A}_{uv} \end{pmatrix},$$

where $\mathbf{A}_{ij} : p_i \times q_j$.

Patterns given by linear structures. Suppose that we have a class of $p \times q$-matrices, say $\mathbb{M}_{p,q}$, where some fixed linear structure between the elements exists (see [**M83**] for a detailed analysis). Typical classes of matrices with linear structures include the symmetric, skew symmetric, diagonal, triangular and symmetric off-diagonal matrices. For all $\alpha, \beta \in \mathbb{R}$ and arbitrary $\mathbf{A}, \mathbf{D} \in \mathbb{M}_{p,q}$, we have

$$\alpha\mathbf{A} + \beta\mathbf{D} \in \mathbb{M}_{p,q},$$

so that $\mathbb{M}_{p,q}$ constitutes a linear space. It is customary to study the vectorized form of any matrix $\mathbf{A} \in \mathbb{M}_{p,q}$, i.e., vec$\mathbf{A}$. Because of the assumed linear structure the vector vec$\mathbf{A}$ belongs to a certain subspace in $\mathbb{R}^{pq}$, say r-dimensional, $r \leq pq$. This space is denoted $\mathbb{M}_{pq}$. The case in which $r = pq$ means that there is no linear structure in the matrix of interest, so we only consider the case $r < pq$. Since the space $\mathbb{M}_{p,q}$ is a linear space, we can always choose a matrix $\mathbf{B} : pq \times r$ which consists of r independent basis vectors such that $\mathbf{B}'\mathbf{B} = \mathbf{I}_r$. Furthermore, r rows of $\mathbf{B}$ are linearly independent. Let these r linearly independent rows of $\mathbf{B}$ form a matrix $\underline{\mathbf{B}}$: $r \times r$ of rank r, with inverse $\underline{\mathbf{B}}^{-1}$.

As noted above the aim is to relate the elements in vec$\mathbf{A}$ with the elements in vec$\mathbf{A}(K)$ so that vec$\mathbf{A}$ can be obtained from vec$\mathbf{A}(K)$ and vice versa. Our strategy is to first find matrices (or transformations) $\mathbf{T}$ and $\mathbf{C}$ such that

$$\mathbf{TB} = \underline{\mathbf{B}}; \qquad \mathbf{C}\underline{\mathbf{B}} = \mathbf{B}. \tag{2.3}$$

This means that $\mathbf{T}$ selects from the basis $\mathbf{B}$ a unique set of row vectors whereas $\mathbf{C}$ creates the original matrix from $\underline{\mathbf{B}}$. It follows immediately, since $\mathbf{B}'\mathbf{B} = \mathbf{I}_r$, that one solution is given by

$$\mathbf{T} = \underline{\mathbf{B}}\mathbf{B}'; \qquad \mathbf{C} = \mathbf{B}\underline{\mathbf{B}}^{-1}.$$

There exist other solutions but it turns out that this is the one which is useful for treating vec$\mathbf{A}$ and vec$\mathbf{A}(K)$. Furthermore, from (2.5), it follows that $\mathbf{TC} = \mathbf{I}$ which implies $\mathbf{TCT} = \mathbf{T}$ and $\mathbf{CTC} = \mathbf{C}$. Since $\mathbf{CT} = \mathbf{BB}'$ is symmetric, $\mathbf{C}$ is the Moore-Penrose inverse, $\mathbf{T}^+$, of $\mathbf{T}$. That is, $\mathbf{C} = \mathbf{T}^+$.

In order to consider the inverse transformation which enables us to "restore" vec$\mathbf{A}$ from vec$\mathbf{A}(K)$ we need a basis $\mathbf{B}$ such that vec$\mathbf{A} = \mathbf{Bq}$ and vec$\mathbf{A}(K) = \underline{\mathbf{B}}\mathbf{q}$. We also need to know which elements have been excluded. This information is given in the form of a pattern identifier induced by the linear structure. However, for simplicity of presentation, from now on we will focus on certain patterned matrices which we call linearly structured matrices. There are no principal difficulties to make necessary modifications for treating the general case.

We say that the matrix $\mathbf{A}$ is *linearly structured* if the only linear structure between the elements is given by $|a_{ij}| = |a_{kl}| \neq 0$ for some $(i,j) \neq (k,l)$ and there exists at least one $(i,j) \neq (k,l)$ so that $|a_{ij}| = |a_{kl}| \neq 0$.

Definition 2.1. *Let $\mathbf{A} : p \times q$ be a linearly structured matrix with r different by absolute value non-zero elements, i.e., $|a_{ij}| \neq |a_{gh}| \neq 0$. A pattern identifier $k(i,j)$ is a function $k : I \times J \longrightarrow H$, with $I = \{1, \dots, p\}$, $J = \{1, \dots, q\}$ and $H = \{0, \dots, r\}$ such that for $a_{ij} \neq 0$, $a_{gh} \neq 0$, $k(i,j) = k(g,h) \iff |a_{ij}| = |a_{gh}|$, and $k(i,j) = 0$ iff $a_{ij} = 0$.*

For a given pattern identifier $k(i,j)$ the definition implies that all possible patterned matrices $\mathbf{A}(K)$, consisting of the different non-zero elements of $\mathbf{A}$, have the same $k(i,j)$.

3. The associated basis of the linear space

In the following the indicator function $1_{\{a,b\}}$ and the sign function $\mathrm{sgn}(a)$ will be used, i.e.,

$$1_{\{a,b\}} = \begin{cases} 1, & a = b, a \neq 0 \\ 0, & \text{otherwise,} \end{cases} \qquad \mathrm{sgn}(a) = \begin{cases} 1, & a > 0 \\ 0, & a = 0 \\ -1, & a < 0. \end{cases}$$

Furthermore, $\otimes$ denotes the Kronecker product. We say that $\mathbf{e}_j : p \times 1$ is a unit base vector if it has 1 in the j-th position and 0 elsewhere, i.e., it is the j-th column of $\mathbf{I}_p$. The next theorem gives a realization of the matrix $\mathbf{B}$ in (2.3).

THEOREM 3.1. *Let $\mathbf{A} : p \times q$ be a linearly structured matrix with r different by absolute value non-zero elements and pattern identifier $k(i,j)$. The basis matrix $\mathbf{B}$ generates* $\mathrm{vec}\mathbf{A}$, *i.e.,* $\mathrm{vec}\mathbf{A} = \mathbf{Bq}$ *for some* $\mathbf{q}$, *if*

$$\mathbf{B} = \sum_{i=1}^{p}\sum_{j=1}^{q}(\mathbf{d}_j \otimes \mathbf{e}_i)\mathbf{f}'_{k(i,j)}\frac{\mathrm{sgn}(a_{ij})}{\sqrt{m(i,j)}}, \tag{3.1}$$

where

$$m(i,j) = \sum_{r=1}^{p}\sum_{s=1}^{q} 1_{\{|a_{ij}|,|a_{rs}|\}} \tag{3.2}$$

and $\mathbf{d}_j : q \times 1$, $\mathbf{e}_i : p \times 1$, $\mathbf{f}_{k(i,j)} : r \times 1$, $k(i,j) \neq 0$, *are unit basis vectors and* $\mathbf{f}_0 = \mathbf{0}$.

PROOF. The statement follows since

$$\begin{aligned} \mathrm{vec}\mathbf{A} &= \sum_{i,j} a_{ij}(\mathbf{d}_j \otimes \mathbf{e}_i) \\ &= \sum_{i,j}\sum_{g,h}(\mathbf{d}_j \otimes \mathbf{e}_i)\mathbf{f}'_{k(i,j)}\mathbf{f}_{k(g,h)}\frac{a_{ij}}{|a_{ij}|}|a_{gh}|\frac{1}{\sqrt{m(i,j)}}\frac{1}{\sqrt{m(g,h)}} \\ &= \mathbf{B}\sum_{g,h}\mathbf{f}_{k(g,h)}|a_{gh}|\frac{1}{\sqrt{m(g,h)}}. \end{aligned}$$

□

THEOREM 3.2. *Let* $\mathbf{B}$ *be given by (3.1). Then, with the notation of Theorem 3.1,*

(i) the column vectors of $\mathbf{B}$ are orthonormal, i.e., $\mathbf{B}'\mathbf{B} = \mathbf{I}_r$;
(ii) $\mathbf{BB}' = \sum_{i,j}\sum_{k,l}(\mathbf{d}_j \otimes \mathbf{e}_i)(\mathbf{d}_l \otimes \mathbf{e}_k)'1_{\{|a_{ij}|,|a_{kl}|\}}\frac{1}{m(i,j)}$.

PROOF. Since $\mathbf{f}_{k(i,j)} = \mathbf{f}_{k(g,h)}$ only if $1_{\{|a_{ij}|=|a_{gh}|\}} = 1$, straightforward calculations show that

$$\mathbf{B}'\mathbf{B} = \sum_{i,j}\mathbf{f}_{k(i,j)}\mathbf{f}'_{k(i,j)}\frac{1}{m(i,j)} = \mathbf{I}_r$$

and thus (i) is verified. Statement (ii) follows in a similar manner. □

Note that for linearly structured matrices $\mathbf{BB}'$ is independent of the pattern identifier $k(i,j)$.

THEOREM 3.3. *For a linearly structured matrix* $\mathbf{A} : p \times q$, *with the pattern identifier* $k(i,j)$ *the matrix* $\underline{\mathbf{B}}$ *in (2.5) is given by*

$$\underline{\mathbf{B}} = \sum_{i=1}^{p}\sum_{j=1}^{q}\sum_{s=1}^{r} \mathbf{f}_s\mathbf{f}_s' 1_{\{k(i,j),s\}} m(i,j)^{-3/2}, \tag{3.3}$$

where $\mathbf{f}_{k(i,j)}$ *and* $m(i,j)$ *are defined in Theorem 3.1.*

REMARK 3.1. *Note that* $\underline{\mathbf{B}}$ *is diagonal and thus*

$$\underline{\mathbf{B}}^{-1} = \sum_{i=1}^{p}\sum_{j=1}^{q}\sum_{s=1}^{r} \mathbf{f}_s\mathbf{f}_s' 1_{\{k(i,j),s\}} m(i,j)^{-1/2}. \tag{3.4}$$

From Theorem 3.1 and Theorem 3.3 it follows that $\mathbf{T}$ and $\mathbf{C} = \mathbf{T}^+$ easily can be established for any linearly structured matrix.

THEOREM 3.4. *Let* $\mathbf{T} = \underline{\mathbf{B}}\mathbf{B}'$ *and* $\mathbf{T}^+ = \mathbf{B}\underline{\mathbf{B}}^{-1}$ *where* $\mathbf{B}$ *and* $\underline{\mathbf{B}}$ *are given by (3.1) and (3.3), respectively. Then,*

$$\mathbf{T} = \sum_{s=1}^{r}\sum_{i,j} \mathbf{f}_s(\mathbf{d}_j \otimes \mathbf{e}_i)' 1_{\{k(i,j),s\}} m(i,j)^{-1}\mathrm{sgn}(a_{ij}), \tag{3.5}$$

$$\mathbf{T}^+ = \sum_{s=1}^{r}\sum_{i,j} (\mathbf{d}_j \otimes \mathbf{e}_i)\mathbf{f}_s' 1_{\{k(i,j),s\}} \mathrm{sgn}(a_{ij}), \tag{3.6}$$

where $m(i,j)$ *is defined by (3.2) and the basis vectors* $\mathbf{e}_i$, $\mathbf{d}_j$, $\mathbf{f}_{k(i,j)}$ *are defined in Theorem 3.1.*

PROOF. From (3.1) and (3.3) we get

$$\begin{aligned}
\mathbf{T} &= \underline{\mathbf{B}}\mathbf{B}' \\
&= \sum_{i_1,j_1}\sum_{s=1}^{r} \mathbf{f}_s\mathbf{f}_s' 1_{\{k(i_1,j_1),s\}} m(i_1,j_1)^{-3/2} \sum_{i,j} \mathbf{f}_{k(i,j)}(\mathbf{d}_j \otimes \mathbf{e}_i)' \frac{\mathrm{sgn}(a_{ij})}{\sqrt{m(i,j)}} \\
&= \sum_{i,j,i_1,j_1}\sum_{s=1}^{r} \mathbf{f}_s(\mathbf{d}_j \otimes \mathbf{e}_i)' 1_{\{k(i_1,j_1),s\}} 1_{\{k(i,j),s\}} m(i_1,j_1)^{-3/2} m(i,j)^{-1/2}\mathrm{sgn}(a_{ij}) \\
&= \sum_{i,j}\sum_{s=1}^{r} \mathbf{f}_s(\mathbf{d}_j \otimes \mathbf{e}_i)' 1_{\{k(i,j),s\}} m(i,j)^{-1}\mathrm{sgn}(a_{ij}),
\end{aligned}$$

since $\sum_{i_1,j_1} 1_{\{k(i_1,j_1),s\}} = m(i_1,j_1)$ and $k(i_1,j_1) = k(i,j)$ implies that $m(i_1,j_1) = m(i,j)$. Thus, the first statement is proved. The second statement follows in the same way from expressions (3.1) and (3.4). □

The following situation has emerged. For a linearly structured matrix $\mathbf{A}$ the structure is described by the pattern identifier $k(i,j)$ with several possible patterns K. The information provided by the $k(i,j)$ can be used to eliminate repeated elements from the matrix $\mathbf{A}$ in order to get a patterned matrix $\mathbf{A}(K)$. We know that $\mathrm{vec}\mathbf{A} = \mathbf{Bq}$ for some vector $\mathbf{q}$, where $\mathbf{B}$ is given in Theorem 3.1. Furthermore, for the same $\mathbf{q}$ we may state that $\mathrm{vec}\mathbf{A}(K) = \underline{\mathbf{B}}\mathbf{q}$. The next results are important consequences of the theorem.

COROLLARY 3.1. *Let* $\mathbf{A}$ *be a linearly structured matrix and* $\mathbf{B}$ *a basis which generates* $\mathrm{vec}\mathbf{A}$. *Let* $\underline{\mathbf{B}}$ *given in Theorem 3.3 generate* $\mathrm{vec}\mathbf{A}(K)$. *Then the transformation matrix* $\mathbf{T}$, *given by (3.5), satisfies the relation*

$$\mathrm{vec}\mathbf{A}(K) = \mathbf{T}\mathrm{vec}\mathbf{A} \tag{3.7}$$

and its Moore-Penrose inverse $\mathbf{T}^+$, *given by (3.6), defines the inverse transformation*

$$\mathrm{vec}\mathbf{A} = \mathbf{T}^+\mathrm{vec}\mathbf{A}(K) \tag{3.8}$$

for any pattern K *which corresponds to the pattern identifier* $k(i,j)$ *which is used in* $\mathbf{T}$.

PROOF. Simple calculations yield that for some $\mathbf{q}$

$$\mathbf{T}\mathrm{vec}\mathbf{A} = \mathbf{T}\mathbf{B}\mathbf{q} = \underline{\mathbf{B}}\mathbf{q} = \mathrm{vec}\mathbf{A}(K).$$

Furthermore, $\mathbf{T}^+\mathrm{vec}\mathbf{A}(K) = \mathbf{T}^+\underline{\mathbf{B}}\mathbf{q} = \mathbf{B}\mathbf{q} = \mathrm{vec}\mathbf{A}$. □

The transformation matrix $\mathbf{T}$ which satisfies (3.7) and (3.8) was called by [**DN**] the *transition* matrix. The next corollary of Theorem 3.4 gives us expressions for any element of $\mathbf{T}$ and $\mathbf{T}^+$, respectively. If we need to point out that the transition matrix $\mathbf{T}$ is applied to a certain matrix $\mathbf{A}$ from the considered class of matrices $\mathbb{M}$, we shall write $\mathbf{A}$ as an argument of $\mathbf{T}$, i.e., $\mathbf{T}(\mathbf{A})$, and sometimes also indicate which class $\mathbf{A}$ belongs to.

COROLLARY 3.2. *Suppose* $\mathbf{A}$ *is a* $p \times q$ *linearly structured matrix with the pattern identifier* $k(i,j)$ *and* $\mathbf{A}(K)$ *denotes a patterned matrix where* K *is one possible pattern which corresponds to* $k(i,j)$ *and* $\mathbf{T}(\mathbf{A})$ *is defined by (3.5). Then the elements of* $\mathbf{T}(\mathbf{A})$ *and* $\mathbf{T}(\mathbf{A})^+$ *are respectively given by*

$$(\mathbf{T}(\mathbf{A}))_{s(j,i)} = \begin{cases} 1/m_s, & \text{if } a_{ij} = (\mathrm{vec}\mathbf{A}(K))_s, \\ -1/m_s, & \text{if } a_{ij} = -(\mathrm{vec}\mathbf{A}(K))_s, \\ 0, & \text{otherwise}, \end{cases}$$

$$(\mathbf{T}^+)_{(j,i)s} = \begin{cases} 1, & \text{if } a_{ij} = (\mathrm{vec}\mathbf{A}(K))_s, \\ -1, & \text{if } a_{ij} = -(\mathrm{vec}\mathbf{A}(K))_s, \\ 0, & \text{otherwise}, \end{cases}$$

where $s = 1,\ldots,r$, $i = 1,\ldots,p$, $j = 1,\ldots,q$, $\mathrm{vec}\mathbf{A}(K) : r \times 1$ *is as in Corollary 3.1, the numeration of elements follows (2.2) and*

$$m_s = \sum_{i=1}^{p}\sum_{j=1}^{q} 1_{\{|(\mathrm{vec}\mathbf{A}(K))_s|,|a_{ij}|\}}.$$

For applications an important special case is the symmetric matrix and its transition matrix. The point of using the transition matrix is that for any choice of elements from the original matrix, the transition matrix will always be the same and described by the so-called duplication matrix which carefully has been examined by [**MN**], as well as others. Let $\mathbf{A}$ be a symmetric $n \times n$-matrix and let the patterned matrix which consists of the elements of its lower triangle be represented by $\mathbf{A}_\triangle$. Denote the corresponding $\frac{1}{2}n(n+1) \times n^2$-transition matrix by $\mathbf{D}_n$:

$$\mathbf{D}_n\mathrm{vec}\mathbf{A} = \mathrm{vec}\mathbf{A}_\triangle. \tag{3.9}$$

Then

$$\mathbf{D}_n^+ \mathrm{vec}\mathbf{A}_\triangle = \mathrm{vec}\mathbf{A}. \tag{3.10}$$

The matrix $\mathbf{D}_n^+$ is called *duplication matrix.*

As we saw in Theorem 3.4, it was possible to characterize transition matrices mathematically. Unfortunately, in general, not many interesting and easily interpretable properties can be found for these matrices. One reason for this is that $\mathbf{T}$ is a function of the pattern identifier which means that results depend on $\mathbf{f}_{k(i,j)}$ and therefore are not easy to look through. However, Nel [**DN**] brought forward the notion of *pattern* matrix which shares some basic properties with $\mathbf{T}^+$. For example, both generate the same subspace. We call it *pattern projection.*

DEFINITION 3.1. *Let* $\mathbf{T}$ *be the transition matrix for a linearly structured* $p \times q$*-matrix* $\mathbf{A}$ *with the pattern identifier* $k(i,j)$. *Then the matrix* $\mathbf{M} : pq \times pq$, *defined by*

$$\mathbf{M} = \mathbf{T}^+\mathbf{T} \tag{3.11}$$

is called the pattern projection matrix of $\mathbf{A}$.

Observe that $\mathbf{M}$ is an orthogonal projector and from a geometrical point of view this is an interesting definition. Some basic properties of the pattern projection matrix $\mathbf{M}$ are collected in the next proposition. The statements, due to [**DN**], all follow in an elementary way from Definition 3.1 and by using the fact that $\mathbf{M}$ is a projector.

PROPOSITION 3.1. *Let* $\mathbf{M}$ *be given by (3.12). Then,* $\mathbf{M}$ *is symmetric,* $\mathbf{M}$ *is idempotent, and if* $\mathbf{A}$ *is* $p \times q$*-matrix with the transition matrix* $\mathbf{T}$, *then* $\mathbf{M}\mathrm{vec}\mathbf{A} = \mathrm{vec}\mathbf{A}$. *Moreover,* $\mathbf{T}$ *is invariant under right multiplication and* $\mathbf{T}^+$ *is invariant under left multiplication by* $\mathbf{M}$, *respectively.*

4. Special classes of matrices

In this section, some widely-used classes of linearly structured square matrices are examined. Suppose that the size of the matrices is $n \times n$. We shall use the following notations:

- s (s_n) for symmetric matrices (s_n if we want to stress the size of the matrix);
- ss (ss_n) – skew-symmetric matrices;
- d (d_n) – diagonal matrices;
- c (c_n) – correlation type matrices or symmetric off-diagonal matrices;
- u (u_n) – upper triangular matrices;
- l (l_n) – lower triangular matrices;
- t (t_n) – Toeplitz matrices.

For instance $\mathbf{M}(s)$ and $\mathbf{T}(s)$ would be the notations for $\mathbf{M}$ and $\mathbf{T}$ for a symmetric $\mathbf{A}$. We will suppose that no other relations between elements exist beside those which define the matrix class. In the following propositions the pattern identifier $k(i,j)$ and matrices $\mathbf{B}$, $\mathbf{T}$, $\mathbf{T}^+$ and $\mathbf{M}$ are presented for all the listed classes of matrices. In all these propositions $\mathbf{e}_i$ and $\mathbf{f}_{k(i,j)}$ stand for the $n-$ and $r-$dimensional basis vectors, respectively, where r depends on the structure. In the subsequent let $\mathbf{e}_i$, $\mathbf{d}_j$ and $\mathbf{f}_s$ denote unit base vectors of proper size and

$$\mathbf{K}_{p,q} = \sum_{i=1}^{p}\sum_{j=1}^{q} \mathbf{e}_i\mathbf{d}_j' \otimes \mathbf{d}_j\mathbf{e}_i'$$

denotes the commutation matrix. The notation $(\mathbf{K}_{p,q})_d$ will stand for the diagonal of $\mathbf{K}_{p,q}$. Let us start with the class of symmetric matrices.

PROPOSITION 4.1. *For the class of symmetric* $\mathbf{A} : n \times n$*-matrices the following equalities hold:*

$$\begin{aligned}
k(i,j) &= n(\min(i,j)-1) - \tfrac{1}{2}\min(i,j)(\min(i,j)-1) + \max(i,j);\\
\mathbf{B}(s_n) &= \frac{1}{\sqrt{2}} \sum_{\substack{i,j=1\\ i\neq j}}^{n} (\mathbf{e}_j \otimes \mathbf{e}_i)\mathbf{f}_{k(i,j)}' \operatorname{sgn}(a_{ij}) + \sum_{i=1}^{n} (\mathbf{e}_i \otimes \mathbf{e}_i)\mathbf{f}_{k(i,i)}' \operatorname{sgn}(a_{ii});\\
\mathbf{T}(s_n) &= \frac{1}{2} \sum_{\substack{i,j=1\\ i<j}}^{n} \sum_{s=1}^{r} \mathbf{f}_s(\mathbf{e}_j \otimes \mathbf{e}_i + \mathbf{e}_i \otimes \mathbf{e}_j)' 1_{\{k(i,j),s\}} \operatorname{sgn}(a_{ij})\\
&\quad + \sum_{i=1}^{n}\sum_{s=1}^{r} \mathbf{f}_s(\mathbf{e}_i \otimes \mathbf{e}_i)' 1_{\{k(i,i),s\}} \operatorname{sgn}(a_{ii}), \qquad r = \frac{1}{2}n(n+1);\\
\mathbf{T}^{+}(s) &= \sum_{\substack{i,j=1\\ i<j}}^{n} \sum_{s=1}^{r} (\mathbf{e}_j \otimes \mathbf{e}_i + \mathbf{e}_i \otimes \mathbf{e}_j)\mathbf{f}_s' 1_{\{k(i,j),s\}} \operatorname{sgn}(a_{ij})\\
&\quad + \sum_{i=1}^{n}\sum_{s=1}^{r} (\mathbf{e}_i \otimes \mathbf{e}_i)\mathbf{f}_s' 1_{\{k(i,i),s\}} \operatorname{sgn}(a_{ii});\\
\mathbf{M}(s) &= \frac{1}{2}(\mathbf{I} + \mathbf{K}_{n,n}).
\end{aligned}$$

PROOF. If we count the unique elements row by row we have n different elements in the first row, $n-1$ non-repeated elements in the second row, $n-2$ elements in the third row, etc. Thus, for a_{ij}, $i \leq j$, we are in the ith row where the jth element should be considered. It follows that a_{ij} is the element with ordering number $n+n-1+n-2+\cdots+n-i+1+j-i$ which equals $n(i-1)-\frac{1}{2}i(i-1)+j = k(i,j)$. If we have an element a_{ij}, $i > j$, it follows by symmetry that the expression for $k(i,j)$ is true and we have verified that the equality for $k(i,j)$ holds. Moreover, $\mathbf{B}(s)$ follows immediately from (3.1). For $\mathbf{T}(s)$ and $\mathbf{T}^{+}(s)$ we apply Theorem 3.4. $\square$

PROPOSITION 4.2. *For the class of skew-symmetric matrices* $\mathbf{A} : n \times n$ *we have* $k(i,i) = 0$,

$$\begin{aligned}
k(i,j) &= n(\min(i,j)-1) - \tfrac{1}{2}\min(i,j)(\min(i,j)-1) + \max(i,j) - \min(i,j), \\
&\qquad\qquad\qquad\qquad\qquad\qquad\qquad\qquad\qquad\qquad i \neq j; \\
\mathbf{B}(ss_n) &= \frac{1}{\sqrt{2}} \sum_{\substack{i,j=1 \\ i \neq j}}^{n} (\mathbf{e}_j \otimes \mathbf{e}_i)\mathbf{f}'_{k(i,j)}\operatorname{sgn}(a_{ij}); \\
\mathbf{T}(ss_n) &= \frac{1}{2} \sum_{\substack{i,j=1 \\ i < j}}^{n} \sum_{s=1}^{r} \mathbf{f}_s(\mathbf{e}_j \otimes \mathbf{e}_i - \mathbf{e}_i \otimes \mathbf{e}_j)' 1_{\{k(i,j),s\}}\operatorname{sgn}(a_{ij}),\ r = \frac{1}{2}n(n-1); \\
\mathbf{T}^+(ss) &= \sum_{\substack{i,j=1 \\ i < j}}^{n} \sum_{s=1}^{r} (\mathbf{e}_j \otimes \mathbf{e}_i - \mathbf{e}_i \otimes \mathbf{e}_j)\mathbf{f}'_s 1_{\{k(i,j),s\}}\operatorname{sgn}(a_{ij}); \\
\mathbf{M}(ss) &= \frac{1}{2}(\mathbf{I} - \mathbf{K}_{n,n}).
\end{aligned}$$

PROOF. Similar calculations to those in Proposition 4.1 show that the relation for $k(i,j)$ holds. The relations for $\mathbf{B}(ss)$, $\mathbf{T}(ss)$ and $\mathbf{T}^+(ss)$ are also obtained in the same way as in the previous proposition. For deriving the last statement we use the fact that $\operatorname{sgn}(a_{ij})\operatorname{sgn}(a_{ij}) = 1$. □

PROPOSITION 4.3. *For diagonal* $n \times n$*-matrices with different diagonal elements we obtain:* $k(i,i) = i$, *and* $k(i,j) = 0$ *for* $i \neq j$;

$$\begin{aligned}
\mathbf{B}(d_n) &= \sum_{i=1}^{n} (\mathbf{e}_i \otimes \mathbf{e}_i)\mathbf{e}'_i; \quad \mathbf{T}(d_n) = \sum_{i=1}^{n} \mathbf{e}_i(\mathbf{e}_i \otimes \mathbf{e}_i)'; \\
\mathbf{T}^+(d_n) &= \sum_{i=1}^{n} (\mathbf{e}_i \otimes \mathbf{e}_i)\mathbf{e}'_i; \quad \mathbf{M}(d_n) = \sum_{i=1}^{n} (\mathbf{e}_i \otimes \mathbf{e}_i)(\mathbf{e}_i \otimes \mathbf{e}_i)' = (\mathbf{K}_{n,n})_d.
\end{aligned}$$

PROOF. In Theorem 3.1 and Theorem 3.4 we may choose $f_{k(i,j)} = \mathbf{e}_i$ which imply the first four equalities. The last statement follows from simple calculations which yield $\mathbf{M}(d) = \mathbf{B}(d)\mathbf{B}'(d) = \sum_{i=1}^{n} (\mathbf{e}_i \otimes \mathbf{e}_i)(\mathbf{e}_i \otimes \mathbf{e}_i)' = (\mathbf{K}_{n,n})_d$. □

PROPOSITION 4.4. *For any* $\mathbf{A} : n \times n$ *which belongs to the class of symmetric off-diagonal matrices it follows that:* $k(i,i) = 0$;

$$\begin{aligned}
k(i,j) &= n(\min(i,j)-1) - \tfrac{1}{2}\min(i,j)(\min(i,j)-1) + \max(i,j) - \min(i,j), \\
&\qquad\qquad\qquad\qquad\qquad\qquad\qquad\qquad\qquad\qquad i \neq j, \\
\mathbf{B}(c_n) &= \frac{1}{\sqrt{2}} \sum_{\substack{i,j=1 \\ i \neq j}}^{n} (\mathbf{e}_j \otimes \mathbf{e}_i)\mathbf{f}'_{k(i,j)}\operatorname{sgn}(a_{ij}); \\
\mathbf{T}(c_n) &= \frac{1}{2} \sum_{\substack{i,j=1 \\ i < j}}^{n} \sum_{s=1}^{r} \mathbf{f}_s(\mathbf{e}_j \otimes \mathbf{e}'_i + \mathbf{e}_i \otimes \mathbf{e}_j)' 1_{\{k(i,j),s\}}\operatorname{sgn}(a_{ij}), \quad r = \frac{1}{2}n(n-1);
\end{aligned}$$

$$
\begin{aligned}
\mathbf{T}^+(c_n) &= \sum_{\substack{i,j=1\\ i<j}}^{n} \sum_{i=1}^{n} (\mathbf{e}_j \otimes \mathbf{e}_i + \mathbf{e}_i \otimes \mathbf{e}_j)\mathbf{f}'_s 1_{\{k(i,j),s\}} \mathrm{sgn}(a_{ij});\\
\mathbf{M}(c_n) &= \frac{1}{2}(\mathbf{I} + \mathbf{K}_{n,n}) - (\mathbf{K}_{n,n})_d.
\end{aligned}
$$

PROOF. The same elements as in the skew-symmetric case are involved. Therefore, the expression for $k(i,j)$ follows from Proposition 4.2. We just prove the last statement since the other follow from Theorem 3.3 and Theorem 3.4. By definition of $\mathbf{B}(c)$, we have $\mathbf{M}(c) = \mathbf{B}(c)\mathbf{B}'(c)$ so that

$$
\begin{aligned}
\mathbf{M}(c) &= \sum_{\substack{i,j=1\\ i\neq j}}^{n} \sum_{\substack{k,l=1\\ k\neq l}}^{n} \frac{1}{2}(\mathbf{e}_j \otimes \mathbf{e}_i) f'_{k(i,j)} f_{k(k,l)} (\mathbf{e}_l \otimes \mathbf{e}_k)' \mathrm{sgn}(a_{ij}) \mathrm{sgn}(a_{kl})\\
&= \frac{1}{2} \sum_{\substack{i,j=1\\ i\neq j}}^{n} (\mathbf{e}_j \otimes \mathbf{e}_i)(\mathbf{e}_j \otimes \mathbf{e}_i + \mathbf{e}_i \otimes \mathbf{e}_j)'\\
&= \frac{1}{2}(\mathbf{I} + \mathbf{K}_{n,n}) - (\mathbf{K}_{n,n})_d.
\end{aligned}
$$

□

PROPOSITION 4.5. *For any* $\mathbf{A} : n \times n$ *from the class of lower triangular matrices*

$$
\begin{aligned}
k(i,j) &= n(j-1) - \tfrac{1}{2}j(j-1) + i, \quad j \leq i,\\
k(i,j) &= 0, \quad i < j;\\
\mathbf{B}(l_n) &= \sum_{\substack{i,j=1\\ i\geq j}}^{n} (\mathbf{e}_i \otimes \mathbf{e}_j)\mathbf{f}'_{k(i,j)} \mathrm{sgn}(a_{ij});\\
\mathbf{T}(l_n) &= \sum_{\substack{i,j=1\\ i\geq j}}^{n} \sum_{s=1}^{r} \mathbf{f}_s (\mathbf{e}_i \otimes \mathbf{e}_j)' 1_{\{k(i,j),s\}} \mathrm{sgn}(a_{ij}), \quad r = \frac{1}{2}n(n+1);\\
\mathbf{T}^+(l_n) &= \sum_{\substack{i,j=1\\ i\geq j}}^{n} \sum_{s=1}^{r} (\mathbf{e}_i \otimes \mathbf{e}_j)\mathbf{f}'_s 1_{\{k(i,j),s\}} \mathrm{sgn}(a_{ij});\\
\mathbf{M}(l_n) &= \sum_{\substack{i,j=1\\ i\geq j}}^{n} (\mathbf{e}_i \otimes \mathbf{e}_j)(\mathbf{e}_i \otimes \mathbf{e}_j)'.
\end{aligned}
$$

PROOF. Here the same elements as in the symmetric case are regarded. The expression for $k(i,j)$ is thus a consequence of Proposition 4.1. The other statements are obtained similarly to the proof of Proposition 4.3, for example. □

The equivalent result for upper triangular matrices follows by symmetry. Moreover, note the similarity between $\mathbf{T}(l_n)$ and $\mathbf{D}_n$, given by (3.9).

PROPOSITION 4.6. *For any Toeplitz matrix* $\mathbf{A} : n \times n$,

$$\begin{aligned}
k(i,j) &= n + j - i; \\
\mathbf{B}(t_n) &= \sum_{i,j=1}^{n} (\mathbf{e}_j \otimes \mathbf{e}_i)\mathbf{f}'_{k(i,j)} \frac{1}{\sqrt{n - |j-i|}} \mathrm{sgn}(a_{ij}); \\
\mathbf{T}(t_n) &= \sum_{i,j=1}^{n} \sum_{s=1}^{r} \mathbf{f}_s (\mathbf{e}_j \otimes \mathbf{e}_i)' 1_{\{k(i,j),s\}} \frac{1}{n - |j-i|} \mathrm{sgn}(a_{ij}), \quad r = 2n-1; \\
\mathbf{T}^+(t_n) &= \sum_{i,j=1}^{n} \sum_{s=1}^{r} (\mathbf{e}_j \otimes \mathbf{e}_i)\mathbf{f}'_s 1_{\{k(i,j),s\}} \mathrm{sgn}(a_{ij}); \\
\mathbf{M}(t_n) &= \sum_{\substack{i,j,k,l=1 \\ j-i=l-k}}^{n} (\mathbf{e}_j \otimes \mathbf{e}_i)(\mathbf{e}_l \otimes \mathbf{e}_k)' \frac{1}{n - |j-i|}.
\end{aligned}$$

PROOF. Since by definition of a Toeplitz matrix $a_{ij} = a_{i-j}$ we have altogether $2n-1$ different elements. Since $j-i$ ranges through

$$-(n-1), -(n-2), \ldots, (n-1)$$

$k(i,j)$ is true. If $j - i = t$ we have of course $j = t + i$. Since $1 \le j \le n$ we have for $t \ge 0$ the restriction $1 \le i \le n - t$. If $t \le 0$, $1 \le i \le n + t$. So, for fixed t we have $n - |t|$ elements. Therefore, $\mathbf{B}(t)$ follows from Theorem 3.1. The other statements are obtained by using Theorem 3.4 and (3.11). □

5. Applications to Jacobians

We will write $d_{k_1}\mathbf{Y}/d_{k_2}\mathbf{X}$ to denote the Jacobian matrix

$$\mathbf{J}(\mathbf{Y} \to \mathbf{X}) = \frac{d\mathrm{vec}\mathbf{Y}(K_1)}{d\mathrm{vec}\mathbf{X}(K_2)},$$

of the one-to-one transformation $\mathbf{Y} \to \mathbf{X}$, where K_1 and K_2 describe the patterns in $\mathbf{Y}$ and $\mathbf{X}$, respectively. As usual, the Jacobian of the transformation is the absolute value of the determinant of $\mathbf{J}(\mathbf{Y} \to \mathbf{X})$, indicated here by $|\mathbf{J}(\mathbf{Y} \to \mathbf{X})|_+$. For many useful results on Jacobians we refer to Deemer and Olkin [**DO**], Olkin and Sampson [**OS**], and the books by Srivastava and Khatri [**SK**], Muirhead [**M**] and Magnus [**M88**], as well as references mentioned in these books. In this section the main purpose is to illustrate how to utilize the notion of patterned matrices and matrix derivatives. The results given below are known but the proofs are new.

THEOREM 5.1. *Let* $\mathbf{A} : n \times n$ *be non-singular,* $\mathbf{V}$ *symmetric and* $\mathbf{W}$ *skew-symmetric.*

(i) $|\mathbf{J}(\mathbf{AVA}' \to \mathbf{V})|_+ = |\mathbf{A}|_+^{n+1}$,

(ii) $|\mathbf{J}(\mathbf{AWA}' \to \mathbf{W})|_+ = |\mathbf{A}|_+^{n-1}$.

PROOF. Without loss of generality we suppose that the eigenvalues of $\mathbf{A}'\mathbf{A}$ are non-zero and distinct. Now

$$\begin{aligned}
\left|\frac{d_k\mathbf{AVA}'}{d_k\mathbf{V}}\right|_+ &= \left|\frac{d\mathbf{V}}{d_k\mathbf{V}}(\mathbf{A}' \otimes \mathbf{A}')\mathbf{T}'(s)\right|_+ = \left|\frac{d_k\mathbf{V}}{d_k\mathbf{V}}(\mathbf{T}^+(s))'(\mathbf{A}' \otimes \mathbf{A}')\mathbf{T}'(s)\right|_+ \\
&= \left|(\mathbf{T}^+(s))'(\mathbf{A}' \otimes \mathbf{A}')\mathbf{T}'(s)\right|_+,
\end{aligned}$$

where $\mathbf{T}(s)$ and $\mathbf{T}^+(s)$ are given in Theorem 3.4 and the index k in the derivative refers to the symmetric structure. By definition of $\mathbf{T}(s)$ this expression can be further explored:

$$\begin{aligned}&\left|\frac{d_k\mathbf{AVA}'}{d_k\mathbf{V}}\right|_+\\&\quad= |\underline{\mathbf{B}}(s)^{-1}\mathbf{B}'(s)(\mathbf{A}'\otimes\mathbf{A}')\mathbf{B}(s)\underline{\mathbf{B}}'|_+ = |\mathbf{B}'(s)(\mathbf{A}'\otimes\mathbf{A}')\mathbf{B}(s)|_+\\&\quad= |\mathbf{B}'(s)(\mathbf{A}'\otimes\mathbf{A}')\mathbf{B}(s)\mathbf{B}'(s)(\mathbf{A}\otimes\mathbf{A})\mathbf{B}(s)|^{1/2} = |\mathbf{B}'(s)(\mathbf{A}'\mathbf{A}\otimes\mathbf{A}'\mathbf{A})\mathbf{B}(s)|^{1/2}\end{aligned}$$

since $\mathbf{B}(s)\mathbf{B}'(s) = \frac{1}{2}(\mathbf{I}+\mathbf{K}_{n,n})$. All different eigenvalues of $\mathbf{A}'\mathbf{A}\otimes\mathbf{A}'\mathbf{A}$ are given by $\lambda_i\lambda_j$, $j\geq i=1,2,\ldots,n$, where λ_k is an eigenvalue of $\mathbf{A}'\mathbf{A}$. However, these eigenvalues are also eigenvalues of $\mathbf{B}'(s)(\mathbf{A}'\mathbf{A}\otimes\mathbf{A}'\mathbf{A})\mathbf{B}(s)$ since, if $\mathbf{u}$ is an eigenvector which corresponds to $\lambda_i\lambda_j$,

$$(\mathbf{A}'\mathbf{A}\otimes\mathbf{A}'\mathbf{A})\mathbf{u} = \lambda_i\lambda_j\mathbf{u}$$

implies that

$$\begin{aligned}&\mathbf{B}'(s)(\mathbf{A}'\mathbf{A}\otimes\mathbf{A}'\mathbf{A})\mathbf{B}(s)\mathbf{B}'(s)\mathbf{u} = \mathbf{B}'(s)\mathbf{B}(s)\mathbf{B}'(s)(\mathbf{A}'\mathbf{A}\otimes\mathbf{A}'\mathbf{A})\mathbf{u}\\&\quad= \mathbf{B}'(s)(\mathbf{A}'\mathbf{A}\otimes\mathbf{A}'\mathbf{A})\mathbf{u} = \lambda_i\lambda_j\mathbf{B}'(s)\mathbf{u}.\end{aligned}$$

Moreover, since

$$r(\mathbf{B}'(s)(\mathbf{A}'\mathbf{A}\otimes\mathbf{A}'\mathbf{A})\mathbf{B}(s)) = r(\mathbf{B}(s)) = \tfrac{1}{2}n(n+1),$$

all eigenvalues of $\mathbf{B}'(s)(\mathbf{A}'\mathbf{A}\otimes\mathbf{A}'\mathbf{A})\mathbf{B}(s)$ are given by $\lambda_i\lambda_j$. Thus,

$$\begin{aligned}|\mathbf{B}'(s)(\mathbf{A}'\mathbf{A}\otimes\mathbf{A}'\mathbf{A})\mathbf{B}(s)|_+^{1/2} &= \prod_{\substack{i,j=1\\ i\leq j}}^{n} |\lambda_i\lambda_j|_+^{1/2} = \prod_{i=1}^{n}|\lambda_i|_+^{\frac{1}{2}(n+1)}\\&= |\mathbf{A}'\mathbf{A}|_+^{\frac{1}{2}(n+1)} = |\mathbf{A}|_+^{n+1}.\end{aligned}$$

Hence, (i) of the theorem is established.

For (ii) it is utilized, as in the symmetric case, that

$$\begin{aligned}\left|\frac{d_k\mathbf{AWA}'}{d_k\mathbf{W}}\right|_+ &= \left|(\mathbf{T}^+(ss))'(\mathbf{A}'\otimes\mathbf{A}')\mathbf{T}'(ss)\right|_+ = |\mathbf{B}'(ss)(\mathbf{A}'\otimes\mathbf{A}')\mathbf{B}(ss)|_+\\&= |\mathbf{B}'(ss)(\mathbf{A}'\mathbf{A}\otimes\mathbf{A}'\mathbf{A})\mathbf{B}(ss)|^{1/2},\end{aligned}$$

where k now indicates the skew-symmetric structure. By definition of $\mathbf{B}(s)$ and $\mathbf{B}(ss)$

$$\begin{aligned}&|\mathbf{A}'\mathbf{A}\otimes\mathbf{A}'\mathbf{A}|_+\\&\quad= |(\mathbf{B}(s)\mathbf{B}'(s)+\mathbf{B}(ss)\mathbf{B}'(ss))(\mathbf{A}'\mathbf{A}\otimes\mathbf{A}'\mathbf{A})(\mathbf{B}(s)\mathbf{B}'(s)+\mathbf{B}(ss)\mathbf{B}'(ss))|_+\\&\quad= |\mathbf{B}(s)\mathbf{B}'(s)(\mathbf{A}'\mathbf{A}\otimes\mathbf{A}'\mathbf{A})\mathbf{B}(s)\mathbf{B}'(s)\\&\qquad\qquad+\mathbf{B}(ss)\mathbf{B}'(ss)(\mathbf{A}'\mathbf{A}\otimes\mathbf{A}'\mathbf{A})\mathbf{B}(ss)\mathbf{B}'(ss)|_+,\end{aligned}\tag{5.1}$$

since $\mathbf{B}'(s)(\mathbf{A}'\mathbf{A}\otimes\mathbf{A}'\mathbf{A})\mathbf{B}(ss)\mathbf{B}'(ss) = \mathbf{B}'(s)\mathbf{B}(ss)\mathbf{B}'(ss)(\mathbf{A}'\mathbf{A}\otimes\mathbf{A}'\mathbf{A}) = \mathbf{0}$. Furthermore, $(\mathbf{B}(s):\mathbf{B}(ss)):n\times n$ is an orthogonal matrix and therefore expression

(5.1) is equivalent to

$$\begin{aligned}
&|\mathbf{A}'\mathbf{A}\otimes\mathbf{A}'\mathbf{A}|_+\\
&=\left|(\mathbf{B}(s):\mathbf{B}(ss))\begin{pmatrix}\mathbf{B}'(s)(\mathbf{A}'\mathbf{A}\otimes\mathbf{A}'\mathbf{A})\mathbf{B}(ss) & \mathbf{0}\\ \mathbf{0} & \mathbf{B}'(ss)(\mathbf{A}'\mathbf{A}\otimes\mathbf{A}'\mathbf{A})\mathbf{B}(ss)\end{pmatrix}\right.\\
&\quad\times\left.\begin{pmatrix}\mathbf{B}'(s)\\ \mathbf{B}'(ss)\end{pmatrix}\right|_+\\
&=|\mathbf{B}'(s)(\mathbf{A}'\mathbf{A}\otimes\mathbf{A}'\mathbf{A})\mathbf{B}(s)|_+|\mathbf{B}'(ss)(\mathbf{A}'\mathbf{A}\otimes\mathbf{A}'\mathbf{A})\mathbf{B}(ss)|_+.
\end{aligned}$$

Thus, it follows from (i) that

$$|\mathbf{B}'(ss)(\mathbf{A}'\mathbf{A}\otimes\mathbf{A}'\mathbf{A})\mathbf{B}(ss)|_+^{1/2}=|\mathbf{A}'\mathbf{A}|^n|\mathbf{A}'\mathbf{A}|^{-\frac{1}{2}(n+1)}=|\mathbf{A}|_+^{n-1}.$$

□

THEOREM 5.2. *Let* $\mathbf{A}$, $\mathbf{B}$ *and* $\mathbf{V}$ *be diagonal matrices, where* $\mathbf{A}$ *and* $\mathbf{B}$ *are constant matrices, and* $\mathbf{W}=\mathbf{AVB}$. *Then,*

$$|\mathbf{J}(\mathbf{W}\to\mathbf{V})|_+=|\mathbf{A}|_+|\mathbf{B}|_+.$$

PROOF. Observe that

$$|\mathbf{J}(\mathbf{W}\to\mathbf{V})|_+=\left|\frac{d_k\mathbf{AVB}}{d_k\mathbf{V}}\right|_+=|\mathbf{B}'(d)(\mathbf{B}\otimes\mathbf{A})\mathbf{B}(d)|_+=|\mathbf{A}|_+|\mathbf{B}|_+,$$

where the subindex k indicates the diagonal structure. □

THEOREM 5.3. *Let* $\mathbf{W}=\mathbf{V}^{-1}:n\times n$.

(i) *If the elements of* $\mathbf{V}$ *are functionally independent, then* $|\mathbf{J}(\mathbf{W}\to\mathbf{V})|_+ =|\mathbf{V}|_+^{-2n}$.

(ii) *If* $\mathbf{V}$ *is symmetric, then* $|\mathbf{J}(\mathbf{W}\to\mathbf{V})|_+=|\mathbf{V}|_+^{-(n+1)}$.

(iii) *If* $\mathbf{V}$ *is skew-symmetric, then* $|\mathbf{J}(\mathbf{W}\to\mathbf{V})|_+=|\mathbf{V}|_+^{-(n-1)}$.

PROOF. Using that

$$\mathbf{0}=\frac{d\mathbf{V}^{-1}\mathbf{V}}{d\mathbf{V}}=\frac{d\mathbf{V}^{-1}}{d\mathbf{V}}(\mathbf{V}\otimes\mathbf{I})+\frac{d\mathbf{V}}{d\mathbf{V}}(\mathbf{I}\otimes\mathbf{V}^{-1})$$

we obtain

$$\left|\frac{d\mathbf{V}^{-1}}{d\mathbf{V}}\right|_+=|-\mathbf{V}^{-1}\otimes(\mathbf{V}^{-1})'|_+=|\mathbf{V}|^{-2n}.$$

Thus (i) is verified. For (ii) we first note that

$$\begin{aligned}
\mathbf{0}&=\frac{d\mathbf{V}^{-1}\mathbf{V}}{d_k\mathbf{V}}=\frac{d\mathbf{V}^{-1}}{d_k\mathbf{V}}(\mathbf{V}\otimes\mathbf{I})+\frac{d\mathbf{V}}{d_k\mathbf{V}}(\mathbf{I}\otimes\mathbf{V}^{-1})\\
&=\frac{d_k\mathbf{V}^{-1}}{d_k\mathbf{V}}(\mathbf{T}^+(s))'(\mathbf{V}\otimes\mathbf{I})+(\mathbf{T}^+(s))'(\mathbf{I}\otimes\mathbf{V}^{-1}),
\end{aligned}$$

where k refers to the symmetric structure. Thus,

$$\frac{d_k\mathbf{V}^{-1}}{d_k\mathbf{V}}=-(\mathbf{T}^+(s))'(\mathbf{V}^{-1}\otimes\mathbf{V}^{-1})\mathbf{T}(s)$$

and the Jacobian is similar to (5.1) which implies that (ii) as well as (iii) with some modifications follow from the proof of Theorem 5.1. □

In the subsequent theorem we represent a symmetric matrix in terms of a triangular matrix. Since the number of non-zero and functionally independent elements are the same in both classes the Jacobian is well-defined.

THEOREM 5.4. *Let* $\mathbf{W} = \mathbf{T}\mathbf{T}'$ *where* $\mathbf{T} : n \times n$ *is a non-singular lower triangular matrix with positive diagonal elements. Then,*

$$|\mathbf{J}(\mathbf{W} \to \mathbf{T})|_+ = 2^n \prod_{j=1}^{n} t_{jj}^{n-j+1}.$$

PROOF. It is noted that

$$\begin{aligned}\frac{d_{k_1}\mathbf{T}\mathbf{T}'}{d_{k_2}\mathbf{T}} &= \frac{d\mathbf{T}}{d_{k_2}\mathbf{T}}(\mathbf{T}' \otimes \mathbf{I})\mathbf{T}'(s) + \frac{d\mathbf{T}}{d_{k_2}\mathbf{T}}\mathbf{K}_{n,n}(\mathbf{I} \otimes \mathbf{T}')\mathbf{T}'(s) \\ &= (\mathbf{T}^+(l))'(\mathbf{T}' \otimes \mathbf{I})(\mathbf{I} + \mathbf{K}_{n,n})\mathbf{T}'(s),\end{aligned}$$

where k_1 and k_2 refer to the symmetric and lower triangular patterns, respectively, and $\mathbf{T}(s)$ and $\mathbf{T}(l)$ are defined in Propositions 4.1 4.5. Since $(\mathbf{I} + \mathbf{K}_{n,n})\mathbf{T}'(s) = 2\mathbf{T}'(s)$, we consider the product

$$\begin{aligned}&(\mathbf{T}^+(l))'(\mathbf{T}' \otimes \mathbf{I})\mathbf{T}'(s) \\ &= \sum_{\substack{i,j=1 \\ i \geq j}}^{n} \sum_{i_1,j_1=1}^{n} \mathbf{f}_{k_2(i,j)}(\mathbf{e}_j \otimes \mathbf{e}_i)'(\mathbf{T}' \otimes \mathbf{I})(\mathbf{e}_{j_1} \otimes \mathbf{e}_{i_1})\mathbf{f}'_{k_1(i_1,j_1)} \\ &= \sum_{\substack{i,j,j_1=1 \\ i \geq j}}^{n} t_{j_1 j}\mathbf{f}_{k_2(i,j)}\mathbf{f}'_{k_1(i,j_1)} \\ &= \sum_{\substack{i,j,j_1=1 \\ j \leq j_1 \leq i}}^{n} t_{j_1 j}\mathbf{f}_{k_2(i,j)}\mathbf{f}'_{k_2(i,j_1)} + \sum_{\substack{i,j,j_1=1 \\ j \leq i < j_1}}^{n} t_{j_1 j}\mathbf{f}_{k_2(i,j)}\mathbf{f}'_{k_1(i,j_1)},\end{aligned}$$

where the last equality follows because $\mathbf{T}$ is lower triangular and $k_2(i,j) = k_1(i,j)$ if $i \geq j$. Moreover, $k_2(i,j) < k_1(i,j_1)$ if $j \leq i < j_1$ and therefore the product $(\mathbf{T}^+(l))'(\mathbf{T}' \otimes \mathbf{I})\mathbf{T}'(s)$ is an upper triangular matrix. Thus,

$$\begin{aligned}\left|\frac{d_{k_1}\mathbf{T}\mathbf{T}'}{d_{k_2}\mathbf{T}}\right|_+ &= |2(\mathbf{T}^+(l))'(\mathbf{T}' \otimes \mathbf{I})\mathbf{T}'(s)|_+ = 2^n \left|\sum_{i \geq j=1}^{n} t_{jj}\mathbf{f}_{k_2(i,j)}\mathbf{f}'_{k_2(i,j)}\right|_+ \\ &= 2^n \prod_{j=1}^{n} |t_{jj}|^{n-j+1}.\end{aligned}$$

□

Acknowledgments

The authors are thankful to the referees and editors whose constructive comments helped to improve considerably the presentation of the paper. The Swedish Natural Research Council (NFR) and Estonian Science Foundation (grant 4366) are gratefully acknowledged for financial support.

References

[A] T. W. Anderson, *An Introduction to Multivariate Statistical Analysis*, Second edition, Wiley, New York, 1984.

[DO] W. L. Deemer and I. Olkin, *The Jacobians of certain matrix transformations useful in multivariate analysis*, Biometrika **38** (1951), 345–367.

[G] F. A. Graybill, *Matrices with Applications in Statistics*, Second edition. Wadsworth, Belmont, 1983.

[M83] J. R. Magnus, *L-structured matrices and linear matrix equations*, Linear and Multilinear Algebra **14** (1983),67–88.

[M88] ——— *Linear Structures*, London and Oxford University Press, New York, 1988.

[MN] J. R. Magnus and H. Neudecker, *Matrix Differential Calculus with Applications in Statistics and Econometrics*, Third edition, Wiley, Chichester, 1999.

[M] R. J. Muirhead, *Aspects of Multivariate Statistical Theory*, Wiley, New York, 1982.

[DN] D. G. Nel, *On matrix differentiation in statistics*, South African Statist. J. **14** (1980), 137–193.

[N] H. Neudecker, *On Jacobians of transformations with skew-symmetric, strictly (lower) triangular or diagonal matrix arguments*, Linear and Multilinear Algebra **14** (1983), 271–295.

[NS] H. Neudecker and A. Satorra, *The algebraic equality of two asymptotic tests for the hypothesis that a normal distribution has a specified correlation matrix*, Statist. Probab. Lett. **30** (1996), 99–103.

[OS] I. Olkin and A. R. Sampson, *Jacobians of matrix transformations and induced functional equations*, Linear Algebra and Its Applications **5** (1972), 257–276.

[P] A. Parring, *General asymptotic distribution of noncomplete multiple correlation coefficients*, Acta et Commentationes Universitatis Tartuensis **541** (1980), 18–26 (in Russian).

[SK] M. S. Srivastava and C. G. Khatri, *An Introduction to Multivariate Statistics*, North Holland, New York, 1979.

Institute of Mathematical Statistics, University of Tartu, Liivi Street 2, 50409 Tartu, Estonia

E-mail address: kollo@ut.ee

Department of Biometry, Swedish University of Agricultural Sciences, Box 7013, SE-750 07 Uppsala, Sweden

E-mail address: dietrich.von.rosen@bi.slu.se

Contemporary Mathematics
Volume **287**, 2001

Random Walks on Regular Languages and Algebraic Systems of Generating Functions

Steven P. Lalley

ABSTRACT. A *random walk on a regular language* is a Markov chain on the set of all finite words from a finite alphabet A whose transition probabilities obey the following rules: (1) Only the last two letters of a word may be modified in one jump, and at most one letter may be adjoined or deleted. (2) Probabilities of modification, deletion, and/or adjunction depend only on the last two letters of the current word. Special cases include (a) reflecting random walks on the nonnegative integers; (b) LIFO queues; (c) finite-range random walks on homogeneous trees; and (d) random walks on the modular group $PSL_2(\mathbb{Z})$. It is shown that the n-step transition probabilities of a random walk on a regular language must obey one of three different types of power laws. The analysis is based on the study of an algebraic system of generating functions related to the Green's function.

1. Introduction

The transition probabilities of random walks on the integer lattices, and more generally on countable Abelian groups, satisfy *local limit theorems* that are intimately related to the *global* limit theorems, such as the CLT, that govern sums of i.i.d. random vectors. For random walks on countable non Abelian groups there are no such global limit theorems. Nevertheless, it is known that for random walks on certain countable non Abelian groups, including the finitely generated free groups [**20, 10, 14**] and the modular group $PSL(2, \mathbb{Z})$ [**22**], and for certain absolutely continuous random walks on semi-simple Lie groups [**3, 17**] the transition probabilities obey local limit theorems; and it is widely suspected that random walks on lattices of semi-simple Lie groups obey similar local laws.

This paper concerns random walks on another class of homogeneous structures, the *regular languages* of computer science [**12**]. Roughly, a *random walk on a regular language* is a Markov chain on the set of all finite words from a finite alphabet A whose transition probabilities obey the following rules: (1) Only the last two letters of a word may be modified in one jump, and at most one letter may be adjoined or deleted. (2) Probabilities of modification, deletion, and/or adjunction depend only on the last two letters of the current word. See Section 2.1 for definitions,

1991 *Mathematics Subject Classification.* Primary 60F05 Secondary 60F99.
Key words and phrases. regular language, algebraic function, Perron-Frobenius theory.
Supported by NSF grant DMS-0071970.

and Section 3 for examples. It will be shown that, under suitable irreducibility and aperiodicity hypotheses, the transition probabilities of such random walks must follow one of three different types of power laws: as $n \to \infty$,

$$P(X_n = y \mid X_0 = x) \sim C_{xy} R^{-n}, \qquad \text{or} \tag{1.1}$$

$$\sim C_{xy} R^{-n} n^{-1/2}, \qquad \text{or} \tag{1.2}$$

$$\sim C_{xy} R^{-n} n^{-3/2} \tag{1.3}$$

for suitable constants $C_{xy} > 0$ and $R \geq 1$. (Here and throughout the paper the symbol $\sim$ indicates that the ratio of the two sides converges to one as $n \to \infty$.) Certain countable non Abelian groups, such as (i) the finitely generated free groups, (ii) free products of two-element groups, and (iii) the modular group, admit descriptions as regular languages, and for random walks on such groups local limit theorems will emerge as special cases. In addition, various other Markov chains which do not have representations as random walks on groups, such as LIFOs (last-in first out queues) with different job types, *do* have representations as random walks on regular languages, and so our results will imply that the transition probabilities of such chains must obey one of the three power laws.

The main tools to be used are Tauberian theorems for generating functions (in particular, Darboux' theorem [**2**]) and techniques for analyzing singularities of generating functions. The key generating functions, the *Green's functions* of the random walks, will be shown to satisfy recursive relations stemming from the recursive nature of a regular language. These recursive relations, in turn, will imply that a related finite system of generating functions satisfy a system of polynomial equations, and thus are algebraic functions. It will then follow that the singularities of the Green's functions are at a branch point on the Riemann surface determined by the algebraic system. Finally, it will be shown that this branch point has ramification two; here the Perron-Frobenius theory of nonnegative matrices will play a crucial role.

The algebraic character of the Green's functions for random walk on a free group was first proved in [**1**]. This result was subsequently generalized to nearest-neighbor random walks on any finitely generated group containing free normal subgroups of finite index in [**23**], and it has been conjectured that these are the *only* groups for which the Green's functions of nearest-neighbor random walks are algebraic. The arguments of papers [**1**] and [**23**] give no information about the nature of the singularities of the Green's functions.

2. Random walk on a regular language

2.1. Definitions and Hypotheses. A *(nearest-neighbor) random walk on a regular language* (abbreviated *RWRL*) is a Markov chain X_n on the set $A^* := \cup_{n \geq 0} A^n$ of all finite words from a finite alphabet A whose transitions obey the following rules:

(a) Only the last two letters of the current word may be modified.
(b) At most one letter may be adjoined or deleted.
(c) Adjunction or deletion may only be done at the *end* of a word.
(d) Probabilities of modification, deletion, and/or adjunction depend only on the last two letters of the current word.

More precisely, a RWRL is a Markov chain whose transition probabilities satisfy:

$$
\begin{aligned}
(2.1) \qquad P(X_{n+1} = x_1x_2\dots x_m a'b' \mid X_n = x_1x_2\dots x_m ab) &= p(a'b'|ab)\\
P(X_{n+1} = x_1x_2\dots x_m a' \mid X_n = x_1x_2\dots x_m ab) &= p(a'|ab)\\
P(X_{n+1} = x_1x_2\dots x_m a'b'c \mid X_n = x_1x_2\dots x_m ab) &= p(a'b'c|ab)\\
P(X_{n+1} = a'b' \mid X_n = a) &= p(a'b'|a)\\
P(X_{n+1} = a' \mid X_n = a) &= p(a'|a)\\
P(X_{n+1} = \emptyset \mid X_n = a) &= p(\emptyset|a)\\
P(X_{n+1} = a \mid X_n = \emptyset) &= p(a|\emptyset) \qquad \text{and}\\
P(X_{n+1} = \emptyset \mid X_n = \emptyset) &= p(\emptyset|\emptyset).
\end{aligned}
$$

Here $\emptyset$ denotes the empty word. We do *not* assume that the transition probabilities are all positive, nor do we assume that there exist positive-probability paths connecting any two words. We shall, however, restrict attention to those random walks satisfying the following aperiodicity and irreducibility hypotheses:

ASSUMPTION 2.1 (Aperiodicity). *For all words $x_1x_2\dots x_m$,*

$$
(2.2) \qquad P(X_{n+1} = x_1x_2\dots x_m \mid X_n = x_1x_2\dots x_m) > 0
$$

Observe that, in view of the homogeneity restrictions implicit in equations (2.1), this is equivalent to assuming that all of the transition probabilities $p(ab \mid ab)$, $p(a \mid a)$, and $p(\emptyset \mid \emptyset)$ are positive.

ASSUMPTION 2.2 (Irreducibility-A). *Let $\mathcal{L} \subset A^*$ be the set of all words that may be reached from $\emptyset$ via positive-probability paths. Then $\forall w \in \mathcal{L}$ there is a positive-probability path from w to $\emptyset$.*

ASSUMPTION 2.3 (Irreducibility-B). *If*

$$
\begin{aligned}
w_1 &= x_1x_2\dots x_m a \in \mathcal{L} \qquad \textit{and}\\
w_2 &= x_1x_2\dots x_m \in \mathcal{L}
\end{aligned}
$$

then there exists a positive-probability path from w_1 to w_2 through words of length at least $m+1$.

ASSUMPTION 2.4 (Irreducibility-C). *If*

$$
\begin{aligned}
w_1 &= x_1x_2\dots x_m a_1a_2\dots a_k \in \mathcal{L} \qquad \textit{and}\\
w_2 &= b_1b_2\dots b_{k'} y_1y_2\dots y_{m'} \in \mathcal{L}
\end{aligned}
$$

then there exists a word $w_3 \in \mathcal{L}$ of the form

$$
w_3 = b_1b_2\dots b_{k'} z_1z_2\dots z_{m''} a_1a_2\dots a_k.
$$

Moreover, for each such triple w_1, w_2, w_3 there exists a positive-probability path from w_2 to w_3 through words of length $\geq k'$.

Henceforth, it is assumed that a RWRL satisfies the aperiodicity and irreducibility assumptions above. A number of interesting examples are discussed in Section 3 below. It should be clear that our results may be extended to RWRLs satisfying somewhat weaker hypotheses. For instance, the hypothesis that changes of state should involve only the last two letters could be replaced by the weaker hypothesis that such changes involve only the last K letters, for some $K < \infty$: see the discussion in paragraph 3.3 below. The aperiodicity hypothesis (2.2) is mainly for ease of exposition — random walks on regular languages that do not satisfy this hypothesis may have transition probabilities that oscillate periodically, with any integer period. See Section 6 below for an example. Finally, the irreducibility assumptions 2.2-2.4 may be weakened; however, the assumptions hold in many interesting examples, including those discussed in Section 3 below.

It is easily proved that the set $\mathcal{L}$ of words accessible from the root $\emptyset$ is a *regular language*, that is, a subset of A^* consisting of all words accepted by a finite-state automaton. See [**12**] for background on regular languages. Since our results and arguments will make no appeal to results from the theory of formal languages, we shall not discuss this point in any further detail; however, we note in passing that it is essentially the recursive structure of $\mathcal{L}$ that lies behind the system of polynomial equations (5.9) below, on which our analysis is based.

2.2. Local Limit Theorem. The main results of this paper are the following theorems, which concern the asymptotic behavior of the transition probabilities $P^x\{X_n = y\} := P(X_n = y \mid X_0 = x)$.

THEOREM 2.5. *Let X_n be a* null recurrent *RWRL. Then for each pair $x, y \in \mathcal{L}$ there exists a constant $0 < C_{xy} < \infty$ such that as $n \to \infty$,*

$$P^x\{X_n = y\} \sim C_{xy} n^{-1/2} \tag{2.3}$$

THEOREM 2.6. *Let X_n be a* transient *RWRL. Then for each pair $x, y \in \mathcal{L}$ there exists a constant $0 < C_{xy} < \infty$ and $1 < R < \infty$ (R is the* spectral radius *of the random walk) such that one of the following holds as $n \to \infty$:*

$$P^x\{X_n = y\} \sim C_{xy} R^{-n}; \quad or \tag{2.4}$$

$$P^x\{X_n = y\} \sim C_{xy} R^{-n} n^{-1/2}; \quad or \tag{2.5}$$

$$P^x\{X_n = y\} \sim C_{xy} R^{-n} n^{-3/2}. \tag{2.6}$$

The proofs of these theorems for the special case where x and y are words of length ≤ 1 will be carried out in Sections 5, 7, and 8 below. The general case, where x, y are arbitrary, may be deduced from this special case by a "recoding" trick — see paragraph 3.3 below. It will be shown in Section 6 that all three cases of Theorem 2.6 may occur. It is known (see, for instance, [**14**] for the proof) that for random walk on a non amenable group the power laws (2.4) and (2.5) cannot occur. Thus, for the examples considered in Sections 3.5 and 3.6 below, only (2.6) is possible.

It is somewhat mysterious that only the three possibilities (2.4)-(2.6) can occur. Although we have no purely probabilistic explanation, we believe that the latter two power laws correspond to different possible limit laws for the "pinned" random walks. Consider the renormalized wordlength process

$$W_n(t) := |X_{\lfloor nt \rfloor}| / \sqrt{n}, \tag{2.7}$$

where $|\cdot|$ denotes wordlength and $\lfloor \cdot \rfloor$ denotes greatest integer. We conjecture that the power laws (2.5) and (2.6) correspond, respectively, to the following possibilities for the processes $\{W_n(t)\}_{0 \leq t \leq 1}$:

$$W_n(t) \xrightarrow{\mathcal{D}} \text{Brownian bridge, and} \tag{2.8}$$

$$W_n(t) \xrightarrow{\mathcal{D}} \text{Brownian excursion.} \tag{2.9}$$

These are, of course, the only possibilities for the reflected random walks discussed in Sections 3.1 and 6 below. Additional evidence for the conjecture is given in a recent paper of BOUGEROL & JEULIN [**4**], where it is shown that the renormalized distance-to-the-identity processes for isotropic random walks on the hyperbolic plane must converge to the Brownian excursion.

2.3. Malyshev's Theorem. Random walks on regular language have previously been studied by MALYSHEV and co-authors in [**18**] and [**9**], in which they are called "random strings". Malyshev proves the following basic theorem:

THEOREM. *The wordlength $L_n := |X_n|$ of a RWRL satisfies a SLLN:*

$$L_n/n \to \mu \qquad \text{almost surely as } n \to \infty, \tag{2.10}$$

and $\mu = 0$ if and only if X_n is recurrent. In all cases, there is "edge stabilization": for any $m \geq 1$, the joint distribution of the last m letters of X_n converges weakly as $n \to \infty$.

Theorems of this type, of course, give no information about the detailed asymptotic behavior of the transition probabilities.

3. Examples of random walks on regular languages

3.1. Nearest-neighbor reflecting random walks on $\mathbb{Z}_+$. The most basic example is the simple nearest-neighbor $p - q$ random walk with reflection at the origin. This is the Markov chain on $\mathbb{Z}_+$ with transition probabilities

$$\begin{aligned} P(X_{n+1} = x+1 \,|\, X_n = x) &= p \qquad &&\text{if } x \geq 1; \\ P(X_{n+1} = x-1 \,|\, X_n = x) &= 1-p \qquad &&\text{if } x \geq 1; \\ P(X_{n+1} = 1 \,|\, X_n = 0) &= r; \\ P(X_{n+1} = 0 \,|\, X_n = 0) &= 1-r \end{aligned} \tag{3.1}$$

Henceforth, we shall refer to this random walk as $RRW(p;r)$. This is clearly equivalent to a random walk on the regular language $\mathcal{L} = \{a^m\}_{m \geq 0}$ with a one-letter alphabet. Notice, though, that this random walk does not satisfy the aperiodicity hypothesis (2.2) and in fact, as we will show in Section 6 below, the transition probabilities do not obey the local limit theorem (2.6), but rather exhibit a periodic oscillation.

3.2. *Non*-nearest-neighbor reflecting random walks on $\mathbb{Z}_+$. Consider a Markov chain on $\mathbb{Z}_+$ with transition probabilities

$$\begin{aligned} P(X_{n+1} = x+k \,|\, X_n = x) &= p_k \qquad &&\text{for } |k| \leq K \text{ and } x \geq K; \\ P(X_{n+1} = x+k \,|\, X_n = x) &= p_{x,k} \qquad &&\text{for } 0 \leq x < K \text{ and } -x \leq k \leq K. \end{aligned} \tag{3.2}$$

Such a process behaves as a (homogeneous) random walk with increments bounded by K outside a finite neighborhood $\{0, 1, \cdots, K-1\}$ of the origin. It is not difficult to see that a Markov chain governed by transition probabilities (3.2) is equivalent to a random walk on the regular language

$$\mathcal{L} = \{a^m b \,|\, m \in \mathbb{Z}_+ \text{ and } b \in \{0, 1, 2, \cdots, K-1\}\}.$$

The word $a^m b$ corresponds to the integer $mK + b$. Since the increments in X_n are of magnitude $\leq K$, only the last two letters of the representing word $aa \cdots ab$ need be modified to account for any possible change of state. Local limit theorems for Markov chains with transition probabilities (3.2) were proved, by Wiener-Hopf methods, in [**15**].

3.3. The "recoding" trick. The hypothesis that transitions in a RWRL involve only the last two letters of the word representing the current state may seem at first glance to be quite restrictive. However, the trick used in the preceding paragraph to represent non-nearest neighbor reflecting random walks on $\mathbb{Z}_+$ as RWRLs may be adapted to various other Markov chains.

Suppose, for instance, that X_n is a Markov chain on A^* whose transition probabilities obey the restrictions (2.1)-(2.1) of Section 2.1, *except* that in any given transition, the last K letters (instead of the last 2) may be modified, and the probabilities of adjunction, deletion, and modification depend on the last K letters (not just the last 2). Although the transition probabilities of the Markov chain X_n do not obey the rules (2.1) (unless $K = 2$), it is not difficult to construct a RWRL Y_n that "represents" X_n as follows: Replace the alphabet A by the larger alphabet $B := \cup_{m=1}^{K} A^m$ consisting of all words of length $\leq K$ from A. For each finite word x in the letters A, let $y = y(x)$ be the word in the letters B obtained from x by grouping letters of x in blocks of K, starting at the beginning of the word. For example, if $K = 3$:

$$\begin{aligned} x &= a_1a_2a_3a_4 \qquad \Longrightarrow \\ y(x) &= (a_1a_2a_3)(a_4). \end{aligned}$$

Observe that the mapping $x \mapsto y(x)$ is bijective. It is routine to verify that the transition probabilities of the Markov chain $Y_n = y(X_n)$ obey the rules (2.1).

Notice that the recoding trick just described also produces a RWRL when the input process X_n is already a RWRL. In this case, the recoding collapses words of length $\leq K$ to words of length 1. Therefore, to prove Theorems 2.5-2.6 for arbitrary x, y, it suffices to prove them for words x, y of length 0 or 1.

3.4. LIFO Queues with multiple job types. A LIFO (last-in first-out) *queue*, also called a *stack*, is a data structure used in many fundamental algorithms of computer science (see, for instance, [**5**], Section 11). The queue is stored as a linked list. New items are adjoined at the top of the linked list ("push"); and when items are removed, they are removed, one at a time, from the top ("pop"). In certain algorithms (for instance, when the stack is used to implement a recursion), the items in the list represent *jobs* to be completed, one at a time. These jobs may be of different types. Assume that the set A of possible job types is finite.

Now suppose that inputs to the algorithm are random. Then during the execution of the algorithm the stack will evolve as a stochastic process whose state space is a subset of the set A^* of finite words over the alphabet A. In algorithms where the state of the computation depends only on the top job in the stack (as is often the case) and inputs to the algorithm are i.i.d., the sequence of successive states of the stack will constitute a random walk on a regular language.

3.5. Random walk on a finitely generated free group. The *free group* $\mathcal{G} = \mathcal{G}(B)$ generated by a finite set $B = \{b_1, b_2, \dots, b_L\}$ consists of all *reduced* finite words in the letters $B^{\pm} = \{b_1^{\pm 1}, b_2^{\pm 1}, \dots, b_L^{\pm 1}\}$. (A word is said to be *reduced* if no letter $b_i^{\pm 1}$ is adjacent to its inverse $b_i^{\mp 1}$.) Multiplication in $\mathcal{G}$ consists of concatenation followed by reduction; the empty word $\emptyset$ is the group identity. A *nearest-neighbor random walk* on $\mathcal{G}$ is a Markov chain X_n that evolves according to the rule

$$X_{n+1} = X_n \xi_{n+1}, \tag{3.3}$$

where $\xi_1, \xi_2, \dots$ are i.i.d. random variables valued in $B^{\pm} \cup \{\emptyset\}$. Clearly, post-multiplication by an element of $B^{\pm}$ can modify only the last letter of a word in $\mathcal{G}$, and so the transition probabilities implied by (3.3) must satisfy the hypotheses (2.1). If each element $\alpha \in B^{\pm} \cup \{\emptyset\}$ is assigned positive probability by the distribution of ξ_n, then the aperiodicity and irreducibility hypotheses of Section 2.1 will be satisfied.

More generally, consider a *non*-nearest-neighbor random walk on $\mathcal{G}$ whose increments are bounded. The law of evolution is again given by (3.3), but now it is assumed that the jumps ξ_n are i.i.d. random variables valued in the set of all reduced words of length $\leq K$. These random walks may also be represented as random walks on a regular language, using the recoding trick of paragraph 3.3 above. The appropriate regular language $\mathcal{L}$ consists of all finite words $a_1 a_2 \dots a_m b$, with $m \geq 0$, such that (i) each $a_i \in \mathcal{G}$ is a word of length K (in the generators of the group $\mathcal{G}$); (ii) $b \in \mathcal{G}$ is a word of length $0 \leq |b| \leq K-1$; and (iii) no cancellations between successive a_i, a_{i+1} (or between a_m, b) are possible. It is easily verified that if the distribution of ξ_1 attaches positive probability to each word of length $\leq K$ then the aperiodicity and irreducibility hypotheses of Section 2.1 are satisfied.

Finally, observe that if the transition probabilities are modified at finitely many elements of $\mathcal{G}$, then the resulting Markov chain, a *perturbed random walk*, although no longer a "random walk" (in the usual sense of the term), is nevertheless still a random walk on a regular language. If the modifications are such that positive-probability transitions remain positive-probability transitions, then the aperiodicity and irreducibility hypotheses will still hold. However, whereas the power laws (2.4) and (2.5) cannot occur for processes of the form (3.3), they *can* occur for perturbed random walks.

3.6. Random walk on the modular group $PSL(2,\mathbb{Z})$. The *modular group* is the quotient group $PSL(2,\mathbb{Z}) = SL(2,\mathbb{Z})/\{\pm I\}$, where $SL(2,\mathbb{Z})$ is the (multiplicative) group of 2×2 matrices with integer entries and determinant 1, and I is the 2×2 identity matrix. (For information about the modular group and its connections with the theory of elliptic modular forms, and also for specific facts used in the subsequent discussion, see [**16**], especially Chapter XI.) A *(right) random walk* on $PSL(2,\mathbb{Z})$ is a sequence

$$X_n = \xi_1 \xi_2 \dots \xi_n, \tag{3.4}$$

where $\xi_1, \xi_2, \dots$ are i.i.d. random variables valued in $PSL(2,\mathbb{Z})$. We shall assume that (a) the distribution of ξ_1 has finite support; (b) $P\{\xi_1 = I\} > 0$; and (c) that for every element $A \in PSL(2,\mathbb{Z})$, the random walk $\{X_n\}_{n \geq 0}$ visits A with positive probability. We shall show that any such random walk has a description as a RWRL. The assumptions (b)-(c) above guarantee that the irreducibility and aperiodicity assumptions 2.1, 2.2, and 2.3 will be satisfied.

The modular group is finitely generated, with generators

$$U = \begin{pmatrix} 1 & 1 \\ 0 & 1 \end{pmatrix}, \qquad U^{-1} = \begin{pmatrix} 1 & -1 \\ 0 & 1 \end{pmatrix}, \qquad \text{and } T = T^{-1} = \begin{pmatrix} 0 & -1 \\ 1 & 0 \end{pmatrix}. \tag{3.5}$$

(More accurately, the group is generated by the *equivalence classes* $\langle T\rangle := \{\pm T\}$ and $\langle U\rangle := \{\pm U\}$. Henceforth, we shall not distinguish between matrices and their equivalence classes in $PSL(2,\mathbb{Z})$.) These generators satisfy the relations

$$T^2 = (TU)^3 = I. \tag{3.6}$$

Thus, every element of $PSL(2,\mathbb{Z})$ may be written as a finite word in the letters T, U, U^{-1} in which no two T's appear consecutively, and no U is adjacent to a U^{-1}. It follows that the modular group is also generated by T, W, W^2, where $W = TU$. The relations (3.6) translate as $T^2 = W^3 = I$. Every element $M \in PSL(2,\mathbb{Z})$ has a representation as a finite word

$$M = T^a W^{n_1} T W^{n_2} T W^{n_3} \cdots T W^{n_k} T^b, \tag{3.7}$$

where $a, b = 0$ or 1 and each n_i is either 1 or 2.

The modular group contains free normal subgroups of finite index, and so it has a representation as a regular language. Following is a sketch of how such a representation may be obtained (see [**16**], Section XI.3E for details). Let Γ' be the *commutator subgroup* of $PSL(2,\mathbb{Z})$, that is, the subgroup consisting of all finite products of *commutators.* (The *commutators* of any group are the elements of the form $aba^{-1}b^{-1}$.) The commutator subgroup of any finite or countable group is a normal subgroup, and so Γ' is a normal subgroup of $PSL(2,\mathbb{Z})$. The group Γ' is free (see [**16**], Theorem XI.3E) , with generators

$$\begin{aligned} X &= TWTW^2, & X^{-1} &= WTW^2T, \\ Y &= TW^2TW, \quad \text{and} & Y^{-1} &= W^2TWT. \end{aligned} \tag{3.8}$$

The index of Γ' in $PSL(2,\mathbb{Z})$ (that is, the number of distinct cosets of Γ') is 6. Thus, since Γ' is a normal subgroup, one may choose 6 distinct elements $A_1, A_2, \dots, A_6$ of $PSL(2,\mathbb{Z})$, one from each coset, such that every element of $PSL(2,\mathbb{Z})$ has a representation

$$M = WA_i, \tag{3.9}$$

where W is a reduced word in the (free) generators X, X^{-1}, Y, Y^{-1}. Observe that there may be some cancellation of letters at the end of W; but since there are only 6 possibilities for A_i, there are only finitely many possible types of cancellation, involving at most a bounded number of letters at the end of W. Thus, there is a finite subset $\mathcal{B} \subset PSL(2,\mathbb{Z})$ such that every element of $PSL(2,\mathbb{Z})$ has a representation as a *reduced* word

$$M = W'B_j, \tag{3.10}$$

where W' is a reduced word in X, X^{-1}, Y, Y^{-1} and $B_j \in \mathcal{B}$. This exhibits $PSL(2,\mathbb{Z})$ as a regular language $\mathcal{L}$, with alphabet $\mathcal{A} = \{X, X^{-1}, Y, Y^{-1}\} \cup \mathcal{B}$.

Now consider a right random walk (3.4) whose jumps ξ_i satisfy the assumptions (a), (b), and (c) above. By (a), the distribution of ξ_i has finite support; consequently, there are only finitely many different ways that multiplication by ξ_{n+1} can modify the end of the word X_n. In particular, there is a finite integer K such that all positive-probability transitions $X_n \to X_{n+1}$ involve only the last K letters of X_n, and such that the transition probabilities depend only on the last K letters of the current state X_n. By replacing the regular language $\mathcal{L}$ by a regular language $\mathcal{L}'$ whose alphabet consists of the words of length $\leq K$ in $\mathcal{L}$, one obtains a representation of the random walk X_n as a random walk on a regular language whose transition probabilities obey the restrictions (2.1) of Section 2.1.

A local limit theorem for random walk on the modular group was proved by Woess [**22**], using different and somewhat more difficult arguments than those employed below. Later [**23**], Woess also proved that, for any nearest-neighbor random walk on a discrete group containing a free normal subgroup of finite index, the Green's function is algebraic; however, his arguments do not allow one to deduce the type of the leading singularity.

4. Darboux' Theorem

The asymptotic behavior of the transition probabilities $P^x(X_n = y)$ is governed by the behavior of the generating function $\sum_n z^n P^x(X_n = y)$ near its smallest positive singularity. To extract the information about the transition probabilities from the behavior of this generating function, we shall appeal to a Tauberian theorem of Darboux [**2**].

Darboux' Theorem. *Let $G(z) = \sum_{n=0}^{\infty} a_n z^n$ be a power series with radius of convergence R. Suppose that G has no singularities in the closed disk $|z| \leq R$ except $z = R$, and that in a neighborhood of $z = R$,*

$$G(z) = A(z)(1 - z/R)^{\alpha} + B(z), \tag{4.1}$$

where $A(z)$ and $B(z)$ are analytic near $z = R$, $A(R) \neq 0$, and $\alpha \in \mathbb{R}\backslash\{0, -1, -2, \ldots\}$. Then as $n \to \infty$,

$$a_n \sim \frac{A(R)}{\Gamma(-\alpha) n^{1+\alpha} R^n} \tag{4.2}$$

More generally, if G has finitely many singularities $\zeta_1, \zeta_2, \ldots, \zeta_m$ on the circle $|z| = R$ of convergence, near each of which it has a representation

$$G(z) = A_i(z)(1 - z/\zeta_i)^{\alpha_i} + B_i(z) \tag{4.3}$$

with $A_i(z)$ and $B_i(z)$ analytic near $z = \zeta_i$ and $\alpha_i \in \mathbb{R} \setminus \{0, -1, -2, \ldots\}$, then as $n \to \infty$,

$$a_n \sim \sum_{i=1}^{m} \frac{A_i(\zeta_i)}{\Gamma(-\alpha_i) n^{1+\alpha_i} \zeta_i^n}, \tag{4.4}$$

provided the right side is not 0.

The hypotheses of Darboux' theorem are tailored to power series that define *algebraic functions.* A complex function $G(z)$ defined for $z \in \Omega$, where Ω is a nonempty open subset of $\mathbb{C}$, is said to be *algebraic* (more precisely, a *branch* of an algebraic function) if there is an irreducible polynomial $P(z, w)$ in two variables such that

$$P(z, G(z)) = 0 \qquad \forall\, z \in \Omega \tag{4.5}$$

Observe that if the polynomial relation (4.5) holds in Ω, then it holds for any analytic continuation of G to a domain containing Ω, by the Law of Permanence of functional equations. See [**11**] for a thorough discussion of algebraic functions, including a proof of the following fundamental result:

If $G(z)$ is an algebraic function, then it has only finitely many singularities in the (extended) complex plane, and near each singularity has the form (4.3), with rational exponent α.

If $G(z) = \sum_{n=0}^{\infty} a_n z^n$ is an algebraic function with a singularity at $z = R$ and no other singularities in $|z| \leq R$, then the conclusion (4.2) holds *provided* the function $A(z)$ in the representation (4.1) is not zero at $z = R$. Rather than verify this directly, we shall find it more convenient to work with leading term of the *Puiseux series* of G at $z = R$. The Puiseux series (see [**11**]) is

$$G(z) = G(R) + \sum_{n=1}^{\infty} b_n (R - z)^{n/k}, \tag{4.6}$$

where $k \geq 1$ is an integer; the series $\sum_{n \geq 1} b_n z^n$ has positive radius of convergence. Using the Darboux theorem, one may easily check that the asymptotic behavior of the coefficients a_n is determined by the exponent of the first (non constant) *nonzero* term in the Puiseux expansion. Thus, we have the following simplification in the hypotheses of Darboux' theorem:

COLLARY 4.1. *Let $G(z) = \sum_{n=0}^{\infty} a_n z^n$ be a power series with nonnegative coefficients a_n and radius of convergence R. Suppose that the function $G(z)$ has no singularities in the closed disk $|z| \leq R$* except *$z = R$, and that as $z \to R-$ (through real arguments less than R) either*

$$G(z) \sim C(1 - z/R)^{\alpha} \qquad or \tag{4.7}$$

$$G(R) - G(z) \sim C(1 - z/R)^{\alpha}, \tag{4.8}$$

with $C > 0$. If $G(z)$ is an algebraic function, then relation (4.2) holds with $A(R)$ replaced by C.

5. Green's functions and related generating functions

5.1. The Green's Functions. Assume now that X_n is a random walk on a regular language $\mathcal{L}$ satisfying the aperiodicity and irreducibility hypotheses of Section 2.1. For $x, y \in \mathcal{L}$, define the *Green's function* $G_{xy}(z)$ as follows:

$$G_{xy}(z) := \sum_{n=0}^{\infty} P^x\{X_n = y\} z^n = E^x \left(\sum_{n=0}^{\infty} z^n \mathbf{1}\{X_n = y\} \right). \tag{5.1}$$

Since the coefficients in the power series are all probabilities, the series converge absolutely and uniformly in the closed unit disc $|z| \leq 1$. Thus, the power series of each of the Green's functions has radius of convergence at least 1. In Section 7.2 below, we shall prove that all of these power series have the same radius R of convergence. Since their coefficients are nonnegative, and, by Assumption 2.2, at least some are strictly *positive*, the following must hold:

(a) $G_{xy}(z)$ and its derivatives are positive for $0 < z < R$.
(b) $G_{xy}(z)$ is strictly increasing and strictly convex in the interval $[0, R)$.
(c) $\lim_{z \to R-} G_{xy}(z) = G_{xy}(R) \in (0, \infty]$.
(d) $G_{xy}(z)$ has a singularity at $z = R$.

Property (d) is PRINGSHEIM's theorem (see [**11**]).

The Green's functions $G_{xy}(z)$ are intimately related to, and in fact completely determined by, a finite system of generating functions $H_{ab,c}(z)$ indexed by words ab and c of lengths 1 and 2, respectively. These are defined as follows:

$$H_{ab,c}(z) := E^{ab} z^T \mathbf{1}\{X_T = c\} = \sum_{n=1}^{\infty} z^n P^{ab}\{T = n \text{ and } X_n = c\}, \tag{5.2}$$

where

$$T := \min\{n : |X_n| < |X_0|\}. \tag{5.3}$$

Observe that some of the generating functions $H_{ab,c}(z)$ may be identically zero, as there may be no positive-probability paths from ab to c through words of length ≥ 2. However, Assumptions 2.2-2.4 imply that at least some of the functions $H_{ab,c}(z)$ are *not* identically zero. We shall refer to these as the *nondegenerate* first-passage generating functions. In Section 7.1 below we will show that the nondegenerate first-passage generating functions $H_{ab,c}(z)$ have a common radius ρ of convergence, and in Section 7.3 that $\rho \geq R$. Because the coefficients in the power series are nonnegative, and, for the nondegenerate generating functions, not all zero, the nondegenerate generating functions $H_{ab,c}(z)$ satisfy properties (a)-(d) above, with R replaced by ρ.

The relationship between the Green's functions and the first-passage generating functions $H_{ab,c}(z)$ is a simple consequence of the Markov property. Observe that the second equality in (5.1) shows that the Green's function $G_{xy}(z)$ may be interpreted as the *expected discounted number of visits to* y when the initial state is x. Let x and y be words of lengths $|x|, |y| \leq 1$. If $X_0 = x$, then X_1 must be a word of length $0, 1$, or 2; if X_1 is a word of length 2, then there can be no more visits to y until after the Markov chain X_n returns from the set of words of length ≥ 2. Hence, by the Markov property, if $|x| \leq 1$ and $|y| \leq 1$, then

$$G_{xy}(z) = \delta_x(y) + z \sum_{w:|w|\leq 1} p(w|x) G_{wy}(z) + z \sum_{ab} \sum_{c} p(ab|x) H_{ab,c}(z) G_{cy}(z), \tag{5.4}$$

where $\sum_{ab}$ is over all two-letter words, and $\sum_c$ is over all one-letter words.

The equations (5.4) may be written compactly in matrix form as follows. Define $\mathbf{G}(z)$ to be the matrix-valued function of z with entries $G_{xy}(z)$, indexed by words x, y of lengths ≤ 1, and define $\mathbf{P}$ to be the matrix of transition probabilities $p(y|x)$, where again x and y are words of length ≤ 1. Define $\mathbf{K}(z)$ to be the matrix-valued function of z with entries $K_{xy}(z)$, indexed by words x, y of length ≤ 1, defined by

$$\begin{aligned} K_{xy}(z) &= \sum_{ab} p(ab|x) H_{ab,y}(z) \qquad && \text{if } |y| = 1; \\ &= 0 && \text{if } |y| = 0. \end{aligned} \tag{5.5}$$

Then equations (5.4) are equivalent to the matrix equations

$$\begin{aligned} \mathbf{G}(z) &= \mathbf{I} + z\mathbf{P}\mathbf{G}(z) + z\mathbf{K}(z)\mathbf{G}(z) \qquad \Longleftrightarrow \\ \mathbf{I} &= (\mathbf{I} - z\mathbf{P} - z\mathbf{K}(z))\,\mathbf{G}(z), \end{aligned} \tag{5.6}$$

where $\mathbf{I}$ is the identity matrix. Notice that for all z of sufficiently small modulus $|z|$, the matrix $(z\mathbf{P} + z\mathbf{K}(z))$ has norm less than one, and so the matrix $(\mathbf{I} - z\mathbf{P} - z\mathbf{K}(z))$

is invertible; thus,

$$\boxed{\mathbf{G}(z) = (\mathbf{I} - z\mathbf{P} - z\mathbf{K}(z))^{-1}} \tag{5.7}$$

for all z in a disc centered at $z = 0$.

5.2. Algebraic relations among the generating functions. Equations (5.6) above show that the generating functions G_{xy} and $H_{w,v}$ are algebraically interrelated. In this section, we will show that the auxiliary generating functions $H_{w,v}$ are themselves algebraically interrelated via a system of quadratic equations that define an algebraic curve. This will imply that these functions are algebraic. This, together with the equations (5.6) above, will imply that the Green's functions $G_{xy}(z)$ are algebraic functions of z.

Recall (see equations 2.1)) that, for a random walk on a regular language, only the last two letters of the current word play a role in the transition to the next word. Consequently, for any initial state $\xi = x_1 x_2 \dots x_m ab \in \mathcal{L}$, the letters $x_1 x_2 \dots x_m$ are ignored up to time T. It follows that for any word $\xi = x_1 x_2 \dots x_m ab \in \mathcal{L}$, and any $c \in A$,

$$H_{ab,c}(z) = E^{\xi} z^T \mathbf{1}\{X_T = x_1 x_2 \dots x_m c\}. \tag{5.8}$$

The homogeneity relations (5.8) lead directly to a system of quadratic equations relating the functions $H_{ab,c}$. These equations derive from the Markov property: Starting from any word w, the random walk must make an initial jump to another state. If this first jump is to a word whose length is less than that of the initial word w, then $T = 1$; otherwise, the initial jump must be either to a word w' of the same length, or to a word w'' with one additional letter. In the latter case, T is the first time after the initial jump that *two* letters are removed; in the former, T is the first time that a single letter is removed. Summing over all possible initial jumps and applying the identity (5.8) yields the following system of equations:

$$\begin{aligned} H_{ab,c}(z) = zp(c|ab) + zp(ab|ab)H_{ab,c}(z) + z \sum_{d,e \in A} p(de|ab) H_{de,c}(z) \\ + z \sum_{d,e,f \in A} p(def|ab) \sum_{g \in A} H_{ef,g}(z) H_{dg,c}(z), \end{aligned} \tag{5.9}$$

We shall abbreviate this system of equations as

$$\boxed{\mathbf{H}(z) = z\mathbf{Q}(\mathbf{H}(z))}, \tag{5.10}$$

where $\mathbf{H}(z)$ is the vector of *nondegenerate* first-passage generating functions $H_{ab,c}(z)$ (that is, those that are *not* identically zero), and $\mathbf{Q}$ is the vector of quadratic polynomials $Q_{ab,c}$ in the variables $H_{ab,c}$ occurring on the right sides of the equations (5.9) above. Observe that $\mathbf{Q}$ is quadratic (that is, each of its components is quadratic) and does not depend on the variable z. Moreover, the coefficients in the polynomials $Q_{ab,c}$ are *positive.* This will be of crucial importance later.

5.3. Algebraic curves and the algebraic character of the Green's functions. An *affine (algebraic) variety* $V \subset \mathbb{C}^n$ is defined to be the set of all zeros $(a_1, a_2, \dots, a_n$ of a finite set of polynomials in n variables over the field $\mathbb{C}$ of complex numbers. (See, for instance **[6]**, Ch. I, for the definition. Some authors use a different terminology, calling what we have defined to be an affine variety an *algebraic set*, and reserving the term *variety* for an *irreducible* algebraic set.) The system of equations (5.10) determines an affine variety V_H, since it consists of polynomial (in fact, quadratic) equations in the variables z and $H_{ab,c}$. Similarly, the system consisting of equations (5.10) and (5.6) determines a variety V_{GH}. (Observe that V_H is obtained by projecting the variety V_{GH} onto the z and $H_{ab,c}$ coordinates.) Notice that the point $(0, \mathbf{H}(0))$ lies on the variety V_H, and that the point $(0, \mathbf{H}(0), \mathbf{G}(0))$ lies on V_{GH}.

A basic theorem of algebraic geometry (see [**6**], Section 4.6) states that any affine variety V has a decomposition $V = \cup_{i=1}^{r} V_i$ into irreducible affine varieties, where a variety W is termed *irreducible* if the only decompositions $W = U_1 \cup U_2$ of W into two varieties are such that $W = U_1$ or $W = U_2$.

PROPOSITION 5.1. *There exists an open neighborhood $\mathcal{N}$ of the point*

$$(0, \mathbf{H}(0), \mathbf{G}(0))$$

such that $\mathcal{N} \cap V_{GH}$ is contained in an irreducible algebraic curve *(a variety of dimension 1).*

PROOF. See [**6**], Chapter 9, for the definition of the (algebraic) dimension of an affine variety. For our purposes, it will only be important that the algebraic dimension coincides with the *transcendence degree* (see the proof of Corollary 5.2 below) of the field of rational functions over $\mathbb{C}$. The system of equations consisting of (5.10) and (5.6) has the form

$$\mathbf{F}(z, \mathbf{w}) = \mathbf{0}, \tag{5.11}$$

where $\mathbf{w} = (w_1, w_2, \ldots, w_m)$ and $\mathbf{F}$ is an $m-$ vector of polynomials in the variables z and $w_1, w_2, \ldots, w_m$. If $(z_*, \mathbf{w}_*)$ is a solution of the equation (5.11), then it is said to be a *nonsingular* point if the Jacobian matrix

$$J := \frac{\partial \mathbf{F}}{\partial \mathbf{w}} \tag{5.12}$$

is nonsingular at $(z, \mathbf{w}) = (z_*, \mathbf{w}_*)$. If $(z_*, \mathbf{w}_*)$ is a nonsingular point of the variety $V_{\mathbf{F}}$ defined by the system (5.11), then, by Theorem 9.6.9 of [**6**], it lies on a unique irreducible component of the variety $V_{\mathbf{F}}$ of dimension 1. For the system consisting of equations (5.10) and (5.6), the function $\mathbf{F}$ may be written $\mathbf{F} = (\mathbf{F_H}, \mathbf{F_G})$, where

$$\mathbf{F_H}(z, \mathbf{H}) = \mathbf{H} - z\mathbf{Q}(\mathbf{H}) \qquad \text{and}$$
$$\mathbf{F_G}(z, \mathbf{H}, \mathbf{G}) = (\mathbf{I} - z\mathbf{P} - z\mathbf{K}(z))\, \mathbf{G}(z).$$

The Jacobian matrix at $(0, \mathbf{H}(0), \mathbf{G}(0))$ has the form

$$J = \begin{pmatrix} I & \mathbf{0} \\ * & I \end{pmatrix},$$

where I denotes the identity matrix and $\mathbf{0}$ the matrix all of whose entries are 0. This is certainly nonsingular, so the point $(0, \mathbf{H}(0), \mathbf{G}(0))$ is a nonsingular point of V_{GH}, and it lies on a unique irreducible component of dimension 1. □

Remark. Observe that the condition (5.12) is also the hypothesis of the (complex) *Implicit Function Theorem* (see, for example, [**13**], Chapter 3). This theorem states that, if the Jacobian matrix J is nonsingular at $(z_*, \mathbf{w}_*)$ then, for each z in a neighborhood of z_*, there is a *unique* solution $\mathbf{w} = \mathbf{w}(z)$ of (5.11), and the mapping $z \mapsto \mathbf{w}(z)$ is analytic. In particular, the locus of solutions to (5.11) near $(z_*, \mathbf{w}_*)$ is analytically parametrized by a single parameter z; in other words, it is a complex manifold of (topological) dimension one. The theorem 9.6.9 of [**6**] cited in the proof above states, in essence, that the *topological* and *algebraic* dimensions of the variety coincide at nonsingular points. □

COLLARY 5.2. *Each of the functions G_{xy} and $H_{ab,c}$ occurring as entries of the Green's function matrix $\mathbf{G}(z)$ and the first-passage generating function vector $\mathbf{H}(z)$ is algebraic.*

PROOF. Let V be the irreducible component the variety V_{GH} containing the point $(0, \mathbf{H}(0), \mathbf{G}(0))$. Each of the coordinate variables z, $H_{ab,c}$, and G_{xy} is a member of the field $\mathbb{C}[V]$ of *rational functions* on the variety V (see [**6**]), Section 5.5). By Theorem 9.5.2 of [**6**], the *transcendence degree* of the field $\mathbb{C}[V]$ (that is, the maximal number of algebraically independent elements) coincides with the algebraic dimension of V, which, by Proposition 5.1, is 1. Thus, there is a polynomial relation between any two elements of $\mathbb{C}[V]$, in particular, between z and $H_{ab,c}$, and between z and G_{xy}. □

In principle, a polynomial $\mathcal{P}(z, w)$ relating z and any one of the Green's functions can be obtained by an "elimination" algorithm, starting from the system (5.9) and (5.6). See [**6**] for an elementary discussion of elimination theory. Unfortunately, in all but the very simplest cases, the system (5.9) is either too large for the computation to be carried out, or leads to a polynomial $\mathcal{P}$ of too high a degree to be solved by quadratures for $G_{xy}(z)$ in terms of z. There is, however, one interesting case where the system may be solved explicitly — this is discussed in the next section.

6. A Special Case: Reflecting Random Walk on $\mathbb{Z}_+$

This special case is instructive, in that explicit formulas for the generating functions may be obtained, and all of the possibilities enumerated in Theorems 2.5 and 2.6 are exhibited. However, the results of this section will play no role in the subsequent sections.

6.1. Explicit expressions for the Green's functions. The Green's functions of the random walks $RRW(p; r)$ defined in Section 3.1 may be written in closed form as algebraic functions of their arguments. Let X_n be the Markov chain on the set of nonnegative integers with transition probabilities given by (3.1). Define the generating functions $G(z), F(z)$ by

$$G(z) = \sum_{n=0}^{\infty} z^n P^0\{X_n = 0\} \qquad \text{and} \tag{6.1}$$

$$F(z) = E^1 z^T \qquad \text{where} \tag{6.2}$$

$$T = \min\{n : X_n = 0\} \tag{6.3}$$

By (5.4) and the Markov property, (see [**7**], Section XIV.4, for similar calculations),

$$G(z) = (1 - z(1 - r) - zrF(z))^{-1} \qquad \text{and} \tag{6.4}$$

$$F(z) = (1 - p)z + pzF(z)^2. \tag{6.5}$$

Notice that the algebraic system (5.10) reduces to the single quadratic equation (6.5). This equation has two solutions; since the function $F(z)$ defined by (5.2) is nondecreasing in z for $0 \leq z \leq 1$, the correct branch is

$$F(z) = \frac{1 - \sqrt{1 - 4p(1-p)z^2}}{2pz}. \tag{6.6}$$

The smallest positive singularity is at $z = \rho$, where

$$\rho = \rho(p) = 1/(2\sqrt{p(1-p)}). \tag{6.7}$$

This is a branch point of order two. Observe that $-\rho$ is also a branch point of order two, but that there are no other singularities on the circle $|z| = \rho$. Notice also that the function $H(\zeta) := \sqrt{\zeta}F(\sqrt{\zeta})$ has only one singularity on its circle of convergence, at $\zeta = (4p(1-p))^{-1}$, and so by Darboux' theorem, as $n \to \infty$ through the *odd* integers,

$$P^1\{T = n\} \sim A(p; r)\rho^{-n}n^{-3/2} \tag{6.8}$$

for a suitable value $A(p; r) \in (0, \infty)$.

6.2. The Case $p = 1/2$. The reflecting random walk $RRW(1/2; r)$ is null recurrent for every value of $r \in (0, 1)$. The equation (6.4) for $G(z)$ simplifies to

$$G(z) = \left(1 - z(1-r) - r + r\sqrt{1 - z^2}\right)^{-1}, \tag{6.9}$$

which shows that G has singularities at $z = \pm 1$. These singularities are of different types, however, as $G(z) \to 0$ as $z \to 1$, while $G(z) \to 2 - r \neq 0$ as $z \to -1$. It is routine to check that

$$G(z) = A_+(z)(1-z)^{-1/2} + B_+(z) \qquad \text{near } z = 1 \qquad \text{and} \tag{6.10}$$

$$G(z) = A_-(z)(1-z)^{+1/2} + B_-(z) \qquad \text{near } z = -1 \tag{6.11}$$

for analytic functions $A_\pm, B_\pm$. Thus, by Darboux' theorem, as $n \to \infty$ the return probabilities satisfy

$$P^0\{X_n = 0\} \sim \frac{C}{\sqrt{n}} \tag{6.12}$$

for a suitable constant $C = C(1/2; r) > 0$.

6.3. The Case $p > 1/2$. When $p > 1/2$ the reflecting random walk $RRW(p; r)$ is transient for every $r \in (0, 1)$. The nature of the leading singularity of the Green's function $G(z)$, and its location *vis a vis* the singularity $z = \rho(p)$ of the first-passage generating function $F(z)$, depend on the value of r. Observe that if $p > 1/2$ then $F(1) < 1$, since the random walk is transient. In fact, by (6.6),

$$F(\rho(p)) = F(1/(2\sqrt{p(1-p)})) = \sqrt{(1-p)/p} < 1, \tag{6.13}$$

and so, by the strict monotonicity of $F(z)$ on the positive real axis, $F(z) < 1$ for all $z \leq \rho(p)$.

Define

$$r_* = r_*(p) = 1 - \frac{\sqrt{1-p}}{\sqrt{p} - \sqrt{1-p}} \tag{6.14}$$

LEMMA 6.1. *Let $z = \zeta = \zeta(p;r)$ be the smallest root of the equation*

$$1 = z(1-r) + rzF(z) \tag{6.15}$$

in the interval $z \in (0, \rho(p)]$, provided there is one. Then for each $p > 1/2$, the zero $\zeta(p;r)$ is continuous and strictly increasing in r for $0 < r \le r_*(p)$. At $r = r_*(p)$,

$$\zeta(p;r) = \rho(p) = 1/(2\sqrt{p(1-p)}). \tag{6.16}$$

For $r > r_*(p)$, there are no zeros of the equation (6.15) in the interval $(0, \rho(p)]$.

PROOF. Since $F(1) < 1$, the zero $\zeta(p;r)$, if it exists, must be strictly greater than 1. Moreover, since $F(z)$ is strictly increasing in $(0,\rho]$, there is at most one zero in $(0,\rho]$. For each $z \in (0,\rho(p)]$, the function $z(1-r)+rzF(z)$ is strictly decreasing in r, since $F(z) < F(\rho(p)) < 1$ by (6.13). Consequently, the function $\zeta(p;r)$ is strictly increasing in r. That $\zeta(p;r)$ is continuous in r follows from the implicit function theorem, as $F'(z) > 0$ for all $z \in (0, \rho(p)]$.

For all r such that $(1-r)^{-1} \le \rho(p)$, the function $z \mapsto 1 - z(1-r) - rzF(z)$ must have a zero in the interval $(1, (1-r)^{-1})$, by the intermediate value theorem. Hence, $\zeta(p;r)$ is defined for all small $r > 0$. By the preceding paragraph, the function $\zeta(p;r)$ is strictly increasing and continuous in r up to the point r_* where $\zeta(p;r_*) = \rho = \rho(p)$. This occurs at the solution $r = r_*$ of the linear equation

$$1 = \rho(1-r) + \rho r F(\rho).$$

Using (6.6), it is easily checked that $r_* = r_*(p)$ is given by (6.14). □

COLLARY 6.2. *Assume that $p > 1/2$. Let $r_* = r_*(p)$, $\zeta = \zeta(p;r)$, and $\rho = \rho(p)$ be as in Lemma 6.1 and equation (6.7). Then for the reflecting random walk $RRW(p;r)$ the transition probabilities satisfy*

$$P^0\{X_n = 0\} \sim C\zeta^{-n} \qquad \textit{if} \qquad 0 < r < r_*; \tag{6.17}$$

$$P^0\{X_n = 0\} \sim C\rho^{-n}n^{-1/2} \qquad \textit{if} \qquad r = r_*; \tag{6.18}$$

$$P^0\{X_n = 0\} \sim (C + (-1)^n C')\rho^{-n}n^{-3/2} \qquad \textit{if} \qquad r_* < r < 1, \tag{6.19}$$

for suitable constants $C = C(p;r) > 0$ and $C' = C'(p;r) \neq 0$.

PROOF. If $r < r_*(p)$ then by Lemma 6.1 the smallest positive zero $\zeta(p;r)$ of the equation (6.15) is strictly less than the radius $\rho(p)$ of convergence of $F(z)$. Hence, by equation (6.4), the smallest positive singularity of $G(z)$ is at $\zeta(p;r) < \rho(p)$. Since $F'(z) > 0$ at $z = \zeta(p;r)$, and since $F(z)$ is analytic in a neighborhood of $\zeta(p;r)$ it follows that $G(z)$ has a simple pole at $z = \zeta(p;r)$. It is not difficult to show, using the explicit expressions (6.4) and (6.5), that $G(z)$ has no other singularities on the circle $|z| = \zeta(p;r)$. The relation (6.17) follows.

Consider next the case $r > r_*(p)$. Since $\rho(1-r) + \rho r F(\rho)$ is strictly decreasing in r (recall that $F(\rho) < 1$, by (6.13)), and takes the value 1 at $r_*(p)$, it follows that, for $r > r_*(p)$,

$$1 > \rho(1-r) + \rho r F(\rho).$$

Consequently, by equation (6.4), $G(z)$ is an analytic function of z and $F(z)$ in a neighborhood of any point $(\xi, F(\xi))$ where $|\xi| = \rho$, and has power series expansion

$$G(z) = G(\xi) + G(\xi)^2\left((z-\xi)(1-r+rF(\xi)) + \xi r(F(z) - F(\xi)) + \cdots\right) + \cdots$$

Since $F(z)$ has branch points of order two at $z = \pm\rho$, so does $G(z)$; and since $F(z)$ has no other singularities on the circle $|z| = \rho$, neither does $G(z)$. Thus, by Darboux' theorem, the asymptotic formula (6.19) holds.

Finally, consider the case $r = r_*(p)$. In this case the Green's function once again has singularities at $z = \pm\rho$, and at no other points on $|z| = \rho$; however, $G(z)$ becomes infinite as $z \to \rho$, but remains finite as $z \to -\rho$. The nature of the singularity at $z = +\rho$ may be deduced from the identity

$$1/G(z) = 1 - z(1-r) - zrF(z);$$

this implies that, in a neighborhood of $z = \rho$,

$$G(z) = A(z)(\rho - z)^{-1/2} + B(z)$$

for suitable analytic function $A(z), B(z)$. Similarly, near $z = -\rho$ the function $G(z)$ satisfies

$$G(z) = C(z)(\rho - z)^{1/2} + D(z)$$

for suitable analytic function $C(z), D(z)$. Thus, by Darboux' theorem, (6.18) holds. □

Remark 1. It may strike the reader as somewhat strange that the period-two oscillation in the transition probabilities occurs only for $r > r_*(p)$. Here is an explanation: Conditional on the event $X_n = 0$, the distribution of the total number N_n of returns to 0 before time n satisfies

$$\begin{aligned} &N_n \xrightarrow{P} \infty && \text{if } r \le r_*(p), \text{ but} \\ &N_n = O_P(1) && \text{if } r > r_*(p) \end{aligned}$$

as $n \to \infty$. Since the state 0 has a positive holding probability, the many returns to 0 destroy any aperiodicity when $r \le r_*(p)$.

Remark 2. The random walks considered in this section do not obey the hypotheses of Theorems 2.5-2.6, and so by themselves do not prove that all the different cases of these theorems actually occur. However, it is a routine (if tedious) matter to obtain explicit formulas for the Green's functions of modified reflecting random walks with holding probability $\alpha > 0$ for all states $x \ge 1$, and to deduce from these expressions local limit theorems similar to those obtained above. The only difference in the results is that the period-two oscillation in formula (6.19) does not occur.

7. Analysis of the leading singularity

We return now to the study of random walks on regular languages in general. Recall from Section 5 that the matrix $\mathbf{G}(z)$ of Green's functions $G_{xy}(z)$ is a rational function (5.6) of the generating functions $H_{ab,c}(z)$, and that these generating functions are interrelated by the system (5.10) of quadratic equations (5.9). As earlier, we denote by $\mathbf{H}$ the vector of nondegenerate first-passage generating functions $H_{ab,c}(z)$.

7.1. Radius of convergence of the system (5.10). In this paragraph we will show that all of the *nondegenerate* first-passage generating functions $H_{ab,c}(z)$ (that is, all those that occur as entries of the vector-valued function $\mathbf{H}(z)$) have the same radius of convergence, and that all have the same *singularity type* at $z = \rho$. In general, two algebraic functions $\varphi(z)$ and $\psi(z)$, both with singularities at $z = \zeta$, will be said to have the same *singularity type* at the point $z = \zeta$ if (a) either $\varphi(z)$ and $\psi(z)$ are both finite at $z = \zeta$, or are both infinite; and (b) the leading terms in the Puiseux expansions of $\varphi(z)$ and $\psi(z)$ at $z = \zeta$ have the same exponent $\alpha > 0$, that is,

$$\varphi(z) \sim C(\zeta - z)^{-\alpha} \quad \text{and} \quad \psi(z) \sim C'(\zeta - z)^{-\alpha} \quad \text{or}$$
$$\varphi(\zeta) - \varphi(z) \sim C(\zeta - z)^{\alpha} \quad \text{and} \quad \psi(\zeta) - \psi(z) \sim C'(\zeta - z)^{\alpha}$$

for suitable constants $C, C' \neq 0$.

PROPOSITION 7.1. *All of the generating functions $H_{ab,c}(z)$ occurring as entries of $\mathbf{H}(z)$ have the same radius ρ of convergence, and all have the same singularity type at $z = \rho$.*

The proof will rely on an irreducibility property of the system (5.10) of quadratic equations governing the generating functions $H_{ab,c}(z)$. The system (5.10) asserts that the function $\mathbf{H}(z)$ is a fixed point of the mapping $\mathbf{F} \mapsto z\mathbf{Q}(\mathbf{F})$; consequently, $\mathbf{H}(z)$ is a fixed point of any iterate of this mapping. In particular,

$$\mathbf{H}(z) = (z\mathbf{Q}) \circ (z\mathbf{Q}) \circ \cdots \circ (z\mathbf{Q})(\mathbf{H}(z)) := z^m \mathbf{Q}^{\circ m}(\mathbf{H}(z)), \tag{7.1}$$

where $\circ$ denotes functional composition. The following lemma concerning this hierarchy of polynomial equations is the key point in the analysis where the Irreducibility Assumptions 2.3 and 2.4 are used.

LEMMA 7.2. *There exists an integer $m \geq 1$ such that, in each component of $(z\mathbf{Q})^{\circ m}(\mathbf{H})$, each variable $H_{ab,c}$ appears in a term with* positive *coefficient.*

PROOF. In each component equation (5.9) of the vector equation (5.10), the right side contains a term that is z times a constant $p(ab|ab)$ times the left side. Consequently, when equation (5.10) is substituted in the mth iterate to obtain the $(m+1)-$th iterate, any product of the variables $H_{ab,c}$ that occurred as a term in the mth iterate will remain as a term in the $(m+1)-$th iterate, albeit with a higher power of z. Therefore, to prove the lemma it suffices to show that, for each pair $H_{ab,c}, H_{a'b',c'}$ of components of $\mathbf{H}$, there exists $m \geq 1$ such that $H_{a'b',c'}$ appears as a factor in some term of the $H_{ab,c}-$component of the $m-$th iterate of (5.10).

Recall that if $H_{ab,c}(z), H_{a'b',c'}(z)$ are entries of $\mathbf{H}(z)$ then they are not identically zero, equivalently, there exist positive-probability paths from ab to c and from $a'b'$ to c' through words of length ≥ 2. Let

$$ab \to w_1 \to w_2 \to \cdots \to w_n \to c$$

be such a path from ab to c. By Assumption 2.4, there is a positive-probability path from w_n to a word $w_n w a'b'$ through the set of words prefaced by w_n. By the homogeneity property (5.8),

$$E^{w_n w a'b'} z^T \mathbf{1}\{X_T = w_n w c'\} = H_{a'b',c'}(z) \not\equiv 0.$$

By Assumption 2.3, for each consecutive pair β, γ of letters in the word wc', there is a positive-probability path from $\beta\gamma$ to β through words prefaced by β. Consequently, for sufficiently large m, the equation for $H_{ab,c}(z)$ in the mth iterate of the system (5.10) contains a term with the factor $H_{a'b',c'}(z)$. □

Proof of Proposition 7.1. That all of the generating functions $H(z)$ have the same radius of convergence follows immediately from Lemma 7.2, because, since $H(1) > 0$ and $H'(z) > 0$ for $z > 0$ and for each function H, Lemma 7.2 implies that for any pair H, H_* there are constants $C, C' > 0$ such that

$$H(z) \geq CH_*(z) \qquad \text{and} \qquad H_*(z) \geq C'H(z) \tag{7.2}$$

for all $z \geq 1$. It also follows from this that either all of the values $H(\rho)$ are $+\infty$, or all are finite (and, of course, positive). In the case where all are $+\infty$, the inequalities (7.2) imply that the leading terms in the Puiseux expansions all have the same exponent α.

Suppose that all of the functions H are finite (and positive) at their common radius of convergence $\rho \geq 1$. By Lemma 7.2, for each integer $k \geq 1$ there exist constants $C_k, C'_k > 0$ such that, for any pair H, H_*, the kth derivatives satisfy

$$H^{(k)}(z) \geq C_k H_*^{(k)}(z) \qquad \text{and} \qquad H_*^{(k)}(z) \geq C'_k H^{(k)}(z). \tag{7.3}$$

It follows that the leading terms in the Puiseux expansions must all have the same exponent α.

Henceforth, we shall refer to ρ as the "radius of convergence" of the system (5.10).

LEMMA 7.3. *Let $H_{ab,c}(z)$ be nondegenerate, and let $0 < r \leq \rho$ be such that $H_{ab,c}(r) < \infty$. Then for any complex number $\zeta \neq r$ such that $|\zeta| = r$,*

$$|H_{ab,c}(\zeta)| < H_{ab,c}(r) \tag{7.4}$$

PROOF. This is a consequence of Assumption 2.1, which implies that, for some $\varepsilon > 0$, $p(ab|ab) \geq \varepsilon$. This, in turn, implies that, for each integer $n \geq 1$,

$$P^{ab}\{T = n+1; X_{n+1} = c\} \geq \varepsilon P^{ab}\{T = n; X_n = c\}. \tag{7.5}$$

Consider any two consecutive terms of the power series expansion of $H_{ab,c}(\zeta)$, with $a_n := P^{ab}\{T = n; X_n = c\}$:

$$\begin{aligned} |a_n\zeta^n + a_{n+1}\zeta^{n+1}| &= r^n|a_n + a_{n+1}\zeta| \\ &< r^n|a_n + a_{n+1}r| \\ &= a_n r^n + a_{n+1}r^{n+1}. \end{aligned}$$

This shows that the sum of any two consecutive nonzero terms in the power series expansion of $H_{ab,c}(\zeta)$ is strictly less, in modulus, than the sum of the corresponding terms in the power series expansion of $H_{ab,c}(r)$. □

7.2. Radius of convergence of the Green's functions $G_{xy}(z)$.

PROPOSITION 7.4. *All of the Green's functions $G_{xy}(z)$ have the same radius of convergence $R \geq 1$. Moreover, all of the Green's functions $G_{xy}(z)$ have the same singularity type at $z = R$.*

PROOF. The argument follows similar lines as the proof of Proposition 7.1, but here Lemma 7.2 is replaced by a "Harnack"-type inequality. By Assumption 2.2, for any two elements u, v of the regular language $\mathcal{L}$, there are positive probability paths from u to v and from v to u. Consequently, for any two pairs $(x, y), (x', y') \in \mathcal{L}^2$ there exist $k \geq 0$ and $\varepsilon > 0$ such that

$$P^x\{X_{n+k} = y\} \geq \varepsilon P^{x'}\{X_n = y'\}. \tag{7.6}$$

This inequality clearly implies that the Green's functions $G_{xy}(z)$ and $G_{x'y'}(z)$ have the same radius of convergence R, and that if $G_{x'y'}(R) = \infty$ then $G_{xy}(R) = \infty$. It follows that the Green's functions $G_{xy}(z)$ are either all finite or all infinite at their common radius of convergence R.

Because each $G_{xy}(z)$ is an algebraic function, it has a Puiseux expansion at $z = R$. Since $G_{xy}(z)$ is strictly increasing in z along the line segment $[0, R]$, the first two terms of the Puiseux series must be of the form

$$G_{xy}(R) - G_{xy}(z) \sim C_{xy}(R - z)^{\alpha_{xy}} \qquad \text{if } G_{xy}(R) < \infty; \qquad \text{or} \tag{7.7}$$

$$G_{xy}(z) \sim C_{xy}(R - z)^{-\alpha_{xy}} \qquad \text{if } G_{xy}(R) = \infty, \tag{7.8}$$

where $\alpha_{xy} > 0$ and $C_{xy} > 0$. Consider first the case where all the Green's functions are infinite at $z = R$: By inequality (7.6) above, for any two pairs $(x, y), (x', y') \in \mathcal{L}$, and any $\alpha > 0$, there exists a constant $\varepsilon > 0$ such that

$$(R - z)^\alpha G_{xy}(z) \geq \varepsilon (R - z)^\alpha G_{x'y'}(z).$$

Consequently, in case (7.8), all of the exponents α_{xy} must be the same.

Now consider the case (7.7). Once again, by inequality (7.6), for any two pairs $(x, y), (x', y') \in \mathcal{L}$, and any $\alpha > 0$, there exist constants $\varepsilon_k > 0$ such that the kth derivatives of the Green's functions satisfy

$$(R - z)^\alpha G_{xy}^{(k)}(z) \geq \varepsilon_k (R - z)^\alpha G_{x'y'}^{(k)}(z).$$

This implies that in case (7.7) all of the exponents α_{xy} are the same. □

7.3. The Perron-Frobenius matrices $M(z) = (\mathbf{P} + \mathbf{K}(z))$. It is not necessarily the case that $R = \rho$: as we have seen in Section 6, it is possible that $R < \rho$, and also that $R = \rho$. Later we will show that $R > \rho$ is impossible. Recall equation (5.6), according to which

$$\mathbf{I} = (\mathbf{I} - z\mathbf{P} - z\mathbf{K}(z))\, \mathbf{G}(z). \tag{7.9}$$

This implies that if z is such that the matrix $(\mathbf{I} - z\mathbf{P} - z\mathbf{K}(z))$ is invertible, and if $\mathbf{K}$ is analytic in a neighborhood of z, then the entries G_{xy} of $\mathbf{G}$ are finite and analytic near z, because (5.7) implies that they may be expressed as convergent Neumann series in the functions K_{xy}. Consequently, if $z = \zeta$ is a singularity of one of the Green's functions $G_{xy}(z)$ then either $z = \zeta$ is a singularity of $\mathbf{H}(z)$ or the matrix $(\mathbf{I} - z\mathbf{P} - z\mathbf{K}(z))$ is *noninvertible*. The following result will be of crucial importance in investigating the invertibility and noninvertibility of this matrix.

LEMMA 7.5. *For $0 < z \leq \rho$ the matrix $M(z) := \mathbf{P} + z\mathbf{K}(z)$ is a Perron-Frobenius matrix.*

PROOF. Recall that a Perron-Frobenius matrix is a square matrix M such that some positive power M^n of M has strictly positive entries. For $z > 0$, the matrices $\mathbf{P}$ and $\mathbf{K}(z)$ have entries

$$\mathbf{P}_{xy} := p(y|x), \tag{7.10}$$

$$K_{xy}(z) := \sum_{a,b} p(ab|x) H_{ab,y}(z), \tag{7.11}$$

where $p(\cdot|\cdot)$ are the transition probabilities in (2.1) and $H_{ab,c}(z)$ is defined by (5.2). The indices x, y are words in $\mathcal{L}$ of length 0 or 1. For any such pair x, y of indices, there is a positive-probability path from x to y, by Assumption 2.2. If there is a positive-probability path of length $n \geq 1$ through words of length ≤ 1, then certainly $\mathbf{P}^n_{xy} > 0$; and since $p(w|w) > 0$ for all elements of $\mathcal{L}$, by Assumption 2.1, it then follows that $\mathbf{P}^{n+k}_{xy} > 0$ for all $k \geq 0$. If there is *not* a positive-probability path from x to y through words of length ≤ 1, then there must be a positive-probability path from x to y that visits a word of length 2. Such a path visits a sequence of words of length ≤ 1, then a word w of length 2, then, after wandering among the words of length ≥ 2, another sequence of words of length 1, etc., and finally arrives at y:

$$\begin{aligned} x \rightarrow x_1 \rightarrow x_2 \rightarrow \cdots x_{k_1} &\rightarrow w_1 \\ \rightarrow x_{k_1+1} \rightarrow x_{k_1+2} \rightarrow \cdots \rightarrow x_{k_2} &\rightarrow w_2 \\ \rightarrow x_{k_2+1} \rightarrow \cdots &\rightarrow \cdots \\ \rightarrow x_{k_r} &\rightarrow w_r \\ &\rightarrow x_m = y \end{aligned}$$

For each pair $w = w_i, v = x_{k_i+1}$ in this sequence, it must be the case that $H_{w,v}(z) > 0$ for $z > 0$, because there is a positive-probability path from w to v through words of length ≥ 2. Moreover, for each pair x_i, x_{i+1} not straddling a w_i, it must be the case that $p(x_{i+1}|x_i) > 0$. Consequently, for all $z > 0$,

$$\begin{aligned} (\mathbf{P} + \mathbf{K}(z))^{r+m}_{xy} > 0 \qquad &\Longrightarrow \\ (\mathbf{P} + \mathbf{K}(z))^{r+m+k}_{xy} > 0 \qquad &\forall\, k \geq 0. \end{aligned}$$

It now follows that, for all sufficiently large $n \geq 1$, the matrix $(\mathbf{P} + \mathbf{K}(z))^n$ has strictly positive entries. ☐

Because $M(z) := (\mathbf{P} + \mathbf{K}(z))$ is a Perron-Frobenius matrix for all $0 < z < \rho$, the conclusions of the Perron-Frobenius theorem (see, for example, [**21**]) apply: In particular, $M(z)$ has a positive eigenvalue $\mu = \mu_z = \mu(z)$ of multiplicity one, and the rest of its spectrum is contained in the open disc $|\zeta| < \mu_z$. Moreover, there are unique right and left eigenvectors $\alpha = \alpha_z = \alpha(z)$ and $\beta = \beta_z = \beta(z)$, with *strictly* positive entries, normalized so that the following system of equations hold:

$$\begin{aligned} M(z)\beta(z) = \mu(z)\beta(z); \qquad &\beta(z)^T\alpha(z) = 1; \\ \alpha(z)^T M(z) = \mu(z)\alpha(z)^T; \qquad &\alpha(z)^T\mathbf{1} \;= 1, \end{aligned} \tag{7.12}$$

where, in the last equation, $\mathbf{1}$ denotes the vector with all entries 1 (thus, the equation asserts that $\alpha(z)$ is a probability vector for each $z \in (0, \rho)$).

LEMMA 7.6. *The functions $\mu(z)$, $\alpha(z)$, and $\beta(z)$ extend to analytic functions in a neighborhood of the line segment $(0, \rho)$, and for $z \in (0, \rho)$, the function $\mu(z)$ is nondecreasing in z. Furthermore, if all of the first-passage generating functions $H_{ab,c}(z)$ are finite at $z = \rho$, then $\mu(z)$, $\alpha(z)$, and $\beta(z)$ are continuous on the* closed *interval $z \in [0, \rho]$.*

PROOF. The entries of $M(z)$, being linear combinations of the first-passage generating functions $H_{ab,c}(z)$, are analytic functions of z for $|z| < \rho$. Moreover, if all of the functions $H_{ab,c}(z)$ are finite at $z = \rho$, then as $z \to \rho-$ the entries of $M(z)$ converge monotonically to those of $M(R)$. Consequently, the analyticity and continuity of $\mu(z)$, $\alpha(z)$, and $\beta(z)$ follow from standard results in regular perturbation theory (see, for example, [**19**], Section XII.1), because the lead eigenvalue $\mu(z)$ has multiplicity one at every $z \in (0, \rho)$. That the eigenvalue $\mu(z)$ is nondecreasing in z for $0 \leq z \leq \rho$ follows because the entries of $M(z)$ are nondecreasing in z. □

LEMMA 7.7. *$z\mu(z) < 1$ for all $z \in [0, R \wedge \rho)$.*

PROOF. Suppose not; then there exists $\zeta \in (0, R \wedge \rho)$ such that $\mu(\zeta) = 1$. By Lemma 7.6, the function $z\mu(z)$ is strictly increasing in z along the line segment $[0, \rho)$, and so $z\mu(z) < 1$ for all $z \in [0, \zeta)$. Because $z\mu(z)$ is the spectral radius of $zM(z)$, it follows that the matrix $(\mathbf{I} - zM(z))$ is invertible for all $z \in (0, \zeta)$. Consequently, the equation (7.9) is solvable for $\mathbf{G}(z)$ along $(0, \zeta)$, via the equation (5.7), and so, by the Law of Permanence of functional equations, the relation (7.9) holds for all z in a neighborhood of the segment $[0, \zeta)$. Taking the inner products with the left eigenvector $\alpha(z)^T$ and the vector $\mathbf{1}$ now gives

$$1 = (1 - z\mu(z))\alpha(z)^T \mathbf{G}(z)\mathbf{1}, \tag{7.13}$$

valid for all $z \in [0, \zeta)$. Now $\zeta < R \wedge \rho$, by hypothesis, and so each of the functions $\mu(z), \alpha(z)$, and $\mathbf{G}(z)$ is continuous at $z = \zeta$. Thus, relation (7.13) persists at $z = \zeta$. But $\zeta\mu(\zeta) = 1$, and so equation (7.13) cannot possibly hold at $z = \zeta$ unless $G(\zeta) = \infty$, which would imply that $\zeta = R$. □

LEMMA 7.8. *For each $r \in (0, \rho)$ and each $\zeta \neq r$ of modulus $|\zeta| = r$, the spectral radius of $M(\zeta)$ is strictly less than $\mu(r)$.*

PROOF. The entries of $M(z) = (\mathbf{P} + \mathbf{K}(z))$ are given by equations (7.10) and (7.11). The nonzero entries of $\mathbf{K}(z)$ are positive linear combinations of the functions $H_{ab,c}(z)$. By Lemma 7.3, these satisfy the inequalities

$$|H_{ab,c}(\zeta)| < H_{ab,c}(r).$$

Consequently, the nonzero entries of $M(\zeta)$ are strictly less, in modulus, than the corresponding entries of $M(r)$. Therefore, the spectral radius of $M(\zeta)$ is strictly less than that of $M(r)$. □

PROPOSITION 7.9. *If $R < \rho$ then $z = R$ is a simple pole of each $G_{xy}(z)$.*

PROOF. Recall that R, being the common radius of convergence of the Green's functions $G_{xy}(z)$, is a singularity of $\mathbf{G}(z)$. Recall also that the functions $H_{ab,c}(z)$ are analytic in the disc $|z| < \rho$, which contains the closed disc $|z| \leq R$. Hence, if $R < \rho$, then, by equation (7.9), the matrix $(\mathbf{I} - M(z))$ must be noninvertible at $z = R$, because if it were invertible then the function $\mathbf{G}(z)$ would extend analytically to a neighborhood of $z = R$, which it does not.

By the preceding lemma, the Perron-Frobenius eigenvalue $\mu(z)$ of $M(z)$ is strictly less than $1/z$ for all $z \in (0, R)$, and, by Lemma 7.6, it is continuous in z for $0 \leq z < \rho$. Since $\mu(z)$ is the spectral radius of $M(z)$, if $R\mu(R) < 1$ then the matrix $(\mathbf{I} - RM(R))$ would be invertible, which it is not. Therefore, if $R < \rho$ then

$$R\mu(R) = 1. \tag{7.14}$$

Now consider the function $F(z) := \alpha(z)^T \mathbf{G}(z)\mathbf{1}$. This is a linear combination of the entries $G_{xy}(z)$ of the Green's matrix $\mathbf{G}(z)$. These entries are multiplied by entries of the left eigenvector $\alpha(z)^T$, which are strictly positive for $0 < z < \rho$ and analytic in z. By equation (5.6), for all $z \in (0, R)$,

$$\begin{aligned} F(z) &= \alpha(z)^T \mathbf{G}(z)\mathbf{1} \\ &= \alpha(z)^T \mathbf{1} + z\alpha(z)^T M(z)\mathbf{G}(z)\mathbf{1} \\ &= 1 + z\mu(z)F(z) \\ &= (1 - z\mu(z))^{-1}, \end{aligned} \tag{7.15}$$

which shows that $F(z)$ has a simple pole at $z = R$, since $z\mu(z) \to 1$ as $z \to R$ and $\mu(z)$ is nondecreasing in z. Since $F(z)$ is a positive linear combination of the entries of the Green's function matrix $\mathbf{G}(z)$, and since all of these entries have the same singularity type at $z = R$ (by Proposition 7.4) it follows that each of these entries has a simple pole at $z = R$. □

COLLARY 7.10. *If $R < \rho$ then for each pair $x, y \in \mathcal{L}$ of words of length* 0 *or* 1*, there exists a constant $C_{xy} > 0$ such that, as $n \to \infty$,*

$$P^x\{X_n = y\} \sim C_{xy} R^{-n}; \tag{7.16}$$

PROOF. By Proposition 7.9, if $R < \rho$ then $G(z)$ has a simple pole at $z = R$. By Lemma 7.8, if $\zeta \neq R$ and $|\zeta| = R$ then the spectral radius of $M(\zeta)$is strictly less than that of $M(R)$, which, by equation (7.14), is $1/R$. Consequently, for each such ζ, the matrix $(\mathbf{I} - \zeta M(\zeta))$ is invertible. Since the entries of $M(z)$ are analytic in $|z| < \rho$, it follows by the implicit function theorem that, for each $\zeta \neq R$ such that $|\zeta| = R$, the Green's function matrix $\mathbf{G}(z)$ is analytic in a neighborhood of $z = \zeta$. Thus, $\mathbf{G}(z)$ has no other singularities in the closed disc $|z| \leq R$. It follows that the coefficients $P^x\{X_n = y\}$ of $G_{xy}(z)$ must obey the asymptotic law (7.16) for some nonzero constant C_{xy}. Since the coefficients are probabilities, the constant must be positive. □

PROPOSITION 7.11. *If $R \not< \rho$, then $R = \rho$. In this case, for each entry $H_{ab,c}$ of* $\mathbf{H}$,

$$H_{ab,c}(\rho) < \infty. \tag{7.17}$$

PROOF. By Lemma 7.7, if $R \geq \rho$ then $z\mu(z) < 1$ for all $z \in [0, \rho)$. It follows that the entries of the matrix $zM(z)$ remain bounded as $z \to \rho-$. Since the entries of $M(z)$ are positive linear combinations of the probabilities $p(y|x)$ and the functions $H_{\beta,c}(z)$ it follows that at least some of the functions $H_{\beta,c}(z)$ remain bounded as $z \to \rho-$. Proposition 7.1 implies that *all* of the functions $H_{\beta,c}(z)$ remain bounded as $z \to \rho-$. □

7.4. The Perron-Frobenius matrix of the system (5.10). Having established the local limit theorem (7.16) in the case $R < \rho$, we now turn our attention to the case

$$R = \rho. \tag{7.18}$$

By Proposition 7.9, if (7.18) holds then all of the entries $H_{ab,c}(z)$ of $\mathbf{H}(z)$ are finite at $z = \rho$. Since all of the generating functions $H_{ab,c}(z)$ occurring as entries of $\mathbf{H}$ have singularities at $z = \rho$, where ρ is the common radius of convergence, the point $z = \rho$ must be a branch point of the algebraic curve (5.10). In particular, by the implicit function theorem, $z = \rho$ must be the smallest positive real number where the *linearization* of the system (5.10) becomes singular. In vector form, the system (5.10) reads $\mathbf{H}(z) = z\mathbf{Q}(\mathbf{H}(z))$; the linearization is obtained by taking the differential of each side:

$$d\mathbf{H} = (dz)\mathbf{Q}(\mathbf{H}) + \left(\frac{\partial \mathbf{Q}}{\partial \mathbf{H}}\right)(d\mathbf{H}) \qquad \Longleftrightarrow$$

$$\left(\mathbf{I} - z\frac{\partial \mathbf{Q}}{\partial \mathbf{H}}\right)(d\mathbf{H}) = (dz)\mathbf{Q}(\mathbf{H}) \tag{7.19}$$

The analysis of the linear system (7.19) is similar to the analysis of the linear system (7.9) in paragraph 7.3 above. Once again, a crucial role is played by a one-parameter family of matrices which are Perron-Frobenius matrices along the line segment $(0, \rho]$. These are defined as follows:

$$\mathbf{L}_z = \mathbf{L}(z) = \left(\frac{\partial \mathbf{Q}}{\partial \mathbf{H}}\right)_{\mathbf{H}=\mathbf{H}(z)} \tag{7.20}$$

LEMMA 7.12. *For each $z \in (0, \rho]$, the matrix $\mathbf{L}_z$ is a Perron-Frobenius matrix.*

PROOF. Since the components of $\mathbf{Q}(\mathbf{H})$ are polynomials in the variables $H_{ab,c}$ with positive coefficients, so are the entries of the matrix $\mathbf{L}_z = \partial\mathbf{Q}/\partial\mathbf{H}$, as are the entries of any positive power $\mathbf{L}_z^m$ of $\mathbf{L}_z$. Since those generating functions $H_{ab,c}(z)$ that occur as entries of the vector $\mathbf{H}(z)$ are strictly positive for $0 < z \le \rho$, it follows that for each $z \in (0, \rho]$ and $m \ge 1$, the entries of $\mathbf{L}_z^m$ are nonnegative. It remains to be shown that, for sufficiently large m, these entries are nonzero.

Let $H = H_{ab,c}$ and $H_* = H_{a'b',c'}$ be entries of $\mathbf{H}$, and, for each $m \ge 1$, let $Q_H^{\circ m}(\mathbf{H})$ be the $H-$component of the vector polynomial $\mathbf{Q}^{\circ m}(\mathbf{H})$. We will show that the $(H, H_*)-$entry of the matrix $\mathbf{L}_z^m$ is nonzero if and only if the variable H_* occurs as a factor in a term of $Q_H^{\circ m}(\mathbf{H})$.The proof is by induction on m. When $m = 1$, the statement is certainly true, because $\mathbf{L}_z = \partial\mathbf{Q}/\partial\mathbf{H}$. Assume, then that the claim is true for all positive integers $\le m$. Consider the $(H, H_*)-$entry of the matrix $\mathbf{L}_z^{m+1}$; this will be nonzero if and only if, for some entry H_{**} of $\mathbf{H}$, the $(H, H_{**})-$entry of $\mathbf{L}_z^m$ and the $(H_{**}, H_*)-$entry of $\mathbf{L}_z$ are both nonzero. By the induction hypothesis, this occurs if and only if H_{**} occurs as a factor in a term of $Q_H^{\circ m}(\mathbf{H})$ and H_* occurs as a factor in a term of $Q_H(\mathbf{H})$. But this occurs if and only if H_* occurs as a factor in a term of $Q_H^{\circ(m+1)}(\mathbf{H})$, because the vector polynomial $\mathbf{Q}^{\circ(m+1)}(\mathbf{H})$ is obtained by substituting $\mathbf{Q}(\mathbf{H})$ for $\mathbf{H}$ in $\mathbf{Q}^{\circ m}(\mathbf{H})$.

Now recall Lemma 7.2, according to which there exists $m \ge 1$ such that, in each component of $\mathbf{Q}^{\circ m}$, *every* entry H of $\mathbf{H}$ occurs as a factor in a term with positive coefficient. In view of the result of the previous paragraph, it follows that, for any such m, every entry of $\mathbf{L}_z^m$ is nonzero. □

Remark. The argument used in the second paragraph of the proof may be modified to prove a somewhat stronger statement. If $m \geq 1$ is so large that every component of $\mathbf{Q}^{\circ m}(\mathbf{H})$ contains every $H = H_{ab,c}$ as a factor in a nonzero term, then, by resubstitution, it follows that every component of $\mathbf{Q}^{\circ 2m}(\mathbf{H})$ contains the *square* of every $H = H_{ab,c}$ as a factor in a nonzero term. Consequently, every entry of the matrix $\mathbf{L}_z^{2m}$ contains every $H = H_{ab,c}$ as a factor in a nonzero term. □

Because $\mathbf{L}_z$ is a Perron-Frobenius matrix when $0 < z \leq \rho$, it is governed by the conclusions of the Perron-Frobenius theorem. Thus, it has a positive eigenvalue $\lambda = \lambda_z = \lambda(z)$ of multiplicity one, and the rest of its spectrum is contained in the open disc $|\zeta| < \lambda_z$. Moreover, there are unique right and left eigenvectors $\mathbf{w} = \mathbf{w}_z = \mathbf{w}(z)$ and $\mathbf{v} = \mathbf{v}_z = \mathbf{v}(z)$, with *strictly* positive entries, satisfying

$$
\begin{aligned}
\text{(7.21)} \qquad \mathbf{L}(z)\mathbf{v}(z) &= \lambda(z)\mathbf{v}(z); \qquad & \mathbf{v}(z)^T\mathbf{w}(z) &= 1; \\
\mathbf{w}(z)^T\mathbf{L}(z) &= \lambda(z)\mathbf{w}(z)^T; \qquad & \mathbf{w}(z)^T\mathbf{1} &= 1,
\end{aligned}
$$

where, in the last equation, $\mathbf{1}$ denotes the vector with all entries 1. Since the matrix $\mathbf{L}(z)$ depends analytically on z for $|z| < \rho$, and since the multiplicity of the eigenvalue $\lambda(z)$ is 1 at every $z \in (0, \rho]$, the functions $\lambda(z), \mathbf{w}(z)$, and $\mathbf{v}(z)$ are all analytic functions of z for z in some open neighborhood of the line segment $(0, \rho]$ (see [**19**], Section XII.1). Since the entries of the matrices $\mathbf{L}_z$ are nondecreasing in z, for $0 \leq z \leq \rho$, the Perron-Frobenius eigenvalue $\lambda(z)$ is also nondecreasing in z.

LEMMA 7.13. $\lambda(\rho) = 1/\rho$.

PROOF. Suppose that $\rho\lambda(\rho) < 1$. Since the spectrum of $\rho\mathbf{L}_\rho$ lies entirely in the disc $|\zeta| \leq \rho\lambda(\rho)$, the matrix $I - \rho\mathbf{L}_\rho$ would be invertible. Consequently, the linearization (7.19) of the algebraic system (5.10)) at $z = \rho$ would be solvable for $d\mathbf{H}$:

$$
d\mathbf{H} = dz\,(\mathbf{I} - z\mathbf{L}_z)^{-1}\, z\mathbf{Q}(\mathbf{H})).
$$

But this would imply, by the implicit function theorem, that $\mathbf{H}(z)$ admits an analytic continuation to a neighborhood of $z = \rho$, which is impossible, since $z = \rho$ is a singularity of $\mathbf{H}(z)$.

Now suppose that $\rho\lambda(\rho) > 1$. Since $z \mapsto z\lambda(z)$ is analytic, and therefore continuous, along the line segment $(0, \rho]$ and takes the value 0 at $z = 0$, it would follow, by the intermediate value theorem, that $r\lambda(r) = 1$ for some $r \in (0, \rho]$. But this would imply that the matrix $(I - r\mathbf{L}_r)$ is singular, and that the system (7.19) cannot be solved for $d\mathbf{H}$ in terms of dz at $z = r$. This, in turn, would imply that $z = r$ is a singularity of the function $\mathbf{H}(z)$, a contradiction, since $z = \rho$ is the smallest positive singularity. □

COLLARY 7.14. *If the RWRL X_n is transient, then $\rho > 1$.*

PROOF. If the RWRL is transient, then at least one of the first-passage generating functions $H_{ab,c}(z)$ must satisfy

$$
H_{ab,c}(1) < 1. \tag{7.22}
$$

□

LEMMA 7.15. *For each $z \neq \rho$ in the disc $|z| \leq \rho$, the spectral radius of the matrix L_z is strictly less than $1/\rho$.*

PROOF. It suffices to show that, for some integer $m \geq 1$, the spectral radius of $\mathbf{L}_z^m$ is less than $1/\rho^m$. Recall that the entries of the matrix $\mathbf{L}_z^m$ are positive linear combinations of products of the variables z and $H_{ab,c}$. Moreover, if m is sufficiently large, then each entry contains every $H_{ab,c}$ as a factor in some term with positive coefficient (see the remark following the proof of Lemma 7.12). Consequently, by Lemma 7.3, if $z \neq \rho$ but $|z| = \rho$ then all of the entries of $\mathbf{L}_z^m$ are strictly less in modulus than the corresponding entries of $\mathbf{L}_\rho^m$. It now follows by a routine argument that the spectral radius of $\mathbf{L}_z^m$ is strictly less than that of $\mathbf{L}_\rho^m$, which, by Lemma 7.13, is $1/\rho^m$. □

LEMMA 7.16. *None of the functions $H(z)$ occurring as entries of* $\mathbf{H}$ *have singularities on the circle $|z| = \rho$ other than those at $z = \rho$.*

PROOF. By Lemma 7.15, if $z \neq \rho$ but $|z| = \rho$ then the spectral radius of $z\mathbf{L}_z$ is strictly less than 1, and so the matrix $\mathbf{I} - z\mathbf{L}_z$ is invertible. Thus, the linearization (7.19) of the system (5.10) is solvable for $d\mathbf{H}$ in terms of dz at z. By the implicit function theorem, $\mathbf{H}$ admits an analytic continuation to a neighborhood of z, and so z is a regular point of $\mathbf{H}$. □

It remains to determine the nature of the singularity at $z = \rho$. By Corollary 5.2, each of the generating functions $H_{ab,c}(z)$ is algebraic; by Proposition 7.1, that all of the *nondegenerate* $H_{ab,c}(z)$ have the same singularity type at $z = \rho$; and by Proposition 7.11, if $\rho = R$ then all of the functions $H_{ab,c}(z)$ are finite at $z = \rho$. Therefore, there exist constants $\alpha > 0$ and $C_H > 0$ such that for each nondegenerate $H = H_{ab,c}$,

$$H(\rho) - H(z) \sim C_H(\rho - z)^\alpha. \tag{7.23}$$

PROPOSITION 7.17. $\alpha = 1/2$.

PROOF. Consider once again the system $\mathbf{H}(z) = z\mathbf{Q}(\mathbf{H}(z))$ of quadratic equations (5.9) for the vector $\mathbf{H}(z)$ of nondegenerate first-passage generating functions $H_{ab,c}(z)$. These equations are valid for all z in the disc $|z| < \rho$, and, by continuity, for all z on the circle of convergence $|z| = \rho$. Taking differences, we have

$$\begin{aligned} \mathbf{H}(\rho) - \mathbf{H}(z) &= \rho\mathbf{Q}(\mathbf{H}(\rho)) - z\mathbf{Q}(\mathbf{H}(z)) \\ &= \rho(\mathbf{Q}(\mathbf{H}(\rho)) - \mathbf{Q}(\mathbf{H}(z))) + (\rho - z)\mathbf{Q}(\mathbf{H}(\rho)) \\ &\quad - (\rho - z)(\mathbf{Q}(\mathbf{H}(\rho)) - \mathbf{Q}(\mathbf{H}(z))). \end{aligned} \tag{7.24}$$

Writing $\Delta\mathbf{H} := \mathbf{H}(\rho) - \mathbf{H}(z)$ and $\Delta z := \rho - z$ and collecting terms, we obtain

$$\begin{aligned} (\mathbf{I} - \rho\mathbf{L}_\rho)\,\Delta\mathbf{H} = (\Delta z)&\mathbf{Q}(\mathbf{H}(\rho)) \\ &- \rho(\text{quadratic terms in } \Delta\mathbf{H}) \\ &- (\Delta z)(\text{linear and quadratic terms in } \Delta\mathbf{H}). \end{aligned} \tag{7.25}$$

Since the quadratic polynomials $\mathbf{Q}(\mathbf{H})$ have nonnegative coefficients, so do the quadratic terms on the right side of (7.25), and at least some of these coefficients are positive. Now take the inner product with the left eigenvector $\mathbf{v}_\rho$ of the Perron-Frobenius matrix $\mathbf{L}_\rho$: since $\rho\lambda(\rho) = 1$, by Lemma 7.13, this gives

$$\begin{aligned} (\Delta z)(\mathbf{v}_\rho^T\mathbf{Q}(\mathbf{H}(\rho))) = &\rho(\text{quadratic form in } \Delta\mathbf{H}) \\ &+ (\Delta z)(\text{linear and quadratic terms in } \Delta\mathbf{H}). \end{aligned} \tag{7.26}$$

Since the eigenvector $\mathbf{v}_R$ has strictly positive entries, the coefficient $\mathbf{v}_\rho^T \mathbf{Q}(\mathbf{H}(\rho))$ of Δz on the left side is positive. Moreover, the coefficients in the quadratic form are nonnegative, and at least some of them are positive; hence, the quadratic form is nonnegative definite and nondegenerate. This implies that the exponent in (7.23) must be $\alpha = 1/2$. □

COLLARY 7.18. *Assume that $R = \rho$. Then there exists $C_\mu > 0$ such that as $z \to R-$,*

$$\mu(R) - \mu(z) \sim C_\mu (R - z)^{1/2}. \tag{7.27}$$

PROOF. When $R = \rho$, the first-passage generating functions $H_{ab,c}(z)$ are finite at $z = R$, by Proposition 7.11. Consequently, the entries of $M(z)$ are finite at $z = R$, because they are positive linear combinations of the functions $H_{ab,c}(z)$. Since the matrix $M(R)$ is a Perron-Frobenius matrix, its leading eigenvalue has multiplicity one, and so, by regular perturbation theory [**19**], the leading eigenvalue is an analytic function of the matrix entries in a neighborhood of $M(R)$. Thus, in particular, the function $\mu(z)$ is given by a convergent power series in the entries of $M(R) - M(z)$, and the linear terms of this power series are not all zero (because locally, near $M(R)$, the spectral radius must increase linearly with any entry). The relation (7.27) therefore follows from Proposition 7.17. □

7.5. Behavior of the Green's functions at $R = \rho$. When $R = \rho$, all of the first-passage generating functions $H_{ab,c}(z)$ are finite at $z = R$, by Proposition 7.11, have Puiseux series of the form (7.23) at R, and have no other singularities on the circle $|z| = R$, by Lemma 7.16. By Lemma 7.7, for every $z \in (0, R)$ the spectral radius $z\mu(z)$ of $zM(z) := z(\mathbf{P} + \mathbf{K}(z))$ is strictly less than one, and so by the continuity of $\mu(z)$, either

$$R\mu(R) < 1 \qquad \text{or} \tag{7.28}$$

$$R\mu(R) = 1 \tag{7.29}$$

The special cases considered in Section 6 show that both possibilities may occur.

PROPOSITION 7.19. *For each entry $G_{xy}(z)$ of the Green's function matrix $\mathbf{G}(z)$, there exists a constant $C_{xy} > 0$ such that, as $z \to R-$,*

$$G_{xy}(z) \sim C_{xy}(R - z)^{-1/2} \qquad \textit{if } R\mu(R) = 1; \textit{ and} \tag{7.30}$$

$$G_{xy}(R) - G_{xy}(z) \sim C_{xy}(R - z)^{1/2} \qquad \textit{if } R\mu(R) < 1. \tag{7.31}$$

PROOF. By Corollary 5.2, all of the Green's functions $G_{xy}(z)$ are algebraic; and by Proposition 7.4, all have the same singularity type at their common radius R of convergence. Consequently, all have convergent Puiseux expansions at $z = R$. It must be shown that the leading terms of these expansions are as implied by (7.30)-(7.31).

Since all of the leading terms have the same exponent, by Proposition 7.4, it will suffice to consider a convenient positive linear combination of the entries of $\mathbf{G}(z)$. Thus, consider the function $F(z) = \alpha(z)^T \mathbf{G}(z)\mathbf{1}$, where $\alpha(z)^T$ is the left Perron-Frobenius eigenvector of $M(z)$ and $\mathbf{1}$ is the vector of all 1s. By equation (7.15),

$$F(z) = (1 - z\mu(z))^{-1}.$$

Hence, by Corollary 7.18,

$$F(z) \sim C_F(R-z)^{-1/2} \qquad \text{if } R\mu(R)=1; \text{ and}$$
$$F(R)-F(z) \sim C_F(R-z)^{1/2} \qquad \text{if } R\mu(R)<1.$$

Since $\lim_{z\to R-}\alpha(z)=\alpha(R)$, it follows that

$$\alpha(R)^T\mathbf{G}(z)\mathbf{1} \sim C_F(R-z)^{-1/2} \qquad \text{if } R\mu(R)=1; \text{ and}$$
$$\alpha(R)^T\mathbf{G}(R)\mathbf{1}-\alpha(R)^T\mathbf{G}(z)\mathbf{1} \sim C_F(R-z)^{1/2} \qquad \text{if } R\mu(R)<1.$$

The relations (7.31)-(7.30) clearly follow. □

COLLARY 7.20. *For any two words x, y of lengths $|x|, |y| \leq 1$, there exists a constant $C_{xy}>0$ such that, as $n\to\infty$,*

$$P^x\{X_n=y\} \sim C_{xy}R^{-n} \qquad \textit{if } R<\rho; \tag{7.32}$$
$$P^x\{X_n=y\} \sim C_{xy}R^{-n}n^{-1/2} \qquad \textit{if } R=\rho \textit{ and } R\mu(R)=1; \textit{ and} \tag{7.33}$$
$$P^x\{X_n=y\} \sim C_{xy}R^{-n}n^{-3/2} \qquad \textit{if } R=\rho \textit{ and } R\mu(R)<1. \tag{7.34}$$

PROOF. The first statement has been proved in Corollary 7.10. In view of Corollary 5.2 and Proposition 7.19, the latter two results are immediate consequences of Corollary 4.1. □

8. Transience and the spectral radius

By Corollary 7.20, the only possible limiting behaviors of the transition probabilities are (7.32), (7.33), and (7.34). Thus, to complete the proofs of Theorems 2.5-2.6, we must show that $R=1$ if and only if the RWRL is recurrent, because

(a) if $R=1$, then (7.32) implies that the transition probabilities converge to nonzero limits, and so the RWRL is *positive* recurrent;
(b) if $R=1$, then (7.33) implies that the transition probabilities converge to zero, and so the RWRL is *null* recurrent;
(c) if $R=1$, then (7.34) implies that the transition probabilities are summable in n, and so the RWRL is *transient*, by a standard test for recurrence; and
(d) if the RWRL is recurrent, then the Green's functions are all infinite at $z=1$, by the same test for recurrence, and so $R=1$.

PROPOSITION 8.1. *If the RWRL is transient, then $R>1$.*

It is possible to prove this in a purely "analytical" fashion, using Lemma 7.13 to show that $\rho=1$ is impossible unless all the nondegenerate first-passage generating functions $H_{ab,c}(z)$ take the value 1 at $z=1$, which implies that the RWRL is recurrent. However, it is simpler to use a probabilistic argument based on a standard large-deviations estimate, which we record in the following lemma.

LEMMA 8.2. *Let $\xi_1, \xi_2, \dots$ be a sequence of Bernoulli random variables adapted to a filtration $\{\mathcal{F}_n\}_{n\geq 0}$. Assume that there exists $p>0$ such that, for every $n\geq 1$,*

$$P(\xi_{n+1}=1 \,\|\, \mathcal{F}_n) \geq p. \tag{8.1}$$

Then for every $\alpha < p$ *there exist* $\beta < 1$ *and* $C < \infty$ *such that* $\forall m \geq 1$,

$$P\{\sum_{j=1}^{n} \xi_j \leq \alpha n \text{ for some } n \geq m\} \leq C\beta^m. \tag{8.2}$$

PROOF. See [**8**]. □

Proof of Proposition 8.1. If the RWRL X_n is transient, then for every $k \geq 1$, with probability one, the first k letters of X_n eventually stabilize. Consequently, there exists a word $w \in \mathcal{L}$ such that

$$P^w\{X_n \text{ is prefixed by } w \; \forall n \geq 0\} := p > 0. \tag{8.3}$$

Define $\mathcal{L}(w)$ to be the set of all words $v \in \mathcal{L}$ that are *suffixed* by w, i.e., $\mathcal{L}(w)$ is the set of all words in $\mathcal{L}$ of the form uw. By Assumption 2.4, for any word $w' \in \mathcal{L}$, there exists a positive-probability path γ from w' to a word $v \in \mathcal{L}(w)$ through words of length $\geq |w'|$. Because only the last two letters in any transition influence the transition probabilities, the probability of such a path γ depends on w' only through its last two letters. Since there are only finitely many two-letter words in $\mathcal{L}$, it follows that, for some $\alpha > 0$ and some positive integer N,

$$P^v\{X_n \in \mathcal{L}(w) \text{ for some } n \leq N\} \geq \alpha \; \forall v \in \mathcal{L}. \tag{8.4}$$

Consider the event $\{X_n = \emptyset\}$. In order that this event occur, it must be the case that, for every visit to the set $\mathcal{L}(w)$ prior to time n, the last letter of the suffix w is erased, or one of the last two letters modified, before time n. By (8.4) and Lemma 8.2, the probability that there are fewer than $\alpha n/2N$ visits to $\mathcal{L}(w)$ by time n decays exponentially in n. By (8.3) and Lemma 8.2, the probability that one of the last two letter of the suffix w is modified $\alpha n/2N$ times before time n decays exponentially in n. Therefore, $P^\emptyset\{X_n = \emptyset\}$ decays exponentially in n, equivalently, $R > 1$.

Concluding Remark. While this paper was being reviewed, the author received a preprint of a paper by T. NAGNIBEDA and W. WOESS in which results similar to Theorems 1 and 2 are proved. The results of Nagnibeda and Woess are proved only for random walks on trees, but by a similar method.

References

[1] K. Aomoto (1984) Spectral theory on a free group and algebraic curves. *J. Fac. Sci. Univ. Tokyo* IA **31**, 297–317.

[2] E. Bender (1974) Asymptotic methods in enumeration. *SIAM Review*, **16**, 485–513.

[3] P. Bougerol (1979) Théorème central limite local sur certains groupes de Lie. *Ann. Sci. cole Norm. Sup.* (4) **14** 403–432.

[4] P. Bougerol & T. Jeulin (1999) Brownian bridge on hyperbolic spaces and on homogeneous trees. *Probability Theory & Related Fields* **115** 95–120.

[5] T. H. Cormen, C. E. Leiserson, & R. L. Rivest (1990) *Algorithms.* MIT Press, Cambridge MA

[6] Cox, Little, & O'Shea (1992) *Ideals, Varieties, and Algorithms.* Springer, NY

[7] W. Feller (196) *An Introduction to Probability Theory and its Applications.* Wiley, NY

[8] D. Freedman (1975) On tail probabilities for martingales. *Annals of Probability* **3** 100–118.

[9] A. Gairat, V. Malyshev, M. Menshikov, & K. Pelikh (1995) Classification of Markov chains that describe the evolution of random strings. (Russian) *Uspekhi Mat. Nauk* **50** (302)5–24; translation in *Russian Math. Surveys* **50** (1995) 237–255.

[10] P. Gerl & W. Woess (1986) Local limits and harmonic functions for nonisotropic random walks on free groups. *Probability Theory & Reated Fields*, **71**, 341–355.
[11] E. Hille (1959) *Analytic Functions*, volumes 1–2. Ginn, Boston.
[12] J. Hopcraft & J. Ullman (1979) *Introduction to Automata Theory, Languages, and Computation.* Addison-Wesley, Reading MA
[13] K. Kendig (1977) *Elementary Algebraic Geometry.* Springer-Verlag, NY
[14] S. Lalley (1993) Finite range random walk on free groups and homogeneous trees. *Annals of Probability* **21** 2087–2130.
[15] S. Lalley (1995) Return probabilities for random walk on a half-line. *J. Theoretical Probability* **8** 571–599.
[16] J. Lehner (1964) *Discontinuous Groups and Automorphic Forms.* American Math. Soc., Providence RI
[17] E. Le Page & M. Peigné (1999) Local limit theorems on some non-unimodular groups. *Rev. Mat. Iberoamericana* **15** 117–141.
[18] V. Malyshev (1994) Stabilization laws in the evolution of a random string. (Russian) *Problemy Peredachi Informatsii* **30** 79–95 translation in *Problems Inform. Transmission* **30** (1995) 260–274.
[19] M. Reed & B. Simon (1978) *Methods of Mathematical Physics IV: Analysis of Operators.* Academic Press, NY
[20] S. Sawyer (1978) Isotropic random walks on a tree. *Z. Warscheinlichkeitstheorie.*, **42**, 279–292.
[21] E. Seneta (1981) *Nonnegative Matrices and Markov Chains*, 2nd ed., Springer, NY.
[22] W. Woess (1982) A local limit theorem for random walks on certain discrete groups. *Probability measures on groups (Oberwolfach, 1981)*, pp. 467–477, Lecture Notes in Math. **928**, Springer, Berlin-New York.
[23] W. Woess (1987) Context-free languages and random walks on groups. *Discrete Math.* **67** 81–87.

University of Chicago, Department of Statistics, USA
E-mail address: lalley@galton.uchicago.edu

Contemporary Mathematics
Volume **287**, 2001

The Normal Quasi-Wishart Distribution

Gérard Letac and Hélène Massam

Abstract. We present in this paper a proof of a result originally given by Geiger and Heckerman [**GH98**]. This result is a characterization of the distributions of (Z_1, Z_2, W), where Z_1, Z_2 are real random variables and $W = \begin{bmatrix} W_1 & W_{12} \\ W_{12} & W_2 \end{bmatrix}$ is a positive definite 2×2 random matrix, subject to the independence conditions

$$(Z_1, W_1 - \frac{W_{12}^2}{W_2}) \text{ is independent of } (Z_2 + \frac{W_{12}}{W_2} Z_1, W_{12}, W_2)$$

and

$$(Z_2, W_2 - \frac{W_{12}^2}{W_1}) \text{ is independent of } (Z_1 + \frac{W_{12}}{W_1} Z_2, W_{12}, W_1).$$

The proof given here is without smoothness hypothesis and relies on a characterization of the generalized inverse Gaussian distribution.

1. Introduction

The origin of the problem is in Bayesian theory. Let us recall some elementary facts on the two-dimensional Gaussian distribution. The mean and variance of the normal distribution will be denoted by capital letters since, in the following sections, they will also be random variables. Let (X_1, X_2) follow a normal $N(Z, W^{-1})$ distribution with mean $Z = (Z_1, Z_2)$ and precision matrix

$$W = \begin{bmatrix} W_1 & W_{12} \\ W_{12} & W_2 \end{bmatrix}.$$

Using its Cholesky decomposition, we can write W as

$$W = \begin{bmatrix} 1 & W_{12}W_2^{-1} \\ 0 & 1 \end{bmatrix} \begin{bmatrix} W_{1\bullet 2} & 0 \\ 0 & W_2 \end{bmatrix} \begin{bmatrix} 1 & 0 \\ W_2^{-1}W_{21} & 1 \end{bmatrix},$$

1991 *Mathematics Subject Classification.* Primary 62H05; Secondary 62C10.
Key words and phrases. Generalized inverse Gaussian, global independence.

where $W_{1\bullet 2} = W_1 - \frac{W_{12}^2}{W_2}$. Then, the density $f_{X_1,X_2}(x_1,x_2)$ of (X_1,X_2) can be written

$$(1.1)\qquad \begin{aligned} f_{X_1,X_2}(x_1,x_2) &= \frac{1}{2\pi}\sqrt{W_1W_2 - W_{12}^2}\exp -\frac{1}{2}(x_1-Z_1)^2W_{1\bullet 2} \\ &\quad \exp -\frac{1}{2}(x_2 - Z_2 + W_{12}W_2^{-1}(x_1-Z_1))^2W_2) \\ &= f_{X_1}(x_1)f_{X_2|X_1}(x_2). \end{aligned}$$

The density of X_1 is Gaussian with mean Z_1 and variance $W_{1\bullet 2}^{-1}$, whereas the density of X_2 given $X_1 = x_1$, is Gaussian with variance W_2^{-1} and mean

$$Z_2 + \frac{W_{12}}{W_2}(Z_1 - x_1).$$

When working in a Bayesian context, is is desirable to have *global independence of the parameters*, as defined by Dawid and Lauritzen [**DL93**]. This means that the prior distributions on (Z,W) should be such that the parameters of the distribution of X_1 should be independent of the parameters of the distribution of X_2 given $X_1 = x_1$, *for all* values x_1 of X_1. The same conditions should be true also if we exchange the roles of X_1 and X_2. From (1.2), it is clear that these independence conditions require that

$$(1.2)\qquad (Z_1, W_{1\bullet 2}) \perp\!\!\!\perp (Z_2 + W_{12}W_2^{-1}Z_1, W_{12}, W_2),$$

$$(1.3)\qquad (Z_2, W_{2\bullet 1}) \perp\!\!\!\perp (Z_1 + W_{21}W_1^{-1}Z_2, W_{12}, W_1).$$

In the special case where it is known that $Z_1 = Z_2 = 0$, these conditions become

$$(1.4)\qquad W_{1\bullet 2} \perp\!\!\!\perp (W_{12}, W_2)$$

$$(1.5)\qquad W_{2\bullet 1} \perp\!\!\!\perp (W_{21}, W_1).$$

It is well known that the Wishart distribution satisfies these conditions. We will see in Section 2 that a more general class of distributions, called quasi-Wishart, will also satisfy (1.4) and (1.5). In the general case, we will see, also in Section 2, that the class of distributions of (Z,W), called normal quasi-Wishart distributions, satisfies (1.2) and (1.3).

The aim of the present paper is to prove that, conversely, if W or (Z,W) fulfill the stated independence conditions, then W is quasi-Wishart and (Z,W) is normal quasi-Wishart.

A characterization of the normal quasi-Wishart distribution was first given by Geiger and Heckerman [**GH98**]. Their proof uses functional equations. In this paper, we give a probabilistic proof of their characterization, without any hypotheses of absolute continuity. Our proof is probably not shorter, but its ingredients are surprising: a remark due to Matsumoto and Yor [**MY01**] from the Brownian motion context, its converse proved by Letac and Wesołowski [**LW00**] and finally a principle of separation of random variables which could be useful in other applications.

In a second paper, Geiger and Heckerman [**GH99**] have extended their bivariate result to Gaussian distributions in $\mathbb{R}^n$ (the proof uses the case $n = 2$ as a crucial step). Essentially, starting from a Gaussian random variable X in $\mathbb{R}^n$, Geiger and Heckerman assume the same constraints for X_1 and X_2, a 2-partition of the original vector X. Our probabilistic proof of Geiger and Heckerman's result is for $n = 2$.

The general case would require an extension of the result by Matsumoto Yor to symmetric matrices without additional smoothness assumptions on the densities.

2. The quasi-Wishart and normal quasi-Wishart distributions

Denote by $\mathcal{P}_n$ the set of symmetric positive definite $n \times n$ real matrices. Let $p > \frac{n-1}{2}$ and σ be in $\mathcal{P}_n$. Recall that the random variable W taking its values in $\mathcal{P}_n$ has a *Wishart* distribution $\gamma_{p,\sigma}$ if its density with respect the Lebesgue measure on the space of symmetric positive definite matrices is

$$C_{p,n}\frac{(\det w)^{p-\frac{n+1}{2}}}{(\det \sigma)^p}e^{-\text{ trace }(\sigma^{-1}w)}\mathbf{1}_{\mathcal{P}_n}(w),$$

where $C_{p,n}$ is the normalizing constant.

DEFINITION 2.1. We say that the random variable (Z, W) taking its values in $\mathbb{R}^n \times \mathcal{P}_n$ has a *normal Wishart* distribution with parameters (p, σ, m, a) if

(1) There exist $p > \frac{n-1}{2}$ and σ in $\mathcal{P}_n$ such that W has the Wishart distribution $\gamma_{p,\sigma}$.
(2) There exist $m \in \mathbb{R}^n$ and $a > 0$ such that the distribution of Z *conditioned* by W is Gaussian with mean m and covariance aW^{-1}.

Let us emphasize the fact that m and a do not depend on W.

The case n=2. If W has density $\gamma_{p,\sigma}$ with $p > 1/2$, the conditional density of (W_1, W_2) knowing $W_{12} = w_{12}$ is

$$K_{w_{12}}(dw_1, dw_2) = C(p, b, w_{12})(w_1 w_2 - w_{12}^2)^{p-\frac{3}{2}}e^{-(b_1 w_1 + b_2 w_2)}dw_1 dw_2, \tag{2.1}$$

where $C(p, b, w_{12})$ is a normalizing constant and $b = (b_1, b_2)$ is the diagonal of the matrix σ^{-1}. The density is concentrated on the set

$$H(w_{12}) = \{(w_1, w_2); w_1 > 0,\ w_1 w_2 - w_{12}^2 > 0\},$$

which is the interior of a hyperbola. Here, it is sufficient to know that the density of W_{12} is absolutely continuous.

DEFINITION 2.2. We say that the probability law of a random variable W in $\mathcal{P}_2$ is *quasi-Wishart* if such law assigns probability zero to the event $W_{12} = 0$ and there exist $p > \frac{1}{2}$ and σ in $\mathcal{P}_2$ such that the *conditional* density of (W_1, W_2) knowing W_{12} is $K_{w_{12}}$, as defined by (2.1).

For the quasi-Wishart distribution, the law of W_{12} on $\mathbb{R} \setminus \{0\}$ is entirely arbitrary . The parameters of the quasi-Wishart distribution are (p, b_1, b_2, w_{12}).

DEFINITION 2.3. We say that the random variable (Z, W) taking its values in $\mathbb{R}^2 \times \mathcal{P}_2$ has a *normal quasi-Wishart* density if

(1) The law of W follows is quasi-Wishart.
(2) There exist $m \in \mathbb{R}^2$ and $a > 0$ such that the law of Z *conditional* on W is Gaussian with mean m and covariance aW^{-1}.

Here again, we insist on the fact that m and a do not depend on W.

PROPOSITION 2.1. *Let W be a random variable taking its values in $\mathcal{P}_2$ and following the quasi-Wishart distribution. Then $W_{1\bullet 2} = W_1 - \frac{W_{12}^2}{W_2}$ is independent of the pair of real random variables (W_{12}, W_2). Let (Z, W) be a random variable* *taking its values in* $\mathbb{R}^2 \times \mathcal{P}_2$ *and following a normal quasi-Wishart distribution. Then the pair* $(Z_1, W_{1\bullet 2})$ *is independent of the triple* $(Z_2+\frac{W_{12}}{W_2}Z_1, W_{12}, W_2)$. *Similar statements hold if we exchange the roles of* 1 *and* 2.

PROOF. For the first part, it suffices to show the results under conditioning by $W_{12} = w_{12}$. The joint density of $S = W_1 - \frac{w_{12}^2}{W_2}$ and W_2, obtained from (2.1), is

$$Cs^{p-3/2}e^{-b_2 s}w_2^{p-3/2}e^{-b_2 w_2 - b_1\frac{w_{12}^2}{w_2}}.$$

Clearly this is a product of two functions depending on s and (w_{12}, w_1) respectively. We proceed in a similar way for the second part. Conditioning on W, the covariance matrix of Z is

$$\frac{1}{a(W_1W_2 - W_{12}^2)}\begin{bmatrix} W_2 & -W_{12} \\ -W_{12} & W_1 \end{bmatrix}$$

the conditional covariance of Z_1 and $Z_2 + \frac{W_{12}}{W_2}Z_1$ is $\operatorname{cov}(Z_1, Z_2) + \frac{W_{12}}{W_2}\operatorname{var}(Z_1) = 0$. The result follows from this and the first part. □

3. The GIG distribution and the Matsumoto-Yor observation

The aim of the present paper is to prove the converse of Proposition 2.1. The following definitions and observations, although apparently unrelated, will be needed in the sequence. Let us recall the definitions of the generalized inverse Gaussian distribution and the Gamma distribution on the positive line.

DEFINITION 3.1. If q, a, and b, are real numbers, with $a > 0$ and $b > 0$, the $GIG(q, a, b)$ density is given by

$$GIG(q,a,b)(dx) = C(q,a,b)x^{q-1}e^{-ax-\frac{b}{x}}\mathbf{1}_{(0,\infty)}(x)dx.$$

If $q > 0$ and $a > 0$ the $\Gamma(q, a)$ density is given by

$$\Gamma(q,a)(dy) = \frac{1}{\Gamma(q)}a^q y^{q-1}e^{-ay}\mathbf{1}_{(0,\infty)}(y)dy.$$

The following theorem will be crucial to our argument.

THEOREM 3.1. *Let* X *and* Y *be independent, non Dirac and positive random variables. Denote*

$$U = \frac{1}{X+Y}, \qquad V = \frac{1}{X} - \frac{1}{X+Y}. \tag{3.1}$$

Then U *and* V *are independent if and only if there exist three positive numbers* q, a, *and* b *such that the densities of* X *and* Y *are* $GIG(-q, a, b)$ *and* $\Gamma(q, a)$ *respectively. Under these circumstances, the densities of* U *and* V *are* $GIG(-q, b, a)$ *and* $\Gamma(q, b)$ *respectively, and the joint density of* $(U_1, X_1) = (1/U, 1/X)$ *is proportional to*

$$(u_1x_1 - 1)^{q-1}e^{-au_1 - bx_1}\mathbf{1}_H(u_1, x_1), \tag{3.2}$$

where $H = \{(u_1, x_1);\ u_1 > 0,\ u_1x_1 - 1 > 0\}$.

PROOF. (1) ⇐ (2). This part of the statement is due to Matsumoto and Yor [**MY01**]. They observed the result first for $a = b$ and $q = 1/2$ in a Brownian motion context and extended it to $a = b$ and $q > 0$. The proof is simple: the joint density of (X, Y) is proportional to

$$\left(\frac{y}{x}\right)^{q-1}\frac{1}{x^2}\,e^{-a(x+y)-bx^{-1}}\mathbf{1}_{(0,\infty)^2}(x, y). \tag{3.3}$$

If $u = (x+y)^{-1}$ and $v = x^{-1}-(x+y)^{-1}$, then $x = (u+v)^{-1}$, $y = u^{-1}-(u+v)^{-1}$ and $dxdy = (u+v)^{-2}u^{-2}dudv$. Noting that $y/x = v/u$ and that the map $(x,y) \mapsto (u,v)$ is a diffeomorphism from $(0,\infty)^2$ to itself, we obtain that the joint density of U and V is proportional to

$$\left(\frac{v}{u}\right)^{q-1}\frac{1}{u^2}\,e^{-b(u+v)-au^{-1}}\mathbf{1}_{(0,\infty)^2}(u,v),$$

which is the product of the $GIG(-q,b,a)$ density for U and the $\Gamma(q,b)$ density for V.

To compute the joint density of (U_1,X_1), let $x = 1/x_1$, $y = u_1 - 1/x_1$ and $dxdy = du_1dx_1/x_1^2$. Noting that $y/x = u_1x_1 - 1$ and that the map $(x,y) \mapsto (x_1,u_1)$ is a diffeomorphism between $(0,\infty)^2$ and H, we see immediately that (3.3) yields the joint density (3.2).

The converse, (1) $\Rightarrow$ (2), is less evident and can be found in Letac and Wesołowski [**LW00**]. □

4. Characterization of the quasi-Wishart distribution

THEOREM 4.1. *Suppose that the probability law of a random variable W in $\mathcal{P}_2$ assigns probability zero to each on of the events $W_{12} = 0$, $W_{1\bullet 2} =$ constant and $W_{2\bullet 1} =$ constant. Then the following two statements are equivalent:*

(1) *$W_{1\bullet 2}$ is independent of (W_{12}, W_2) and $W_{2\bullet 1}$ is independent of (W_{12}, W_1).*
(2) *The law of W is quasi-Wishart.*

PROOF. The part (1) $\Leftarrow$ (2) is the first part of Proposition 2.1. We prove (1) $\Rightarrow$ (2) with the help of Theorem 3.1 as follows. We write

$$X = \frac{|W_{12}|}{W_2},\quad U = \frac{|W_{12}|}{W_1},\quad Y = \frac{W_{1\bullet 2}}{|W_{12}|},\quad V = \frac{W_{2\bullet 1}}{|W_{12}|}.$$

It follows from (1) that, conditional on W_{12}, X and Y are independent of each other and that U and V are also independent of each other. Moreover, we observe that

$$U = (X+Y)^{-1} \quad\text{and}\quad V = X^{-1} - (X+Y)^{-1}.$$

By Theorem 3.1, there exist three constants q, a and b, all positive, such that X is $GIG(-q,a,b)$, Y is $\Gamma(q,a)$, U is $GIG(-q,b,a)$ and V is $\Gamma(q,b)$. Note that at this point we use the hypothesis of non constancy of $W_{1\bullet 2}$ and $W_{2\bullet 1}$, for Theorem 3.1 requires non Dirac distributions. Since the independence of X and Y and of U and V is conditional on $W_{12} = w_{12}$, the constants q, a and b may vary with W_{12}. Let us see how. If Y is $\Gamma(q,a)$, then $|W_{12}|Y$ is $\Gamma(q, a|w_{12}|^{-1})$. However $|W_{12}|Y$ is independent of (W_{12}, W_2) and in particular of W_{12}. It follows that q and $b_1 = a|W_{12}|^{-1}$ are constants independent of W_{12}. Similarly , $|W_{12}|V = W_{2\bullet 1}$ is $\Gamma(q, b|W_{12}|^{-1})$ with q and $b_2 = b|W_{12}|^{-1}$ constants. Therefore, $W_{1\bullet 2}$ and $W_{2\bullet 1}$ are respectively $\Gamma(q,b_1)$ and $\Gamma(q,b_2)$. It remains to find the joint distribution of (W_1,W_2) conditional on $W_{12} = w_{12}$. But $(W_1,W_2) = |w_{12}|(X+Y, X^{-1})$ and this distribution has been essentially given in (3.2) and is equal to

$$\frac{1}{C}(w_1w_2 - w_{12}^2)^{q-1}e^{-b_1w_1-b_2w_2}\mathbf{1}_{\{w_1>0,\ w_2>0,\ w_1w_2-w_{12}^2>0\}}(w_1,w_2). \tag{4.1}$$

This distribution has the required form (2.1), provided we denote $p = q+1/2 > 1/2$, as required for a quasi-Wishart distribution. This proves the theorem. □

5. Characterization of the normal quasi-Wishart distribution

THEOREM 5.1. *Suppose that the probability law of the random variable* (Z, W) *has support* $\mathbb{R}^2 \times \mathcal{P}_2$ *and assigns probability zero to each of the events* $W_{12} = 0$, $W_{1\bullet 2} =$ *constant and* $W_{2\bullet 1} =$ *constant. Then the following two statements are equivalent:*

(1) $(Z_1, W_{1\bullet 2})$ *is independent of* $(Z_2 + \frac{W_{12}}{W_2} Z_1, W_{12}, W_2)$ *and* $(Z_2, W_{2\bullet 1})$ *is independent of* $(Z_1 + \frac{W_{12}}{W_1} Z_2, W_{12}, W_1)$.
(2) *The law of* (Z, W) *is normal quasi-Wishart.*

PROOF. The part $\Leftarrow$ is the second part of Proposition 2.1. The rest of the paper is devoted to the proof of $\Rightarrow$. It follows immediately from Theorem 4.1 that W is quasi-Wishart. Thus we may assume that there exists $p > 1/2$, $b_1 > 0$, and $b_2 > 0$ such that (2.1) holds. We first condition by W and prove that there exist two random variables M and A which are W measurable and such that Z, given W, has the Gaussian distribution $N(M, (AW)^{-1})$.

Denote

$$X_1 = Z_1,\ X_2 = Z_2 + W_2^{-1} W_{12} Z_1,\ Y_2 = Z_2,\ Y_1 = Z_1 + W_1^{-1} W_{12} Z_2,$$

$$a_{11} = 1 - \frac{W_{12}^2}{W_1 W_2},\ a_{12} = \frac{W_{12}}{W_1},\ a_{21} = -\frac{W_{12}}{W_2},\ a_{22} = 1.$$

With this notation, we can say that a_{ij}, $i, j = 1, 2$ are different from 0 with probability 1, that $a_{11}a_{22} - a_{12}a_{21} = 1$, and that

$$Y_1 = a_{11}X_1 + a_{12}X_2, \quad Y_2 = a_{21}X_1 + a_{22}X_2.$$

Moreover, the hypothesis implies that, conditional on W, X_1 and X_2 are independent and Y_1 and Y_2 are also independent.

Following a similar argument by Geiger and Heckerman [**GH98**], it then follows from the characterization of the normal distribution stated in Theorem A.1 (Appendix) that X_1 and X_2 are normal, with respective variances σ_1^2 and σ_2^2 satisfying

$$\sigma_1^2(W_1W_2 - W_{12}^2) = \sigma_2^2 W_2^2.$$

In terms of Z, this means that conditionally on W, Z is Gaussian with covariance proportional to W^{-1}, say equal to $(AW)^{-1}$ and with mean M. It remains to show that A and M denoted respectively $a(W)$ and $m(W)$ as functions of W, are actually constants.

To prove the constancy of A, we use Proposition A.1 (Appendix). In fact, since $(Z_1, W_{1\bullet 2})$ is independent of (W_{12}, W_2), and since, when conditioned on W, the random variable Z_1 is Gaussian with variance $(a(W)W)^{-1}$, we deduce that a actually depend on $W_{1\bullet 2}$, only. Let us write $a(W) = a_1(W_{1\bullet 2})$. Symmetrically, $a(W) = a_2(W_{2\bullet 1})$. Now, to show that a is a constant, we are going to show that conditional on W_{12}, a is a constant and therefore depends only upon W_{12} but since W_{12} is independent of $W_{1\bullet 2}$ and of $W_{2\bullet 1}$, this will show that a is a constant. So, *we now fix* W_{12} and we use the random variables X, Y, U and V as defined in (4). Consider $f_1(Y) = a_1(|W_{12}|Y)$ and $f_2(V) = a_2(|W_{12}|V)$. We now apply Proposition A.1 with $E_1 = E_2 = \mathbb{R}$, with λ_1 and λ_2 the Lebesgue measure and with $\mathbf{P}$ as the distribution of (X, V). Indeed, since the map between H and $(0, \infty)^2$ defined by $(U_1, X_1) = (\frac{1}{U}, \frac{1}{X}) \mapsto (Y, V) = (\frac{1}{U} - X, \frac{1}{X} - U)$ is clearly a bijection, it follows from (3.2) that the density of (Y, V) is strictly positive. Thus A is a constant.

To see that the mean $m(W) = (m_1(W), m_2(W))$ of (Z_1, Z_2) does not depend on W we condition first by W_{12}. Let ϵ be the sign of W_{12}. We keep the notations (4). Since (Z_1, Y) and $(Z_2 + \epsilon X Z_1, X)$ are independent, $m_1(W)$ is a function $f_1(Y)$. Symmetrically, $m_2(W)$ is a function $f_2(V)$. The independence of (Z_1, Y) and $(Z_2 + \epsilon X Z_1, X)$ implies that $(f_2(V) + \epsilon X f_1(Y), X)$ is independent of Y.

Let $K(ds)$ be the distribution of $f_1(Y)+\frac{\epsilon}{X} f_2(V)$. The independence of $(f_2(V)+\epsilon X f_1(Y), X)$ and Y shows that the distribution $K_y(ds)$ of

$$S(y) = f_1(y) + \frac{\epsilon}{X} f_2(\frac{1}{X} - \frac{1}{X+y})$$

is equal to $K(ds)$, dy almost everywhere.

Next, let us show that, for $j = 1, 2$, f_j is equal, almost everywhere, to a continuous function. The characteristic function of $S(y)$ is

$$\psi_y(t) = \exp(itf_1(y))\mathbb{E}(\exp(it\frac{\epsilon}{X} f_2(\frac{1}{X} - \frac{1}{X+y}))), \tag{5.1}$$

and coincides dy almost everywhere with the characteristic function $\psi(t)$ of $K(ds)$. We note that the set $A = \{y : \psi_y(t) = \psi(t)\}$ on which, given t, ψ_y and ψ coincide, does not, in fact, depend upon t and that its complement A' has Lebesgue measure 0. Thus, for a constant and sufficiently small,

$$\frac{1}{\psi(t)}\mathbb{E}(\exp(it\frac{\epsilon}{X(V,y)} f_2(V))) = \exp -itf_1(y), \quad \forall |t| \le a, \forall y \in A. \tag{5.2}$$

From its definition, $y \mapsto X(V, y)$ is continuous. Since

$$\exp(it\frac{\epsilon}{X(V,y)} f_2(V))$$

is bounded, this implies that the left hand side of (5.2), as a function of y, is continuous for all t. Hence $y \mapsto \exp(-itf_1(y))$ is also continuous on A. Applying Lemma A.1 (Appendix) to $u_n = f_1(y + h_n) - f_1(y)$ where $h_n \to 0$, we deduce immediately that there exists a continuous function which coincides with f_1 on A.

At this point, we could try to refine the previous argument to show that $f_j, j = 1, 2$ is C^1. However, similarly to Geiger and Heckerman [**GH98**, p. 123, formula (9)], we prefer to rely on Jarai [**J86**] to assert that, since $F(x) = \epsilon x f_1(y) + f_2(v(x, y))$ depends only on x and since f_1 and f_2 are continuous, then f_1 and f_2 are infinitely differentiable. We can therefore take the derivative of $S(y)$ with respect to y and we obtain

$$f_1'(y) + \frac{\epsilon}{x(x+y)^2} f_2'(\frac{1}{x} - \frac{1}{x+y}) = 0. \tag{5.3}$$

Introducing $g_1(y) = y^2 f_1'(y)$ and $g_2(v) = v^2 f_2'(v)$, with the obvious notation

$$v = \frac{1}{x} - \frac{1}{x+y}, \quad u = \frac{1}{x+y},$$

(5.3) becomes $g_1(y) + \epsilon x g_2(v) = 0$. Symmetrically, $g_2(y) + \epsilon u g_1(y) = 0$. From the definition of u, this can be rewritten as $(x + y)g_2(y) + g_1(y) = 0$. Subtracting from $g_1(y) + x g_2(v) = 0$, we obtain that $g_2 = 0$. Similarly $g_1 = 0$ and therefore f_1 and f_2 are constants. These two constants may still depend on W_{12}. But Z_1 is independent of W_{12}, thus its mean Z_1 conditioned by W is an absolute constant after all. The same holds, of course, for the mean of Z_2 and Theorem 5.1 is proved. □

Appendix A

THEOREM A.1 (Characterization of the normal distribution). *Let X_1 and X_2 be independent non Dirac real random variables. Let a_{ij} $i,j = 1,2$ be non zero real numbers such that $a_{11}a_{22} - a_{12}a_{21} \neq 0$. Define*

$$Y_1 = a_{11}X_1 + a_{12}X_2, \quad Y_2 = a_{21}X_1 + a_{22}X_2.$$

Then Y_1 and Y_2 are independent if and only if there exist numbers m_1, m_2, $\sigma_1 > 0$ and $\sigma_2 > 0$ such that X_j has Gaussian distribution N_{m_j,σ_j^2} for $j = 1,2$ and such that $\sigma_1^2 a_{11}a_{21} + \sigma_2^2 a_{12}a_{22} = 0$.

PROPOSITION A.1 (Principle of separation of random variables). *Let $(E_1, \mathcal{E}_1, \lambda_1)$ and $(E_2, \mathcal{E}_2, \lambda_2)$ be two measured spaces. Let (X_1, X_2) be a random variable taking its values in $E_1 \times E_2$ with distribution $\mathbf{P}$ and suppose that $\mathbf{P}$ has a density with respect to $\lambda_1 \otimes \lambda_2$ which is strictly positive almost everywhere. Let f_1 and f_2 be two measurable functions on E_1 and E_2 such that $f_1(X_1) = f_2(X_2)$, $\mathbf{P}$ almost surely. Then f_1 and f_2 are constants, λ_1 and λ_2 almost everywhere, respectively.*

PROOF. Denote $F = f_1(X_1) = f_2(X_2)$. If F is not a constant, there exists c such that $0 < \mathbf{P}(F \leq c) < 1$. We write

$$A_1 = \{x_1 \in E_1; f_1(x_1) \leq c\}, \quad A_2 = \{x_2 \in E_2; f_2(x_2) \leq c\}.$$

Let A_1' and A_2' be their complementary sets in E_1 and E_2. Since

$$\{F \leq c\} = \{X_1 \in A_1\} = \{X_2 \in A_2\},$$

it follows that

$$\mathbf{P}(X_1 \in A_1; X_2 \in A_2') = \mathbf{P}(X_1 \in A_1'; X_2 \in A_2) = 0.$$

The absolute continuity of $\lambda_1 \otimes \lambda_2$ with respect to $\mathbf{P}$ and the previous equality implies $0 = \lambda_1(A_1)\lambda_2(A_2') = \lambda_1(A_1')\lambda_2(A_2)$. We then have two possible cases: either $\lambda_1(A_1) = 0$ or $\lambda_1(A_1) \neq 0$. In the first case, since $\mathbf{P}$ is absolutely continuous with respect to $\lambda_1 \otimes \lambda_2$, then $\mathbf{P}(X_1 \in A_1) = 0$. In the second case, $\lambda_1(A_1) \neq 0$ implies $\lambda_2(A_2') = 0$, which itself implies that $\lambda_2(A_2) \neq 0$ and therefore $\lambda_1(A_1') = 0$. By the absolute continuity of $\mathbf{P}$ with respect to $\lambda_1 \otimes \lambda_2$, this implies $\mathbf{P}(X_1 \in A_1') = 0$. From the consideration of these two possible cases, we see that we always have either $\mathbf{P}(X_1 \in A_1) = 0$ or $\mathbf{P}(X_1 \in A_1') = 0$, that is, for any constant c, $\mathbf{P}(F \leq c)$ is equal either to 0 or to 1. It follows immediately that F is a constant, which proves the proposition. □

LEMMA A.1. *Let $(u_n)_{n \geq 0}$ be a real sequence such that*

$$\forall |t| \leq a, \lim_{n \to \infty} \exp itu_n = 1.$$

Then $u_n \to 0$.

PROOF. By the dominated convergence theorem,

$$\frac{\sin(au_n)}{au_n} = \frac{1}{2a} \int_{-a}^{a} e^{itu_n} dt \to_{n \to \infty} 1.$$

Thus $u_n \to 0$. □

References

[DL93] Dawid, P. and Lauritzen, S. *Hyper Markov laws in statistical analysis of decomposable graphical models*, Ann. of Stat. **21** (1993), 1272-1317.

[GH98] Geiger, D. and Heckerman, D. *A characterization of the bivariate Wishart distribution*, Probab. and Math. Statist. **18** (1998), 119-131.

[GH99] ______ *Parameters Priors for Directed Acyclic Graphical Models and the Characterization of Several Probability Distributions*, Preprint, (1999).

[J86] Jarai, A. *On regular solutions of functional equations*, Aequationes Math. **30** (1986), 21-54.

[LW00] Letac, G. and Wesołowski, J. *An independence property of the GIG and gamma distributions*, Ann. Prob. **28** (2000), 1371-1383.

[MY01] Matsumoto, H. and Yor, M. *An analogue of Pitman's 2M-X theorem for exponential Wiener functionals, Part II: the role of the generalized inverse Gaussian laws*, Nagoya Math. J. (to appear).

Laboratoire de Statistique et Probabilités, Université Paul Sabatier, 31062, Toulouse, France

E-mail address: `letac@cict.fr`

Department of Statistics, University of Virginia, Halsey Hall, Charlottesville, VA 22904, U.S.A.

E-mail address: `massamh@mathstat.yorku.ca`

Contemporary Mathematics
Volume **287**, 2001

Rank Score Statistics for Spherical Data

Teresa Neeman and Ted Chang

ABSTRACT. We introduce rank score statistics for nonparametric inference on the modal location when the data comes from a rotationally symmetric distribution on the sphere. The Pitman efficacy of such a rank score statistic is calculated and a theory of optimal score functions developed.

The spherical sign test is the special case of a constant score function. We introduce a new distribution, the "spherical exponential" distribution for which the spherical sign test is optimal (in the same sense that the usual univariate sign test is optimal for a double exponential distribution).

Lower bounds on the asymptotic relative efficiency of the spherical sign test to a spherical t-test are found. Due to the equivalency of ARE for tests and estimators, these lower bounds are also lower bounds on the asymptotic relative efficiency for the spherical median, as introduced by Fisher, relative to the directional mean.

Similar lower bounds are also found for the ARE of the spatial mean, introduced by Brown, to the sample mean for bivariate Euclidean space data from a rotationally symmetric distribution.

1. Introduction

Suppose $X_1, ..., X_n \in R^p$ are i.i.d. with a distribution which depends only upon $\|x - \theta\|$ for some $\theta \in R^p$. Given a nondecreasing score generating function ϕ, Neeman (1995) and Hössjer and Croux (1995) define a rank score statistic to test $H_0 : \theta = \theta_0$ as

$$V_\phi(\theta_0) = \sum\nolimits_i \phi(\frac{R_i}{n+1}) \frac{X_i - \theta_0}{\|X_i - \theta_0\|}.$$

Here $R_1, ..., R_n$ are the rankings of $\|X_1 - \theta_0\|, ..., \|X_n - \theta_0\|$ and the unit length vector

$$S(X; \theta_0) \equiv (X - \theta_0)/\|X - \theta_0\|$$

plays the role that $sign(X - \theta_0)$ plays for univariate data. Under reasonable and standard regularity conditions, Neeman (1995) and Hōssjer and Croux (1995) calculate the efficacy of tests which reject H_0 when $\|V_\phi(\theta_0)\|$ is too large.

1991 *Mathematics Subject Classification.* Primary 62H11, 62G05; Secondary: 62G20, 62G30.

Key words and phrases. nonparametric inference, directional data, inference on the sphere, spherical median, spatial median.

Given a vector subspace $\mathfrak{V}$ of Euclidean space, we shall denote by $\Omega(\mathfrak{V})$ the unit sphere of $\mathfrak{V}$, that is the vectors of $\mathfrak{V}$ which have unit length. The unit length vector $S(X;\theta)$ gives the direction that X deviates from θ and this vector is independent, under the assumed symmetry of the distribution of X, from $\|X-\theta\|$.

We use these ideas here to develop rank score statistics for spherical data following the work in Neeman (1995). We assume that $X_1, ..., X_n \in \Omega(R^p)$ are i.i.d. with a density (with respect to surface measure on $\Omega(R^p)$) which depends only upon $x^t\theta$ for some $\theta \in \Omega(R^p)$. A popular such distribution for use with spherical data is the Fisher-von Mises-Langevin distribution whose density is of the form $c(\kappa)\exp(\kappa x^t\theta)$.

Since X is confined to $\Omega(R^p)$, its deviation from θ cannot have an arbitrary direction and hence, for spherical data we define the "sign" of X to be

$$S_\Omega(X;\theta) \equiv \frac{X-(X^t\theta)\theta}{\|X-(X^t\theta)\theta\|}.$$

Note that $S_\Omega(X;\theta)$ lies in $\Omega(\theta^\perp)$, the unit length vectors which are perpendicular to θ.

For Euclidean space data in R^p, Brown (1983) defined the spatial median as the solution to the estimating equation $\sum_i S(X_i;\theta)=0$. Similarly, Fisher (1985) defined the spherical median for data in $\Omega(R^3)$ as the solution to the estimating equation $\sum_i S_\Omega(X_i;\theta)=0$. Since both these statistics solve estimating equations defined by rank scores statistics (with $\phi\equiv 1$), the calculations of Neeman (1995) and Hössjer and Croux (1995) immediately yield, as a corollary, the calculations of Brown (1983) of the asymptotic distribution of the spatial median. Similarly, the calculation in Neeman (1995) of the asymptotic distribution of the spherical median becomes a special case of the results presented here.

It is tempting to define ranks for spherical data by ranking the spherical distances of the data from the hypothesized direction θ_0, that is to rank the values of $\arccos(X_i^t\theta_0)$. This procedure would assign high ranks to points close to $-\theta_0$, and we believe such an approach is inherently unstable for diffuse distributions. For example, suppose θ_0 is the North Pole and X is the point with latitude 89^oS and longitude 0°. This point is essentially no different from the point $\widetilde{X}$ with latitude 89^oS and longitude 180°. However $S_\Omega(X;\theta_0) = -S_\Omega(\widetilde{X};\theta_0)$ and it would seem prudent for a ranking procedure to assign low ranks to X and $\widetilde{X}$.

Accordingly, we propose to rank the X_i based upon values of

$$\|X_i-(X_i^t\theta_0)\theta_0\| = \sqrt{1-(X_i^t\theta_0)^2}$$

and to study rank score statistics of the form

$$V_\phi^\Omega(\theta_0) = \sum_i \phi(\frac{R_i}{n+1})S_\Omega(X_i;\theta_0). \tag{1.1}$$

We acknowledge that statistics of the form (1.1) cannot distinguish between θ_0 and $-\theta_0$, however this property is shared by several common spherical statistical procedures.

For example, when $\phi\equiv 1$, one gets the rank score statistic V_1^Ω associated to the spherical median. We shall refer to V_1^Ω as the "spherical sign test". Watson (1983) bases a test for location on

$$t^\Omega(\theta_0) = \sum_i (X_i-(X_i^t\theta_0)\theta_0) = \sum_i \|X_i-(X_i^t\theta_o)\theta_0\|\, S_\Omega(X_i;\theta_0), \tag{1.2}$$

which, although not a rank score statistic of the form (1.1), also has the feature that $t^{\Omega}(\theta_0) = t^{\Omega}(-\theta_0)$. If $X_1, ..., X_n$ are i.i.d. Fisher-von Mises-Langevin, the m.l.e. $\widehat{\theta}$ satisfies $t^{\Omega}(\widehat{\theta}) = 0$.

Ducharme and Milasevic (1987) studied the "normalized spatial median" which is defined to be $\tilde{\theta} = \tilde{\eta}/\,\|\tilde{\eta}\|$ where $\tilde{\eta}$ is the spatial median of the X_i. The normalized spatial median is distinct from the spherical median and indeed we do not believe it is the associated estimator to any of the rank score statistics considered here: that is, $\tilde{\theta}$ cannot be defined as the solution to an estimating equation $V_{\phi}^{\Omega}(\theta) = 0$ where $V_{\phi}^{\Omega}(\theta)$ has the form (1.1). Thus the results given here do not include as a subcase those of Ducharme and Milasevic.

Section 2 will calculate the Pitman efficacy of spherical rank score statistics. Optimal rank score statistics will be calculated there. We introduce a new distribution, which we call the "spherical exponential" distribution which has the spherical sign test as its optimal rank score test. A numerical example on remnant magnetization is given in Section 3. In Section 4 we will give lower bounds for the asymptotic relative efficiency, for rotationally symmetric data in R^2, of

$$V_1(\theta_0) = \sum\nolimits_i S(X_i;\theta_0) \tag{1.3}$$

to

$$t(\theta_0) = \sum\nolimits_i (X_i - \theta_0)\,. \tag{1.4}$$

It is well known (see Bickel and Lehmann (1975)) that for symmetric unimodal distributions on the real line, the asymptotic relative efficiency of V_1 to t is bounded below by 1/3. In Section 5, we derive lower bounds on the asymptotic relative efficiency of V_1^{Ω} to t^{Ω} for data in $\Omega(R^3)$.

2. Pitman efficacy for rank score statistics on the sphere

Let $\omega_p = 2\pi^{p/2}/\Gamma(p/2)$ be the (surface) area of $\Omega(R^p)$ and let $d\omega_p$ be (unnormalized) surface measure on $\Omega(R^p)$, so that $\omega_p = \int_{\Omega(R^p)} d\omega_p$. It is well known (see Watson (1983)) that

$$d\omega_p = (1-t^2)^{(p-3)/2} dt d\omega_{p-1},$$

where $t = x^t\theta$ and, by minor abuse of notation, $d\omega_{p-1}$ is the surface measure on $\Omega(\theta^{\perp})$. Suppose $X \in \Omega(R^p)$ has a distribution of the form $f(x^t\theta)d\omega_p/\omega_{p-1}$, so that the distribution of t is $f(t)(1-t^2)^{(p-3)/2}dt$. It also follows that t and $\xi = S_{\Omega}(X;\theta)$ are independent and that ξ is uniformly distributed on $\Omega(\theta^{\perp})$. In particular, $E(\xi) = 0$ and, using $Tr(\xi\xi^t) = 1$, it can easily be seen that $E(\xi\xi^t) = (p-1)^{-1}(I - \theta\theta^t)$.

Thus under the null distribution,

$$\begin{aligned} E(V_{\phi}^{\Omega}(\theta_0)) &= 0, \\ Var(V_{\phi}^{\Omega}(\theta_0)) &= (p-1)^{-1}(I-\theta_0\theta_0^t)\sum\nolimits_i \phi(i/(n+1))^2. \end{aligned}$$

The asymptotic distribution of $n^{-1/2}V_{\phi}^{\Omega}(\theta_0)$ is multivariate normal with covariance $\Sigma = (p-1)^{-1}(I-\theta_0\theta_0^t)\int_0^1 \phi^2(u)du$.

Ko and Chang (1993) have shown that Fisher information $I(f)$ can be interpreted as a quadratic form on $\Omega(\theta^{\perp})$ given by

$$I(f)(\delta_1, \delta_2) = E\left[\left(\frac{f'(t)}{f(t)}\right)^2 (1-t^2)\right] \frac{\delta_1^t \delta_2}{p-1}, \tag{2.1}$$

where $\delta_1, \delta_2 \in \Omega(\theta^{\perp})$. In what follows we assume that

- f is twice continuously differentiable,
- $0 < E\left[\left(\frac{f'(t)}{f(t)}\right)^2 (1-t^2)\right] < \infty$,
- The score generating function ϕ is non-negative, non-decreasing and square integrable.

Recalling that ranking is done on $\sqrt{1-t^2}$, let $F_+(t) = \Pr(|X^t\theta| > t)$ for $t \in [0,1]$. Fix a sequence of alternatives to $H_0 : \theta = \theta_0$,

$$H_{1n} : \theta_n = (1 - \frac{\delta^t\delta}{2n})\theta_0 + \delta/\sqrt{n} + O(n^{-3/2}),$$

where $\delta \in \theta_0^{\perp}$ and the $O(n^{-3/2})$ term is adjusted to insure that $\theta_n \in \Omega(R^p)$.

THEOREM 2.1. *Under the sequence of alternatives* $\{\theta_n\}$

$$n^{-1/2} V_\phi^{\Omega}(\theta_0) \xrightarrow{d} N(\mu, \Sigma),$$

where

$$\begin{aligned}
\mu &= \frac{\delta}{p-1}\int_{-1}^{1} \phi(F_+(|t|)) f'(t)(1-t^2)^{(p-2)/2} dt, \\
\Sigma &= \frac{I - \theta_0\theta_0^t}{p-1}\int_0^1 \phi^2(u) du.
\end{aligned}$$

PROOF. The proof follows the general outlines of the proofs in Section A.2 of Hettmansperger and McKean (1998) or in Hössjer and Croux (1995) and uses LeCam's third lemma. We will outline only the computations needed by LeCam's third lemma.

Let q_n and p_n be the density of $X_1, ..., X_n$ under H_{1n} and H_0 respectively and let $L_n = \log(q_n/p_n)$. We will further let $t_i = X_i^t\theta_0$, $\xi_i = S_\Omega(X_i; \theta_0)$, and

$$W_\phi^{\Omega}(\theta_0) = \sum\nolimits_i \phi(F_+(|t_i|))\xi_i. \tag{2.2}$$

To apply LeCam's third lemma, we calculate the asymptotic mean and variance of $(n^{-1/2}W_\phi^{\Omega}(\theta_0), L_n)$ under p_n.

Under p_n, $E(n^{-1/2}W_\phi^{\Omega}(\theta_0)) = 0$ and $Cov(n^{-1/2}W_\phi^{\Omega}(\theta_0)) = \Sigma$. Furthermore

$$f(X^t\theta_n) = f(t) + f'(t)(\frac{X^t\delta}{\sqrt{n}} - \frac{t\delta^t\delta}{2n}) + f''(t)\frac{(X^t\delta)^2}{2n} + O_p(n^{-3/2}),$$

so that

$$\begin{aligned} L_n &= \sum_i \log \left(\frac{f(t_i) + f'(t_i)(\frac{X_i^t \delta}{\sqrt{n}} - \frac{t_i \delta^t \delta}{2n}) + f''(t_i)\frac{(X_i^t\delta)^2}{2n} + O_p(n^{-3/2})}{f(t_i)} \right) \\ &= \sum_i \{ \frac{1}{\sqrt{n}} \frac{f'(t_i)\sqrt{1-t_i^2}\xi_i^t \delta}{f(t_i)} \\ &+ \frac{1}{2n} \frac{(f(t_i)f''(t_i) - f'(t_i)^2)(1-t_i^2)(\xi_i^t\delta)^2 - f(t_i)f'(t_i)t_i\delta^t\delta}{f(t_i)^2} \\ &+ O_p(n^{-3/2})\}. \end{aligned}$$

Notice that $E(\xi) = 0$ and

$$E((\xi^t\delta)^2) = \delta^t E(\xi\xi^t)\delta = \frac{1}{p-1}\delta^t(I - \theta_0\theta_0^t)\delta = \frac{1}{p-1}\delta^t\delta.$$

Furthermore

$$\begin{aligned} E\left[\frac{f''(t)(1-t^2)}{(p-1)f(t)}\right] &= \int \frac{1}{p-1} f''(t)(1-t^2)^{(p-1)/2}dt = \int f'(t)t(1-t^2)^{(p-3)/2}dt \\ &= E\left[\frac{f'(t)t}{f(t)}\right], \end{aligned}$$

and hence

$$\begin{aligned} E(L_n) &= -\frac{\delta^t\delta}{2(p-1)} E\left[\left(\frac{f'(t)}{f(t)}\right)^2 (1-t^2)\right] + O(n^{-1/2}) \\ &= -\frac{1}{2} I(f)(\delta,\delta) + O(n^{-1/2}). \end{aligned}$$

In addition

$$\begin{aligned} Var(L_n) &= E\left[\left(\frac{f'(t)\sqrt{1-t^2}(\xi^t\delta)}{f(t)}\right)^2\right] + O(n^{-1/2}) \\ &= I(f)(\delta,\delta) + O(n^{-1/2}). \end{aligned}$$

$$\begin{aligned} Cov(n^{-1/2}W_\phi^\Omega(\theta_0), L_n) &= Cov\left(\phi(F_+(|t|))\xi, \frac{f'(t)\sqrt{1-t^2}(\xi^t\delta)}{f(t)}\right) + O(n^{-1/2}) \\ &= \frac{\delta}{p-1} E\left[\phi(F_+(|t|))\frac{f'(t)\sqrt{1-t^2}}{f(t)}\right] + O(n^{-1/2}) \\ &= \mu + O(n^{-1/2}). \end{aligned}$$

The remainder of the proof, including the asymptotic multivariate normality of $(n^{-1/2}W_\phi^\Omega(\theta_0), L_n)$, is similar to the proof given in Hettmansperger and McKean (1998). □

COROLLARY 2.2. The square Pitman efficacy of $V_\phi^\Omega(\theta_0)$ in the direction $\delta \in \theta_0^\perp$ is

$$c_\phi^2(\delta) = \frac{\left(\int_{-1}^1 \phi(F_+(|t|))f'(t)(1-t^2)^{(p-2)/2}dt\right)^2}{(p-1)\int_0^1 \phi^2(u)du}\delta^t\delta. \tag{2.3}$$

Given two scoring functions ϕ and ψ, the asymptotic relative efficiency of $V_\phi^\Omega(\theta_0)$ and $V_\psi^\Omega(\theta_0)$ is

$$ARE(V_\phi^\Omega, V_\psi^\Omega) = \frac{\left(\int_{-1}^1 \phi(F_+(|t|))f'(t)(1-t^2)^{(p-2)/2}dt\right)^2 \left(\int_0^1 \psi^2(u)du\right)}{\left(\int_0^1 \phi^2(u)du\right)\left(\int_{-1}^1 \psi(F_+(|t|))f'(t)(1-t^2)^{(p-2)/2}dt\right)^2}.$$

COROLLARY 2.3. Given a distribution $f(x^t\theta)d\omega_p/\omega_{p-1}$ on $\Omega(R^p)$, define a score generating function ϕ_f by

$$\phi_f(u) = \frac{f'(F_+^{-1}(u)) + f'(-F_+^{-1}(u))}{f(F_+^{-1}(u)) + f(-F_+^{-1}(u))}(1-[F_+^{-1}(u)]^2)^{1/2}. \tag{2.4}$$

Then $V_{\phi_f}^\Omega$ is optimal in the sense that, for any other score generating function ψ, $ARE(V_{\phi_f}^\Omega, V_\psi^\Omega) \geq 1$. Furthermore the efficacy of $V_{\phi_f}^\Omega$ is

$$\begin{aligned} c_{\phi_f}^2(\delta) &= (p-1)^{-1}\int_0^1 \phi_f^2(u)du\ \delta^t\delta \\ &= (p-1)^{-1}\int_0^1 \phi_f^2(F_+(t))(f(t)+f(-t))(1-t^2)^{(p-3)/2}dt\ \delta^t\delta. \end{aligned} \tag{2.5}$$

REMARK 2.4. In Euclidean space, the efficacy of the optimal score generating function is Fisher information. This is not generally true for spherical rank score statistics, as we have defined them, because of the requirement that the ranks be antipodally symmetric.

PROOF OF COROLLARY 2.3. Let ψ be an arbitrary score generating function. Making the substitution $u = F_+(t)$, $du = -[f(t)+f(-t)](1-t^2)^{(p-3)/2}dt$

$$\begin{aligned} \int_{-1}^1 \psi(F_+(|t|))f'(t)(1-t^2)^{(p-2)/2}dt &= \\ \int_0^1 \psi(F_+(t))(f'(t)+f'(-t))(1-t^2)^{(p-2)/2}dt &= \int_0^1 \psi(u)\phi_f(u)du. \end{aligned}$$

and the corollary follows by the Cauchy-Schwarz inequality. □

COROLLARY 2.5. For the Fisher-von Mises-Langevin distribution, the optimal score generating function is $\phi_f(F_+(t)) \propto (1-t^2)^{1/2}$. In addition, ϕ_f is fully efficient, that is, its efficacy is Fisher information.

PROOF. Substituting $f(t) = c(\kappa)\exp(\kappa t)$ into (2.4), we see that $\phi_f(F_+(t)) = \kappa(1-t^2)^{1/2}$. It is now easily seen that (2.1) and (2.5) coincide. □

REMARK 2.6. For the Fisher-von Mises-Langevin distribution, it is easily seen that $V_{\phi_f}^\Omega(\theta_0)$ is asymptotically equivalent to $t^\Omega(\theta_0)$ as defined in equation (1.2). Indeed the proof of Theorem 2.1 (see Hettmansperger and McKean (1998), page 404) implies that $n^{-1/2}(V_{\phi_f}^\Omega(\theta_0) - W_{\phi_f}^\Omega(\theta_0)) \xrightarrow{P} 0$ where $W_{\phi_f}^\Omega(\theta_0)$ is defined in equation

(2.2) and is easily seen to be, for Fisher-von Mises Langevin f, proportional to $t^{\Omega}(\theta_0)$.

COROLLARY 2.7. The spherical sign test is optimal for distributions of the form $f(t) \propto \exp(-\gamma \arccos(t))$.

PROOF. The corollary can be easily proven by checking that with $f(t)$ as given $(f(t)+f(-t))^{-1}(f'(t)+f'(-t))(1-t^2)^{1/2}$ is a constant. □

We shall call this distribution a "spherical exponential distribution". It is analogous to the double exponential distribution in R^1 for which the usual sign test is optimal. While the double exponential distribution and the normal distribution on the real line are fundamentally different in the tails of the distribution, the corresponding distributions on the sphere are similar when the data is somewhat disperse, because of the compactness of the sphere. To illustrate this, following Watson (1983), for various values of γ, we matched a spherical exponential distribution with a Fisher-von Mises-Langevin distribution by choosing κ to equalize $E(t)$. After matching the two distributions, we calculated the maximum deviation between the two cumulative distributions of t. Table 1 below lists the matched concentrations, the approximate angle $\theta = \arccos(t)$ at which the maximum deviation between occurred, and the value of the maximum deviation

TABLE 1

γ	κ	θ	max. dev.
0.01	0.0118	0.816	0.00000786
0.1	0.118	0.816	0.000785
1.0	1.203	0.785	0.0692
2.0	2.570	0.660	0.209
10.0	34.666	0.141	0.861

Our results indicate that when the data are somewhat disperse, either distribution can be used. In this case, the spherical sign statistic is preferable to t^{Ω} because it is more robust. On the other hand, when the data are concentrated, one of the two families of distributions will be preferable.

3. A Numerical example

Fisher et al. (1987) consider a data set of $n = 17$ measurements of magnetic remanence from specimens collected from the Tumblagooda Sandstone of Australia. The data is plotted in Figure 1. They conclude that although there is no evidence of a lack of rotational symmetry in the data, the data does not seem to follow a Fisher distribution. Indeed the data has two points (the most northwestern and southeastern points in Figure 1) which are too far from the center to be accounted for by a Fisher distribution. This lack of fit together with a somewhat small sample size makes the data a good candidate for a nonparametric analysis.

We consider four different score functions ϕ: $\phi \equiv 1$ (sign test), $\phi(u) = u$ (Wilcoxon scoring), and optimal Fisher scoring when $\kappa = 2$ and when $\kappa = 40$. The

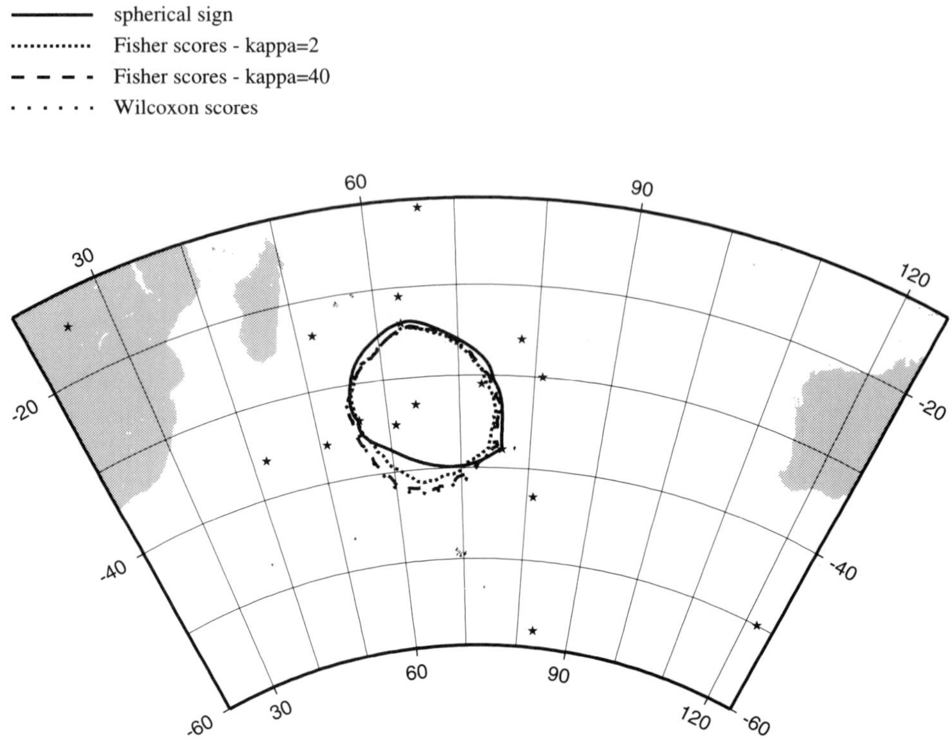

FIGURE 1. 95% confidence regions for modal direction

value $\kappa = 40$ is chosen because the analysis in Fisher et al. (1987) shows that if a Fisher distribution were fit to the data, the value of κ would be approximately 40 or 50. The value $\kappa = 2$ was chosen to see the effect of a wildly unrealistic choice of κ. Although the data is not Fisherian, the nonparametric analysis with Fisher scoring is still valid, although perhaps inefficient.

The null distribution of V_ϕ^Ω is identical to that of $U = \sum_i \phi(i/(n+1))\xi_i$ where $\xi_1, ..., \xi_n$ are i.i.d. uniform on $\Omega(\theta_0^\perp)$. Thus $E(U) = 0$ and $Var(U) = (p-1)^{-1}(\sum_i \phi(i/(n+1))^2)(I - \theta_0\theta_0^T)$. A natural approximation to the null distribution of $\left\|V_\phi^\Omega\right\|^2$ is $(p-1)^{-1}(\sum_i \phi(i/(n+1))^2)\chi^2_{p-1}$. Exact critical points for the distribution of $\left\|V_\phi^\Omega\right\|^2$ can, with a very modest investment in computer time, be obtained by simulation. Indeed an S-Plus algorithm to do that is:

```
V2CRIT ¡- function(ai,nsim1000=10,alpha=c(.5,.75,.9,.95,.99)) {
    #Monte Carlo simulation of critical points of —V—ˆ2
    #scores in ai, 1000*nsim1000 simulated data sets
    n ¡- length(ai)
    v2 ¡- matrix(NA,nrow=1000,ncol=nsim1000)
    for (i in 1:nsim1000) { theta ¡- 2*pi*runif(1000*n)
     x ¡- matrix(ai*cos(theta),nrow=n)
```

```
y ¡- matrix(ai*sin(theta),nrow=n)
v2[,i] ¡- (apply(x,2,sum)^2+apply(y,2,sum)^2) }
v2 ¡- sort(as.vector(v2))
qalpha ¡- round(alpha*1000*nsim1000)
v2[qalpha] }
```

Table 2 gives the values of $a_i = \phi(i/(n+1))$ together with 95% critical points for $\left\|V_\phi^\Omega\right\|^2$ calculated using simulation (with 100,000 simulated data sets $\xi_1, ..., \xi_n$) and χ_2^2 approximation.

TABLE 2

	sign test	Fisher $\kappa = 2$	Fisher $\kappa = 40$	Wilcoxon
a_1	1.000	1.000	1.000	1.000
a_2	1.000	1.423	1.435	2.000
a_3	1.000	1.754	1.785	3.000
a_4	1.000	2.039	2.094	4.000
a_5	1.000	2.294	2.382	5.000
a_6	1.000	2.529	2.658	6.000
a_7	1.000	2.749	2.927	7.000
a_8	1.000	2.957	3.196	8.000
a_9	1.000	3.155	3.468	9.000
a_{10}	1.000	3.343	3.749	10.000
a_{11}	1.000	3.522	4.042	11.000
a_{12}	1.000	3.692	4.355	12.000
a_{13}	1.000	3.849	4.698	13.000
a_{14}	1.000	3.992	5.083	14.000
a_{15}	1.000	4.114	5.538	15.000
a_{16}	1.000	4.210	6.117	16.000
a_{17}	1.000	4.273	6.984	17.000
95%quantile	50.441	495.693	788.952	5198.417
χ^2approximation	50.927	506.396	803.771	5347.382

If c is the appropriate critical point $\{\theta \mid V_\phi^\Omega(\theta) < c\}$ is a 95% confidence region for the modal vector θ. The four confidence regions corresponding to the four scoring functions are also plotted in Figure 1. Examining them we see that they are quite similar however the sign test and Fisher scoring for $\kappa = 2$ yield slightly smaller regions than those for Fisher scoring with $\kappa = 40$ and Wilcoxon scoring.

(Figure 1 is mapped with an Albers conic equal area projection, so that spherical areas are accurately portrayed.) Clearly the scoring methods designed for more diffuse distributions do better than those designed for concentrated distributions. There also seems to be a slight systematic movement of the region towards the southwest as the weighting of the spherical signs of the higher ranked observations increases.

Examining the data more carefully, one can see that the two highest ranked points are in almost opposite directions (northwest and southeast) from the center of the data. The same is true for the next two highest ranking points which are north and slightly east of south from the center. Thus the highest ranking points come with spherical signs which almost cancel out. These fortuitous circumstances explain why the different scoring schemes yield such similar results.

The spherical median satisfies $V_1^{\Omega}(\widehat{\theta}) = 0$. In general given a scoring function ϕ, there is an associated estimator defined by $V_{\phi}^{\Omega}(\widehat{\theta}) = 0$. For this data the four estimates of θ are essentially identical: with latitudes -32.996, -32.681, -32.997, -32.931 and longitudes 64.001, 64.701, 64.805, 65.199 respectively. Note that, as usual for rank score procedures, the confidence regions have a coverage probability that is nonparametrically correct, but the estimates are not centered in the confidence regions and their covariances depend upon the underlying distribution.

Because of this, standard errors for $\hat{\theta}$ are not possible without additional parametric assumptions. Nevertheless, adopting the idea of Scheffé simultaneous inference, one can take the regions shown in Figure 1 and calculate for each of them their smallest enclosing rectangle in lat-long space. In this way, a pair of simultaneous 95% confidence intervals (derived from the sign test) for the latitude and longitude is calculated as $(-39.86, -23.83)$ and $(55.19, 75.97)$ respectively. The other three regions yield very similar results.

4. The ARE of the spatial sign test to the t-test for rotationally symmetric data on R^2

We assume here that the distribution of the data $X_1, ..., X_n \in R^2$ is rotationally symmetric; that is $X_1, ..., X_n$ are i.i.d. with a density of the form $f(r)/(2\pi)$ where $r = \|x - \theta\|$ for some $\theta \in R^2$. With this convention, the density of $R = \|X - \theta\|$ is $rf(r)$. We calculate lower bounds on $ARE(V_1, t)$ where V_1 and t are defined by equations (1.3) and (1.4) respectively. Brown (1983) calculated the asymptotic covariance of the spatial median defined as the solution to the equation $V_1(\widehat{\theta}) = 0$. In the notation of the present paper this asymptotic covariance is $\frac{2}{n}\left(\int_0^\infty f(r)dr\right)^{-2} I$ where I is a 2×2 identity matrix. Note that $\bar{X}$ satisfies $t(\bar{X}) = 0$ and has asymptotic covariance $\frac{1}{2n}\left(\int_0^\infty f(r)r^3 dr\right) I$. By the well known equivalence of ARE for test statistics and their corresponding estimators (see, for example, Hettmansperger and McKean (1998))

$$ARE(V_1, t) = \frac{1}{4}\left(\int_0^\infty f(r)dr\right)^2 \left(\int_0^\infty f(r)r^3 dr\right).$$

THEOREM 4.1. *If the distribution of the data is rotationally symmetric, then*

$$ARE(V_1, t) \geq 1/4.$$

PROOF. By the Cauchy-Schwarz inequality

$$\begin{aligned}\left(\int_0^\infty f(r)rdr\right)^4 &\le \left(\int_0^\infty f(r)r^2dr\right)^2\left(\int_0^\infty f(r)dr\right)^2\\ &\le \left(\int_0^\infty f(r)r^3dr\right)\left(\int_0^\infty f(r)rdr\right)\left(\int_0^\infty f(r)dr\right)^2.\end{aligned}$$

Since $\int_0^\infty f(r)rdr = 1$, the theorem follows. □

THEOREM 4.2. *If f is monotonically decreasing on $[0,\infty)$, then*

$$ARE(V_1, t) \ge 1/2.$$

PROOF. Let $f(r)/(2\pi)$ be a given decreasing density and let $u(r)/(2\pi)$ be a uniform density on a disk of radius r_0, where r_0 is chosen so that $\int_0^\infty f(r)dr = \int_0^\infty u(r)dr = 2/r_0$. That is, $u(r) = 2/r_0^2$ for $r \in [0, r_0]$ and 0 otherwise. Then $ARE_u(V_1, t) = 1/2$. It suffices to show that $\int_0^\infty f(r)r^3dr \ge \int_0^\infty u(r)r^3dr$. We claim that if F and U are the anti derivatives of f and u respectively, then, for any $s > 0$.

$$\int_0^s (F(\infty) - F(r))dr \le \int_0^s (U(\infty) - U(r))dr,$$

Given the claim, we can complete the proof of Theorem 4.2 as follows:

$$\begin{aligned}\int_0^\infty f(r)r^3dr &= 3\int_0^\infty (F(\infty) - F(r))r^2dr\\ &= -6\int_0^\infty r\int_0^r (F(\infty) - F(s))dsdr\\ &\ge -6\int_0^\infty r\int_0^r (U(\infty) - U(s))dsdr = \int_0^\infty u(r)r^3dr.\end{aligned}$$

To prove the claim, let $G_F(r) = F(\infty) - F(r)$ and $G_U(r) = U(\infty) - U(r)$. Then G_F and G_U are continuous, monotonically decreasing, and satisfy $G_F(0) = G_U(0) = 2/r_0$, $\int_0^\infty G_F(r)dr = \int_0^\infty f(r)rdr = 1 = \int_0^\infty G_U(r)dr$, and $G_F''(r) = -f'(r) \ge 0$.

We claim that there exists an $r_1 > 0$, $r_1 < r_0$ with $f(r_1) > 0$ and $G_F(r_1) = G_U(r_1)$. Suppose not. Let $r_f = \sup\{r \mid f(r) > 0\}$ (r_f can be infinite). Suppose $r_f \le r_0$. If $G_U(r) > G_F(r)$ for all $r < r_f$, then $\int_0^{r_f} G_U(r)dr > \int_0^{r_f} G_F(r)dr = 1$, a contradiction. On the other hand, if $G_U(r) < G_F(r)$ for all $r < r_f$, we have $G_U(r_f) \le G_F(r_f) = 0$ and hence $r_f = r_0$. Then $1 = \int_0^{r_f} G_U(r)dr < \int_0^{r_f} G_F(r)dr = 1$, again a contradiction. Similar contradictions apply if $r_f \ge r_0$ and therefore r_1 must exist.

For $0 < r < r_1$, $G_U'(r)$ is a constant and negative. By the concavity of G_F, $G_F'(r) \le G_U'(r)$ and hence $G_F(r) \le G_U(r)$. Thus

$$\int_0^r (F(\infty) - F(s))ds \le \int_0^r (U(\infty) - U(s))ds.$$

Now consider $r_0 > r > r_1$. By the concavity of G_F, $G_F'(r) \ge G_U'(r)$ and hence $G_F(r) \ge G_U(r)$. $G_F(r) \ge G_U(r) = 0$ is trivially true for $r \ge r_0$. Thus for $r \ge r_1$,

$$\begin{aligned}\int_0^r (F(\infty) - F(s))ds &= 1 - \int_r^\infty G_F(s)ds\\ &\le 1 - \int_r^\infty G_U(s)ds = \int_0^r (U(\infty) - U(s))ds,\end{aligned}$$

and the lemma is proven. □

REMARK 4.3. The proof of Theorem 4.2 shows that the lower bound of 1/2 is sharp. The lower bound of 1/4 in Theorem 4.1 is also sharp as can be shown by taking a sequence of densities converging to a distribution which makes $R = 1$ almost surely.

5. The ARE of the spherical sign test to the spherical t-test for rotationally symmetric data on $\Omega(R^3)$

In this section, we study the asymptotic relative efficiency of the spherical sign test statistic $V_1^\Omega(\theta_0)$ to the statistic $t^\Omega(\theta_0)$ as defined in equation (1.2). For $\Omega(R^3)$, Fisher (1985) calculated the $ARE(V_1^\Omega(\theta_0), t^\Omega(\theta_0))$ for the Fisher-von Mises-Langevin distributions. These results were generalized to arbitrary dimensions p and arbitrary rotationally symmetric distributions by Ko and Chang (1993) (see also the discussion in Chang and Ko (1995)). We note that Fisher, Ko and Chang actually calculate the asymptotic variances of the spherical median (defined by $V_1^\Omega(\widehat{\theta})) = 0$) and the spherical mean (defined by $t^\Omega(\widehat{\theta}) = 0$).

Using Theorem 4.2 of Ko and Chang (1993), or by direct computation of the Pitman efficacy of t^Ω together with (2.3), it is readily shown that

$$(5.1)\quad ARE(V_1^\Omega, t^\Omega) = \frac{\left(\int_{-1}^1 f(t)(1-t^2)^{(p-1)/2}dt\right)\left(\int_{-1}^1 f'(t)(1-t^2)^{(p-2)/2}dt\right)^2}{\left(\int_{-1}^1 f'(t)(1-t^2)^{(p-1)/2}dt\right)^2}.$$

We are interested in the case $p = 3$. Let $g(\psi) = f(\cos(\psi))$. Then with a little calculus, equation (5.1) becomes

$$(5.2)\qquad ARE_g(V_1^\Omega, t^\Omega) = \frac{\left(\int_0^\pi g(\psi)\cos\psi d\psi\right)^2\left(\int_0^\pi g(\psi)\sin^3\psi d\psi\right)}{\left(\int_0^\pi g(\psi)\sin 2\psi d\psi\right)^2\left(\int_0^\pi g(\psi)\sin\psi d\psi\right)}.$$

Here we have used $\int_0^\pi g(\psi)\sin\psi d\psi = \int_{-1}^1 f(t)dt = 1$.There are many examples of densities for which $ARE_g(V_1^\Omega, t^\Omega) = 0$. One example is $g(\psi) \propto (5/8) + (3/8)\cos 2\psi + (1/3)\cos 3\psi$. On the other hand, if we require $g(\psi)$ to be monotonically decreasing, we show below that $ARE_g(V_1^\Omega, t^\Omega) \geq .5$.

This lower bound is strict. Letting

$$g_n(\psi) \propto \begin{cases} 1 & 0 \leq \psi \leq 1/n \\ 0 & \text{otherwise} \end{cases}$$

it is easy to see that $\lim_{n\to\infty} ARE_g(V_1^\Omega, t^\Omega) = .5$. On the other hand, letting

$$g_n(\psi) \propto \begin{cases} 1 & 0 \leq \psi \leq \pi - 1/n \\ 0 & \text{otherwise} \end{cases}$$

one can see that $ARE_g(V_1^\Omega, t^\Omega)$ is not bounded above.

THEOREM 5.1. *If $g(\psi)$ is monotonically decreasing on $[0, \pi]$, then*

$$ARE(V_1^\Omega, t^\Omega) \geq 1/2.$$

Before proving the theorem, we state three lemmas.

LEMMA 5.2. *Let $f(\psi)$ be a monotonically decreasing nonnegative function on $[0,\pi/2]$. Define a new function $g(\psi)=\left[\int_0^{\pi/2} f(\xi)\cos\xi d\xi\right]^{-1} f(\psi)\cos\psi$. Then*

$$\frac{\int_0^{\pi/2} g(\psi)\sin^3\psi d\psi}{\int_0^{\pi/2} g(\psi)\sin\psi d\psi} \le \frac{\int_0^{\pi/2} f(\psi)\sin^3\psi d\psi}{\int_0^{\pi/2} f(\psi)\sin\psi d\psi}.$$

PROOF. Without loss of generality, we can assume $\int_0^{\pi/2} f(\psi)d\psi = 1$. Then f/g is nondecreasing, so that, in the notation of Capéraà (1988), the distribution functions of f and g satisfy $F >_{(+)} G$. The lemma follows from the Corollary to the main theorem of Capéraà (1988). □

LEMMA 5.3. *Suppose f is nonnegative and monotonically decreasing on $[0,\pi]$. Then*

$$\frac{\int_0^{\pi/2} f(\psi)\sin^3\psi d\psi}{\int_0^{\pi/2} f(\psi)\sin\psi d\psi} \le \frac{\int_{\pi/2}^{\pi} f(\psi)\sin^3\psi d\psi}{\int_{\pi/2}^{\pi} f(\psi)\sin\psi d\psi}.$$

PROOF. Define

$$g_1(\psi)=[\int_0^{\pi/2} f(\xi)d\xi]^{-1}f(\psi),\quad g_2(\psi)=[\int_0^{\pi/2} f(\pi-\xi)d\xi]^{-1}f(\pi-\psi).$$

Clearly $\int_0^{\pi/2} g_1(\psi)d\psi = 1 = \int_0^{\pi/2} g_2(\psi)d\psi$, g_1 is monotonically decreasing and g_2 is monotonically increasing on $[0,\pi/2]$, so that the lemma again follows from Capéraà (1988). □

LEMMA 5.4. *Suppose f is nonnegative and monotonically decreasing on $[0,\pi]$. Then*

$$\frac{\int_0^{\pi/2} f(\psi)\sin^3\psi d\psi - \int_{\pi/2}^{\pi} f(\psi)\sin^3\psi d\psi}{\int_0^{\pi/2} f(\psi)\sin\psi d\psi - \int_{\pi/2}^{\pi} f(\psi)\sin\psi d\psi} \le \frac{\int_0^{\pi/2} f(\psi)\sin^3\psi d\psi + \int_{\pi/2}^{\pi} f(\psi)\sin^3\psi d\psi}{\int_0^{\pi/2} f(\psi)\sin\psi d\psi + \int_{\pi/2}^{\pi} f(\psi)\sin\psi d\psi}.$$

PROOF OF THEOREM 5.1. We resume the notation of the beginning of this section: $f(t)$ represents the density of $t = X^t\theta_0$ and $g(\psi) = f(\cos(\psi))$. We consider two cases.

First suppose that the support of g is contained in $[0,\pi/2]$. As in Lemma 5.2, let $g_1(\psi)=\left[\int_0^{\pi/2} g(\xi)\cos\xi d\xi\right]^{-1} g(\psi)\cos\psi$. Then

$$\begin{aligned}
ARE_g(V_1^{\Omega}, t^{\Omega}) &= \frac{\left(\int_0^{\pi} g(\psi)\cos\psi d\psi\right)^2 \left(\int_0^{\pi} g(\psi)\sin^3\psi d\psi\right)}{\left(\int_0^{\pi} g(\psi)\sin 2\psi d\psi\right)^2 \left(\int_0^{\pi} g(\psi)\sin\psi d\psi\right)} \\
&\ge \frac{\left(\int_0^{\pi} g(\psi)\cos\psi d\psi\right)^2 \left(\int_0^{\pi} g_1(\psi)\sin^3\psi d\psi\right)}{\left(\int_0^{\pi} g(\psi)\sin 2\psi d\psi\right)^2 \left(\int_0^{\pi} g_1(\psi)\sin\psi d\psi\right)} \\
&= \frac{\left(\int_0^{\pi} g(\psi)\cos\psi d\psi\right)^2 \left(\int_0^{\pi} g(\psi)\sin^3\psi\cos\psi d\psi\right)}{\left(\int_0^{\pi} g(\psi)\sin 2\psi d\psi\right)^2 \left(\int_0^{\pi} g(\psi)\sin\psi\cos\psi d\psi\right)} \\
&= \frac{\left(\int_0^1 h(r)dr\right)^2 \left(\int_0^1 h(r)r^3 dr\right)}{4\left(\int_0^1 h(r)r dr\right)^3},
\end{aligned}$$

where $r = \sin\psi$ and $h(r) = g(\arcsin r)$. By Theorem 4.2 $ARE_g(V_1^\Omega, t^\Omega) \geq 1/2$.

In the general case, let $g_1(\psi) = g(\psi) - g(\pi - \psi)$ for $0 \leq \psi \leq \pi/2$ and 0 elsewhere. Then $ARE_{g_1}(V_1^\Omega, t^\Omega) \geq 1/2$. Then

$$\begin{aligned} ARE_g(V_1^\Omega, t^\Omega) &= \frac{\left(\int_0^\pi g(\psi)\cos\psi d\psi\right)^2 \left(\int_0^\pi g(\psi)\sin^3\psi d\psi\right)}{\left(\int_0^\pi g(\psi)\sin 2\psi d\psi\right)^2 \left(\int_0^\pi g(\psi)\sin\psi d\psi\right)} \\ &= \frac{\left(\int_0^\pi g_1(\psi)\cos\psi d\psi\right)^2 \left(\int_0^\pi g(\psi)\sin^3\psi d\psi\right)}{\left(\int_0^\pi g_1(\psi)\sin 2\psi d\psi\right)^2 \left(\int_0^\pi g(\psi)\sin\psi d\psi\right)} \\ &\geq \frac{\left(\int_0^\pi g_1(\psi)\cos\psi d\psi\right)^2 \left(\int_0^\pi g_1(\psi)\sin^3\psi d\psi\right)}{\left(\int_0^\pi g_1(\psi)\sin 2\psi d\psi\right)^2 \left(\int_0^\pi g_1(\psi)\sin\psi d\psi\right)} \\ &= ARE_{g_1}(V_1^\Omega, t^\Omega) \geq 1/2, \end{aligned}$$

where the inequality in the third line above follows from Lemma 5.4. □

References

[1] P. Bickel and E. L. Lehmann, *Descriptive statistics for nonparametric models, II: Location*, Ann. Statist. **3** (1975), 1045–1069.

[2] B. M. Brown, *Statistical uses of the spatial median*, J. R. Statist. Soc. **B45** (1983), 25–30.

[3] P. Capéraà, *Tail ordering and asymptotic efficiency of rank tests*, Ann. Statist. **16** (1988), 470–478.

[4] T. Chang and D. Ko, *M-estimates of rigid body motion on the sphere and in Euclidean space*, Ann. Statist. **23** (1995), 1823–1847.

[5] G. R. Ducharme and P. Milasevic, *Spatial median and directional data*, Biometrika **74** (1987), 212–215.

[6] N. I. Fisher, *Spherical medians.* J. R. Statist. Soc. **B47** (1985), 342–348.

[7] N. I. Fisher, T. Lewis, and B. J. J. Embleton, *Statistical Analysis of Spherical Data*, Cambridge University Press, Cambridge, 1987.

[8] O. Hössjer and C. Croux, *Generalized univariate signed rank statistics for testing and estimating a multivariate location parameter*, Nonparametric Statistics **4** (1995), 293–308.

[9] T. P. Hettmansperger and J. W. McKean, *Robust Nonparametric Statistical Methods*, Arnold, London, 1998.

[10] D. Ko and T. Chang, *Robust M-Estimators on Spheres*, J. Mult. Anal. **45** (1993), 104–136.

[11] T. Neeman, *Rank Statistics for Spherical Data*, University of Virginia dissertation, 1995.

[12] G. S. Watson, *Statistics on Spheres*, John Wiley & Sons, New York, 1983.

Australian Bureau of Agricultural and Resource Economics, GPO Box 1563, Canberra, ACT 2601

E-mail address: `Teresa.Neeman@abare.gov.au`

Department of Statistics, University of Virginia, Charlottesville, VA 22904-4135

E-mail address: `tcc8v@pitman.stat.virginia.edu`

Contemporary Mathematics
Volume **287**, 2001

Graphical Model Search Via Essential Graphs

Michael D. Perlman

ABSTRACT. Searching the space of graphical Markov models determined by acyclic digraphs (ADG $\equiv$ DAG) is facilitated by the use of essential graphs, each of which uniquely represents a Markov equivalence class of ADGs. Existing Bayesian and non-Bayesian model search procedures attempt to move through the space by altering at most two edges from the current essential graph at each step, but may require many attempts before actually reaching another essential graph. This paper provides the theoretical foundation of a new method for searching the space of essential graphs, wherein the characterizing properties of essential graphs are used to ensure greater likelihood of success in actually reaching another essential graph at each step.

1. Introduction

Graphical Markov models (GMMs) use graphs to succinctly specify dependences in multivariate distributions. GMMs determined by acyclic digraphs (ADG) (also called directed acyclic graphs $\equiv$ DAG) admit especially simple statistical analysis. The joint probability density admits a recursive factorization that allows efficient updating algorithms for Bayesian analysis (see [**LDLL**], [**L**], [**SL**]), and, in expert system applications, simple causal interpretations ([**LS**], [**SDLC**]). Indeed, the vibrant UAI (uncertainty in artificial intelligence) community has concentrated much of its effort on ADG models $\equiv$ Bayesian networks.

In the Bayesian setting, experts specify a prior distribution on competing ADG models, then these prior distributions combine with likelihoods (typically integrated over parameters) to give posterior model probabilities. Model selection algorithms then seek out the ADG models with highest posterior probability, and inference proceeds conditionally on these selected models ([**CH**], [**B**], [**HGC**], [**MR**], [**MY**]). Non-Bayesian model selection methods are similar, replacing posterior model probabilities by, for example, penalized maximum likelihoods ([**C95**]).

[**HGC**] highlighted a fundamental problem with this general approach: the non-uniqueness of the graph associated with the model. Unlike undirected graphs, two or more ADGs may determine the same GMM, i.e., may determine the same set of conditional independence (CI) restrictions among a given set of random variates.

2000 *Mathematics Subject Classification.* Primary 62M45, 60K99; Secondary 68R10, 68T30.

Key words and phrases. Graphical Markov model, acyclic digraph, directed acyclic graph, Markov equivalence, essential graph, model search.

This research was supported in part by the National Science Foundation.

$b \to a \to c \qquad b \leftarrow a \to c \qquad b \leftarrow a \leftarrow c \qquad b \to a \leftarrow c$

FIGURE 1. Four acyclic digraphs with the vertex set $V \equiv \{a, b, c\}$.

For the simplest example, the two ADGs $a \to b$ and $a \leftarrow b$ represent the same (unrestricted $\equiv$ saturated) bivariate model.

[**P**] and [**VP90**], [**VP92**] showed that two ADGs are Markov equivalent iff they have the same *skeleton* ($\equiv$ underlying UG) and same *immoral* configurations: $a \to c \leftarrow b$, with a and b non-adjacent (also see [**AMP97**]). The collection of all possible ADGs for a given collection of variates coalesces into classes of Markov-equivalent ADGs, where all ADGs within a Markov-equivalence class determine the same statistical model. For example, the first three graphs in Figure 1 comprise one equivalence class – each specifies the single CI $X_b \perp\!\!\!\perp X_c | X_a$, abbreviated as $b \perp\!\!\!\perp c|a$ – while the fourth graph specifies the independence $b \perp\!\!\!\perp c$ and is the sole member of another equivalence class.

Model selection algorithms that ignore Markov equivalence face several difficulties:

- Repeating analyses for equivalent ADGs leads to significant computational inefficiencies: in some cases, the Markov equivalence class $[D]$ containing a given ADG D may be superexponentially large.
- Ensuring that equivalent ADGs have equal posterior probabilities severely constrains prior distributions over the space of ADGs – cf. [**HGC**], [**MAPV**].
- Weighting individual ADGs in Bayesian model averaging procedures to achieve specified weights for all Markov-equivalence classes is impractical without an explicit representation of these classes.

Treating each Markov-equivalence class as a single model would overcome these difficulties. In principle, one can conduct model selection and model averaging by searching over the space of Markov equivalence classes rather than over the larger space of individual ADGs. Of course, this requires a tractable representation of the space of Markov-equivalence classes. *Essential graphs,* reviewed in Section 2, provide this representation. In Section 3 we show how their characterizing properties may be used to devise efficient search procedures.

2. The essential graph for an ADG model

The reader is referred to [**L**], [**AMP97**], [**AMP01**], or [**AMPT**] for the graph-theoretic terminology used in this paper. We use standard abbreviations such as *pa* for *parents*, *ch* for *children*, *nb* for *neighbors*, *cl* for *closure*, *de* for *descendants*, and *De(A)* for *descendants*$(A) \cup A$.

A *chain graph* $\equiv$ *adicyclic graph* $G \equiv (V, E)$ is a graph with both undirected and directed edges (indicated by lines and arrows, respectively) and no semi-directed cycles. [A semi-directed cycle has at least one arrow and no opposing arrows.] If all arrows of G are removed, the *chain components* of G are the connected components of the resulting undirected graph. The family of chain components is denoted by $\mathcal{T} \equiv \mathcal{T}(G)$. These chain components comprise the vertex of the ADG $\mathcal{D} \equiv \mathcal{D}(G)$, where, for any $\sigma, \tau \in \mathcal{T}$, $\sigma \to \tau$ in $\mathcal{D}$ iff $s \to t$ for at least one pair $s \in \sigma$, $t \in \tau$.

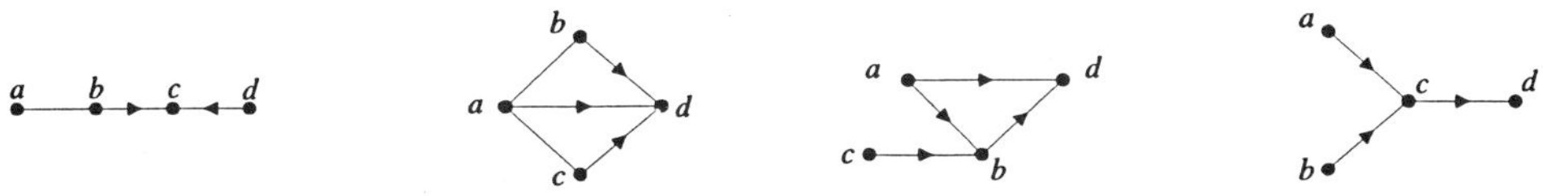

FIGURE 2. Four essential graphs D^* for 4 variables.

[**AMP97**] showed that the Markov equivalence class $[D]$ containing any ADG D can be uniquely represented by a certain Markov-equivalent chain graph

$$D^* := \cup\{D' | D' \in [D]\}, \tag{2.1}$$

the *essential graph* associated with $[D]$. (The essential graph was called the *completed pattern* in [**VP90**].) Here, the *union* $\cup$ of ADGs with the same skeleton is again a graph with the same skeleton, but any directed edge ($\equiv$ arrow) that occurs with opposite orientations in two ADGs is replaced by an undirected edge ($\equiv$ line) in the union. Thus *an arrow occurs in the essential graph D^* iff it occurs with the same orientation in every $D' \in [D]$*. Clearly, every arrow involved in an immorality in D is essential, but other arrows may be essential as well, for example, the arrow $a \to d$ in the second graph in Figure 2. ([**PV**], [**SGS**], and [**C95**] note that, under certain additional assumptions, the *essential arrows* of an ADG D, i.e. those directed edges of D that occur in D^*, may indicate causal influences.)

Four examples of essential graphs D^* for 4 variables appear in Figure 2. In the first example, D can be taken to be the ADG $a \to b \to c \leftarrow d$; in the second example, D can be taken to be the ADG with arrows $a \to b$ and $a \to c$. In the third and fourth examples, $D = D^*$. [**AMP97**] exhibit all 185 essential graphs for 4 variables, compared to 543 ADGs; for 5 variables these counts are 8,782 and 29,281, respectively. A computer study by [**GP**] for up to ten variables suggests an asymptotic value of approximately 0.27 for the ratio of essential graphs to ADGs.

Because $[D]$ can be superexponentially large (a complete graph admits $n!$ acyclic orderings), direct calculation of D^* via (2.1) is computationally infeasible. To overcome this obstacle, [**AMP97**], Theorem 4.1, gave the following explicit characterization of essential graphs:

THEOREM 2.1. *A graph $G \equiv (V, E)$ is the essential graph D^* for some ADG D with vertex set V iff G satisfies the following four conditions:*

(1) *G is a chain graph.*
(2) *For each chain component $\tau \in \mathcal{T}(G)$, the UG G_τ is chordal.*
(3) *G has no flags, i.e., no induced subgraphs of the form $a \to b — c$.*
(4) *Each arrow $a \to b$ in G is "strongly protected", that is, occurs in at least one of the following four configurations as an induced subgraph of G:*

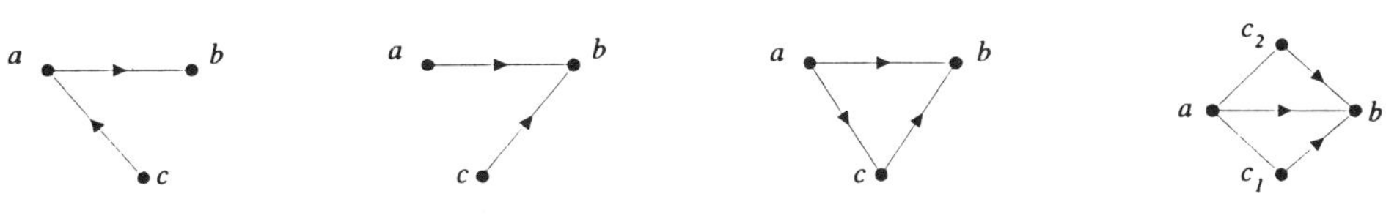

(a) (b) (c) (d)

[**AMP97**] applied Theorem 2.1 to obtain a polynomial-time algorithm for constructing D^* from D and to establish the irreducibility of certain Markov chains used for Monte Carlo search procedures over the space of essential graphs – see Proposition 3.1. Efficient algorithms for determining D^* appear in [**C95**] and [**M**].

3. ADG model search via essential graphs

Procedures for ADG model selection and ADG model averaging based on essential graphs were implemented by [**C96**] and [**MAPV**]; also see [**DF**] and [**GG**]. The approach of [**MAPV**] is based on the following proposition ([**AMP97**], Proposition 4.5)):

PROPOSITION 3.1. *Suppose that $G, H \in \mathcal{E} \equiv \mathcal{E}(V)$, the set of all essential graphs with vertex set V. Then there exists a sequence $G \equiv G_1, \ldots, G_n \equiv H$ in $\mathcal{E}$ such that the members of each consecutive pair G_i, G_{i+1} differ by either:*

(i) *adding or deleting one line;*
(ii) *adding or deleting one arrow; or*
(iii) *adding or deleting one immorality. (Other immoralities may be affected.)*

The "moves" in $\mathcal{E}$ allowed by the search procedure of [**MAPV**] consist of all possible edge additions and deletions described in (i), (ii), (iii). After each such move, the new graph is tested to see that it satisfies conditions (1), (2), (3), and (4) and therefore remains in $\mathcal{E}$. Proposition 3.1 implies that any Markov chain with state space $\mathcal{E}$ that assigns positive probabilities to all moves of these six types must be irreducible, a necessary condition for the application of Markov Chain Monte Carlo procedures (see Remark 3.5).

The search procedure of [**MAPV**] is inefficient, however, because most of its moves produce a new graph for which one or more of (1), (2), (3), (4) fails. Greater efficiency might be achieved by more fully exploiting the characterization of essential graphs given in Theorem 2.1. The main purpose of the present paper is to describe a new class of search procedures that are more efficient because their moves are more likely to remain in $\mathcal{E}$ – this is summarized in Theorem 3.4 and Remark 3.5.

Most notably, property (3) implies that if $a \to b \in G \in \mathcal{E}$, then $a \to b' \in G\ \forall b' \in \tau(b)$ (the unique chain component containing b). Thus if the arrow $a \to b$ is added to (removed from) $G \in \mathcal{E}$, then all arrows $a \to b'$, $b' \in \tau(b)$, must also be added to (removed from) G in order to produce a new graph that is again in $\mathcal{E}$. Furthermore, if G is any graph that satisfies (1) and (3), then it is straightforward to show that

(3.1) $a \to b$ is strongly protected in G via (a) (resp. (b), (c), (d))
$\Rightarrow a \to b'$ is strongly protected in G via (a) (resp. (b), (c), (d)) $\forall b' \in \tau(b)$.

These considerations motivate the following definition:

DEFINITION 3.2. For each pair $a \in V$, $\tau \in \mathcal{T}(G)$ such that $a \notin \tau$, the *arrow bundle* $a \to \tau$ is the collection of arrows $\{a \to t \mid t \in \tau\}$.

Thus, if $a \to t \in G$ for one $t \in \tau$, then $a \to t \in G$ for all $t \in \tau$. Therefore it makes sense to say that either $a \to \tau \in G$ or $a \to \tau \notin G$ and, by (3.1), that

(3.2) *either:* $a \to \tau$ is strongly protected in G via (a) (resp. (b), (c), (d));
or: $a \to \tau$ is not strongly protected in G via (a) (resp. (b), (c), (d)).

To verify that an arrow bundle $a \to \tau$ is strongly protected in G, a local search involving the parents and neighbors of a and the parents of τ in $\mathcal{D}$ suffices:

$a \to \tau$ is strongly protected in G via (a), (b), or (c) iff

$$\mathrm{pa}_G(a) \neq \mathrm{pa}_G(\tau) \setminus (\{a\} \cup \mathrm{nb}_G(a)); \tag{3.3}$$

$a \to \tau$ is strongly protected in G via (d) iff

$$\mathrm{pa}_G(\tau) \cap \mathrm{nb}_G(a) \text{ is not complete in } G. \tag{3.4}$$

The allowable moves in our new class of search procedures are now described. Let G denote the essential graph at the current state of the search, and let G' denote the graph obtained from G after altering one or more edges or immoralities according to one of the moves described in A, B, and C below. All quantities below are defined relative to G, e.g., $\mathcal{T} \equiv \mathcal{T}(G)$, $\mathcal{D} \equiv \mathcal{D}(G)$.

A. Select $\tau \in \mathcal{T}$ and two distinct points $s, t \in \tau$.

(Aα) If s—$t \in G_\tau$, *delete* s—t. Then G' satisfies (1), (3), and (4). [Note that (4d) in G may become (4b) in G'.] Also, G' will satisfy (2) iff G'_τ is decomposable iff $\mathrm{nb}_{G_\tau}(s) \cap \mathrm{nb}_{G_\tau}(t)$ is complete. (Also see Theorem 1 of [**GG**].) Verification of this condition requires only a *local* search among the neighbors of s and t in G. [G'_τ may remain connected, or may now comprise exactly two connected components of G'.]

(Aβ) If s—$t \notin G_\tau$, *add* s—t. Then G' satisfies (1) and (3). G' will satisfy (2) *unless* there exists a chordless undirected path with three or more edges between s and t in G_τ. (Also see Theorem 2 of [**GG**].) The following paragraph facilitates the verification of (4):

An arrow bundle that is strongly protected in G via (a) (resp. (c)) will remain strongly protected in G' via (a) (resp. (c)). If $\mathrm{ch}_G(s) \cap \mathrm{ch}_G(t) = \emptyset$, an arrow bundle that is strongly protected in G via (b) (resp., (d)) will remain strongly protected in G' via (b) (resp., (d)). If $\mathrm{ch}_G(s) \cap \mathrm{ch}_G(t) \neq \emptyset$, strong protection in G' of each arrow bundle $a \to \tilde{\tau}$ with $a \in \{s,t\} \cup (\mathrm{nb}_{G_\tau}(s) \cap \mathrm{nb}_{G_\tau}(t))$ and $\tilde{\tau} \subseteq \mathrm{ch}_G(s) \cap \mathrm{ch}_G(t)$ must be verified directly. [By applying (3.3) and (3.4) with G replaced by G', it can be seen that this verification can be accomplished without updating G by means of a local search involving the parents and neighbors of a in G, the children of a in $\mathcal{D}$, and the parents of $\tilde{\tau}$ in $\mathcal{D}$.]

B. Select distinct $\sigma, \tau \in \mathcal{T}$. Next, determine whether σ and τ are comparable (one is an ancestor of the other) in $\mathcal{D}$ or non-comparable in $\mathcal{D}$.

First assume that σ and τ are comparable; without loss of generality assume that $\sigma \leq_{\mathcal{D}} \tau$, which is equivalent to

$$\mathrm{De}_{\mathcal{D}}(\sigma) \cap \mathrm{pa}_{\mathcal{D}}(\tau) \neq \emptyset, \tag{3.5}$$

and which implies that $\tau \not\leq_{\mathcal{D}} \sigma$. Select $s \in \sigma$.

(Bα) If $s \to \tau \in G$, *delete* $s \to \tau$. Then G' satisfies (1), (2), and (3), but (4) must be verified. The following paragraph facilitates this verification:

If $\mathrm{ch}_{\mathcal{D}}(\tau) \setminus \mathrm{ch}_G(s) = \emptyset$, an arrow bundle that is strongly protected in G via (a) will remain strongly protected in G' via (a). If $\mathrm{ch}_{\mathcal{D}}(\tau) \setminus \mathrm{ch}_G(s) \neq \emptyset$, strong protection in G' of each arrow bundle $a \to \tilde{\tau}$ with $a \in \tau$ and $\tilde{\tau} \in \mathrm{ch}_{\mathcal{D}}(\tau) \setminus \mathrm{ch}_G(s)$ must be verified directly. If $\mathrm{pa}_G(\tau) \setminus (\{s\} \cup \mathrm{ch}_G(s)) = \emptyset$, an arrow bundle that is strongly protected in G via (b) (resp., (c)) will remain strongly protected in G' via (b) (resp., (b) or (c)). If $\mathrm{pa}_G(\tau) \setminus (\{s\} \cup \mathrm{ch}_G(s)) \neq \emptyset$, strong protection in G' of each arrow bundle

$a \to \tau$ with $a \in \mathrm{pa}_G(\tau) \setminus (\{s\} \cup \mathrm{ch}_G(s))$ must be verified directly. If

$$(3.6) \qquad \mathrm{nb}_{G_\sigma}(s) \cap \mathrm{pa}_G(\tau) \cap \mathrm{nb}_{G_\sigma}(\mathrm{pa}_G(\tau) \setminus \mathrm{cl}_{G_\sigma(s)})$$

$= \emptyset$, an arrow bundle that is strongly protected in G via (d) will remain strongly protected in G' via (d). If (3.6) $\neq \emptyset$, strong protection in G' of each arrow bundle $a \to \tau$ with $a \in$ (3.6) must be verified directly. [Again, local search suffices for the direct verifications.]

(Bβ) If $s \to \tau \notin G$, *add* $s \to \tau$. Then G' satisfies (1), (2), and (3), but (4) must be verified. The following paragraph facilitates this verification:

By (3.5), $s \to \tau$ is strongly protected in G' via (a), (b), (c), or (d) (see Remark 3.3 below). Any other arrow bundle that is strongly protected in G via (c) (resp., (d)) will remain strongly protected in G' via (c) (resp., (d)). If $\mathrm{ch}_G(s) \cap \mathrm{pa}_G(\tau) = \emptyset$, an arrow bundle that is strongly protected in G via (a) will remain strongly protected in G' via (a). If $\mathrm{ch}_G(s) \cap \mathrm{pa}_G(\tau) \neq \emptyset$, strong protection in G' of each arrow bundle $a \to \tau$ with $a \in \mathrm{ch}_G(s) \cap \mathrm{pa}_G(\tau)$ must be verified directly. If $\mathrm{ch}_G(s) \cap \mathrm{ch}_{\mathcal{D}}(\tau) = \emptyset$, an arrow bundle that is strongly protected in G via (b) will remain strongly protected in G' via either (b) or (c). If $\mathrm{ch}_G(s) \cap \mathrm{ch}_{\mathcal{D}}(\tau) \neq \emptyset$, strong protection in G' of each arrow bundle $a \to \tilde{\tau}$ with $a \in \tau$ and $\tilde{\tau} \in \mathrm{ch}_G(s)$ must be verified directly. [Again, local search suffices for the direct verifications.]

Next assume that σ and τ are non-comparable in $\mathcal{D}$. Select $s \in \sigma$.

(Bγ) If $\mathrm{pa}_G(\sigma) \neq \mathrm{pa}_G(\tau)$, *add* $s \to \tau$. Then G' satisfies (1), (2), and (3), but (4) must be verified. The following paragraph facilitates this verification:

Since $\mathrm{pa}_G(\sigma) \neq \mathrm{pa}_G(\tau)$, $s \to \tau$ is strongly protected in G' via (a) or (b). Any other arrow bundle that is strongly protected in G via (a) (resp., (c), (d)) will remain strongly protected in G' via (a) (resp., (c), (d)). If $\mathrm{ch}_G(s) \cap \mathrm{ch}_{\mathcal{D}}(\tau) = \emptyset$, an arrow bundle that is strongly protected in G via (b) will remain strongly protected in G' via either (b) or (c). If $\mathrm{ch}_G(s) \cap \mathrm{ch}_{\mathcal{D}}(\tau) \neq \emptyset$, strong protection in G' of each arrow bundle $a \to \tilde{\tau}$ with $a \in \tau$ and $\tilde{\tau} \in \mathrm{ch}_G(s)$ must be verified directly. [Again, a local search suffices for this direct verification.]

(Bδ) If $\mathrm{pa}_G(\sigma) = \mathrm{pa}_G(\tau)$, select $t \in \tau$ and *add* s—t. This merges the two chain components σ and τ of G into a single chain component $\sigma \cup \tau$ of G'. Then G' satisfies (1), (2), and (3), the latter because $\mathrm{pa}_G(\sigma) = \mathrm{pa}_G(\tau)$, but (4) must be verified. The following facilitates this verification:

An arrow bundle that is strongly protected in G via (a) (resp., (c), (d)) will remain strongly protected in G' via (a) (resp., (c), (d)). If $\mathrm{ch}_G(s) \cap \mathrm{ch}_G(t) = \emptyset$, an arrow bundle that is strongly protected in G via (b) will remain strongly protected in G' via (b). If $\mathrm{ch}_G(s) \cap \mathrm{ch}_G(t) \neq \emptyset$, strong protection in G' of each arrow bundle $a \to \tilde{\tau}$ with $a \in \{s, t\}$ and $\tilde{\tau} \subseteq \mathrm{ch}_G(s) \cap \mathrm{ch}_G(t)$ must be verified directly. [Again, local search suffices.]

C. Select distinct $\rho, \sigma, \tau \in \mathcal{T}$ such that τ is maximal $\equiv$ terminal in $\mathcal{D}$, $|\rho| = |\sigma| = 1$, ρ and σ are non-adjacent in $\mathcal{D}$, and the arrow bundles $r \to \tau$ and $s \to \tau$ are either both present or both absent in G, where $\rho = \{r\}$ and $\sigma = \{s\}$. Thus, either $r \to \tau \leftarrow s$ occurs in G or r, s, τ are mutually non-adjacent in G.

(Cα) If $r \to \tau \leftarrow s$ occurs in G, *delete both* $r \to \tau$ *and* $s \to \tau$. Then G' satisfies (1), (2), and (3), but (4) must be verified. The following facilitates this verification:

By the maximality of τ (resp., the restrictions $|\rho| = |\sigma| = 1$), an arrow bundle that is strongly protected in G via (a) (resp., (d)) will remain strongly protected in G' via (a) (resp., (d)). If $\mathrm{pa}_G(\tau) \setminus (\{r, s\} \cup \mathrm{ch}_G(r) \cup \mathrm{ch}_G(s)) = \emptyset$, an arrow bundle that is strongly protected in G via (b) (resp., (c)) will remain strongly protected in G' via (a) or (b) (resp., (a) or (c)). If $\mathrm{pa}_G(\tau) \setminus (\{r, s\} \cup \mathrm{ch}_G(r) \cup \mathrm{ch}_G(s)) \neq \emptyset$, strong protection in G' of each arrow bundle $a \to \tau$ with $a \in \mathrm{pa}_G(\tau) \setminus (\{r, s\} \cup \mathrm{ch}_G(r) \cup \mathrm{ch}_G(s))$ must be verified directly. [Again, local search suffices.]

(Cβ) If r, s, τ are mutually non-adjacent in G, *add both* $r \to \tau$ *and* $s \to \tau$. Then G' satisfies (1) (by the maximality of τ), (2), and (3), but (4) must be verified. The following facilitates this verification:

By the maximality of τ, an arrow bundle that is strongly protected in G via (b) (resp., (c), (d)) will remain strongly protected in G' via (b) (resp., (c), (d)). If $(\mathrm{ch}_G(r) \cup \mathrm{ch}_G(s)) \cap \mathrm{pa}_G(\tau) = \emptyset$, an arrow bundle that is strongly protected in G via (a) will remain strongly protected in G' via (a). If $(\mathrm{ch}_G(r) \cup \mathrm{ch}_G(s)) \cap \mathrm{pa}_G(\tau) \neq \emptyset$, strong protection in G' of each arrow bundle $a \to \tau$ with $a \in (\mathrm{ch}_G(r) \cup \mathrm{ch}_G(s)) \cap \mathrm{pa}_G(\tau)$ must be verified directly. [Again, local search suffices.]

REMARK 3.3. In (Bβ) it was assumed that $s \to \tau \notin G$ and $s \to \tau$ was *added*, producing G'. It was then asserted that, by (3.5), $s \to \tau$ is strongly protected in G' via (a), (b), (c), or (d). The proof is given here.

By (3.5), either $\mathrm{de}_{\mathcal{D}}(\sigma) \cap \mathrm{pa}_{\mathcal{D}}(\tau) \neq \emptyset$ or $\sigma \cap \mathrm{pa}_{\mathcal{D}}(\tau) \neq \emptyset$. If the former, choose $\rho \in \mathrm{de}_{\mathcal{D}}(\sigma) \cap \mathrm{pa}_{\mathcal{D}}(\tau)$ and select $r \in \rho \cap \mathrm{pa}_G(\tau)$. Because $s \leftarrow r \notin G$ (recall that $s \in \sigma$) and therefore $s \leftarrow r \notin G'$, either $s \to r \in G, G'$, in which case $s \to \tau$ is strongly protected in G' via (c), or else $s \cdot\!\!/\!\!\cdot r$ in G, G', in which case $s \to \tau$ is strongly protected in G' via (b).

Next, suppose that $\sigma \cap \mathrm{pa}_{\mathcal{D}}(\tau) \neq \emptyset$ and choose $s' \in \sigma \cap \mathrm{pa}_{\mathcal{D}}(\tau)$, so $s' \neq s$ and $s' \to \tau \in G, G'$. Either $s \cdot\!\!/\!\!\cdot s'$ in G_σ, in which case $s \to \tau$ is strongly protected in G' via (b), or else $s - s' \in G_\sigma, G'_\sigma$. Assume that the latter holds and select any $t \in \tau$.

Because $G \in \mathcal{E}$, $s' \to t$ is strongly protected in G via (a), (b), (c), or (d). If (a), then $c \to s' \to t$ occurs as an induced subgraph in G for some $c \in V$, so by (3) $c \to s \to t$ occurs as an induced subgraph in G', hence $s \to t$ is strongly protected in G' via (a).

If $s' \to t$ is strongly protected in G via (b), then the immorality $s' \to t \leftarrow c$ occurs in G, G' for some $c \neq s \in V$, hence $s' \cdot\!\!/\!\!\cdot c$ in G, G' and, by (3), $s \leftarrow c \notin G, G'$. If $s \to c \in G, G'$, then $s \to t$ is strongly protected in G' via (c); if $s \cdot\!\!/\!\!\cdot c$ in G, G', then $s \to t$ is strongly protected in G' via (b); if$s - c \in G_\sigma, G'_\sigma$, then $s \to t$ is strongly protected in G' via (d) (take $c_1 = s', c_2 = c$).

If $s' \to t$ is strongly protected in G via (c), then both $s' \to c \to t$ and $s' \to t$ occur in G, G' for some $c \notin \sigma, \tau$. By (1), $s \leftarrow c \notin G, G'$. Thus either $s \to c \in G, G'$, whence $s \to t$ is strongly protected in G' via (c), or else $s \cdot\!\!/\!\!\cdot c$ in G, G', whence $s \to t$ is strongly protected in G' via (b).

If $s' \to t$ is strongly protected in G via (d), then configuration (d) occurs as an induced subgraph of G, G' with $a = s'$, $b = t$, and $c_1, c_2 \in \sigma \setminus \{s, s'\}$. If $s \cdot\!\!/\!\!\cdot c_i$ in G_σ, G'_σ for $i = 1$ or 2, then the immorality $s \to t \leftarrow c_i$ occurs in G' so $s \to t$ is strongly protected in G' via (b). If $s - c_1$ and $s - c_2$ in G_σ, G'_σ, then configuration (d) occurs as an induced subgraph of G' with $a = s$, $b = t$, and the given c_1, c_2, hence $s \to t$ is strongly protected in G' via (d). This establishes the desired assertion.

THEOREM 3.4. *Let G, H be two essential graphs with common vertex set V. Then there exists a sequence $G \equiv G_1, \ldots, G_n \equiv H$ in $\mathcal{E}$ such that each consecutive pair G_i, G_{i+1} differs by exactly one of the "moves" described in (Aα), (Aβ), (Bα), (Bβ), (Bγ), (Bδ), (Cα), (Cβ) above.*

PROOF. The proof of Proposition 3.1 (see Section 4) implies that there exists a sequence $G \equiv G_1, \ldots, G_n \equiv H$ in $\mathcal{E}$ such that for each pair G_i, G_{i+1}, either:

(i) G_{i+1} is obtained from G_i by the addition or deletion of one line; or

(iv) G_i is an ADG and G_{i+1} is obtained from G_i by the addition or deletion of one arrow (hence G_{i+1} is also an ADG); or

(v) G_i is an ADG and G_{i+1} is obtained from G_i by the addition or deletion of one immorality $r \to t \leftarrow s$, where t is terminal in G_i (hence G_{i+1} is also an ADG).

But (i) is identical to (Aα), (Aβ), and (Bδ) combined, (iv) is a special case of (Bα), (Bβ), and (Bγ) combined, and (v) is a special case of (Cα) and (Cβ) combined, so the desired result is established. □

REMARK 3.5. [**MAPV**] describe stochastic search procedures for model selection and model averaging, based on the Metropolis-Hastings algorithm, that act directly on essential graphs rather than ADGs. Such procedures move through the space of essential graphs according to a Markov chain whose transition probabilities are chosen to achieve a desired stationary distribution. Convergence to the stationary distribution requires that the Markov chain be irreducible and aperiodic. By Theorem 3.4, irreducibility will hold whenever the chain assigns positive probability to each available move of the types (Aα), (Aβ), (Bα), (Bβ), (Bγ), (Bδ), (Cα), and (Cβ). [Some of these moves may not be available: for example, (Aα) is unavailable if the current essential graph has no lines ($\equiv$ undirected edges).] Aperiodicity can be guaranteed, for example, by ensuring that the chain has positive probability of remaining in its current state. Implementation of this new class of search procedures is currently under study.

4. Proof of Proposition 3.1

The proof of Proposition 4.5 in [**AMP97**] ($\equiv$ Proposition 3.1 here) is incomplete. A complete proof is now presented. (The terminology and notation of [**AMP97**] is maintained here.)

It suffices to establish the result when H has no edges, i.e., $H = (V, \emptyset)$. First assume that G contains at least one line ($\equiv$ undirected edge), so $G \equiv G_1$ has at least one chain component τ with at least two vertices. Since G_τ is chordal, it has at least one simplicial vertex a (cf. [**BP**], Lemma 2.2); since G_τ is connected, $\mathrm{nb}_{G_\tau}(a) \neq \emptyset$. Choose any $b \in \mathrm{nb}_{G_\tau}(a)$, so that $a — b \in G_\tau$, then remove this line to produce a graph G_2. Because a was simplicial in G_τ, $(G_2)_\tau$ is also chordal. Because G_1 is an essential graph, it is now straightforward to verify that G_2 satisfies the conditions of Theorem 2.1, hence G_2 is also an essential graph. Continue this process of single line removal until reaching an essential graph G_j with no lines. (A related argument appears in Lemma 5 of [**FL**].)

If G_j has no arrows (that is, no directed edges), then $G_j = H$ and we are done. Otherwise, we can reach $H \equiv (V, \emptyset)$ by removing arrows from the ADG G_j according to (iv) or (v) as follows. Let $B (\neq \emptyset)$ be the set of all *terminal* vertices of G_j, that is, the set of all $b \in V$ such that b is maximal in V with respect to the partial ordering $\preceq$ determined by the ADG G_j. Since G_j has at least one arrow, there must exist at least one $b \in B$ such that $A := \mathrm{pa}_G(b) \neq \emptyset$. We shall show that we can remove either one arrow $a \to b$, $a \in A$, or one immorality $a_1 \to b \leftarrow a_2$, $a_1, a_2 \in A$, such that the resulting digraph $G_{j+1} \in \mathcal{E}$, i.e., is an essential graph.

Assume, for the moment, that such a removal has occurred. By Corollary 4.2 of [**AMP97**], a digraph G is an essential graph iff it is an ADG and each arrow is *protected*, that is, occurs in at least one of the three configurations (a), (b), (c) as an induced subgraph of G. Clearly G_{j+1} is an ADG. Because each arrow of G_j is protected in G_j, if an arrow of G_{j+1} is unprotected in G_{j+1} then, by the maximality of b in G_j, this arrow must be of the form $a^* \to b$ for some $a^* \in A$.

We now show that the desired removal is possible:

Step 1. Define $A_1 := A$ and choose a_1 to be minimal in A_1 with respect to the partial ordering $\preceq$. Then A_1 can be expressed as the disjoint union

$$A \equiv A_1 = \{a_1\} \dot\cup A_2 \dot\cup A_1',$$

where $A_2 := A_1 \cap \mathrm{ch}_{G_j}(a_1)$ and $A_1' := A_1 \setminus (\{a_1\} \dot\cup A_2)$. By the minimality of a_1, $a_1 \cdot\!\!/\!\!\cdot a'$ in G_j $\forall a' \in A_1'$.

1.1. If $A_1' = \emptyset$, remove $a_1 \to b$. Then any remaining arrow of the form $a^* \to b$, $a^* \in A$, must have $a^* \in A_2$, hence occurs in configuration (a) in G_{j+1} with $a = a^*$ and $c = a_1$, hence is protected in G_{j+1}.

1.2. If $|A_1'| = 1$ and $A_2 = \emptyset$, remove the immorality $a_1 \to b \leftarrow a'$, where $A_1' = \{a'\}$. Here no arrows of the form $a^* \to b$, $a^* \in A$, remain in G_{j+1}.

1.3. If $|A_1'| \geq 2$ and $A_2 = \emptyset$, remove any one arrow of the form $a' \to b$, $a' \in A_1'$. Then all remaining arrows of the form $a^* \to b$, $a^* \in A$, are protected in G_{j+1}: if $a^* = a_1$ then $a^* \to b$ occurs in configuration (b) in G_{j+1} with $a = a^*$ and $c = a''$ for any $a'' \in A_1' \setminus \{a'\}$; if $a^* = a'' \in A_1' \setminus \{a'\}$ then $a^* \to b$ occurs in configuration (b) in G_{j+1} with $a = a^*$ and $c = a_1$.

Step 2. If $A_1' \neq \emptyset$ and $A_2 \neq \emptyset$, choose a_2 to be minimal in A_2 with respect to the partial ordering $\preceq$. Then A_2 can be expressed as the disjoint union $A_2 = \{a_2\} \dot\cup A_3 \dot\cup A_2'$, where $A_3 := A_2 \cap \mathrm{ch}_{G_j}(a_2)$ and $A_2' := A_2 \setminus (\{a_2\} \dot\cup A_3)$. By the minimality of a_2, $a_2 \cdot\!\!/\!\!\cdot a'$ in G_j $\forall a \in A_2'$. Note that

$$A = \{a_1\} \dot\cup \{a_2\} \dot\cup A_3 \dot\cup A_2' \dot\cup A_1'.$$

2.1. If $A_2' = \emptyset$, remove $a_2 \to b$. Then any remaining arrow of the form $a^* \to b$, $a^* \in A$, is protected in G_{j+1}: if $a^* \in A_3$ then $a^* \to b$ occurs in configuration (a) in G_{j+1} with $a = a^*$ and $c = a_2$; if $a^* \in \{a_1\} \dot\cup A_1'$ then $a^* \to b$ occurs in configuration (b) in G_{j+1} as in 1.3.

2.2. If $|A_2'| = 1$ and $A_3 = \emptyset$, remove the immorality $a_2 \to b \leftarrow a'$, where $A_2' = \{a'\}$. Any remaining arrow of the form $a^* \to b$, $a^* \in A$, must have $a^* \in \{a_1\} \dot\cup A_1'$, hence occurs in configuration (b) in G_{j+1} as in 1.3.

2.3. If $|A_2'| \geq 2$ and $A_3 = \emptyset$, remove any one arrow of the form $a' \to b$, $a' \in A_2$. Then all remaining arrows of the form $a^* \to b$, $a^* \in A$, are protected in G_{j+1}: if $a^* = a_2$ then $a^* \to b$ occurs in configuration (b) in G_{j+1} with $a = a^*$ and $c = a''$ for any $a'' \in A_2' \setminus \{a'\}$; if $a^* = a'' \in A_2' \setminus \{a'\}$ then $a^* \to b$ occurs in configuration (b) in G_{j+1} with $a = a^*$ and $c = a_2$; if $a^* \in \{a_1\} \dot\cup A_1'$ then $a^* \to b$ occurs in configuration (b) in G_{j+1} as in 1.3.

Step k. If $A'_{k-1} \neq \emptyset$ and $A_k \neq \emptyset$, choose a_k to be minimal in A_k with respect to the partial ordering $\preceq$. Then A_k can be expressed as the disjoint union $A_k = \{a_k\}\dot{\cup}A_{k+1}\dot{\cup}A'_k$, where $A_{k+1} := A_k \cap \mathrm{ch}_{G_j}(a_k)$ and $A'_k := A_k \setminus (\{a_k\}\dot{\cup}A_{k+1})$. By the minimality of a_k, $a_k \cdot\!\!/\!\!\cdot a'$ in G_j $\forall a' \in A'_k$. Note that

$$A = \{a_1\}\dot{\cup}\cdots\dot{\cup}\{a_k\}\dot{\cup}A_{k+1}\dot{\cup}A'_k\dot{\cup}\ldots\dot{\cup}A'_1.$$

k.1. If $A'_k = \emptyset$, remove $a_k \to b$. Then any remaining arrow of the form $a^* \to b$, $a^* \in A$, is protected in G_{j+1}: if $a^* \in A_{k+1}$ then $a^* \to b$ occurs in configuration (a) in G_{j+1} with $a = a^*$ and $c = a_k$; if $a^* \in \{a_1\}\dot{\cup}\cdots\dot{\cup}\{a_{k-1}\}\dot{\cup}A'_{k-1}\dot{\cup}\ldots\dot{\cup}A'_1$ then $a^* \to b$ occurs in configuration (b) in G_{j+1} as in $(k-1).3$.

k.2. If $|A'_k| = 1$ and $A_{k+1} = \emptyset$, remove the immorality $a_k \to b \leftarrow a'$, where $A'_k = \{a'\}$. Any remaining arrow of the form $a^* \to b$, $a^* \in A$, must have

$$a^* \in \{a_1\}\dot{\cup}\cdots\dot{\cup}\{a_{k-1}\}\dot{\cup}A'_{k-1}\dot{\cup}\ldots\dot{\cup}A'_1,$$

hence occurs in configuration (b) in G_{j+1} as in $(k-1).3$.

k.3. If $|A'_k| \geq 2$ and $A_{k+1} = \emptyset$, remove any one arrow of the form $a' \to b$, $a' \in A_k$. Then all remaining arrows of the form $a^* \to b$, $a^* \in A$, are protected in G_{j+1}: if $a^* = a_k$ then $a^* \to b$ occurs in configuration (b) in G_{j+1} with $a = a^*$ and $c = a''$ for any $a'' \in A'_k \setminus \{a'\}$; if $a^* = a'' \in A'_k \setminus \{a'\}$ then $a^* \to b$ occurs in configuration (b) in G_{j+1} with $a = a^*$ and $c = a_k$; if $a^* \in \{a_1\}\dot{\cup}\cdots\dot{\cup}\{a_{k-1}\}\dot{\cup}A'_{k-1}\dot{\cup}\ldots\dot{\cup}A'_1$ then $a^* \to b$ occurs in configuration (b) in G_{j+1} as in $(k-1).3$.

Because A is finite, this process must terminate after finitely many steps, so the desired removal is always possible.

Acknowledgments

We thank Steen Andersson, Robert Castelo, Max Chickering, Steve Gillispie, Paolo Giudici, Michael Levitz, David Madigan, Chris Meek, Thomas Richardson, Milan Studený, the Editors, and two referees for many helpful comments.

References

[AMP97] S. A. Andersson, D. Madigan, and M. D. Perlman, *A characterization of Markov equivalence classes for acyclic digraphs*, Ann. Statist. **25** (1997), 505–541.

[AMP01] S. A. Andersson, D. Madigan, and M. D. Perlman, *Alternative Markov properties for chain graphs*, Scand. J. Statist. **28** (2001), 33–85.

[AMPT] S. A. Andersson, D. Madigan, M. D. Perlman, and C. M. Triggs, *A graphical characterization of lattice conditional independence models*, Ann. Math. and Artificial Intelligence **21** (1997), 27–50.

[BP] J. R. S. Blair and B. Peyton, *An introduction to chordal graphs and clique trees*, Graph Theory and Sparse Matrix Computation (A. George, J. R. Gilbert, and J. W. H. Liu, eds.), Springer-Verlag, NY, 1993, pp. 1–29.

[B] W. Buntine, *Operations for learning with graphical models*, J. Artificial Intelligence Research **2** (1994), 159–225.

[C95] D. M. Chickering, *A transformational characterization of equivalent Bayesian network structures*, Uncertainty in Artificial Intelligence: Proceedings of the Eleventh Conference (P. Besnard and S. Hanks, eds.), Morgan Kaufman, San Francisco, 1995, pp.87–98.

[C96] D. M. Chickering, *Learning equivalence classes of Bayesian network structures*, Uncertainty in Artificial Intelligence: Proceeding of the Twelfth Conference (F. Jensen and E. Horvitz, eds.), Morgan Kaufmann, San Francisco, 1996, pp.150–157.

[CH] G. F. Cooper and E. Herskovits, *A Bayesian method for the induction of probabilistic networks from data*, Machine Learning **9** (1992), 309–347.

[DF] P. Dellaportas and J. J. Forster, *Markov chain Monte Carlo model determination for hierarchical and graphical log-linear models*, Biometrika **86** (1999), 616–633.

[FL] M. Frydenberg and S. L. Lauritzen, *Decomposition of maximum likelihood in mixed graphical interaction models*, Biometrika **72** (1989), 539–555.

[GP] S. B. Gillispie and M. D. Perlman, *Enumerating Markov equivalence classes of acyclic digraph models*, Uncertainty in Artificial Intelligence: Proceedings of the Seventeenth Conference (J. Breese and D. Koller, eds.), Morgan Kaufman, San Francisco, 2001, pp.171–177.

[GG] P. Giudici and P. Green, *Decomposable graphical Gaussian model determination*, Biometrika **86** (1999), 785–801.

[HGC] D. Heckerman, D. Geiger, and D. M. Chickering, *Learning Bayesian networks: the combination of knowledge and statistical data*, Machine Learning **20** (1995), 197–243.

[L] S. L. Lauritzen, *Graphical Models*, Oxford University Press, Oxford, 1993.

[LDLL] S. L. Lauritzen, A. P. Dawid, B. N. Larsen, and H.-G. Leimer, *Independence properties of directed Markov fields*, Networks **20** (1990), 491–505.

[LS] S. L. Lauritzen and D. J. Spiegelhalter, *Local computations with probabilities on graphical structures and their application to expert systems (with discussion)*, J. Royal Statist. Soc. (Series B) **50** (1988), 157–224.

[MAPV] D. Madigan, S. A. Andersson, M. D. Perlman, and C. M. Volinsky, *Bayesian model averaging and model selection for Markov equivalence classes of acyclic digraphs*, Commun. Statist. Thy. Meth. **25** (1996), 2493–2519.

[MR] D. Madigan, A. E. Raftery, *Model selection and accounting for model uncertainty in graphical models using Occam's window*, J. Amer. Statist. Assoc. **89** (1994), 1535–1546.

[MY] D. Madigan, J. York, *Bayesian graphical models for discrete data*, Int. Statist. Review **63** (1995), 215–232.

[M] C. Meek *Causal inference and causal explanation with background knowledge*, Proceedings of the Eleventh Annual Conference on Uncertainty in Artificial Intelligence (Philippe Besnard and Steve Hanks, eds.), Morgan Kaufmann, San Mateo, (1995), pp. 403–410.

[P] J. Pearl, *Probabilistic Reasoning in Intelligent Systems: Networks of Plausible Inference.* Morgan Kaufmann, San Mateo, (1988).

[PV] J. Pearl and T. Verma, *A theory of inferred causation*, Proceedings of the Second International Conference on Principles of Knowledge Representation (J. Allen, R. Fikes, E. Sandewall, eds.), Morgan Kaufmann, San Francisco, (1991), pp.441–452.

[SL] D. J. Spiegelhalter and S. L. Lauritzen, *Sequential updating of conditional probabilities on directed graphical structures*, Networks **20** (1990), 579–605.

[SDLC] D. J. Spiegelhalter, A. P. Dawid, S. L. Lauritzen, and R. G. Cowell, *Bayesian analysis in expert systems (with discussion)*, Statist. Science **8** (1993), 219–283.

[SGS] P. Spirtes, C. Glymour, and R. Scheines, Causation, Prediction, and Search, Springer, New York, 1993.

[VP90] T. Verma and J. Pearl, *Equivalence and synthesis of causal models*, Uncertainty in Artificial Intelligence: Proceedings of the Sixth Conference (M. Henrion, R. Shachter, L. Kanal, J. Lemmer, eds.), Morgan Kaufman, San Francisco, 1990, pp.220-227.

[VP92] T. Verma and J. Pearl, *An algorithm for deciding if a set of observed independencies has a causal explanation*m Uncertainty in Artificial Intelligence: Proceedings of the Eighth Conference (D. Dubois, M. P. Wellman, B. DAmbrosio, and P. Smets, eds.), Morgan Kaurman, San Francisco, 1992, pp.323–330.

DEPARTMENT OF STATISTICS, UNIVERSITY OF WASHINGTON, SEATTLE, WASHINGTON 98195
E-mail address: `michael@ms.washington.edu`

Contemporary Mathematics
Volume **287**, 2001

Computational Commutative Algebra in Discrete Statistics

Giovanni Pistone, Eva Riccomagno, and Henry P. Wynn

ABSTRACT. We show that exponential probability models with lattice support are algebraic varieties within which computational and conceptual tools from commutative algebra can be applied. Log-linear models for contingency tables and conditional independence probability models on trees are discussed within this algebraic framework.

1. Introduction

In the last few years the subject of Computational Commutative Algebra (CCA) has attracted much attention in a number of applied fields. This interest is proved in particular by the number of textbooks and monographs recently published. In this paper we refer especially to the textbooks [**7**] and [**13**]; see also [**5**] for a collection of more advanced short reviews. In the monograph [**18**, Chapter 6] a fundamental methodology is given which uses CCA at various levels in the construction and analysis of statistical models and sub-models. We give a condensed version of this construction together with results pertaining statistical models on lattices. This line of research was originated by work on simulation [**9**] and on design of experiments [**19**], and developed in [**21**].

Our use of CCA can be described heuristically as falling in three stages. We start with a finite set of points $D \subset \mathbb{R}^d$, called an experimental design or support. The first stage refers to the construction of the ring of real random variables on D as a quotient space of the polynomial ring in d indeterminates, modulo the set of all polynomials whose zeros include D. Each random variable is represented by an equivalence class of interpolating polynomials. The second stage considers discrete probability models with support D, described as random variables through the density p. There are two principal cases, $y = p(x)$ and $y = \log p(x)$. As polynomials are parametrized by their coefficients, the two cases lead to different classes of statistical models. Suppose that in the second case the sought-after

1991 *Mathematics Subject Classification.* Primary 62B15, 68W30; Secondary 14Q99.

G.P. was supported by the grant MURST *Methods for planning experiments and their applications* (A. Giovagnoli) and by UFR de Mathématiques et d'Informatique, Université Louis Pasteur, Strasbourg. This work was done while E. R. was a post-doctoral research fellow at Eurandom, NL..

interpolating polynomial is $t(x)$. Then,

$$p(x) = \exp(t(x)), \quad x \in D,$$

is an exponential model.

If D is a subset of a non-negative integer lattice, $D \subset \mathbb{Z}^d_{\geq 0}$, a third stage is possible in which, by reparametrization, $p(x)$ can be written as a "power product." We shall see below that the algebraic equations satisfied by the probabilities $p(x)$, $x \in D$, considered as indeterminates, are seen to belong to a toric ideal (see [**2**]).

In each of the above algebraic descriptions of statistical models, statistical submodels are set up by considering further algebraic restrictions to be satisfied by parameters. For example, a full description of an independence model

$$p(x_1, x_2) = p_1(x_1)p_2(x_2)$$

is possible if a full description of the marginal probabilities is available. CCA allows us to compute such descriptions in polynomial form, given a polynomial description of D. We shall see below how CCA methods apply to more general statistical models described by marginal restrictions, or sufficiency (conditional independence) assumptions. When the exponential representation of probabilities is used, CCA methods include the familiar log-linear model theory.

Once the probabilistic quantities of interest, whether the support of a probability function or a toric statistical model for the probabilities, are expressed as polynomials, then one can take advantage of the structure of polynomial rings and quotient spaces and use the polynomial algebraic tools to manipulate these quantities. Section 2 presents the relevant results from CCA and Gröbner basis theory. Section 3 describes the link to design theory and non-saturated models. In Section 4 the algebraic notion of a toric ideal and its relation with the model residual space are introduced. This is developed for exponential models in Section 5. Sections 6 and 7 describe the implication of the polynomial approach to modeling of independence and conditional independence on trees.

2. Commutative algebra in statistics

Traditionally, the algebraic language of statistics has been linear algebra. Computational linear algebra is a powerful tool, but not fully adapted to tackle some problems, such as the computation of sufficient statistics and the specification of non-regression statistical models. The approaches have been various. Recent theoretical work by McCullagh [**16**] has suggested the framework of category theory. Our approach starts from the remark that the class of random variables on a finite sample space has the algebraic structure of a ring. We stress here the fact that *computational* commutative polynomial algebra is the relevant tool in such framework, the connection being that the ring of random variables can be shown to be isomorphic to a quotient polynomial ring.

In the following, we list definitions and properties of objects in commutative algebra we need, and refer to [**7**], [**8**] and [**13**] for proofs and in-depth analysis. The actual computational feasibility of our methods is important. For the purpose of illustration in this paper we use the system CoCoA, see [**4**] and [**13**, 275–304]. Many other symbolic computation systems are available, see [**7**, Appendix C] and [**18**, Section 1.2] for short reviews. The use of Gröbner bases in statistics is just one of the possible applications of symbolic computation in statistics. Compare [**1**] for a different use of symbolic computation in this field of application.

The key notion allowing the actual computation over polynomial rings is that of total ordering on terms as defined below. Here a *term* is a product of indeterminates (or variables), a *monomial* is a constant (typically a real number) multiplied by a term, a *polynomial* is a sum of monomials. The ring of polynomials with variables $x_1, \ldots, x_d$ and real coefficients is denoted by $\mathbb{R}[x_1, \ldots, x_d]$. In the univariate case, there is a unique such ordering: $\ldots \succ x^2 \succ x \succ 1$. In the multivariate case the ordering allows to perform computations among polynomials in a way similar to the usual algebra of polynomials in one indeterminate.

DEFINITION 2.1. Let d be a positive integer. A *term-ordering* is a total ordering relation $\succ$ over the terms in d indeterminates such that

(1) $x^\alpha \succ 1$ for all $x^\alpha = x_1^{\alpha_1} \ldots x_d^{\alpha_d}$ with $\alpha_i \in \mathbb{Z}_{\geq 0}$ for all $i = 1, \ldots, d$,
(2) if $x^\alpha \succ x^\beta$, then $x^{\alpha+\gamma} \succ x^{\beta+\gamma}$ for all $\alpha, \beta, \gamma \in \mathbb{Z}^d_{\geq 0}$.

The *leading term* $\mathrm{LT}_\succ(f)$ of a polynomial $f \in \mathbb{R}[x_1, \ldots, x_d]$ with respect to the term-ordering $\succ$ is the largest of its terms with respect to $\succ$.

Point (1) corresponds to the requirement that the relation $\succ$ is a well-ordering and Point (2) expresses compatibility with simplification of monomials, that is, if x^β divides x^α, then $x^\alpha \succ x^\beta$. Alternative notations for a term-ordering are $\succ_\tau$ and τ.

To a finite set of points, $D \subset \mathbb{R}^d$, one can associate (in a non-unique way) a set of $N = \#D$ terms, so that a real function over D can be expressed as a real linear combination of these terms. These bases are called *monomial bases.* Theorem 2.4 below gives an automatic technique to compute a special class of monomial bases. This method uses Gröbner bases (see Definition 2.3) and assumes a term-ordering. A so-obtained basis has the property that if x^α is in the basis and x^β divides it, then x^β is in the basis. A set of terms with this property is called in the algebraic language *order ideal*; see [**24**]. In the statistical literature, terms like *functional marginality* (see [**15**]) and *hierarchical models* are used.

A term in a monomial basis has order equal to the number of its indeterminates. A single term of order larger than two in a monomial basis is usually called in statistics an *interaction* term.

A support $D \subset \mathbb{R}^d$ is seen as an (affine) *algebraic variety*, that is, the set of zeros of a system of polynomial equations.

DEFINITION 2.2. Let $D \subset \mathbb{R}^d$ be a finite set of points, that is a support. The set $\mathrm{Ideal}\,(D) \subset \mathbb{R}[x_1, \ldots, x_d]$ is the set of all polynomials whose real zeros include the points in D.

We recall that a *polynomial ideal* I is a subset of $\mathbb{R}[x_1, \ldots, x_d]$ such that (i) $f + g \in I$ for all $f, g \in I$ and (ii) $sf \in I$ for all $f \in I$ and $s \in \mathbb{R}[x_1, \ldots, x_d]$. For example, $\mathrm{Ideal}\,(D)$ is a polynomial ideal. An ideal I is *finitely generated* if there exist (non-unique) $f_1, \ldots, f_s \in \mathbb{R}[x_1, \ldots, x_d]$ such that for all $f \in I$, then $f = \sum_{i=1}^s h_i f_i$, where $h_i \in \mathbb{R}[x_1, \ldots, x_d]$. We write $I = \mathrm{Ideal}\,(f_1, \ldots, f_s)$. The important *Hilbert basis theorem* states that every polynomial ideal (except $\mathrm{Ideal}\,(0)$) is finitely generated. An ideal I is *radical* if $f^m \in I$ implies $f \in I$.

Some properties of $\mathrm{Ideal}\,(D)$ are

(1) $\mathrm{Ideal}\,(D)$ is a radical ideal.
(2) $\mathrm{Ideal}\,(D)$ is generated by $f_1, \ldots, f_s \in \mathbb{R}[x_1, \ldots, x_d]$ if and only if the complex solutions of the system of polynomial equations $f_1 = \ldots = f_s = 0$ are of multiplicity one and are the points in D.

DEFINITION 2.3. Given a polynomial ideal $I \subset \mathbb{R}[x_1, \ldots, x_d]$ and a term-ordering τ, a finite subset G of I is called a *Gröbner basis* for I with respect to τ if

$$\text{Ideal}\,(\text{LT}_\tau\,(g) : g \in G) = \text{Ideal}\,(\text{LT}_\tau\,(f) : f \in I)\,.$$

A Gröbner basis G is *reduced* if for all $g \in G$ (i) the coefficient of $\text{LT}_\tau\,(g)$ is 1 and (ii) no term of g belongs to the ideal $\text{Ideal}\,(\text{LT}\,(f) : f \in G \setminus \{g\})$.

For every term-ordering there exist Gröbner bases. Moreover for a term-ordering τ, there exists a unique *reduced* Gröbner basis.

Given an ideal $I \subset \mathbb{R}[x_1, \ldots, x_d]$ the equivalence relation, $\sim_I$ over $\mathbb{R}[x_1, \ldots, x_d]$ modulo I is defined as: for all $f, g \in \mathbb{R}[x_1, \ldots, x_d]$ then $f \sim_I g$ if and only if $f - g \in I$. The quotient space is defined as

$$\mathbb{R}[x_1, \ldots, x_d]/I = \{[f] : f \sim_I g \text{ for } f, g \in \mathbb{R}[x_1, \ldots, x_d]\}\,.$$

THEOREM 2.4. *Let I be a polynomial ideal in $\mathbb{R}[x_1, \ldots, x_d]$ and τ a term-ordering.*

(1) *The quotient space $\mathbb{R}[x_1, \ldots, x_d]/I$ is a vector space over the real numbers.*
(2) *Let G be a Gröbner basis of I with respect to τ.*
   (a) *A vector space basis of $\mathbb{R}[x_1, \ldots, x_d]/I$ is given by*

$$\begin{aligned}\text{Est}_\tau\,(D) &= \{x^\alpha : x^\alpha \textit{ is not divisible by } \text{LT}_\tau(g),\ g \in G\}\\ &= \{x^\alpha : \alpha \in L\}\,.\end{aligned}$$

   (b) *The full design matrix, $Z = [x^\alpha]_{x \in D; \alpha \in L}$, is invertible.*
(3) *Let D be a support and $I = \text{Ideal}\,(D)$ the corresponding support ideal. Then*
   (a) *$f \sim_I g$ if and only if $f(a) = g(a)$ for all $a \in D$.*
   (b) *The dimension of $\mathbb{R}[x_1, \ldots, x_d]/I$ as $\mathbb{R}$-vector space is equal to the number of points in D.*
   (c) *The design matrix for the regression model*

$$y = \sum_{\alpha \in M} \theta_\alpha x^\alpha + \varepsilon$$

   with $M \subseteq L$ is full rank.

For a proof see, for example, [**18**, Chapter 3]. Note that the monomial bases obtained via Gröbner bases are just instances of all possible monomial bases; see [**3**]. For a discussion of monomial bases arising in binary regular and non-regular designs; see [**11**].

3. Experimental designs and residual space

In this section we consider sub-models specified by a polynomial model of the type considered in Theorem 2.4(3)(c). We apply the algebraic theory of the previous section to experimental designs in order to show the link with "statistical thinking." Thus, let a set

$$\{x^\alpha : \alpha \in M\}$$

of linearly independent terms on D be given. The linear independence is always true if $M \subset L$, that is, the model is a subset of the exponent list in the estimable set Est_τ. Let

$$Z_1 = \left[x^\alpha\right]_{x\in D;\alpha\in M}$$

be the design matrix corresponding to the statistical sub-model M. It is very common, for example in statistical analysis, to fit sub-models such as linear (planar) models or quadratic models. Now extend Z_1 to a full basis for the column space (range) of Z in an orthogonal way. With abuse of notation we can replace Z by this partially orthogonal basis and write Z in block form as

$$Z = [Z_1 : Z_2],$$

where $Z_1^t Z_2 = 0$. In statistical modeling we consider the model

$$y = Z_1\theta + \varepsilon,$$

with ε a mean zero error. The least squares estimate of θ is $\hat{\theta} = \left(Z_1^t Z_1\right)^{-1} Z_1^t y$. The fit is $\hat{y} = Py$, where P is the projector $Z_1\left(Z_1^t Z_1\right)^{-1} Z_1^t$ and the orthogonal projector is

$$I - P = Z_2\left(Z_2^t Z_2\right)^{-1} Z_2^t.$$

The invertibility of $Z_1^t Z_1$ and $Z_2^t Z_2$ comes from the fact that Z_1 and Z_2 are full rank by construction. In statistical jargon we might consider Z_2 (columns) as "spanning the space of residuals" and thus we call Z_2 a *residual matrix*.

Despite this straightforward explanation, the structure of Z_2 is not so transparent. We shall see that special instances, such as integer valued matrices, are useful for clarification.

EXAMPLE 3.1 (One-dimensional polynomial regression). Consider the set of points $D = \{0, 1, \dots, n-1\} \subset \mathbb{Z}$. Construct orthogonal univariate polynomials on D of increasing degree

$$1, p_1(x), , \dots, p_{n-1}(x).$$

Let the sub-model consist of the terms

$$1, x, x^2, \dots, x^q,$$

where $q < n-1$. Then, it is clear that a candidate for Z_2 is

$$\left[p_j(x)\right]_{x\in D; j=q+1,\dots,n-1}.$$

Notice that since D consists of integers the $p_j(x)$ have rational coefficients. These can be converted to integer by multiplication of a suitable integer. Then, Z_2 as constructed can be replaced by an integer matrix.

EXAMPLE 3.2. Consider the full factorial design 2^d with levels -1 and 1. For all term-orderings the Est set of Theorem 2.4 is composed of all the multi-linear terms, that is, in L there are all the vectors $\alpha = (\alpha_1, \dots, \alpha_d)$ with $\alpha_i = 0, 1$ for all $i = 1, \dots, d$. The choice of -1 and 1 as levels implies that the full design matrix, Z, is orthogonal. Thus, for every sub-model $M \subset L$ automatically $Z_2 = Z_{|\overline{M}}$ is orthogonal to $Z_1 = Z_{|M}$, where $\overline{M} = L \setminus M$.

EXAMPLE 3.3. Let D be the two-dimensional support with the particular structure given below

that is, $D = \{(0,2),(0,1),(0,0),(1,0),(2,0)\}$. This is called an echelon design. Echelon designs are defined by the property that if a point $a = (a_1, \dots, a_d)$ is in D, then also $(y_1, \dots, y_d)$ is in D for $0 \le y_i \le a_i$, $i = 1, \dots, d$. Also in this case, for all term-orderings there is only one Est, namely $\{x^a : a \in D\}$ (see [**18**]).

Let M be a subset of L of echelon form. For example $M = \{(0,0),(1,0),(0,1)\}$. Then, the full design matrix is

$$Z = \begin{bmatrix} 1 & 0 & 2 & 0 & 4 \\ 1 & 0 & 1 & 0 & 1 \\ 1 & 0 & 0 & 0 & 0 \\ 1 & 1 & 0 & 1 & 0 \\ 1 & 2 & 0 & 4 & 0 \end{bmatrix},$$

where the first three columns refer to M and give Z_1. The last two columns are the evaluations of x_1^2 and x_2^2 at D. Now we may chose Z_2 so that its columns are orthogonal to the first three columns of Z. For example we can take

$$Z_2 = \begin{bmatrix} 1 & -1 \\ -2 & 2 \\ 1 & 0 \\ 0 & -2 \\ 0 & 1 \end{bmatrix}.$$

We shall see that the notion of residual space of a model in the vector space sense is not enough for a fully satisfying theory of exponential models on a lattice. We shall need a more refined structure according to the following definition.

DEFINITION 3.4. Let $D \subset \mathbb{Z}^d_{\ge 0}$ be a finite lattice support and let $\{x^\alpha : \alpha \in L\}$ be a monomial basis (for example obtained with the Gröbner basis method). If $M \subseteq L$ specifies the terms of a model and $Z_1 = \left[x^\beta\right]_{x \in D; \beta \in M}$ the model matrix, we call *integer orthogonal* a vector such that

(1) the vector q has integer components, $q(x) \in \mathbb{Z}$, $x \in D$.
(2) The integer values $q(x)$, $x \in D$, are relatively prime.
(3) The vector q is orthogonal to the model, that is,

$$\sum_{x \in D} x^\beta q(x) = 0, \quad \beta \in M.$$

In the lattice case we can always construct a Z_2 matrix whose columns are integer orthogonal vectors by using any orthogonalization procedure. We shall denote by $\mathcal{Z}_2$ the set of all integer orthogonal vectors. If Z_2 is any residual matrix, then $\mathcal{Z}_2 \subset \operatorname{Span}(Z_2)$

The *projection* of the polynomial ideal $I \subset \mathbb{R}[x_1, \dots, x_d]$ on the sub-ring of the indeterminates $x_{p+1}, \dots, x_d$, $\mathbb{R}[x_{p+1}, \dots, x_d]$, for $p < d$, is

$$I_p = I \cap \mathbb{R}[x_{p+1}, \dots, x_d].$$

It is a polynomial ideal called also the *p-elimination* ideal of I. We talk of elimination of $x_1, \ldots, x_p$ from I. If the ideal I is the ideal of a support D, then the elimination produces the ideal of the corresponding marginal support. The space of marginal responses is associate with the marginal ring and the elimination ideal.

A special type of term-orderings, called *elimination orderings*, is used to compute the Gröbner basis of the elimination ideal; see [**18**, Section 2.9].

4. Toric ideals

We recall briefly the notion of a toric ideal as presented in [**13**, Tutorial 38] and [**2**]; see also [**24**], [**8**], [**22**] [**20**]. The relevance of toric ideals for the analysis of contingency tables with fixed margins and the application to simulation for exact tests has been singled out in [**9**]; see also [**10**]. We present below our application to exponential models on a finite lattice, $D \subset \mathbb{Z}^d_{\geq 0}$. The reader will recognize, for example, one of the main themes in the classical theory of log-linear models for contingency tables; see [**12**].

Toric ideals are related to algebraic varieties defined parametrically by monomials.

DEFINITION 4.1. Let $\mathbb{R}[\zeta] = \mathbb{R}[\zeta_0, \ldots, \zeta_s]$ be a polynomial ring and for $\alpha_i \in \mathbb{Z}^{s+1}_{\geq 0}$, $i = 1, \ldots, n$. Let $\zeta_0^{\alpha_{i0}} \cdots \zeta_s^{\alpha_{is}}$ be a finite list of terms. Consider the extended polynomial ring $\mathbb{R}[p, \zeta] = \mathbb{R}[p_1, \ldots, p_n, \zeta_0, \ldots, \zeta_s]$ and the ideal J generated by the binomials $p_i - \zeta_0^{\alpha_{i0}} \cdots \zeta_s^{\alpha_{is}}$. If $\mathbb{R}[p] = \mathbb{R}[p_1, \ldots, p_n]$, the elimination ideal $I = J \cap R[p]$ is called the *toric ideal* of the matrix $Z_1 = [\alpha_{ij}]_{j=1,\ldots,n; i=0,\ldots,s}$.

We outline the main facts about toric ideals with two non-statistical examples, both taken from the literature. The theory relevant for statistical models will be described in Section 5.

EXAMPLE 4.2 ([**7**, Section 4.6]). In $\mathbb{R}[p_1, p_2, p_3, \zeta]$ consider the identities $p_1 = \zeta^3$, $p_2 = \zeta^4$, $p_3 = \zeta^5$. These can be considered in several ways: a) a parametric description of a curve in the three dimensional space of coordinates p_1, p_2, p_3; b) a homeomorphism of the ring $\mathbb{R}[p_1, p_2, p_3]$ into the ring $\mathbb{R}[\zeta]$ derived uniquely from the rules $p_1 \mapsto \zeta^3$, $p_2 \mapsto \zeta^4$, $p_1 \mapsto \zeta^5$; c) the ideal $J \subset \mathbb{R}[p_1, p_2, p_3, \zeta]$ generated by the polynomials $p_1 - \zeta^3$, $p_2 - \zeta^4$, $p_1 - \zeta^5$. The kernel of the mapping defined in b) is the ideal I consisting of all polynomials $f(p_1, p_2, p_3)$ such that $f(\zeta_0^3, \zeta_0^4, \zeta_0^5) = 0$. This ideal is easily seen to be equal to the elimination ideal $J \cap \mathbb{R}[p_1, p_2, p_3]$. The curve a) is the variety of I. The computation using CoCoA shows that I is generated by the following polynomials $-p_1^3 + p_2p_3$, $-p_2^2 + p_1p_3$, $p_1^2p_2 - p_3^2$. The corresponding equations are the implicit equations for the curve a). Note that the elements of this basis are binomials.

Now consider the same example in a linear algebra setting. Assume $\zeta > 0$ and rewrite the initial identities as $\log p_1 = 3 \log \zeta$, $\log p_2 = 4 \log \zeta$, $p_3 = 5 \log \zeta$, that is, the vector $\log[p]$, $[p] = (p_1, p_2, p_3)$, is proportional to the vector $u = (3, 4, 5)$. By introducing the vectors $v = (1, -2, 1)$ and $w = (3, -1, -1)$ which span $u^\perp$ (the orthogonal space of u), we can write $v^t \log[p] = 0$, $w^t \log[p] = 0$, that is, $p_1p_3 = p_2^2$, $p_1^3 = p_2p_3$. Comparison with the previous results shows that one equation is missing, and, more important, that the variety now contains the axis $p_1 = p_2 = 0$ which is not part of the original curve a). One of the integer orthogonal vectors in Definition 3.4, namely, $(2, 1, -2)$ is missing.

We describe now the theory relevant to fill the gap between the residual matrix and the toric ideal. Given two polynomial ideals $I, J \subset \mathbb{R}[x_1, \ldots, x_d]$, the *colon ideal* $I : J$ is defined as

$$I : J = \{g \in \mathbb{R}[x_1, \ldots, x_d] : fg \in I \text{ for all } f \in J\}.$$

If J is the ideal generated by a single polynomial f, then we write $I : f$.

The saturation ideal of the ideal $I \subset \mathbb{R}[x_1, \ldots, x_d]$ with respect to the polynomial $f \in \mathbb{R}[x_1, \ldots, x_d]$ is defined as

$$I : f^\infty = \{g \in \mathbb{R}[x_1, \ldots, x_d] : f^m g \in I \text{ for some } m > 0\}.$$

THEOREM 4.3. *Let $I \subset \mathbb{R}[x_1, \ldots, x_d]$ be an ideal and let $f \in \mathbb{R}[x_1, \ldots, x_d]$ be a polynomial. Then*

(1) *$I : f^\infty$ is an ideal.*
(2) *$I : f \subset I : f^2 \subset I : f^3 \subset \ldots$.*
(3) *There exists $m > 0$ such that $I : f^m = I : f^\infty$.*
(4) *Suppose that I is generated by $f_1, \ldots, f_s$ and define the ideal $\tilde{I}$ in the extended polynomial ring $\mathbb{R}[x_1, \ldots, x_d, y]$ generated by $f_1, \ldots, f_s$ and $1 - fy$. Then $I : f^\infty = \tilde{I} \bigcap \mathbb{R}[x_1, \ldots, x_d]$.*

Saturation is a mean to "localize" around f, that is, to make f invertible "locally." In the next section this construction will be related to the choice of the Z_2 matrix above ($Z_1^t Z_2 = 0$).

EXAMPLE 4.4 ([**23**]). Consider the following system of equations

$$\begin{cases} p_1 = st \\ p_2 = s^3t^2 \\ p_3 = st^3 \\ p_4 = s^5t^2. \end{cases} \tag{4.1}$$

The elimination of s and t from System (4.1) is equivalent to determining the kernel of the following homomorphism of $\mathbb{R}$-algebras

$$\begin{array}{rccc} \phi: & \mathbb{R}[p_1, p_2, p_3, p_4] & \longrightarrow & \mathbb{R}[s, t] \\ & (p_1, p_2, p_3, p_4) & \longmapsto & (st, s^3t^2, st^4, s^5t^2). \end{array}$$

To System (4.1) we can associate the Z_1 matrix

$$Z_1 = \begin{bmatrix} 1 & 1 \\ 3 & 2 \\ 1 & 3 \\ 5 & 2 \end{bmatrix},$$

and the system of linear equations

$$\begin{cases} z_1 + 3z_2 + z_3 + 5z_4 = 0 \\ z_1 + 2z_2 + 3z_3 + 2z_4 = 0, \end{cases}$$

whose solutions include the integer orthogonal vectors. There is an interesting connection of toric ideals to Diophantine linear systems, first pointed out in [**6**]. Two linearly independent solutions for this system are $(z_1, z_2, z_3, z_4) = (7, -2, -1, 0)$ and $(4, -3, 0, 1)$. Then, a possible Z_2 matrix is

$$Z_2 = \begin{bmatrix} -7 & 4 \\ 2 & -3 \\ 1 & 0 \\ 0 & 1 \end{bmatrix},$$

and the binomials $p_1^7 - p_2^2 p_3$ and $p_1^4 p_4 - p_2^3$ are in the kernel of ϕ. Also, any other polynomials in $\text{Ideal}\left(p_1^7 - p_2^2 p_3, p_1^4 p_4 - p_2^3\right)$ belongs to the kernel of ϕ. Actually there are other polynomials in $\ker(\phi)$ which do not belong to J, for example $p_1^3 p_2 - p_3 p_4$. It can be proved that $\ker(\phi)$ is the saturation ideal of $\text{Ideal}\left(p_1^7 - p_2^2 p_3, p_1^4 p_4 - p_2^3\right)$ with respect to the polynomial $p_1 p_2 p_3 p_4$.

A method based on Theorem 4.3 to compute the saturation ideal is to consider the ideal

$$\text{Ideal}\left(p_1^7 - p_2^2 p_3, p_1^4 p_4 - p_2^3, p_1 p_2 p_3 p_4 t - 1\right) \subset \mathbb{R}[p_1, p_2, p_3, p_4, t]$$

in the extended space and intersect it with $\mathbb{R}[p_1, p_2, p_3, p_4]$. Definition 4.1 itself suggests another method to compute the sought-after toric ideal, namely, as elimination ideals of s and t from the system

$$\begin{cases} p_1 - st \\ p_2 - s^3 t^2 \\ p_3 - st^3 \\ p_4 - s^5 t^2. \end{cases}$$

5. Exponential models and toric ideals

In this section we apply the results of previous sections to exponential models. Let $D \subset \mathbb{Z}^d_{\geq 0}$ be a finite (lattice) support. Let $\mathcal{M} = \{x^\beta : \beta \in M\}$ be a set of monomial functions on D, linearly independent over the real numbers. We call $\mathcal{M}$ a *model of terms*. We assume $0_d \in M$ and consider the exponential model

$$p(x; \psi) = \exp\left(\sum_{\beta \in M} \psi_\beta x^\beta\right). \tag{5.1}$$

As $x \in Z^d_{\geq 0}$, then x^β is a non-negative integer for all $\beta \in M$, and we introduce the ζ-parameters defined by

$$\zeta_\beta = \exp\left(\psi_\beta\right), \quad \beta \in M. \tag{5.2}$$

The p-parameters, that is, the vector [p] with components $p(x; \psi)$, $x \in D$, depend on the ζ-parameters according to Equations (5.1) and (5.2). The dependence between the p-parameters and the ζ-parameters is a relation among monomials

$$p(x) = \prod_{\beta \in M} \zeta_\beta^{x^\beta}. \tag{5.3}$$

The normalizing conditions are algebraic in both parameters, $\sum_{x \in D} p(x) = 1$ or $\sum_{x \in D} \prod_{\beta \in M} \zeta_\beta^{x^\beta} = 1$.

As before, we consider the model matrix $Z_1 = \left[x^\beta\right]_{x \in D; \beta \in M}$, a residual integer valued matrix, Z_2, and the set of all integer orthogonal vectors, $\mathcal{Z}_2$. Recall that for an ideal, $I \subset \mathbb{R}[x_1, \dots, x_n]$,

$$\text{Variety}\,(I) = \{a \in \mathbb{R}^n : f(a) = 0 \text{ for all } f \in I\}.$$

THEOREM 5.1. *Consider the polynomial rings* $\mathbb{R}[p, \zeta] = \mathbb{R}[p(x) : x \in D, \zeta_\beta : \beta \in M]$, $\mathbb{R}[p] = \mathbb{R}[p(x) : x \in D]$, $\mathbb{R}[\zeta] = \mathbb{R}[\zeta_\beta : \beta \in M]$. *Let* M *be a model of terms as above and let* J *be the ideal generated in* $\mathbb{R}[p, \zeta]$ *by Equations (5.3). Let* $\mathcal{Z}_2$ *be the set of all integer orthogonal vectors to* Z_1. *Let* $I = J \cap R[p]$ *be the toric ideal correspondent to* M.

(1) *The probability parameters* $[p]$ *of the exponential model in Equation (5.1) belong to the irreducible variety* Variety $(I) \cap$ Variety $\left(\sum_{x \in D} p(x) - 1\right)$.
(2) *The toric ideal* I *has a binomial homogeneous basis. This basis can be computed by elimination of the* ζ*-variables in* J.
(3) *For each column* f *of the residual matrix,* Z_2*, denote by* f_+ *and* f_- *respectively the positive and the negative part. Then, the binomial*

$$\prod_{x \in D} p(x)^{f_+(x)} - \prod_{x \in D} p(x)^{f_-(x)} \tag{5.4}$$

belongs to the toric ideal I.
(4) *Let* I_0 *be the ideal generated by the binomials in Equation (5.4). Then,* $I_0 \subset I$*, and* I *is the saturation ideal of* I_0 *for the monomial* $\prod_{\beta \in M} \zeta_\beta$. *In particular, a basis of* I *can be computed by the elimination of* t *and* ζ *in the ideal generated by the polynomials in Equation (5.4) and* $t \prod_{\beta \in M} \zeta_\beta - 1$.
(5) *The set of binomials*

$$\prod_{x \in D} p(x)^{g_+(x)} - \prod_{x \in D} p(x)^{g_-(x)}, \qquad \text{for all } g \in \mathcal{Z}_2,$$

is a basis of the toric ideal I.

PROOF. Point 1 follows from the exponential model expressed in the ζ-parameters and the normalization condition. Note that the existence of a polynomial parametric representation for this variety implies that the variety is irreducible; see [**7**, Section 4.5]. Point 2 is a well known result in the theory of toric ideals. Point 3 follows from considering positive values for the parameters, and observing that the vector of the log-probabilities $\log[p]$ is a linear combination of the columns of the model matrix, Z_1, or, equivalently, orthogonal to the columns of the Z_2 matrix. Point 4 follows from the general theory. Point 5 follows from the fact that any vector in $\mathcal{Z}_2$ can be a column in Z_2 and the logarithms of the binomials in a basis of the toric ideal form an integer orthogonal vector. □

There is a very close connection between the toric ideal constructed by exhibiting the $p(x)$ as power products and the residual space construction on Z_2. To rephrase Theorem 5.1 (3), suppose that D is lattice and M a model of terms linearly independent over D. We can always construct an orthogonal matrix, Z_2 ($Z_1^t Z_2 = 0$), with integer coefficients. Any standard construction, such as Gram-Schmidt or the Cholesky orthogonal polynomial construction, will lead to a rational Z_2 which can be made integer by suitable multiplication by an integer. Now Z_2 gives a basis for the toric ideal as follows: For each column j of Z_2, divide the row indices in three groups according as to whether the entry ζ_{ij} is positive, zero or negative. Call the three set of indices so obtained J_j^+, J_j^0 and J_j^-, respectively.

Then, the saturation ideal of the ideal generated by

$$\prod_{i\in J_j^+} p_i^{\zeta_{ij}} - \prod_{i\in J_j^-} p_i^{\zeta_{ij}} \qquad \text{for all } j$$

is the sought-after toric ideal.

One may also obtain different Gröbner bases of the toric ideal by varying the term-ordering, but there is only a finite number of such bases. Their union is still a Gröbner basis and is called the universal Gröbner basis of the toric ideal (see [**24**]).

6. Modeling independence

In statistical inference many of the probability models used are based on assumptions of independence or conditional independence. In the present section we examine the implications for independence and conditional independence of our assumption of finite lattice support, $D \subset \mathbb{Z}^d_{\geq 0}$. Consider the following example.

EXAMPLE 6.1. Let X_1 and X_2 be two independent Bernoulli (0, 1 valued) random variables with marginal probabilities $p_1 = \mathrm{P}(X_1 = 1)$ and $p_2 = \mathrm{P}(X_2 = 1)$. Independence says that

$$\mathrm{P}(\{X_1 = 1\} \cap \{X_2 = 1\}) = p_1 p_2.$$

Now consider the exponential formulation. By independence this is

$$\log p(x_1, x_2; \psi) = \exp(\psi_0 + \psi_1 x_1 + \psi_2 x_2).$$

Then, we have

$$Z_1 = \begin{bmatrix} 1 & 0 & 0 \\ 1 & 1 & 0 \\ 1 & 0 & 1 \\ 1 & 1 & 1 \end{bmatrix}, \quad Z_2 = \begin{bmatrix} 1 \\ -1 \\ -1 \\ 1 \end{bmatrix}.$$

It is easy to check that the Z_2 column corresponds to the unique integer orthogonal vector to Z_1. Then, the toric ideal in obvious notation is

$$p_{00}p_{11} - p_{10}p_{01} = 0.$$

This condition is very familiar to students of the 2×2 contingency table.

The attractive feature of the exponential interpolation is that essentially any conditional structure can be captured by the structure of the terms $X^\alpha Y^\beta Z^\gamma$, $(\alpha, \beta, \gamma) \in M$ appearing in the sub-model

$$p(X, Y, Z; \psi) = \exp\left(\sum_{(\alpha,\beta,\gamma)\in M} \psi_{\alpha\beta\gamma} X^\alpha Y^\beta Z^\gamma\right). \tag{6.1}$$

Here capital letters denote blocks of variables.

It is enough for the purpose of the present paper to model a single conditional independence structure. Thus, for three vectors of random variables X, Y, Z, let X and Y be conditional independent given Z, written as $X \coprod Y | Z$. Then, the following proposition holds. It is a version of a well known result in graphical models; see [**14**, Chapter 4]. We partially follow the presentation given in [**17**].

LEMMA 6.2. *Let X, Y and Z be disjoint blocks of variables and let D_A be the support of the vector of random variables A, where $A \subseteq 2^{\{X,Y,Z\}}$. Let us assume the following:*

(1) *for each $z \in D_Z$ the corresponding section of the sample space is a product*

$$D_{X,Y,Z=z} = D_{X,Z=z} \times D_{Y,Z=z};$$

(2) *The joint quotient ring* $\mathbb{R}[X,Y,Z]/\operatorname{Ideal}(D_{X,Y,Z})$ *has a monomial vector space basis that contains the terms in the model* $X^\alpha Y^\beta Z^\gamma$, $(\alpha,\beta,\gamma) \in M$, *and a monomial vector space basis of both the marginal quotient rings*

$$\mathbb{R}[X,Z]/\operatorname{Ideal}(D_{X,Z}) \text{ and } \mathbb{R}[Y,Z]/\operatorname{Ideal}(D_{Y,Z}).$$

(The assumption is in particular true if $D_{X,Y,Z} = D_X \times D_Y \times D_Z$.*) Then, under the distribution in Equation (6.1),* $X \coprod Y \mid Z$ *if and only if the model structure* M *does not contain any* (α,β,γ) *with* α, β *both nonzero, that is, any* X, Y *interaction is excluded.*

PROOF. Under Assumption (1) the conditional independence is equivalent to the factorization, or, taking the logarithms, the representation of

$$\sum_{\alpha\beta\gamma \in M} \psi_{\alpha\beta\gamma} X^\alpha Y^\beta Z^\gamma$$

as a sum of a function in the marginal space $\mathbb{R}[X,Z]/\operatorname{Ideal}(D_{X,Z})$ and a function in the marginal space $\mathbb{R}[Y,Z]/\operatorname{Ideal}(D_{Y,Z})$. Under Assumption (2) this is possible only in absence of the indicated interactions. □

The omission of the terms $X^\alpha Y^\beta Z^\gamma$ with α and β both non-zero, $(\alpha \neq 0)(\beta \neq 0) = 1$, from the model affects the form of the residual matrix, Z_2, and of $\mathcal{Z}_2$ in the analysis of the previous sections and equivalently the toric ideals. The following lemma links the model structure implied by $X \coprod Y \mid Z$ to the structure of the toric ideal.

LEMMA 6.3. *In the set up of Lemma 6.2, in the case* $D_{X,Y,Z} = D_X \times D_Y \times D_Z$ *construct separate series of orthogonal polynomials with integer coefficients* $\{p_\alpha\}$, $\{q_\beta\}$, $\{r_\gamma\}$ *on* D_X, D_Y *and* D_Z, *respectively. Then, one residual matrix is*

$$Z_2 = \left[p_\alpha q_\beta r_\gamma\right]_{(\alpha \neq 0)(\beta \neq 0)=1},$$

and the toric ideal can be constructed by saturation.

PROOF. It follows directly from the fact that the product of integer orthogonal bases is an orthogonal integer basis for the product space. □

The above can be generalized to a very important group of models based on trees.

7. Conditional independence on trees

We refer to [**14**] for graph theory in graphical models. Recall that a tree is a graph $G(E,V)$ with no cycles where V is the set of vertices and E the set of edges, indicated as pairs of vertices. By identifying a source (root) vertex $v \in V$ and giving the edges directions (arrows) away from the source, the tree becomes a special case of a directed graph. Directed graphs are at the foundation of Bayesian influence diagrams and hence of modern probabilistic artificial intelligence, or learning. Moreover trees are an important subclass of models which we now explain by mean of an example.

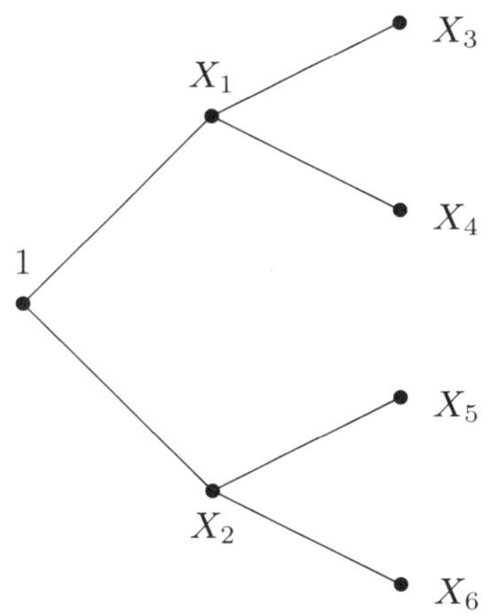

FIGURE 1. A tree of random variables.

EXAMPLE 7.1. Consider six binary random variables $X_1, \dots, X_6$ and the following tree in which each node holds one of the random variables except for the root to which we simply attach the symbol 1. The tree in Figure 1 holds the information about conditional independence as follows

$$\begin{gathered} X_1 \coprod X_2 \\ X_3 \coprod X_4 | X_1 \\ X_5 \coprod X_6 | X_2. \end{gathered}$$

The support is taken to be $D = \{0,1\}^6$ and the joint density $p(x) > 0$, $x \in D$, can be factorized using the chain rule for conditional probability. Thus, with $x = (x_1, \dots, x_6)$

$$p(x) = p_{31}(x_3|x_1)p_{41}(x_4|x_1)p_{52}(x_5|x_2)p_{62}(x_6|x_2)p_1(x_1)p_2(x_2).$$

Now switch to the exponential interpolation. The full $\mathrm{Est}_\tau(D)$ is all 2^6 multilinear terms

$$1, x_1, x_2, \dots, x_1x_2x_3x_4x_5x_6,$$

for any term-ordering, τ. The sub-model represented by the tree is

$$\begin{array}{llllllll} M: & 1, & & & & & \\ & x_1, & x_2, & x_3, & x_4, & x_5, & x_6, \\ & x_1x_3, & x_1x_4, & x_2x_5, & x_2x_6. & & \end{array}$$

Notice that, with care, M can be read directly off the tree. Below we shall give a systematic way of doing this.

DEFINITION 7.2. Let $D = \{0,1\}^d$. Let the binary random variables $X_1, \dots, X_d$ have a joint distribution with density $p(x) > 0$, $x \in D$. The random variables are said to be a *tree model* with respect to a directed tree $T(E, V)$ with $\#V = 1 + d$ and some vertex v_0 if "children" are conditionally independent with respect to "parents"

$$\coprod_{e(i,j) \in E(i)} X_j \mid X_i,$$

where $E(i)$ is the set of directed edges out of vertex i.

The corresponding factorization of $p(x)$ takes the form

$$p(x) = \prod_{i,j:\, i\in V,\, e(i,j)\in E(i)} p(x_j|x_i).$$

DEFINITION 7.3. For any tree $T(E, V)$ define a *maximal chain* as a connected path from the root to a final branch, that is, a vertex such that $E(v)$ is the empty set.

The maximal chains together with all their subsets form a simplicial complex. These subsets are themselves sometimes referred to as chains, but need not be connected. Call the simplicial complex generated in this way $\mathcal{C}(T)$. In the Example 7.1 above, the maximal chains are x_1x_3, x_1x_4, x_2x_5, x_2x_6.

THEOREM 7.4. *Let $X_1, \dots, X_d$ be a tree model with respect to the tree $T(E, V)$ and let $\mathcal{C}(T)$ be the simplicial complex generated by the maximal chains of T. Let the distribution be $p(x) > 0$, $x \in D = \{0,1\}^d$ and $\sum_{x\in D} p(x) = 1$. Define the exponential interpolator*

$$p(x) = \exp\left(\sum_{\alpha\in M} \psi_\alpha x^\alpha\right),$$

with $M \subseteq \{\alpha : \alpha \in D\}$. Then

$$M = \{\alpha : x^\alpha \in \mathcal{C}(T)\}.$$

PROOF. It is straightforward by repeated applications of Lemma 6.2. □

We are now in a position to give explicit generators for the toric ideal for the probabilities $p(x)$ and hence to set up algebraic formulae for the probabilistic model. Consider the setup of Theorem 7.4. The following theorem is a direct consequence of the "automatic" orthogonality of Example 3.2.

THEOREM 7.5. *Let L be the full list for the (saturated) model consisting of all multi-linear terms and let M be like in Theorem (7.4). Define*

$$\overline{M} = L \setminus M$$

namely the complementary list of M in L. Then, the toric ideal corresponding to the model of terms generated by M, is the saturation ideal of the ideal generated by the binomials

$$\left\{\prod_{x:x^\alpha=1} p(x) - \prod_{x:x^\alpha=0} p(x) : \alpha \in \overline{M}\right\}.$$

PROOF. By computation of the relevant Z_2 matrix and the result on saturation in Theorem 5.1. □

EXAMPLE 7.6. For Example 7.1 the list $\{x^\alpha : \alpha \in \overline{M}\}$ is

$$\begin{array}{lll}
x_1x_2, & x_1x_5, & x_1x_6, \\
x_2x_3, & x_2x_4, & \\
x_3x_4, & x_3x_5, & x_3x_6, \\
x_4x_5, & x_6x_6, & \\
x_5x_6, & & \\
x_1x_2x_3, & \dots, & x_4x_5x_6, \\
x_1x_2x_3x_4, & \dots, & x_3x_4x_5x_6, \\
x_1x_2x_3x_4x_5, & \dots, & x_2x_3x_4x_5x_6, \\
x_1x_2x_3x_4x_5x_6. & &
\end{array}$$

Each of these terms leads to a generator of the toric ideal. For x_3x_4 we have

$$x_3x_4 : \quad \Pi_{x:x_3x_4=1}p(x) - \Pi_{x:x_3x_4=0}p(x) \ .$$

The algebraic conditions for the $p(x)$ are given by

$$\left\{\Pi_{x:x^\alpha=1}p(x) - \Pi_{x:x^\alpha=0}p(x) = 0 | \alpha \in \overline{M}\right\}.$$

In the case exhibited here $\overline{M}$ can itself be seen to have its own simplicial structure.

The binary case can be extended in a straightforward way to the case when each X_i is a vector of binary random variable by using Lemmas 6.2 and 6.3.

This more general case arises as an important reduction of directed graphs which are the main component of Bayesian influence diagrams (for details see [**14**]). These can be explained simply as follows. We associate to each vertex v_i of a directed graph $G(E, V)$ a random variable X_i. Then, the $X_1, \dots, X_n$ (where n is the number of vertices) are said to have the Markov property with respect to $G(E, V)$ if (again) "children" are conditionally independent given "parents" and in such a way that "parents" separate "children" from "ancestors" (vertices further back but reverse reachable from children). However, when parents are grouped together, so that any two parents which have the same children are merged ("moralization" to use the quaint term), the graph becomes a tree in which the merged parents are grouped together at a super-node (called a "clique"). Thus, the general tree formula can be seen, in a sense, as holding together the maximal amount of first order Markov resolution.

Returning to the first case of interpolation in Section 1 we can call the raw polynomial through the probabilities in the saturated case

$$p(x) = \sum_{\alpha \in L} \theta_\alpha x^\alpha.$$

In matrix terms we have

$$\begin{aligned}
[p] &= Z[\theta] \\
[\theta] &= Z^{-1}[p].
\end{aligned}$$

Adjoining to this the toric ideal conditions derived from the model M and using Theorem 5.1, we can produce polynomial interpolators parametrized by θ which satisfy the model conditions. We can compute, at least in principle, as possibly rational forms in θ, all the conditional probabilities in the tree case.

References

1. David F. Andrew and James E. Stafford, *Symbolic computation for statistical inference*, Oxford University Press, Oxford, 2000.
2. Anna M. Bigatti and Lorenzo Robbiano, *Toric ideals*, Matemática Contemporânea (2000), to appear.
3. Massimo Caboara and Lorenzo Robbiano, *Families of estimable terms*, Proceedings ISSAC, UWO, Canada, ACM (2001), to appear.
4. Antonio Capani, Gianfranco Niesi, and Lorenzo Robbiano, CoCoA, *a system for doing computations in commutative algebra*, Available via anonymous ftp from `cocoa.dima.unige.it`.
5. Arjeh M. Cohen, Hans Cuypers, and Hans Sterk (eds.), *Some tapas of computer algebra*, Springer-Verlag, Berlin, 1999.
6. Pasqualina Conti and Carlo Traverso, *Buchberger algorithm and integer programming*, Applied algebra, algebraic algorithms and error-correcting codes (New Orleans, LA, 1991), Springer, Berlin, 1991, pp. 130–139.
7. David Cox, John Little, and Donal O'Shea, *Ideals, varieties, and algorithms: An introduction to computational algebraic geometry and commutative algebra*, second ed., Springer-Verlag, New York, 1997.
8. ______, *Using algebraic geometry*, Springer-Verlag, New York, 1998.
9. Persi Diaconis and Bernt Sturmfels, *Algebraic algorithms for sampling from conditional distributions*, The Annals of Statistics **26** (1998), no. 1, 363–397.
10. Ian H. Dinwoodie, *The Diaconis-Sturmfels algorithm and rules of succession*, Bernoulli **4** (1998), no. 3, 401–410.
11. Roberto Fontana, Giovanni Pistone, and Maria Piera Rogantin, *Classification of two-level factorial fractions*, J. Statist. Plann. Inference **87** (2000), no. 1, 149–172.
12. Shelby J. Haberman, *The analysis of frequency data*, The University of Chicago Press, Chicago, Ill.-London, 1974, Statistical Research Monographs, Vol. IV.
13. Martin Kreuzer and Lorenzo Robbiano, *Computational commutative algebra. 1*, Springer-Verlag, Berlin, 2000.
14. Steffen L. Lauritzen, *Graphical models*, The Clarendon Press Oxford University Press, New York, 1996, Oxford Science Publications.
15. P. McCullagh and J. A. Nelder, *Generalised linear models*, second ed., Chapman & Hall, London, 1989.
16. Peter McCullagh, *Invariance and factorial models. With discussion and a reply by the author*, J. R. Stat. Soc. Ser. B Stat. Methodol. **62** (2000), no. 2, 209–256.
17. Mauro Piccioni, *Graphical models and structural zeros*, GROSTAT IV Workshop, IUT STID Menton, 2000.
18. Giovanni Pistone, Eva Riccomagno, and Henry P. Wynn, *Algebraic statistics: Computational commutative algebra in statistics*, Chapman&Hall/CRC, Boca Raton, 2000.
19. Giovanni Pistone and Henry P. Wynn, *Generalised confounding with Gröbner bases*, Biometrika **83** (1996), no. 3, 653–666.
20. Fabio Rapallo, *Algebraic Markov bases and MCMC for two-ways contingency tables*, submitted, 2001.
21. Eva Riccomagno, *Algebraic geometry in experimental design and related fields*, Phd thesis, Department of Statistics, University of Warwick, August 1997.
22. Tiziana Rizzo, *Markov chains and toric ideals*, GROSTAT IV Workshop, IUT STID Menton, 2000.
23. Lorenzo Robbiano, *Toric ideals*, GROSTAT IV Workshop, IUT STID Menton, 2000.
24. Bernd Sturmfels, *Gröbner bases and convex polytopes*, American Mathematical Society, Providence, RI, 1996.

Dipartimento di Matematica, Politecnico di Torino, Italy
E-mail address: `piston@calvino.polito.it`

Department of Statistics, University of Warwick
E-mail address: `emr@stats.warwick.ac.uk`

Department of Statistics, University of Warwick
E-mail address: `hpw@stats.warwick.ac.uk`

Contemporary Mathematics
Volume **287**, 2001

Maximum Covariance Difference Test for Equality of Two Covariance Matrices

Akimichi Takemura and Satoshi Kuriki

ABSTRACT. We propose a test of equality of two covariance matrices based on the maximum standardized difference of scalar covariances of two sample covariance matrices. We derive the tail probability of the asymptotic null distribution of the test statistic by the tube method. However the usual formal tube formula has to be suitably modified, because in this case the index set, around which the tube is formed, has zero critical radius.

1. Introduction

Consider the null hypothesis of equality of two covariance matrices

$$H_0 : \Sigma_1 = \Sigma_2$$

of two Wishart populations against the alternative $\Sigma_1 \neq \Sigma_2$. In this paper we propose a test statistic based on maximizing the standardized difference of sample scalar covariances from two populations. Let $p\times p$ ($p \geq 2$) random matrices W_1, W_2 be independently distributed according to Wishart distributions $W_p(n_i, \Sigma_i)$, $i = 1, 2$. n_1 and n_2 are real numbers greater than $p-1$. Consider the difference of the scalar covariances $a'S_1b - a'S_2b$, where $S_i = W_i/n_i$, $i = 1, 2$, are sample covariance matrices and a, b are $p \times 1$ vectors. Under the null hypothesis $H_0 : \Sigma_1 = \Sigma_2 = \Sigma$, the variance of the difference for fixed a, b is written as

$$\text{Var}(a'S_1b - a'S_2b) = \left(\frac{1}{n_1} + \frac{1}{n_2}\right)\left((a'\Sigma a)(b'\Sigma b) + (a'\Sigma b)^2\right). \tag{1.1}$$

See Appendix A. Therefore our proposed statistic is

$$T = \max_{a,b\in R^p} \frac{a'S_1b - a'S_2b}{\sqrt{(1/n_1 + 1/n_2)((a'Sa)(b'Sb) + (a'Sb)^2)}}, \tag{1.2}$$

where S is the pooled sample covariance matrix

$$S = \frac{W_1 + W_2}{n_1 + n_2}.$$

In the case that maximizing vectors a and b in (1.2) coincide, our proposed statistic detects the difference in the scalar variances of two populations. However as discussed in the next section, maximizing vectors a and b may be different. In

---

1991 *Mathematics Subject Classification.* Primary 62H10; Secondary 62H15.

this case our statistic detects the difference in scalar covariances. In this sense our statistic is different from Roy's maximum-minimum roots test ([**8**]), which compares only the scalar variances. In fact we shall show that with overwhelming probability the maximizing values of a and b are different for large p (see Table 1 below) or large value of our statistic T (see Proposition 3.2). Therefore the behavior of our statistic is quite different from that of Roy's test. It is clearly different from omnibus type test procedures such as likelihood ratio test or Nagao's trace test ([**5**]). Comparisons of various test statistics mainly from the viewpoint of power behavior have been given in [**7**], [**6**], [**1**] among others.

We have described our statistic in the setting of two-sample problem. One-sample version is even simpler. Let a $p \times p$ matrix $W = nS$ be distributed according to Wishart distribution $W_p(n, \Sigma)$. For testing

$$H_0 : \Sigma = I \quad \text{(the identity matrix)}$$

our test statistic is

$$T = \max_{a,b \in R^p} \frac{\sqrt{n}a'(S-I)b}{\sqrt{(a'a)(b'b) + (a'b)^2}}. \tag{1.3}$$

The advantage of maximum type test, including our test statistic and Roy's maximum-minimum roots test, is that when the null hypothesis is rejected, the test statistic itself suggests the direction of departure from the null hypothesis. On the other hand the drawback of the maximum type test is that the asymptotic null distribution is difficult to evaluate, because the limiting distribution is not a χ^2 distribution in general. Recently it has been recognized that the tube formula or the Euler characteristic method provide a general methodology for evaluating the tail probability of the asymptotic null distribution of maximum type statistic in various testing problems in classical multivariate analysis. The tube method can be applied to our proposed statistic as well.

However there is a substantial difficulty in formal application of the tube formula to the present problem. For the application of the tube formula we set up a Gaussian random field $Z(u) = \langle u, Z \rangle$, $u \in M$, corresponding to the asymptotic null distribution of Wishart matrices. The index set M of the Gaussian field is a subset of the $p(p+1)/2 - 1$ dimensional unit sphere $S^{p(p+1)/2-1}$. It turns out that the critical radius of M for our present problem is zero and the formal expansion of the tail probability based on tube formula is only partly valid. In our previous work ([**11**]) we have investigated the validity of tube formula and found that the asymptotic expansion is valid up to certain degrees of freedom in the case of zero critical radius. Employing our previous argument it will be shown that for our present problem the asymptotic expansion is valid for $[p/2]$ (the integral part of $p/2$) terms.

The organization of this paper is as follows. In Section 2 we explicitly solve the maximization in (1.2) and (1.3) and give some intuitive meaning of our statistic. Section 3 is the main section of this paper and we investigate the geometry of the index set M and evaluate coefficients in the tube formula for M. By studying singularities of M, we will show that the asymptotic expansion based on the tube formula is valid for $[p/2]$ terms. In Section 4 we give some simulation results to confirm the accuracy of approximation by tube formula. From the simulation we see that for the present problem the tube formula approximation is practical, but not remarkably good as in the cases with positive critical radius. Satisfactory approximation is achieved by explicit evaluation of the term of the order, where the formal tube formula breaks down. We also give some simulation results of power behavior of our statistic. Note that the theoretical investigation of the power behavior of our statistic is difficult, because at the present the tube formula is not applicable under contiguous alternative hypotheses.

2. Some simple interpretations of proposed statistic

In this section we explicitly solve the maximization in (1.2) and (1.3). This leads to a clear understanding of the meaning of our proposed test statistic. Throughout this paper for a real symmetric matrix A, we denote the positive (non-negative, non-positive, negative) definiteness by $A > 0$ ($A \geq 0$, $A \leq 0$, $A < 0$).

The maximization is entirely equivalent in (1.2) and (1.3). In (1.2) let

$$Z = \sqrt{\frac{n_1 n_2}{2(n_1+n_2)}} S^{-1/2}(S_1 - S_2)S^{-1/2} \tag{2.1}$$

and replace a by $S^{1/2}a$ and b by $S^{1/2}a$. Then the maximization in (1.2) reduces to the maximization in (1.3), where

$$Z = \sqrt{\frac{n}{2}}(S - I). \tag{2.2}$$

Therefore we consider the maximization problem

$$\max_{a,b\in R^p} \frac{\sqrt{2}a'Zb}{\sqrt{(a'a)(b'b) + (a'b)^2}},$$

where Z is a real symmetric matrix. It is convenient for later discussion to include the factor $1/\sqrt{2}$ in the definition of Z. Since the maximum is always non-negative, we can equivalently solve

$$\max_{a,b\in R^p} \frac{2(a'Zb)^2}{(a'a)(b'b) + (a'b)^2} = \max_{a,b\in R^p} \frac{2a'Zbb'Za}{a'((b'b)I + bb')a},$$

which is a ratio of quadratic forms in a for fixed b. Therefore for fixed b, the maximizing a is proportional to

$$(I + bb'/(b'b))^{-1}Zb = (I - bb'/(2b'b))Zb = Zb - \frac{b'Zb}{2b'b}b$$

and the maximum value is given by

$$\frac{2b'Z(I + bb'/(b'b))^{-1}Zb}{b'b} = \frac{2b'b \cdot b'Z^2b - (b'Zb)^2}{(b'b)^2}. \tag{2.3}$$

We now maximize this with respect to b under the restriction $b'b = 1$. Note that by diagonalizing Z, we can reduce the problem to the case of diagonal $Z = \mathrm{diag}(l_1, \ldots, l_p)$, $l_1 \geq \cdots \geq l_p$. Write $b = (b_1, \ldots, b_p)'$ and $x_i = b_i^2 \geq 0$, $i = 1, \ldots, p$, with $\sum_{i=1}^p x_i = 1$. Then (2.3) is written as

$$Q = 2\sum x_i l_i^2 - \left(\sum x_i l_i\right)^2.$$

Suppose that $l_1 \geq l_i > 0$. Fixing $x_j, j \neq 1, i$, consider changing x_1, x_i to $x_1 + c, x_i - c$, respectively. Differentiating Q by c we have

$$\frac{1}{2}\frac{\partial Q}{\partial c} = (l_1^2 - l_i^2) - \left(\sum x_j l_j\right)(l_1 - l_i) = (l_1 - l_i)\left(l_1 + l_i - \sum x_j l_j\right).$$

This is non-negative because $l_1 \geq \sum x_j l_j$. Therefore we increase Q by changing x_1, x_i to $x_1 + x_i, 0$, respectively. Similarly if $0 > l_i \geq l_p$ we increase Q by changing x_p, x_i to $x_p + x_i, 0$, respectively. It now follows that for positive definite Z we maximize Q by taking $x_1 = 1, x_2 = \cdots = x_p = 0$. Similarly for negative definite Z we maximize Q by $x_p = 1, x_1 = \cdots = x_{p-1} = 0$. It remains to consider the case $l_1 \geq 0 \geq l_p$. The above argument shows that at the maximum $x_2 = \cdots = x_{p-1} = 0$. Write $x_1 = c$, $x_p = 1 - c$. $\partial Q/\partial c = 2(l_1 - l_p)(l_1 + l_p - cl_1 - (1-c)l_p) = 0$ yields

$$x_1 = c = \frac{l_1}{l_1 - l_p}, \quad x_p = 1 - c = \frac{-l_p}{l_1 - l_p}.$$

a and b are given as

$$b = \left(\frac{\sqrt{l_1}}{\sqrt{l_1 - l_p}}, 0, \ldots, 0, \pm\frac{\sqrt{-l_p}}{\sqrt{l_1 - l_p}}\right)',$$

$$(2.4) \qquad a = \left(\frac{\sqrt{l_1}}{\sqrt{l_1 - l_p}}, 0, \ldots, 0, \mp\frac{\sqrt{-l_p}}{\sqrt{l_1 - l_p}}\right)'.$$

(see Appendix A for a.) Summarizing the above derivation in terms of original Z, we have proved the following theorem.

THEOREM 2.1. *Define Z by (2.1) for the two-sample problem or by (2.2) for the one-sample problem. Write the spectral decomposition of Z as $Z = \sum_{i=1}^p l_i h_i h_i'$, $l_1 \geq \cdots \geq l_p$. Then the value of the statistic T in (1.2) or (1.3) is given by*

$$T = \begin{cases} l_1, & \text{if } Z > 0, \\ \sqrt{l_1^2 + l_p^2}, & \text{if } l_1 \geq 0 \geq l_p, \\ -l_p, & \text{if } Z < 0. \end{cases}$$

The set of maximizing vectors $\{S^{1/2}a, S^{1/2}b\}$ for the two-sample problem or $\{a, b\}$ for the one-sample problem is given by

$$\{S^{1/2}a, S^{1/2}b\} \text{ or } \{a, b\} = \begin{cases} \{h_1, h_1\}, & \text{if } Z > 0, \\ \left\{\frac{\sqrt{l_1}h_1 + \sqrt{-l_p}h_p}{\sqrt{l_1 - l_p}}, \frac{\sqrt{l_1}h_1 - \sqrt{-l_p}h_p}{\sqrt{l_1 - l_p}}\right\}, & \text{if } l_1 \geq 0 \geq l_p, \\ \{h_p, -h_p\}, & \text{if } Z < 0. \end{cases}$$

Note that for the case of non-negative definite or non-positive definite Z, $a = \pm b$ and our proposed statistic T corresponds to maximized difference in scalar variances. However for the case $l_1 > 0 > l_p$, a and b are not parallel and T corresponds to difference in scalar covariances. The situation is clearly understood by considering the simple case of one-sample problem with $p = 2$. In Figure 1 we have depicted the concentration ellipse $x'S^{-1}x = 1$ for 3 cases of Theorem 2.1. In Case 1, $S - I_2$ is positive definite. In this case $a = b$ and we are looking at the maximum variance. Similarly in Case 3, where $S - I_2$ is negative definite, we are looking at the minimum variance. Case 2 depicts the intermediate case. For simplicity consider

$$S = \begin{pmatrix} \mathrm{Var}(x_1) & \mathrm{Cov}(x_1, x_2) \\ \mathrm{Cov}(x_1, x_2) & \mathrm{Var}(x_2) \end{pmatrix} = \begin{pmatrix} 1 & r \\ r & 1 \end{pmatrix}, \quad r \neq 0.$$

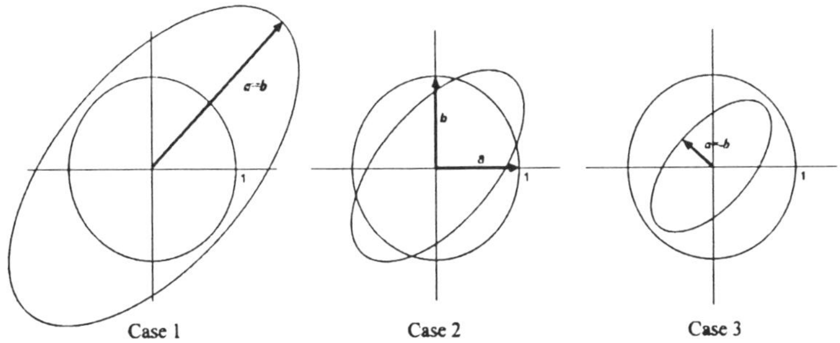

FIGURE 1. Three cases of maximization.

Then $a = (1,0)'$, $b = (0,1)'$ and

$$T = |\operatorname{Cov}(x_1, x_2)| = |r|.$$

In this case our statistic T detects the deviation of the covariance from 0.

3. Evaluation of tail probability of asymptotic null distribution by tube formula

In this section we evaluate the tail probability of the asymptotic null distribution of our statistic using the tube formula for general dimension. For the case $p = 2$, the exact form of the asymptotic null distribution is easy to evaluate and given in Appendix D.

Considering the asymptotic distribution of Z in (2.1) or (2.2), in this section let the distribution of Z be the standard symmetric matrix normal distribution, i.e., the elements z_{ij}, $1 \le i \le j \le p$, of Z are independent normal random variables with mean 0 and variance

$$\operatorname{Var}(z_{ij}) = \begin{cases} 1, & \text{if } i = j, \\ 1/2, & \text{if } i < j. \end{cases}$$

Then by the invariance principle the asymptotic null distribution of T in (1.2) or (1.3) is written as

$$\lim_{n \to \infty} P(T > x) = P\left(\max_{a,b \in R^p} \frac{\sqrt{2}a'Zb}{\sqrt{(a'a)(b'b) + (a'b)^2}} > x \right),$$

where $n = \min(n_1, n_2)$ for the two-sample problem. Therefore the problem of evaluating the asymptotic null distribution is reduced to the evaluation of the distribution of the maximum of a Gaussian random field, for which the tube formula ([**9**], see also [**3**], [**10**]) can be employed.

The sample space $\mathcal{S}$ of the standard symmetric matrix normal variable Z is the set of $p \times p$ real symmetric matrices with the inner product

$$\langle A, B \rangle = \operatorname{tr} AB, \quad A, B \in \mathcal{S}.$$

By the correspondence

$$A = (a_{ij}) \leftrightarrow (a_{11}, \dots, a_{pp}, \sqrt{2}a_{12}, \dots, \sqrt{2}a_{p-1,p})$$

$\mathcal{S}$ can be identified with $R^{p(p+1)/2}$ as is done in [**4**]. Let $K \subset \mathcal{S}$ denote the cone

$$K = \{ab' + ba' \mid a, b \in R^p\}$$

and define

$$M = K \cap S^{p(p+1)/2-1} = \left\{ \frac{ab' + ba'}{\|ab' + ba'\|} \mid a, b \in R^p \right\},$$

where

$$\|ab' + ba'\| = \langle ab' + ba', ab' + ba' \rangle^{1/2} = \sqrt{2((a'a)(b'b) + (a'b)^2)}.$$

Then

$$\max_{a,b \in R^p} \frac{\sqrt{2}a'Zb}{\sqrt{(a'a)(b'b) + (a'b)^2}} = \max_{u \in M} \langle u, Z \rangle$$

and our problem is reduced to the canonical form suitable for application of the tube formula. We first determine the global geometry of K and M.

PROPOSITION 3.1. *Let K_2 denote the set of $p \times p$ real symmetric matrices of rank less than or equal to 2. Let $\mathcal{S}_+$ and $\mathcal{S}_-$ denote the set of positive definite matrices and negative definite matrices, respectively. Then*

$$K = K_2 \cap \mathcal{S}_+^C \cap \mathcal{S}_-^C,$$

where A^C denotes the complement of A.

PROOF. If $a \propto b$ then $ab' + ba' \propto aa'$ is of rank 0 or 1. If a and b are linearly independent then

$$ab' + ba' = \frac{1}{2}\left((a + b)(a + b)' - (a - b)(a - b)'\right).$$

Here $a + b$ and $a - b$ are linearly independent. By Sylvester's law $ab' + ba'$ has 1 positive and 1 negative characteristic root. Furthermore note that all matrices of rank 2 with 1 positive and 1 negative characteristic root can be written as $ab' + ba'$ again by Sylvester's law. This proves the proposition. □

From this proposition we see that $M = K \cap S^{p(p+1)/2-1}$ is a manifold with boundary

$$\partial M = \{A \in M \mid \operatorname{rank} A = 1\} = \{hh' \mid h \in S^{p-1}\} \cup \{-hh' \mid h \in S^{p-1}\}.$$

Note that each of the two components of ∂M forms a manifold of dimension $p - 1$. We shall show that except for $p = 2$ the boundary of M has a singularity in the sense that at $u \in \partial M$ the tangent cone (supporting cone) $S_u(M)$ of M is not convex. Because of this singularity the critical radius of M is 0 for $p \geq 3$ as discussed in [**11**]. For the case $p = 2$, $\mathcal{S}$ is identified with R^3 and we can fully describe K and M. For illustrative purpose this is done in Appendix B. The relative interior M^O of M consists of matrices with one positive and one negative root and forms a manifold of dimension $2p - 2$. In the following we denote the smallest cone containing M^O and ∂K by $K^O = \bigcup_{c>0} cM^O$, $\partial K = \bigcup_{c>0} c\, \partial M$, respectively.

Let $\bar{G}_m$ denote the upper probability of χ^2 distribution with m degrees of freedom. From the standard argument of the tube formula, we know that the order of the tail probability of the projection of Z onto K^O, is of the order $O(\bar{G}_{2p-1})$, i.e.,

$$P\left(\sup_{u \in M^O} \langle u, Z \rangle > t\right) = P\left(\max_{u \in M} \langle u, Z \rangle > t\right) = O(\bar{G}_{2p-1}(t^2)) \quad \text{as } t \to \infty. \tag{3.1}$$

TABLE 1. Probability of positive definiteness of Z.

p	$P(Z>0)$
2	0.1464466
3	0.0249209
4	0.0024567
5	0.0001401
6	0.0000046
7	8.8×10^{-8}

Here $2p-1$ is the dimension of K^O. Similarly the order of the tail probability of the projection of Z onto ∂K is of the order $O(\bar{G}_p)$:

$$P\left(\max_{u\in\partial M}\langle u,Z\rangle > t\right) = O(\bar{G}_p(t^2)) = O\big(t^{-(p-1)}\bar{G}_{2p-1}(t^2)\big) \quad \text{as } t\to\infty. \tag{3.2}$$

Note that (3.1) corresponds to the second case of Theorem 2.1 and (3.2) corresponds to the cases 1,3 of Theorem 2.1. Therefore simply by counting the dimensions of K^O and ∂K we have the following proposition.

PROPOSITION 3.2. Under the asymptotic null distribution

$$P(Z>0 \text{ or } Z<0 \mid T>t) = O\big(t^{-(p-1)}\big) \quad \text{as } t\to\infty.$$

This proposition shows that as far as small P-values are concerned, the case of $a=\pm b$ in the maximization of (1.2) or (1.3) can be ignored. The statement of this proposition will be strengthened in Proposition 3.8

Actually the unconditional probability of the case $a=\pm b$ becomes very small as p becomes large. Table 1 lists the probabilities of Case 1, i.e., the probability of Z being positive definite, for dimensions $p=2,3,\dots,7$. We see that for $p\geq 4$, the unconditional probability is negligible at the usual significance levels of 5% or 1%. The probabilities in Table 1 were calculated by a recurrence formula given in Lemma 3.5 below.

We now determine various differential geometric quantities of K and M. Actually most of the calculations have been done in [4]. The local geometry of ∂M is given by the following proposition. Since the two components of ∂M are entirely similar, we only consider the non-negative definite part of ∂M. Since the relation between the geometries of M and K is trivial we only state results in terms of K.

PROPOSITION 3.3. Let $y=h_1h_1'\in\partial M$. Choose $h_2,\dots,h_p$ such that $\{h_1,\dots,h_p\}$ is an orthonormal basis of R^p. The tangent cone $S_y(K)$ of K at y is given by

$$S_y(K) = \operatorname{span}\{h_1h_j'+h_jh_1',\ j=1,\dots,p\}\oplus\{-qq' \mid h_1'q=0\},$$

where $\oplus$ denotes the orthogonal sum. The convex hull $C(S_y(K))$ of $S_y(K)$ is

$$C(S_y(K)) = \operatorname{span}\{h_1h_j'+h_jh_1',\ j=1,\dots,p\}\oplus\{A\leq 0 \mid h_1'Ah_1=0\}$$

and the normal cone $N_y(K)=C(S_y(K))^*$ is

$$N_y(K)=C(S_y(K))^* = \{A\geq 0\mid h_1'Ah_1=0\}.$$

The set of non-zero characteristic roots of the second fundamental form of K at y with respect to $A\in N_y(K)$ is given by $\{-l_j,\ j=2,\dots,p\}$, where $l_2,\dots,l_p$ are characteristic roots of A.

PROOF. All the results are given in Section 2 of [**4**] except for the description of the tangent cone $S_y(K)$. Differentiating $y = h_1 h_1'$ with respect to h_1 gives the first term of $S_y(K)$. Now consider $x = h_1 h_1' - \epsilon q q' \in K^O$, $h_1' q = 0$, $\epsilon > 0$, in a neighborhood of y. This gives the second term of $S_y(K)$. □

Note that at $y \in \partial M$, $S_y(K)$ and its convex hull $C(S_y(K))$ are different and $S_y(K)$ is not convex. As discussed in [**11**], this implies that the critical radius of M is zero and the tube formula applied to M^O is not entirely valid. We shall discuss this point below in more detail.

M^O and K^O are smooth and we summarize results on M^O and K^O from [**4**].

PROPOSITION 3.4. *Let $y \in M^O$ and write the spectral decomposition of y as $y = l_1 h_1 h_1' + l_p h_p h_p'$, $l_1 > 0 > l_p$, $l_1^2 + l_p^2 = 1$. Choose $h_2, \ldots, h_{p-1}$ such that $\{h_1, \ldots, h_p\}$ is an orthonormal basis of R^p. The tangent space of $S_y(K)$ of K at y is*

$$S_y(K) = \operatorname{span}\left\{h_i h_j' + h_j h_i',\ i = 1, p,\ j = 1, 2, \ldots, p\right\}.$$

The normal space $N_y(K)$ of K at y is

$$N_y(K) = S_y(K)^* = \operatorname{span}\left\{h_i h_j' + h_j h_i',\ 2 \le i \le j \le p-1\right\}.$$

The set of non-zero characteristic roots of the second fundamental form of K at y with respect to $A \in N_y(K)$ is given by

$$\left\{-\frac{l_j}{l_1}, -\frac{l_j}{l_p},\ j = 2, \ldots, p-1\right\},$$

where $l_2, \ldots, l_{p-1}$ are characteristic roots of A.

Given the above geometrical quantities we can now employ the tube formula for approximating the tail probability of asymptotic null distribution of our statistic T. Calculations of the tube formula for this case is very similar to those in [**4**] and in Section 3.2 of [**11**].

Let $l_1 \ge \cdots \ge l_p$ denote the characteristic roots of Z. The exact tail probability of T is given

$$\begin{aligned} P(T > t) &= 2P(l_1 > t, Z > 0) + P\left(l_1^2 + l_p^2 > t^2,\ l_1 > 0 > l_p\right) \\ &= 2F_1(t) + F_2(t) \quad \text{(say)}. \end{aligned} \tag{3.3}$$

The joint density function of $l_1, \ldots, l_p$ is given by

$$\begin{aligned} f(l_1, \ldots, l_p) &= d(p) \exp\left(-\frac{1}{2}\sum_{i=1}^{p} l_i^2\right) \prod_{1 \le i < j \le p} (l_i - l_j) \\ &= d(p) \exp\left(-\frac{1}{2}\sum_{i=1}^{p} l_i^2\right) \det\left(l_i^{p-j}\right)_{1 \le i,j \le p}, \end{aligned} \tag{3.4}$$

where

$$d(p) = \frac{1}{2^{p/2} \prod_{i=1}^{p} \Gamma(i/2)}. \tag{3.5}$$

Let

$$\begin{aligned} B_1(t) &= \{l_1 > t,\ \infty > l_2 > \cdots > l_p > 0\}, \\ B_2(t) &= \left\{l_1^2 + l_p^2 > t^2,\ l_1 > 0 > l_p,\ \infty > l_2 > \cdots > l_{p-1} > -\infty\right\}. \end{aligned}$$

Following the derivation in [4] and the example of Section 3.2 of [11], it is shown that the tube formula approximation to $F_1(t)$ and $F_2(t)$ is obtained by ignoring the order constraint $l_1 \geq l_2$ and $l_{p-1} \geq l_p$ in integrating the joint density:

$$
\begin{aligned}
(3.6)\quad P(T>t) \simeq \tilde{P}(T>t) &= \left\{ 2\int_{B_1(t)} + \int_{B_2(t)} \right\} f(l_1,\dots,l_p)\,dl_1\dots dl_p \\
&= 2\tilde{F}_1(t) + \tilde{F}_2(t).
\end{aligned}
$$

Now we employ the techniques of [2] for evaluating $\tilde{F}_1(t)$ and $\tilde{F}_2(t)$. Define

$$
(3.7)\quad U_k(q_1,\dots,q_k) = \int\cdots\int_{\infty>l_1>\cdots>l_k>0} \exp\left(-\frac{1}{2}\sum_{i=1}^k l_i^2\right) \det(l_i^{q_j})_{1\leq i,j\leq k} \prod_{i=1}^k dl_i,
$$

$$
(3.8)\quad V_k(q_1,\dots,q_k) = \int\cdots\int_{\infty>l_1>\cdots>l_k>-\infty} \exp\left(-\frac{1}{2}\sum_{i=1}^k l_i^2\right) \det(l_i^{q_j})_{1\leq i,j\leq k} \prod_{i=1}^k dl_i,
$$

where $q_1,\dots,q_k$ are non-negative integers. U_k was introduced in Section 2 of [2] and can be evaluated by the following recurrence formula.

LEMMA 3.5 (Theorem 2.2 of [2]). *$U_k(q_1,\dots,q_k)$ satisfies the following recurrence relation:*

$$
\begin{aligned}
&U_k(q_1,\dots,q_k) \\
&\quad = (-1)^{k-1}U_{k-1}(q_2,\dots,q_k)I(q_1=1) \\
&\quad\quad +(q_1-1)U_k(q_1-2,q_2,\dots,q_k) \\
&\quad\quad +2\sum_{j=2}^k(-1)^j\frac{1}{2^{\frac{1}{2}(q_1+q_j)}}U_1(q_1+q_j-1)U_{k-2}(q_2,\dots,q_{j-1},q_{j+1},\dots,q_k)
\end{aligned}
$$

$(k\geq 2,\ q_1\geq 1)$, with the initial condition

$$
\begin{aligned}
U_1(q) &= I(q=1)+(q-1)U_1(q-2) \qquad (q\geq 1), \\
U_1(0) &= \sqrt{\pi/2},
\end{aligned}
$$

where $I(\cdot)$ denotes the indicator function.

Similarly $V_k(q_1,\dots,q_k)$ can be evaluated by the the following recurrence formula. The proof is entirely the same as Theorem 2.2 of [2] and omitted.

LEMMA 3.6. *$V_k(q_1,\dots,q_k)$ satisfies the following recurrence relation:*

$$
\begin{aligned}
&V_k(q_1,\dots,q_k) \\
&\quad = (q_1-1)V_k(q_1-2,q_2,\dots,q_k) \\
&\quad\quad +2\sum_{j=2}^k(-1)^j\frac{1}{2^{\frac{1}{2}(q_1+q_j)}}V_1(q_1+q_j-1)V_{k-2}(q_2,\dots,q_{j-1},q_{j+1},\dots,q_k)
\end{aligned}
$$

$(k\geq 2,\ q_1\geq 1)$, with the initial condition

$$
\begin{aligned}
V_1(q) &= (q-1)V_1(q-2) \qquad (q\geq 1), \\
V_1(0) &= \sqrt{2\pi}.
\end{aligned}
$$

Using U_k and V_k the tube formula approximations $\tilde{F}_1(t),\tilde{F}_2(t)$ to $F_1(t),F_2(t)$ are evaluated as follows.

PROPOSITION 3.7. For $k = 1, \dots, p$, let

$$\tau_{p-k+1} = d(p)(-1)^{k-1}2^{(p-k-1)/2}\Gamma\left(\frac{p-k+1}{2}\right) \times U_{p-1}(p-1,\dots,p-k+1,p-k-1,\dots,0)$$

and for $k = 3, 5, \dots, 2p-1$, let

(3.9)

$$\omega_{2p-k+2} = d(p) \sum_{1\le l<m\le p,\, l+m=k} (-1)^{l-1}2^{(2p-k)/2} \times\Gamma\left(\frac{p-l+1}{2}\right)\Gamma\left(\frac{p-m+1}{2}\right) \times V_{p-2}(p-1,\dots,p-l+1,p-l-1,\dots,p-m+1,p-m-1,\dots,0),$$

where $d(p)$ is given in (3.5) and U_k, V_k are given in (3.7), (3.8). Then $\tilde{F}_1(t), \tilde{F}_2(t)$ in the formal tube formula approximation $\tilde{P}(T > t) = 2\tilde{F}_1(t) + \tilde{F}_2(t)$ to the tail probability $P(T > t)$ are written as

$$\tilde{F}_1(t) = \sum_{k=1}^{p} \tau_{p-k+1}\bar{G}_{p-k+1}(t^2),$$

$$\tilde{F}_2(t) = \sum_{k=3,\, k:\mathrm{odd}}^{2p-1} \omega_{2p-k+2}\,\bar{G}_{2p-k+2}(t^2).$$

PROOF. Consider the determinant term $\det(l_i^{p-j})_{1\le i,j\le p}$ in the joint density (3.4) of the characteristic roots of Z. For $\tilde{F}_1$ expand this determinant as

$$\det\left(l_i^{p-j}\right)_{1\le i,j\le p} = \sum_{k=1}^{p}(-1)^{k-1}l_1^{p-k}\det\begin{pmatrix} l_2^{p-1} & \cdots & l_2^{p-k+1} & l_2^{p-k-1} & \cdots & 1 \\ \vdots & & \vdots & \vdots & & \vdots \\ l_p^{p-1} & \cdots & l_p^{p-k+1} & l_p^{p-k-1} & \cdots & 1 \end{pmatrix}$$
$$= \sum_{k=1}^{p}(-1)^{k-1}l_1^{p-k}\det\left(l_{i+1}^{p-j-I(k\le j)}\right)_{1\le i,j\le p-1}.$$

Now l_1 and $l_2, \dots, l_p$ can be separately integrated out. Integration with respect to l_1 gives

$$\int_t^\infty l_1^{p-k}e^{-l_1^2/2}\,dl_1 = 2^{(p-k-1)/2}\Gamma\left(\frac{p-k+1}{2}\right)\bar{G}_{p-k+1}(t^2).$$

Integration with respect to $l_2, \dots, l_p$ gives U_{p-1}. This proves the expansion of $\tilde{F}_1$.

For $\tilde{F}_2$, expand the determinant $\det(l_i^{p-j})_{1\le i,j\le p}$ as

$$\det\left(l_i^{p-j}\right)_{1\le i,j\le p} = \sum_{1\le l<m\le p}(-1)^{l+m+p-1} \times\det\begin{pmatrix} l_1^{p-l} & l_1^{p-m} \\ l_p^{p-l} & l_p^{p-m}\end{pmatrix}\det\left(l_{i+1}^{p-j-I(l\le j)-I(m-1\le j)}\right)_{1\le i,j\le p-2}.$$

Now l_1, l_p and $l_2, \ldots, l_{p-1}$ can be separately integrated out. Integration with respect to l_1, l_p gives

$$\int_{l_1^2+l_p^2>t^2,\, l_1>0>l_p} \left(l_1^{p-l} l_p^{p-m} - l_1^{p-m} l_p^{p-l}\right) e^{-(l_1^2+l_p^2)/2}\, dl_1 dl_p$$
$$= \{(-1)^{p-m} - (-1)^{p-l}\} 2^{(2p-l-m-2)/2}$$
$$\times \Gamma\left(\frac{p-l+1}{2}\right) \Gamma\left(\frac{p-m+1}{2}\right) \bar{G}_{2p-l-m+2}(t^2).$$

Note that

$$(-1)^{p-m} - (-1)^{p-l} = \begin{cases} 0 & \textit{if } m+l \textit{ is even}, \\ 2(-1)^{p-m} & \textit{if } m+l \textit{ is odd}. \end{cases}$$

Integration with respect to $l_2, \ldots, l_{p-1}$ gives V_{p-2}. Letting $k = m+l$ and rearranging terms according to the value of k gives the expansion of $\tilde{F}_2$. □

In Appendix C we list $\tilde{F}_1(t), \tilde{F}_2(t)$ for $p = 2, \ldots, 5$.

It is of considerable interest to explicitly evaluate the leading term of the tube formula approximations for general p, because the leading term of the tube formula is always valid as shown in [**11**].

PROPOSITION 3.8. The order of $F_1(t)$ and $F_2(t)$ (as $t \to \infty$) in (3.3) is the same as the order of $\tilde{F}_1(t)$ and $\tilde{F}_2(t)$, respectively, and given by

$$F_1(t) \sim 2^{(p-3)/2} P(Z_{p-1} > 0)\, \bar{G}_p(t^2), \tag{3.10}$$

$$F_2(t) \sim 2^{p-5/2}\, \bar{G}_{2p-1}(t^2), \tag{3.11}$$

where $P(Z_{p-1} > 0)$ denotes the probability that $(p-1) \times (p-1)$ symmetric matrix normal random variable is positive definite.

PROOF. The leading term of $\tilde{F}_1(t)$ or $F_1(t)$ is

$$d(p)\, 2^{(p-2)/2}\, \Gamma(p/2)\, U_{p-1}(p-2, \ldots, 0)\, \bar{G}_p(t^2).$$

Now $U_{p-1}(p-2, \ldots, 0) = P(Z_{p-1} > 0)/d(p-1)$, $d(p)/d(p-1) = 1/\{2^{1/2}\Gamma(p/2)\}$. This proves (3.10). Next, the leading term of $\tilde{F}_2(t)$ or $F_2(t)$ is

$$d(p)\, 2^{(2p-3)/2}\, \Gamma(p/2)\, \Gamma((p-1)/2)\, V_{p-2}(p-3, \ldots, 0)\, \bar{G}_{2p-1}(t^2).$$

Now $V_{p-2}(p-3, \ldots, 0) = 1/d(p-2)$, $d(p)/d(p-2) = 1/\{2\Gamma(p/2)\Gamma((p-1)/2)\}$. This proves (3.11). □

Proposition 3.8 provides a more precise statement of Proposition 3.2.

If M has positive critical radius, then the tube formula approximation is valid in the sense that in (3.6), $P(T > t) = \tilde{P}(T > t)(1 + R(t))$, where the remainder term $R(t)$ is of exponentially small order in t as $t \to \infty$. An exponential bound for the remainder term $R(t)$ in terms of the critical radius was given in [**3**].

However in our case M has zero critical radius and the tube formula is only partly valid. From [**11**] we see that the terms of the tube formula $\tilde{F}_2$ for K^O is valid for the degrees of freedom greater than $p = \dim \partial K$. In other words, the error $F_2(t) - \tilde{F}_2(t)$ is of the same order as the order of $F_1(t)$ or $\tilde{F}_1(t)$. Therefore if we use the tube formula approximation $\tilde{F}_2(t)$ for $F_2(t)$, then $F_1(t)$ or $\tilde{F}_1(t)$ is of no use. We state this fact as the following theorem.

THEOREM 3.9. *Under the asymptotic null distribution, the tail probability of our statistic T of (1.2) or (1.3) is evaluated as*

$$P(T > t) = \begin{cases} \omega_{2p-1}\bar{G}_{2p-1}(t^2) + \omega_{2p-3}\bar{G}_{2p-3}(t^2) + \cdots + \omega_{p+2}\bar{G}_{p+2}(t^2) \\ \qquad + R_p(t), \qquad p : \text{odd}, \\ \omega_{2p-1}\bar{G}_{2p-1}(t^2) + \omega_{2p-3}\bar{G}_{2p-3}(t^2) + \cdots + \omega_{p+1}\bar{G}_{p+1}(t^2) \\ \qquad + R_p(t), \qquad p : \text{even}, \end{cases}$$

where ω_m is defined in (3.9), the order of the remainder term $R_p(t)$ is $O(\bar{G}_p(t^2))$.

It is of interest to investigate the term of order $O(\bar{G}_p(t^2))$ in $F_2(t) - \tilde{F}_2(t)$. Since the formal tube formula breaks down at this order, we need to examine the multiple integral in $F_2(t)$ and $\tilde{F}_2(t)$ much more closely.

For $p \geq 3$ define

(3.12)

$$\eta_p = -d(p)\, 2^{p/2}\, \Gamma\left(\frac{p}{2}\right) \int_{0>l_p>l_{p-1}}^{\infty>l_2>\cdots>l_{p-1}} \exp\left(-\frac{1}{2}\sum_{i=2}^{p-1} l_i^2\right) \prod_{2\leq i<j\leq p} (l_i - l_j) \prod_{i=2}^{p} dl_i.$$

Then we have the following proposition.

PROPOSITION 3.10. Let $p \geq 3$. As $t \to \infty$

$$F_2(t) - \tilde{F}_2(t) = \eta_p\, \bar{G}_p(t^2) + o(\bar{G}_p(t^2)), \tag{3.13}$$

where $d(p)$ is given in (3.5).

We give an outline of a proof of Proposition 3.10 in Appendix E. Combining (3.13) with the leading term of $F_1(t)$ we can strengthen Theorem 3.9 as follows.

THEOREM 3.11. *The term of order $\bar{G}_p(t^2)$ in the remainder term $R_p(t)$ of Theorem 3.9 is evaluated as*

(3.14)

$$R_p(t) = \begin{cases} \{\omega_p + \eta_p + 2^{(p-1)/2}\, P(Z_{p-1} > 0)\}\bar{G}_p(t^2) + o(\bar{G}_p(t^2)), & p : \text{odd}, \\ \{\eta_p + 2^{(p-1)/2}\, P(Z_{p-1} > 0)\}\bar{G}_p(t^2) + o(\bar{G}_p(t^2)), & p : \text{even}, \end{cases}$$

where ω_m is defined in (3.9).

This theorem recovers the term of order $\bar{G}_p(t^2)$ in the formal tube formula approximation for $P(T > t)$ in Theorem 3.9. However from a practical viewpoint there is a difficulty in applying (3.14) for large p, because exact evaluation of the constant η_p for large p seems to be difficult. For $p = 3, 4, 5$, the values of η_p in (3.12) are evaluated as

$$\eta_3 = \frac{1}{2\sqrt{2}}, \qquad \eta_4 = \frac{3}{2} + \frac{4}{3\pi}, \qquad \eta_5 = \frac{135}{16\sqrt{2}} + \frac{25}{8\pi} - \frac{81}{8\sqrt{2}\pi}\tan^{-1}\left(\frac{1}{\sqrt{2}}\right).$$

Evaluation of η_5 is already quite laborious.

4. Some simulation results

In this section we present some simulation results concerning our proposed statistic.

First we investigate accuracy of our tail probability approximation to the asymptotic null distribution by simulation. We do not consider approximation of (finite degrees of freedom) Wishart distribution by the symmetric matrix normal distribution, because this approximation is not specific to our statistic.

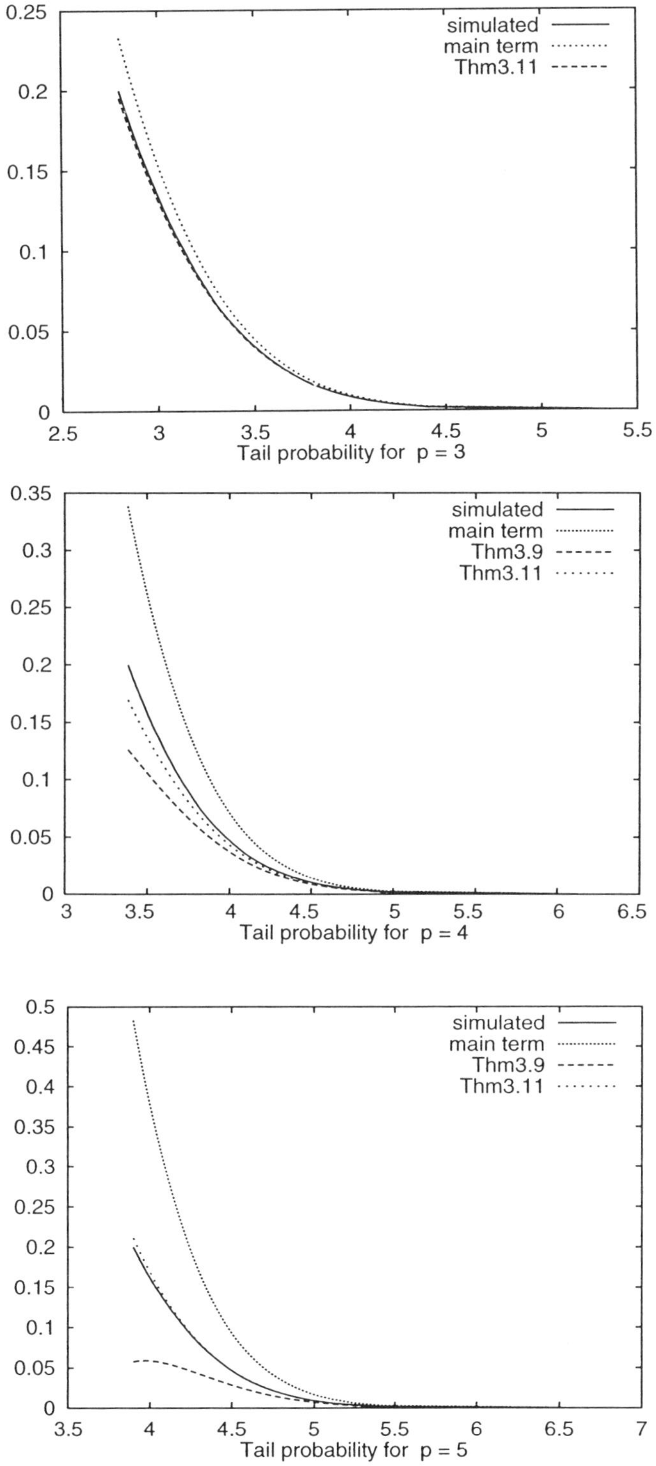

FIGURE 2. Tube formula approximation for $p = 3, 4, 5$.

TABLE 2. Limiting power behavior of three tests.

$p = 4$, Level of significance = 5%

$(\psi_1,\psi_2,\psi_3,\psi_4)$	LR	Roy	TK
(2, 2, 0, 0)	0.435	0.436	0.323
(2, 0, 0, −2)	0.435	0.352	0.451
(1.5, 1.5, −1.5, −1.5)	0.490	0.332	0.446
(2, 0, −1, −1)	0.323	0.261	0.332

Figure 2 shows the tube formula approximation for the dimensions $p = 3, 4, 5$. In Figure 2 "simulated" (solid line) shows the simulated true tail probability of our statistic based on simulation of size 100,000. "main term" (dotted line) is the approximation using only the main term (3.11). "Thm3.9" and "Thm3.11" (dashed lines) are based on the approximation in Theorems 3.9 and 3.11. Note that for $p = 3$ the approximation in Theorem 3.9 consists of only the main term. From the figures we see that the approximation by Theorem 3.9 is practical for P-values below 5% range but is not as good as our previous studies given in [**3**]. This suggests that the case of zero critical radius poses some difficulty from the numerical viewpoint as well. On the other hand the addition of the d.f. p term in Theorem 3.11 greatly improves the approximation and is very satisfactory.

We now very briefly investigate the power behavior of our statistic. We only compare the first order limiting power behavior of our statistic against the likelihood ratio test and Roy's maximum-minimum roots test. For more extensive power comparison of existing test procedures against two-sided alternatives see [**1**]. [**7**] and [**6**] give detailed power comparisons of existing tests against one-sided alternatives.

As the limit of contiguous alternatives, let Z be distributed according to the symmetric matrix normal distribution with non-zero mean matrix Ψ. The covariance structure of Z is the same as the null case. In the first order, the likelihood ratio test and other omnibus type test procedures are equivalent and have non-central χ^2 distribution with $p(p+1)/2$ degrees of freedom and non-centrality parameter $\operatorname{tr}\Psi^2$. Roy's maximum-minimum roots test is $T_R = \max(l_1, -l_p)$. The significance level is taken to be 5%. We obtain the upper 5 percentile of Roy's maximum-minimum roots test by generating 100,000 T_R's under the asymptotic null distribution. The upper 5 percentile of our statistic was already obtained by the simulation study of the previous paragraph. Then the power is computed by generating T_R and our static 100,000 times each under the limiting alternative distribution and by counting the number of times these statistics exceed the 5 percentiles.

Note that Ψ can be assumed to be diagonal $\operatorname{diag}(\psi_1, \dots, \psi_p)$ without loss of generality. We only present the results for the case of $p = 4$. In Table 2, TK stands for our proposed statistic. The entries of the table are the values of the asymptotic power.

From Table 2 our statistic tends to do better than other procedures when Ψ has one large root and one small (i.e., negative but large in absolute value) root. On the other hand Roy's test is good when Ψ is positive definite or negative definite. The likelihood ratio test seems to be a reasonable overall test.

Appendix A. Derivation of some formulae

Derivation of (1.1).

Note that $\mathrm{Var}(a'S_1b - a'S_2b) = \mathrm{Var}(a'W_1b)/n_1^2 + \mathrm{Var}(a'W_2b)/n_2^2$. Let $W = (w_{ij})$ be distributed according to Wishart distribution $W_p(n, \Sigma)$, $\Sigma = (\sigma_{ij})$. It suffices to show that $\mathrm{Var}(a'Wb) = n(a'\Sigma a)(b'\Sigma b) + n(a'\Sigma b)^2$. Using the fact that $\mathrm{Cov}(w_{ij}, w_{kl}) = n(\sigma_{ik}\sigma_{jl} + \sigma_{il}\sigma_{jk})$ we have

$$\begin{aligned}
\mathrm{Var}(a'Wb) &= \mathrm{Cov}\Big(\sum_{i,j} a_i b_j w_{ij}, \sum_{k,l} a_k b_l w_{kl}\Big) \\
&= n \sum_{i,j,k,l} a_i a_k b_j b_l (\sigma_{ik}\sigma_{jl} + \sigma_{il}\sigma_{jk}) \\
&= n\Big(\sum_{i,k} a_i a_k \sigma_{ik}\Big)\Big(\sum_{j,l} b_j b_l \sigma_{jl}\Big) + n\Big(\sum_{i,l} a_i b_l \sigma_{il}\Big)\Big(\sum_{j,k} b_j a_k \sigma_{jk}\Big) \\
&= n(a'\Sigma a)(b'\Sigma b) + n(a'\Sigma b)^2.
\end{aligned}$$

Derivation of (2.4).

The factor $1/\sqrt{l_1 - l_p}$ is only for normalization. Therefore we can take

$$b = (\sqrt{l_1}, 0, \dots, 0, \pm\sqrt{-l_p})'.$$

Then

$$\begin{aligned}
a &= Zb - \frac{b'Zb}{2b'b} b \\
&= (l_1\sqrt{l_1}, 0, \dots, 0, \pm l_p\sqrt{-l_p})' - \frac{l_1^2 - l_p^2}{2(l_1 - l_p)}(\sqrt{l_1}, 0, \dots, 0, \pm\sqrt{-l_p})' \\
&= \frac{l_1 - l_p}{2}(\sqrt{l_1}, 0, \dots, 0, \mp\sqrt{-l_p})'.
\end{aligned}$$

Normalizing this we obtain a in (2.4).

Appendix B. The cone K for $p = 2$

For $p = 2$ the sample space $\mathcal{S}$ is identified with R^3 and almost all $Z \in \mathcal{S}$ has rank 2. Therefore we only need to identify $\mathcal{S}_+$ and $\mathcal{S}_-$. For real symmetric $Z = (z_{ij})_{1\le i,j\le 2}$ write $u = z_{11}, v = z_{22}, w = \sqrt{2}z_{12}$. Then $Z > 0$ if and only if

$$u > 0, \quad 2uv - w^2 > 0.$$

If we make 45 degrees rotation in (u, v) plane and define $s = (u + v)/\sqrt{2}$, $t = (u - v)/\sqrt{2}$. Then $2uv = s^2 - t^2$ and

$$\mathcal{S}_+ = \{(s, t, w) \mid s + t > 0,\ s^2 > t^2 + w^2\}$$

is a circular cone. $\mathcal{S}_- = -\mathcal{S}_+$ is the circular cone opposite to $\mathcal{S}_+$. Therefore K is the cone obtained by subtracting two circular cones from R^3. We see that for $p = 2$, $M = K \cap S^2$ is a 2-dimensional manifold with smooth boundaries.

Appendix C. Explicit form of tube formula approximations for small dimensions

Here we give a brief list of tube formula approximations of Proposition 3.7. These following formulas were obtained by the applying the recurrence relations of Lemmas 3.5 and 3.6.

$p = 2:$

$$\begin{aligned}\tilde{F}_1 &= \frac{1}{2\sqrt{2}}(\bar{G}_2 - \bar{G}_1),\\ \tilde{F}_2 &= \frac{1}{\sqrt{2}}\bar{G}_3.\end{aligned}$$

$p = 3:$

$$\begin{aligned}\tilde{F}_1 &= \left(-\frac{1}{2\sqrt{2}} + \frac{1}{2}\right)\bar{G}_3 - \frac{1}{\sqrt{2}\pi}\bar{G}_2 + \left(\frac{1}{2\sqrt{2}} + \frac{1}{4}\right)\bar{G}_1,\\ \tilde{F}_2 &= \sqrt{2}(\bar{G}_5 - \bar{G}_3).\end{aligned}$$

$p = 4:$

$$\begin{aligned}\tilde{F}_1 &= \left(\frac{1}{2\sqrt{2}} - \frac{1}{\pi}\right)\bar{G}_4 + \left(\frac{1}{4\sqrt{2}} - \frac{1}{4}\right)\bar{G}_3 + \left(-\frac{3}{8\sqrt{2}} + \frac{1}{\pi}\right)\bar{G}_2\\ &\quad + \left(-\frac{3}{8\sqrt{2}} + \frac{1}{4}\right)\bar{G}_1,\\ \tilde{F}_2 &= 2\sqrt{2}\bar{G}_7 - \frac{7}{\sqrt{2}}\bar{G}_5 + \frac{7}{2\sqrt{2}}\bar{G}_3.\end{aligned}$$

$p = 5:$

$$\begin{aligned}\tilde{F}_1 &= \left(-\frac{1}{4\sqrt{2}} - \frac{1}{\pi} + \frac{1}{2}\right)\bar{G}_5 + \left(-\frac{1}{\sqrt{2}\pi} + \frac{2}{3\pi}\right)\bar{G}_4 + \left(\frac{1}{8\sqrt{2}} + \frac{4}{3\pi} - \frac{1}{2}\right)\bar{G}_3\\ &\quad + \left(\frac{11}{12\sqrt{2}\pi} - \frac{2}{3\pi}\right)\bar{G}_2 + \left(\frac{1}{8\sqrt{2}} - \frac{2}{3\pi} + \frac{1}{8}\right)\bar{G}_1,\\ \tilde{F}_2 &= 4\sqrt{2}\bar{G}_9 - 9\sqrt{2}\bar{G}_7 + \frac{15}{\sqrt{2}}\bar{G}_5 - \frac{5}{\sqrt{2}}\bar{G}_3.\end{aligned}$$

Appendix D. Exact asymptotic tail probability for $p = 2$

For the case $p = 2$, we can easily evaluate the exact tail probability of the asymptotic null distribution. Note that for $p = 2$, $\langle Z, Z\rangle = l_1^2 + l_2^2$ and the event $\{l_1 > 0 > l_2\}$ are independent. Furthermore $P(l_1 > 0 > l_2) = 1/\sqrt{2}$. Therefore for $p = 2$

$$F_2(t) = \frac{1}{\sqrt{2}}\bar{G}_3(t^2).$$

Now we evaluate $F_1(t) = \int_{l_1 > t,\, l_1 > l_2 > 0} f(l_1, l_2)\, dl_1 dl_2$. Integrating with respect to l_2 first then with respect to l_1 gives

$$\int_0^{l_1} f(l_1, l_2)\, dl_2 = \frac{l_1}{\sqrt{2}} e^{-l_1^2/2}\left(\Phi(l_1) - \frac{1}{2}\right) - \frac{1}{2\sqrt{\pi}} e^{-l_1^2/2} + \frac{1}{2\sqrt{\pi}} e^{-l_1^2},$$

$$\int_{l_1=t}^{\infty}\int_{l_2=0}^{l_1} f(l_1,l_2)\,dl_2 dl_1 = \frac{e^{-t^2/2}}{2\sqrt{2}} - \frac{1-\Phi(t)}{\sqrt{2}} - \frac{1-\Phi(t)}{\sqrt{2}}e^{-t^2/2} + (1-\Phi(\sqrt{2}t)),$$

where Φ is the cumulative distribution function of the standard normal distribution. Noting that $\bar{G}_2(t^2) = e^{-t^2/2}$, $\bar{G}_1(t^2) = 2(1-\Phi(t))$ we obtain

$$P(T>t) = \frac{1}{\sqrt{2}}\bar{G}_3(t^2) + \frac{1}{\sqrt{2}}\bar{G}_2(t^2) - \frac{1}{\sqrt{2}}\bar{G}_1(t^2) - \frac{1}{\sqrt{2}}\bar{G}_2(t^2)\bar{G}_1(t^2) + \bar{G}_1(2t^2).$$

Note that this exact formula coincides with the tube formula $\tilde{F}_2+2\tilde{F}_1$ for $p=2$ given in Appendix C, except for remainder terms which is negligible with exponentially small order. This should be the case because M has positive critical radius for $p=2$.

Appendix E. Outline of a Proof of Proposition 3.10

Here we give an outline of a proof of Proposition 3.10. Since full justification of all the steps of the approximations takes too much space, we omit some justifications of neglecting remainder terms. A complete memo of justifications can be obtained from the authors upon request.

Note that the difference between $\tilde{F}_2$ and F_2 is due to the difference of the ranges of integration. Let

$$\Omega = \{\,(l_1,\ldots,l_p) \mid l_1 > 0 > l_p,\ l_2 > \cdots > l_{p-1},\ l_1^2 + l_p^2 > t^2\,\}$$

Then

$$\text{(E.1)}\qquad \frac{1}{d(p)}\left(\tilde{F}_2(t) - F_2(t)\right) = \int_{\Omega_1} \exp\left(-\frac{1}{2}\sum_{i=1}^{p} l_i^2\right)\det(l_i^{p-j})\prod_{i=1}^{p} dl_i = \int_{\Omega_2} + \int_{\Omega_3} - \int_{\Omega_4},$$

where

$$\begin{aligned}
\Omega_1 &= \Omega\cap\{\,(l_1,\ldots,l_p) \mid l_2 > l_1 \text{ or } l_{p-1} < l_p\,\},\\
\Omega_2 &= \Omega\cap\{\,(l_1,\ldots,l_p) \mid l_2 > l_1\,\},\\
\Omega_3 &= \Omega\cap\{\,(l_1,\ldots,l_p) \mid l_{p-1} < l_p\,\},\\
\Omega_4 &= \Omega_2\cap\Omega_3 = \Omega\cap\{\,(l_1,\ldots,l_p) \mid l_2 > l_1 \text{ and } l_{p-1} < l_p\,\}.
\end{aligned}$$

It can be easily shown that the component of (E.1) corresponding to Ω_4 is negligible with exponentially small order. Intuitively this is the case because the probability of both l_2 and l_{p-1} being greater than t in absolute value should be exponentially smaller than the probability that just one of them is greater than t in absolute value. By symmetry the first term and the second term on the right hand side are equal. Therefore

$$\frac{1}{d(p)}\left(\tilde{F}_2(t) - F_2(t)\right) \sim 2\int_{\Omega_3} \exp\left(-\frac{1}{2}\sum_{i=1}^{p} l_i^2\right)\det(l_i^{p-j})\prod_{i=1}^{p} dl_i.$$

We now argue that the main contribution to this integral comes from the region where $|l_p|/l_1$ is close to 0. In fact consider the range $-\sqrt{t} > l_p$. Fix $0 < \epsilon < 1$.

Since $l_{p-1}^2 > l_p^2 > t$ on the range of integration we have

$$\exp\left(-\frac{1}{2}l_{p-1}^2\right) \le \exp\left(-\frac{1-\epsilon}{2}t\right)\exp\left(-\frac{\epsilon}{2}l_{p-1}^2\right).$$

From this it easily follows that there exists C_ϵ such that

$$\int_{\Omega_3\cap\{-\sqrt{t}>l_p\}} \exp\left(-\frac{1}{2}\sum_{i=1}^p l_i^2\right)\left|\det\left(l_i^{p-j}\right)\right|\prod_{i=1}^p dl_i \le C_\epsilon \exp\left(-\frac{1-\epsilon}{2}t\right)\bar{G}_1(t^2)$$

for all $t > 0$. Therefore this range of integration is negligible with exponentially small order. Similarly we can show that the range $l_i^2 > t^2$, $i = 2, \ldots, p-1$, can be ignored.

It follows that

$$\frac{1}{d(p)}\left(\tilde{F}_2(t) - F_2(t)\right) \sim 2\int_{\Omega_3\cap\{\sqrt{t}\ge l_2\}\cap\{-\sqrt{t}\le l_{p-1}\}} \exp\left(-\frac{1}{2}\sum_{i=1}^p l_i^2\right)\det\left(l_i^{p-j}\right)\prod_{i=1}^p dl_i.$$

Note that on this range of integration l_1 is large:

$$l_1 > \sqrt{t^2 - l_p^2} \ge \sqrt{t^2 - t}.$$

This implies that on our range of integration the main term of $\det(l_i^{p-j})$ is the term with the highest degree in l_1, i.e.,

$$\det\left(l_i^{p-j}\right) \sim l_1^{p-1}\prod_{2\le i<j\le p}(l_i - l_j).$$

Now we integrate l_1 out. Using

$$\int_{\sqrt{t^2-l_p^2}}^\infty l_1^{p-1}\exp\left(-\frac{1}{2}l_1^2\right)dl_1 \sim t^{p-2}\exp\left(-\frac{t^2}{2}\right)\exp\left(\frac{l_p^2}{2}\right)$$

we have

$$\frac{1}{d(p)}\left(\tilde{F}_2(t) - F_2(t)\right) \sim 2t^{p-2}\exp\left(-\frac{t^2}{2}\right)$$
$$\times\int_{0>l_p>l_{p-1}\ge-\sqrt{t}}^{\sqrt{t}\ge l_2>\cdots>l_{p-1}} \exp\left(-\frac{1}{2}\sum_{i=2}^{p-1} l_i^2\right)\prod_{2\le i<j\le p}(l_i - l_j)\prod_{i=2}^p dl_i.$$

Proposition 3.10 follows from this because $2t^{p-2}e^{-t^2/2} \sim \Gamma(p/2)2^{p/2}\bar{G}_p(t^2)$.

References

[1] S. S. Chu and K. C. S. Pillai, Power comparison of two-sided tests of equality of two covariance matrices based on six criteria, *Ann. Inst. Stat. Math.* **31** (1979), 182–205.

[2] S. Kuriki, One-sided test for the equality of two covariance matrices, *Ann. Statist.* **21** (1993), 1379–1384.

[3] S. Kuriki and A. Takemura, Tail probabilities of the maxima of multilinear forms and their applications, *Ann. Statist.* **29** (2001), to appear.

[4] ———, Some geometry of the cone of nonnegative definite matrices and weights of associated $\bar{\chi}^2$ distribution, *Ann. Inst. Stat. Math.* **52** (2000), 1–14.

[5] H. Nagao, On some test criteria for covariance matrix, *Ann. Statist.* **1** (1973), 700–709.

[6] K. C. S. Pillai and S. Al-Ani, Power comparisons of tests of equality of two covariance matrices based on individual characteristic roots, *J. Amer. Statist. Assoc.* **65** (1970), 438–446.

[7] K. C. S. Pillai and K. Jayachandran, Power comparisons of tests of equality of two covariance matrices based on four criteria, *Biometrika* **55** (1968), 335–342.

[8] S. N. Roy, On a heuristic method of test construction and its use in multivariate analysis, *Ann. Math. Statist.* **24** (1953), 220–238.

[9] J. Sun, Tail probabilities of the maxima of Gaussian random fields, *Ann. Probab.* **21** (1993), 34–71.

[10] A. Takemura and S. Kuriki, Maximum of Gaussian field on piecewise smooth domain: Equivalence of tube method and Euler characteristic method, Discussion Paper CIRJE-F-54, Faculty of Economics, Univ. of Tokyo (1999). Submitted for publication.

[11] ———, Tail probability via tube formula and Euler characteristic method when critical radius is zero, Discussion Paper CIRJE-F-59, Faculty of Economics, Univ. of Tokyo (1999). Submitted for publication.

FACULTY OF ECONOMICS, UNIVERSITY OF TOKYO, 7-3-1 HONGO, BUNKYO-KU, TOKYO 113-0033, JAPAN.

E-mail address: takemura@e.u-tokyo.ac.jp

THE INSTITUTE OF STATISTICAL MATHEMATICS, 4-6-7 MINAMI-AZABU, MINATO-KU, TOKYO 106-8569, JAPAN.

E-mail address: kuriki@ism.ac.jp

Contemporary Mathematics
Volume **287**, 2001

The Covariance Structure of Random Permutation Matrices

Marlos A. G. Viana

ABSTRACT. Random permutation matrices (U) arise naturally in the stochastic representation of vectors of order statistics, induced order statistics and associated ranks. When the probability law of U is uniform, the covariance structure among the entries of U is derived explicitely, and a constructive derivation for the covariance in the general case (U^k) is described and related to the distribution of cycles in the symmetric group. It is shown that the covariance structure of vectors resulting from the multiplicative action Ux of U on a vector x requires averaging the conjugates $Uxx'U'$ of xx' over the group of permutation matrices. The mean conjugate is a projection operator which leads to a trace decomposition with the usual ANOVA interpretation. Numerical examples of these decompositions are discussed in the context of the analysis of circularly symmetric data.

1. Introduction

Permutation matrices are square matrices characterized by having all entries either 0 or 1 and exactly one 1 in each row and column. The collection (G) of all permutation matrices (g) of order n is in one-to-one correspondence (ρ) with the collection (S_n) of all permutations (π) on $\{1, 2, \ldots, n\}$. The permutation matrix associated with a permutation π is the matrix representing the linear transformation that maps the canonical basis $\{e_1, \ldots, e_n\}$ of R^n into the basis $\{e_{\pi(1)}, \ldots e_{\pi(n)}\}$. The collection S_n together with the operation of function composition defines the symmetric group on n objects. The correspondence ρ, defined in S_n, has the homomorphic property $\rho(\pi_1\pi_2) = \rho(\pi_1)\rho(\pi_2)$ for all π_1, π_2 in S_n. This characterizes ρ as a n-dimensional linear representation of S_n (the permutation representation). If a permutation π is selected according to a probability law in S_n, then $\rho(\pi)$ is a random permutation matrix (U) in G. In the present paper we concentrate on random permutation matrices in G, keeping in mind the fact that those are just one of the many (random) linear representations for random events defined in the symmetric group S_n. We will write G to indicate both S_n and the group (M_n) of corresponding permutation matrices.

Random permutation matrices arise naturally, for example, in the linear representation of order statistics. This is illustrated in the introductory examples below,

1991 *Mathematics Subject Classification.* 15A52, 20B30.
Key words and phrases. random matrices, symmetric group, moments.

where the determination of the uniform mean $E(U)$ and the uniform mean conjugate $E(UHU')$ of a given matrix H are required. The uniform mean is simply the average of all elements in G, whereas the mean conjugate of H is defined as the average of the matrix conjugates gHg' of H over G. More generally, as the examples will show, we need to describe the corresponding uniform higher-order means $E(U^k)$ and mean conjugates $E(U^kHU^{k'})$. This is equivalent to deriving the covariance structure among the entries of U^k.

The covariance structure of U, a uniform random permutation matrix, is studied in detail in Section 2. Proposition 2.2 in Section 2 states that the uniform mean conjugate of H decomposes as $v_0ee'/n + v_1(I - ee'/n)$, where $v_0 = e'He/n$ and $v_0 + (n-1)v_1 = \text{tr } H$. This result, when applied to the linear representation of order statistics, for example, leads to the covariance structure of ranks of permutation-symmetric observations, or to the mean and variance of the usual Wilcoxson rank-sum statistics. Proposition 2.2 is a direct consequence of the (n-1)-dimensional irreducible linear representation of G, which leads to a decomposition of tr $E(UHU')$ into the weighted average of corresponding orthogonal projections with weights proportional to the dimensions of the irreducible subspaces, and is carried out by the intertwining action of $E(UHU')$. The ANOVA interpretation of such decomposition is discussed in Section 3, where an application to circularly symmetric data is outlined. Finally, in Section 4, we derive the covariance structure of moments of random permutations. The main result (Theorem 4.1) shows that, when the law of U is uniform in G, the trace of U^k is in average equal to the number of positive integer divisors of k not greater than n.

EXAMPLE 1.1. Representation of order statistics. Let X_i indicate the rank of the i-th component Y_i of a random vector Y, defined on a linear subspace V of $\mathbb{R}^n$, when these components are ordered from the smallest to the largest value, and assume for the moment that either the components of Y are distinct or that there is a fixed rule assigning distinct ranks to tied observations. Let $\pi \in G$ indicate the permutation taking X_i to i, that is, $X_{\pi(i)} = i$, $i = 1, \ldots, n$. Given the standard basis $\{e_1, \ldots, e_n\}$ of V, define the $n \times n$ matrix $\rho(\pi)$ by $e_{\pi(j)} = \sum_i \rho(\pi)_{ij} e_i$, $j = 1, \ldots, n$. Equivalently, $\rho(\pi)_{ij} = 1$ if and only if π takes j into i, that is $\pi(j) = i$. The homomorphic map $\pi \to \rho(\pi)$ is the *permutation representation* introduced above. By virtue of the randomness in the ranks of Y, the matrix $U = \rho(\pi)$ is a random permutation matrix, and $U'Y$ linearly represents the ordered version of Y. For example, if the vector of ranks of an observed vector y is $x' = (3, 4, 2, 1)$, then π is the permutation taking x' into $r' = (1, 2, 3, 4)$. In cycle notation, we write π as (1423) to indicate that $1 \to 4$, $4 \to 2$, $2 \to 3$ and $3 \to 1$. The corresponding linear representation is

$$u = \rho(\pi) = \begin{bmatrix} 0 & 0 & 1 & 0 \\ 0 & 0 & 0 & 1 \\ 0 & 1 & 0 & 0 \\ 1 & 0 & 0 & 0 \end{bmatrix}.$$

If we denote by $\mathcal{Y}$ the ordered version of Y, then the random permutation matrix U $(= \rho(\pi))$ is such that $X = Ur$ and $\mathcal{Y} = U'Y$. If (Y,Z) are concomitant random vectors of corresponding dimensions, then $U'Z$ is the vector of concomitants of order (or induced order) statistics. If the distribution of Y is permutation symmetric (that is Y and gY are equally distributed for all $g \in G$), then the probability law of U is uniform in G. The covariance structure of ranks, order statistics and concomitants may then be expressed as multiplicative actions of uniformly distributed random permutations. In particular, to obtain the covariance structure of ranks $X = Ur$, we need to evaluate the uniform expected value $\sum_g gr/n!$ of $U'r$, indicated by $E(Ur)$, and the uniform expected value $\sum_g grr'g'/n!$ of $U'rr'U$, denoted by $E(Urr'U')$. The characterization and interpretation of uniform averages over G, such as those in $E(U)$ and $E(Urr'U')$ will be obtained in Section 2.

Other formulations leading to random linear representations can be obtained in terms of random walks or random graphs, or more generally as iterated random functions (e.g., [**DF99**]).

2. Means and mean conjugates

Let U indicate a random permutation matrix distributed according to $w \equiv \{w(g); g \in G\}$. We refer to

$$E_w(UHU') = \sum_{g\in G} gHg'w(g)$$

as the mean conjugate of H when U is distributed according to w on G. The mean conjugate is a particular invariant mean $\sum_g c(g)/|G|$ on G, where c is a function defined in G. [e.g., [**Nai82**, p.60]]. When U is uniformly distributed in G we write $E(UHU')$ and refer to the uniform mean conjugate of H. When an operator M commutes with every element of G we say that M *has the symmetry of* G.

PROPOSITION 2.1. *$E_w(UHU')$ has the symmetry of G if and only $E_w(UHU')$ is the uniform mean conjugate of H.*

PROOF. If $E_w(UHU')$ is the uniform mean conjugate of H, the transitivity of (the action of) G (on itself) shows that $tE(UHU')t' = E(UHU')$ for all $t \in G$. Conversely, if U is assigned the value g with probability $w(g)$, and $E_w(UHU')$ commutes with every element t of G, we have $E_w(UHU') = t\sum_g w(g)gHg't' = \sum_g w(g)tgH(tg)' = \sum_v w(t'v)vHv'$, where the last equality uses the transitivity of the corresponding action on G. The same fact shows that $\sum_t w(t'v) = 1$, for all t, so that

$$E_w(UHU') = \frac{1}{n!}\sum_t\sum_v w(t'v)vHv' = \sum_v \frac{1}{n!}\sum_t w(t'v)vHv' = \sum_v \frac{1}{n!}vHv'.$$

□

The following result gives the explicit evaluation of $E(UHU')$ (see also [**Dan62**] for an heuristic proof).

PROPOSITION 2.2. *(a) The uniform mean $E(U)$ of U in G is ee'/n and (b) the uniform mean conjugate $E(UHU')$ of H on G decomposes as $v_0\frac{ee'}{n} + v_1(I - \frac{ee'}{n})$, where $v_0 = e'He/n$ and $v_0 + (n-1)v_1 = tr\ H$.*

PROOF. Let V be the one-dimensional subspace of $\mathbb{R}^n$ spanned by $e' = (1, ..., 1)$, W its orthogonal complement in $\mathbb{R}^n$, $M = E(U) - \frac{ee'}{n}$ and $\mathcal{M} = E(UHU') - v_0\frac{ee'}{n} - v_1(I - \frac{ee'}{n})$, with v_0 and v_1 as above. Allowing for complex solutions, if all eigenvalues of M or $\mathcal{M}$ are zero then the proposition holds. Otherwise, let λ be a non-zero eigenvalue of M, ℓ a non-zero eigenvalue of $\mathcal{M}$, $K = kern\ \{M - \lambda I\}$ and $\mathcal{K} = kern\ \{\mathcal{M} - \ell I\}$. We will show that both K and $\mathcal{K}$ are subspaces of W stable under G, and use the fact[e.g., [**Ser77**, p.17], [**Dia88**, p.6] and Appendix A] that

V and W are irreducible subspaces of the linear representation ρ. Consequently, both K and $\mathcal{K}$ must reduce to null subspaces, and this will complete the proof.

(1) $K \subset W$. First note that $e'M = e'(\frac{1}{n!}\sum g - \frac{ee'}{n}) = 0$, so that for $v \in K$ we have $(M - \lambda I)v = 0 = e'M - \lambda e'v = \lambda e'v$, and because $\lambda \neq 0$ we obtain $e'v = 0$, or $v \in W$.

(2) M has the symmetry of G. In fact, from the transitivity of G, $tMt' = \frac{1}{n!}\sum tgt' - \frac{tee't'}{n} = \frac{1}{n!}\sum_g g - \frac{ee'}{n} = M$, $\forall t \in G$.

(3) K is stable under G. Take $v \in K$ and $t \in G$, then, by (2), $Mtv = tMv = t\lambda v = \lambda tv$, that is $tv \in K$, for all $t \in G$. This completes the proof of part (a).

(4) $\mathcal{K} \subset W$. We use part (a) to show, similarly to (1) above, that $e'\mathcal{M} = 0$. In fact, $e'\mathcal{M} = \frac{1}{n!}\sum_g e'gHg' - v_0\frac{e'ee'}{n} - v_1(e' - \frac{e'ee'}{n}) = e'H\frac{1}{n!}\sum_g g - v_0 e' = e'H\frac{1}{n}ee' - v_0 e' = (\frac{1}{n}e'He - v_0)e' = 0$, by the definition of v_0. Now for $v \in \mathcal{K}$ we have $(\mathcal{M} - \ell I)v = 0 = e'\mathcal{M} - \ell e'v = \ell e'v$, and because $\ell \neq 0$ we obtain $e'v = 0$, or $v \in W$.

(5) $\mathcal{M}$ has the symmetry of G. Again, from the transitivity of the action on G, $t\mathcal{M}t' = \sum_g tgHg't' - v_0\frac{tee't'}{n} - v_1(tt' - \frac{tee't'}{n}) = t\mathcal{M}t'$, $\forall t \in G$.

(6) $\mathcal{K}$ is stable under G. Take $v \in \mathcal{K}$ and $t \in G$, then, by (5), $\mathcal{M}tv = t\mathcal{M}v = t\ell v = \ell tv$. That is, $tv \in \mathcal{K}$, for all $t \in G$. This completes the proof of part (b).

□

Equivalently, the decomposition of $E(UHU')$ may be expressed as $a_0 ee' + a_1 I$, for some scalars a_0 and a_1 determined by $n(a_0 + a_1) = \text{tr } H$ and $n(n-1)a_0 = e'He - \text{tr } H$.

EXAMPLE 2.1. Let U indicate the order statistics representation introduced in Section 1 associated with a random vector Y, and let $r' = (1, 2, \dots, n)$. The vector $U'r$ indicates the vector of ranks associated with Y, when the components of Y are all distinct. Average ranks with tied observations are discussed in Example 2.3. If the distribution of Y is permutation symmetric then the distribution of U is uniform. The vector of means, $E(Ur)$, from Proposition 2.2, is

$$E(Ur) = E(U)r = \frac{\mathbf{e}\mathbf{e}'}{n}r = \frac{n(n+1)}{2n}\mathbf{e} = \frac{n+1}{2}\mathbf{e},$$

whereas the covariance of Ur is given by $E(Urr'U') - E(Ur)E(Ur)'$. To evaluate the mean conjugate of rr' in G, following Proposition 2.2, we note that the trace of rr' is $n(n+1)(2n+1)/6$, the sum of the squares of the first n positive integers, whereas the sum $e'rr'e$ of all entries of rr' is $(\mathbf{e}'r)^2 = [n(n+1)/2]^2$. Therefore,

$$n(a_0 + a_1) = \frac{n(n+1)(2n+1)}{6},$$

$$n(n-1)a_0 = [\frac{n^2(n+1)^2}{4} - \frac{n(n+1)(2n+1)}{6}] = \frac{1}{12}n(n+1)(n-1)(3n+2),$$

from which we obtain

$$\text{Cov}(Ur) = [a_0 - \frac{(n+1)^2}{4}]\mathbf{e}\mathbf{e}' + a_1\mathbf{I} = \frac{-(n+1)}{12}\mathbf{e}\mathbf{e}' + \frac{n(n+1)}{12}\mathbf{I}. \tag{2.1}$$

The result shows that the common variance among the ranks is $(n+1)(n-1)/12$, and the common covariance between any two ranks is $-(n+1)/12$. The resulting common correlation is $-1/(n-1)$. The mean and variance of the usual Wilcoxson rank-sum statistics W follows from fixing (say, the first) m components of R. Writing $f' = (1,...,1,0,...,0)$ with $1 \leq m < n$ components equal to 1, then $W = f'R$ and $EW = f'E(Ur) = m(n+1)/2$, whereas $var(W) = f'\text{Cov}(Ur)f = m(n-m)(n+1)/12$. Similar arguments can be applied to rank correlations (e.g., [**BE73**]).

EXAMPLE 2.2. Covariance structure of order statistics. Let $\Gamma_1 = \{y \in \mathbb{R}^n : y_1 \leq \ldots \leq y_n\}$, denote by Γ_g the set image $g\Gamma_1$ of Γ_1 under the permutation matrix g, and set $w(g) = P(Y \in \Gamma_g)$. Let also $E_g Y$ indicate the expectation of Y conditional to Γ_g, that is, $E_g Y = \int y I_g(y) F(dy)/w(g)$, where I_g is the indicator function of Γ_g and F is the probability law of Y. The expectation of $\mathcal{Y} = U'Y$ is then

$$E(U'Y) = \sum_{g\in G} w(g) g' E_g Y = \sum_{g\in G} \int g' y I_g(y) F(dy) = \sum_{g\in G} \int x I_g(gx) F(gdx).$$

which reduces to

$$E(\mathcal{Y}) = E(U'Y) = \int x I_1(x) F^*(dx), \tag{2.2}$$

where $F^*(dx) = \sum_{g\in G} F(gdx)$ is the symmetrized form of $F(dx)$. If, in addition, the integral withe respect to F is invariant under $y \to gy$, then $E(U'Y)$ reduces to the usual form $\int x I_1(x) n! F(dx)$. Similar argument leads to an expression for $E(\mathcal{Y}\mathcal{Y}')$. Note that equation (2.2) has many useful extensions. For example, consider the case of conditioning Y on a concomitant (eventually random) vector X. Given $X = x$, we observe Y_x according to the probability law of $Y|x$ and represent its ordered version by $U'_{Y_x} Y_x$, as before. By definition, we have $E(U'Y \mid x) = E(U'_{Y_x} Y_x)$, $w_x(g) = P(Y \in \Gamma_g | x)$, so that

$$E(U'Y) = E_x E(U'Y \mid x) = E_x \sum_{g\in G} w_x(g) g' E_g(Y_x).$$

This construction works, for example, when the probability law of Y_x is symmetrically generated, e.g., $Y_x = Mx + V$ where M and V have the symmetry of G. The covariance structure of ordered observations from symmetrically dependent observations is described in [**LV99**], [**Via98**], [**VO97**], [**OV95**].

EXAMPLE 2.3. The representation of ordered tied observations. Given $y \in \mathbb{R}^n$, define the equivalence relation $i \sim j$ if and only if $y_i = y_j$, and let $C_1, \ldots, C_m$ be the resulting equivalent classes, the C_i class with k_i elements and block-rank i. The k_i rank-equivalent elements in C_i are assigned to labels $c_{i1}, \ldots, c_{ik_i}$, for example, monotonically increasing with j. Index the canonical basis accordingly by e_{ij}, $i = 1, \ldots, m$ and $j = 1, \ldots, k_i$. Let π_i be a permutation in $S_{k_i} \equiv S_i$, $i = 1, \ldots, m$ and define the action $e_{ij} \to e_{\pi_i(c_{ij})}$ of the product group $\Pi_i S_i$ on the standard basis. Let $\rho(\pi_1, \ldots, \pi_m)$ be the corresponding permutation representation. Then, for any choice of $\pi_1, \ldots, \pi_m$, $\rho(\pi_1, \ldots, \pi_m)$ is an order representation, that is, $\rho(\pi_1, \ldots, \pi_m)Y = \mathcal{Y}$, whereas the resulting vector R of average ranks is obtained from the multiplicative action of the average inverse order representations on $r' = (1, 2, \ldots, n)$, that is,

$$R = \frac{1}{\Pi_i k_i!} \sum_{S_1 \times \cdots \times S_m} \rho(\pi_1, \ldots, \pi_m)' r.$$

To illustrate, let $y' = (1,0,0,1,-1)$. Then $C_1 = \{5\}$, $C_2 = \{2,3\}$ and $C_3 = \{1,4\}$, so that $c_{11} = 5$, $c_{21} = 2$, $c_{22} = 3$, $c_{31} = 1$ and $c_{32} = 4$. The resulting actions

$$e_{11} \to e_{\pi_1(c_{11})}, e_{21} \to e_{\pi_2(c_{21})}, e_{22} \to e_{\pi_2(c_{22})}, e_{31} \to e_{\pi_3(c_{31})}, e_{32} \to e_{\pi_3(c_{32})},$$

of $S_1 \times S_2 \times S_2$ on the standard basis leads to the order representations. Corresponding to the identity, for example, a linear order representation U is

$$U = \begin{bmatrix} 0 & 0 & 0 & 0 & 1 \\ 0 & 1 & 0 & 0 & 0 \\ 0 & 0 & 1 & 0 & 0 \\ 1 & 0 & 0 & 0 & 0 \\ 0 & 0 & 0 & 1 & 0 \end{bmatrix},$$

with the effect that $UY = \mathcal{Y}$. There are $1!2!2! = 4$ of those and their sum is

$$\begin{bmatrix} 0 & 0 & 0 & 0 & 4 \\ 0 & 2 & 2 & 0 & 0 \\ 0 & 2 & 2 & 0 & 0 \\ 2 & 0 & 0 & 2 & 0 \\ 2 & 0 & 0 & 2 & 0 \end{bmatrix}.$$

Taking the transpose, it follows that,

$$\frac{1}{4}\begin{bmatrix} 0 & 0 & 0 & 2 & 2 \\ 0 & 2 & 2 & 0 & 0 \\ 0 & 2 & 2 & 0 & 0 \\ 0 & 0 & 0 & 2 & 2 \\ 4 & 0 & 0 & 0 & 0 \end{bmatrix}\begin{bmatrix} 1 \\ 2 \\ 3 \\ 4 \\ 5 \end{bmatrix} = \begin{bmatrix} 4.5 \\ 2.5 \\ 2.5 \\ 4.5 \\ 1 \end{bmatrix}.$$

EXAMPLE 2.4. Familial/Bilateral Index. If familial factors do not influence the risk of (certain types of) cancer beyond their effects on sex, age at treatment and dose, then the residuals (on a proportional hazards model, say) should be distributed independently of family membership [**PGL**$^+$**88**]. The proposed statistics is the sum of the products of the residuals $\{x_i, y_i\}$ for pairmates $i = 1, \dots, n$. Large values of $T = \sum_i^n x_i y_i = x'y$ are indicative of familial effects. To assess the expected variance of T, given the data x, y, under the assumption of random pairing of siblings, define the random outcome $T_u = x'Uy$, where U is a random permutation matrix, uniformly distributed in G. Then, $var(T_u) = x'E(Uyy'U')x$, and direct application of Proposition 2.2 shows that $E(T_u) = n\bar{x}\bar{y}$, and that $var(T_u) = (n-1)S_x^2 S_y^2$, where S_x^2 and S_y^2 are the corresponding usual estimates of the lateral variability in the data.

3. Trace decompositions

The proof of Proposition 2.2 is based on the existence of an irreducible $V \oplus W$ representation of G, which leads to a decomposition of tr H into the weighted average of corresponding orthogonal projections with weights proportional to dim V and dim W, respectively. This is carried out by the intertwining action of $E(UHU')$. Table 1 illustrates the decomposition and its ANOVA interpretation ($v_0 \neq 0$).

EXAMPLE 3.1. The following cases will justify the interpretation:

(1) $H = mI$, $m \neq 0$. Then $v_0 = v_1 = m$ and $\mathcal{F} = 1$;

TABLE 1. Trace decomposition for the $(n-1)$ irreducible representation of G.

effect	components	df	mss	$\mathcal{F}$
W	$(n-1)v_1$	n-1	v_1	v_1/v_0
V	v_0	1	v_0	
total	tr H	n		

(2) Let $H \in G$, uniformly. Then $v_0 = e'He/n = 1$ and $\mathcal{F} = v_1 = (\text{tr }(H) - 1)/(n-1)$. One shows that $\mathcal{F}$ has mean zero and standard deviation $1/(n-1)$ under uniform sampling in G. Moreover, $\mathcal{F} = 1$ if and only if $H = I$ and the whole n-dimensional $\mathbb{R}^n$ is fixed by H, whereas $\mathcal{F} = -1/(n-1)$ if and only if H is cyclic, in which case the largest invariant subspace is the 1-dimensional subspace generated by $e' = (1, ...1)$. Here, larger values of $\mathcal{F}$ reflect a larger invariant subspace;

(3) Let H be doubly stochastic with non-negative entries. It can be written as a convex combinations of permutation matrices, say, $H = w_1 g_1 + \ldots + w_r g_r$. Direct computation shows that
$$\mathcal{F} = \frac{\text{tr } H - 1}{n-1}.$$
Again, because H is non-negative and doubly stochastic, large values of $\mathcal{F}$ imply that the off-diagonal entries must be very small and hence H must be concentrated on its diagonal. In fact $\mathcal{F} = 1$ if and only if $H = I$.

(4) Let $y \in \mathbb{R}^n$ and $H = yy'$. Then $(n-1)v_1 = ||y - \bar{y}e||^2$ and $v_0 + (n-1)v_1 = ||y||^2$, so that
$$\mathcal{F}^{-1/2} = \sqrt{n}\frac{|\bar{y}|}{\sqrt{\sum_i (y_i - \bar{y})^2/(n-1)}}.$$

(5) Let $H = \sum_i (x_i - \bar{x})(y_i - \bar{y})'$. Then,
$$\mathcal{F} = \frac{1-r}{1+(n-1)r},$$
where r is the usual sample correlation coefficient under the permutation symmetric assumption for the underlying covariance matrix, that is, under the assumption that the covariance matrix has the symmetry of G. More precisely,
$$r = \frac{e'He - \text{tr } H}{(n-1)\text{tr } H}.$$

(6) Let $y \in \{0,1\}^n$, $H = yy'$ and $\bar{y}$ the proportion of ones in y. Then one shows that
$$\mathcal{F} = (n-1)\frac{\bar{y}}{1-\bar{y}},$$
that is, $\mathcal{F}/(n-1)$ is the observed odds on "1"-components in y.

EXAMPLE 3.2. Cyclic decomposition. Let g in G be an element of order n, and let C_n indicate the corresponding cyclic group generate by g. In direct analogy to Proposition 2.2, we obtain that the uniform mean conjugate $E(UHU')$ of H on C_n decomposes as $v_0 I + v_1 g + \ldots + v_{n-1} g^{n-1}$, where the weights v_i are given by

$v_i = \text{tr } g^{n-i}H/n$ (there are n one-dimensional irredicible representations). When H is symmetric, we have, in addition, that $v_i = v_{n-i}$. The corresponding ANOVA interpretation, for $n = 4$ (and similarly for any even n) and $H = yy'$ is illustrated on Table 2. The case $n = 5$ (and similarly for any odd n) is illustrated on Table 3.

TABLE 2. Analysis of Variance for Cyclic Decomposition, $n = 4$, $H = yy'$.

i	components	df	mss (v_i)
0	$\text{tr } H/n = y'y/4$	1	$y'y/4$
1	$2\text{tr } gH/n = 2y'gy/4$	2	$y'gy/4$
2	$\text{tr } g^2H = 1y'g^2y/4$	1	$y'g^2y/4$
total	$e'He/n = 4\bar{y}^2$	4	

TABLE 3. Analysis of Variance for Cyclic Decomposition, $n = 5$, $H = yy'$.

i	components	df	mss (v_i)
0	$\text{tr } H/n = y'y/5$	1	$y'y/5$
1	$2\text{tr } gH/n = 2y'gy/5$	2	$y'gy/5$
2	$2\text{tr } g^2H = 2y'g^2y/5$	2	$y'g^2y/5$
total	$e'He/n = 5\bar{y}^2$	5	

EXAMPLE 3.3. ANOVA for cyclic decompositions (Keratometry Data). Keratometry is the measurement of corneal curvature of a small area using a sample of four reflected points of light along an annulus 3 to 4 mm. in diameter, centered about the line of sight. For normal cornea this commonly approximates the apex of the cornea [**VOM93**]. The fundamental principle of computerized keratometry is similar in that the relative separation of reflected points of light along concentric rings are used to calculate the curvature of the measured surface. Using a pattern of concentric light-reflecting rings and sampling at specific circularly equidistant intervals, a numerical model of the measured surface may be obtained. Sampling takes place at equally-spaced ring-semimeridian intersections (e.g., [**KW89**]). Figure 1 shows the apex (scaled) distances measured along 360 equally-spaced semimeridians (9-degree separation, n=40), based on the reflected image of a fixed ring. Similar data are generated for the actual surface curvature. The resulting ANOVA is shown in Table 4. The total *cyclic* variability $40 \times \bar{y}^2 = 0.2498454423\ 10^8$ is decomposed into the cyclic components $y'g^iy/40$. More specifically, $40 \times \bar{y}^2 = \sum_{i=0}^{40} v_i \times df_i$. Figure 2 illustrates the size of the semimeridian *effects*, with maximum effect sizes orthogonally located. The analysis also picks up the direction (corresponding to steep and flat curvatures) of the corresponding effects.

TABLE 4. Analysis of Variance for Circular Data with decomposition $40 \times \bar{y} = 0.2498454423\ 10^8 = \sum_{i=0}^{40} y'g^iy$.

meridian (i)	$y'g^iy$	df_i	meridian (i)	$y'g^iy$	df_i
0	2501.44	1	10, 30	2496.61	2
1, 21	2501.19	2	11, 31	2496.63	2
2, 22	2500.56	2	12, 32	2496.72	2
3, 23	2499.70	2	13, 33	2496.93	2
4, 24	2498.80	2	14, 34	2497.30	2
5, 25	2498.00	2	15, 35	2497.88	2
6, 26	2497.39	2	16, 36	2498.66	2
7, 27	2496.99	2	17, 37	2499.55	2
8, 28	2496.76	2	18, 38	2500.39	2
9, 29	2496.65	2	19, 39	2501.00	2
			20	2501.23	1
			total	$0.2498454423\ 10^8$	40

FIGURE 1. Keratometry data with 9-degree meridian separation.

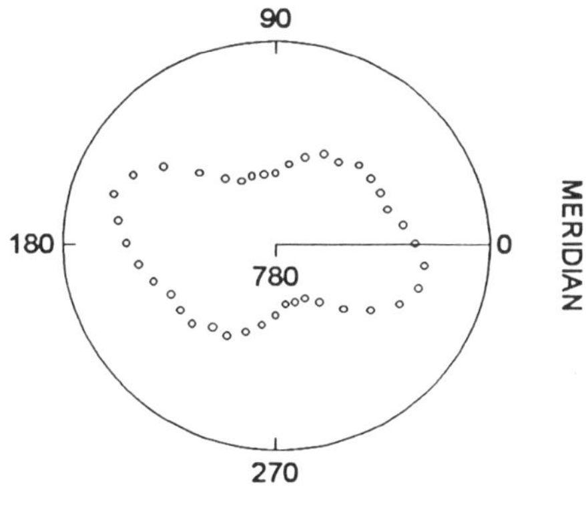

4. The covariance structure and transition probabilities

Recall that given two random matrices U and V, the (ij,st)-entry of Cov (UV) is given by $\mathrm{Cov}(UV)_{ij;st} = e_i'\mathrm{E}(UV'e_je_t'VU')e_s - e_i'\mathrm{E}(UV)e_je_t'\mathrm{E}(UV)'e_s$. Taking $m = n$ and $\mathbf{V}$ constant and equal to $\mathbf{I}$, the (ij,st) entry of $\mathrm{Cov}(U^k)$ is expressed as

$$\text{(4.1)} \qquad \mathrm{Cov}(U^k)_{ij,st} = E(e_i'U^ke_je_s'U^ke_t) - E(e_i'U^ke_j)E(e_s'U^ke_t), \quad k = 1, \ldots .$$

The ij-entry $U_{ij,k} = e_i'U^ke_j = e_i'e_{U^k(j)}$ of U^k indicates whether or not U^k moves j into i. That is, observed $U = g$, the value of $U_{ij,k}$ is 1 if $g^k(j) = i$ and is 0 otherwise. Then, under the uniform law, the average $E(U^k)$ of U^k over G is the matrix in which the ij-component is the probability $P_k(j \to i)$ that U^k moves j into i. Similarly, $U_{ij,k}U_{st,k}$ indicates whether or not U^k moves (j,t) into (i,s).

FIGURE 2. Analysis of Variance for Circular Data, $MSQ = y'g^iy/10^6$, for each semimeridian angle $9\,i$, $i = 1, \ldots, 40$.

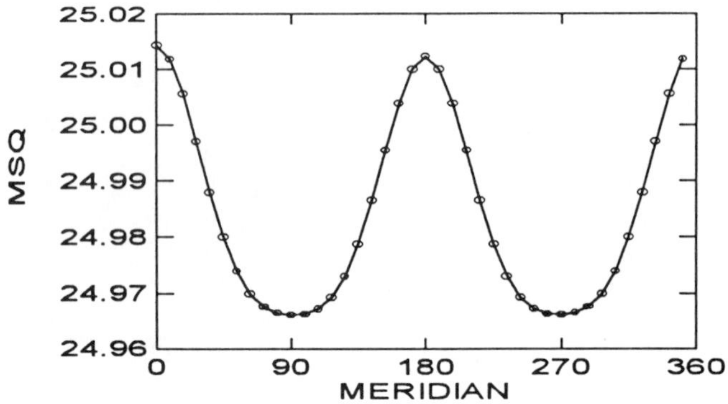

The corresponding probability $\mathcal{P}_k((j,t) \to (i,s))$ is the average $E(U_{ij,k}U_{st,k})$ of $U_{ij,k}U_{st,k}$ over G under the uniformly probability law. It then follows that the (ij,st)-entry of $\text{Cov}(U^k)$, indicated in expression (4.1), has the interpretation

$$\text{Cov}(U^k)_{ij,st} = \mathcal{P}_k[(j,t) \to (i,s)] - P_k(j \to i)P_k(t \to s).$$

Denoting the associated random walk on $\{1, \ldots, n\}$ by L (with law P_k) and the random walk on $\{1, \ldots, n\}^2$ by L^2 (with law $\mathcal{P}_k$), we write, shortly, $\text{Cov}(U^k) = \mathcal{P}_k(L^2) - P_k(L)^2$. If the entries of $\mathcal{P}_k(L^2)$ are assembled in lexicographic order $11, \ldots, 1n, \ldots, n1, \ldots, nn$, and $C_{l,k}$ indicates the l-th column of U^k, and C_k the $1 \times n$ block matrix $(C_{1,1}, \ldots, C_{n,k})$, then

$$\text{Cov}(U^k) = E(C_kC_k') - E(C_k)E(C_k)', \tag{4.2}$$

with the (l,m) block corresponding to $\text{Cov}(C_{l,k}, C_{m,k})$, $l, m = 1 \ldots, n$, displays the covariance structure of U^k in lexicographic form.

4.1. The covariance structure of U. The following proposition describes the covariance structure of a random permutation matrix.

PROPOSITION 4.1. *If U is uniformly distributed in G, then,*

$$Cov(U)_{ij;st} = \frac{1}{n^2(n-1)}[1 + n(n\delta_{jt}\delta_{is} - \delta_{jt} - \delta_{is})]. \tag{4.3}$$

PROOF. Taking expression (4.1) with k=1, $\text{E}(Ue_je_t'U')$ is the uniform mean conjugate of e_je_t', and hence, from Proposition 2.2, it has the form $a_0ee' + a_1I$, with

$$n(n-1)a_0 = \text{tr }(ee' - I)e_je_t' = 1 - \delta_{jt}, \quad n(a_0 + a_1) = \text{tr }e_je_t' = \delta_{jt}.$$

Substituting the expressions for $E(Ue_je_t'U')$ and $E(U) = \frac{1}{n}ee'$ into (4.1), the result follows. □

EXAMPLE 4.1. To illustrate, with $n = 3$, the covariance structure of U is determined by the covariance between any two columns i and s;

$$\operatorname{Cov}(C_i, C_s) = \frac{1}{18}\left[\begin{array}{rrr|rrr} 4 & -2 & -2 & -2 & 1 & 1 \\ -2 & 4 & -2 & 1 & -2 & 1 \\ -2 & -2 & 4 & 1 & 1 & -2 \\ \hline -2 & 1 & 1 & 4 & -2 & -2 \\ 1 & -2 & 1 & -2 & 4 & -2 \\ 1 & 1 & -2 & -2 & -2 & 4 \end{array}\right].$$

In general, the covariance matrix between any two columns i and s of U with uniform distribution in G has the form

$$\operatorname{Cov}(C_i, C_s) = \begin{cases} \frac{1}{n^2(n-1)}\mathbf{ee}' + \frac{-1}{n(n-1)}\mathbf{I}, & \text{when } i \neq s, \\ \frac{-1}{n^2}\mathbf{ee}' + \frac{1}{n}\mathbf{I}, & \text{when } i = s. \end{cases}$$

4.2. The covariance structure of U^k. Here it will be convenient to assemble the entries of $\operatorname{Cov}(U^k)$ using the matrix direct-product $U \otimes V$ notation. We also write $\otimes^2 V$ to indicate $V \otimes V$. First we note that both $E(\otimes^2 U^k)$ and $\otimes^2 E(U^k)$ are block matrices, with respective blocks $E(U_{ij,k}U^k)$ and $E(U_{ij,k})E(U^k)$ indexed by ij. Within each block the entries are indexed by s (rows) and t (columns). Clearly, $\operatorname{Cov}(U^k) = E(\otimes^2 U^k) - \otimes^2 E(U^k)$. For example, the sums of the rows in lexicographic notation, as in equation (4.2), correspond to block sums in $\otimes$-notation. To illustrate, with $n = 2$, we have (omitting the k-index),

$$U = \begin{bmatrix} U_{11} & U_{12} \\ U_{21} & U_{22} \end{bmatrix}, \quad \otimes^2 U = \left[\begin{array}{cc|cc} U_{11}U_{11} & U_{11}U_{12} & U_{12}U_{11} & U_{12}U_{12} \\ U_{11}U_{21} & U_{11}U_{22} & U_{12}U_{21} & U_{12}U_{22} \\ \hline U_{21}U_{11} & U_{21}U_{12} & U_{22}U_{11} & U_{22}U_{12} \\ U_{21}U_{21} & U_{21}U_{22} & U_{22}U_{21} & U_{22}U_{22} \end{array}\right].$$

The associated random walk L^2 visits four points labeled by (1,1),(1,2),(2,1) and (2,2). The transition from (j,t) to (i,s) occurs with probability $E(U_{ij}U_{st})$. The diagonal of the matrix $E(\otimes^2 U)$ contains the *resting* probabilities $[(i,j) \to (i,j)]$ and the average trace of $E(\otimes^2 U)$, for example, is the walk's resting probability when the past state distribution is uniform. Similarly, $E(U)$ describes the marginal transition probabilities associated with the marginal walk L. In this section we will derive $E(U^k)$, $E(\otimes^2 U^k)$ and $\otimes^2 E(U^k)$, thus leading to the assembling of $\operatorname{Cov}(U^k)$.

4.2.1. *Derivation of $E(U^k)$.*

PROPOSITION 4.2. *If U is uniformly distributed in G, then $E(U^k) = a_0 ee' + a_1 I$, where a_0 and a_1 are determined by $n(a_0 + a_1) = E(tr\, U^k)$ and $n(n-1)a_0 = n - E(tr\, U^k)$.*

PROOF. Let $m_{ji} = \{g \in G; g^k(j) = i\}$, so that $|G|E(U^k)_{ij} = |m_{ji}|$. If $j \neq i$ then m_{ij} and any other m_{uv} with $u \neq v$ are isomorphic via $g \to t'gt$ where t is (any permutation) such that $t(v) = j$ and $t(u) = i$; take g in m_{ij}, then $(t'gt)^k(v) = t'g^k t(v) = t'g^k(j) = t'(i) = u$, so that $t'gt$ is in m_{uv}, and hence $|m_{ij}| = |m_{uv}|$ (off-diagonal homogeneity). Similarly, if $j = i$ then m_{ii} and any other m_{ff} are isomorphic via $g \to t'gt$ where t is any permutation satisfying $t(f) = i$, and hence $|m_{ii}| = |m_{ff}|$ (diagonal homogeneity). Therefore, $E(U^k)$ has the form $a_0 ee' + a_1 I$, for constants a_0 and a_1, determined by taking the trace and overall sum on both sides of the equality $E(U^k) = a_0 ee' + a_1 I$. □

To determine the coefficients a_0 and a_1, we need the following: Given positive integers n and k let $D_{k,n}$ indicate the set of all positive integer divisors of k not greater than n, with $|D_{k,n}|$ elements in it. For computational purposes, note that $|D_{k,n}|$ may be expressed as $\sum_{r=1}^{n} I_{\{0\}}(k \mod r)$, where $I_{\{0\}}$ is the indicator function with base the set $\{0\}$.

THEOREM 4.1. *If U is uniformly distributed in G then $E(\text{tr } U^k) = |D_{k,n}|$.*

PROOF. Let $c_{m,n}(U)$ indicate the random number of cycles of length m associated with a random permutation U in G, $m = 1, \ldots, n$. Then, noting that $\text{tr } g^k = \sum_{m \in D_{k,n}} m c_{m,n}(g)$, it follows that

$$E(\text{tr } U^k) = \sum_{m \in D_{k,n}} m E(c_{m,n}(U)).$$

However, because there are in average $1/m$ cycles of length m in a uniform random permutation in G, for all n, (that is $E(c_{m,n}(U)) = 1/m$) [see also Appendix B], the proposed result follows. □

From proposition 4.1, we obtain,

$$a_0 + a_1 = \frac{|D_{k,n}|}{n}, \quad a_0 = \frac{n - |D_{k,n}|}{n(n-1)}, \quad n a_0 + a_1 = 1. \tag{4.4}$$

The corresponding ANOVA decomposition corresponding to Table 1 leads to $v_1 = (|D_{k,n}| - 1)/(n-1)$ and $v_0 = 1$, so that $\mathcal{F} = v_1 = (|D_{k,n}| - 1)/(n-1)$.

EXAMPLE 4.2. Calculations based on Theorem 4.1. In the following tables 5 and 8, a permutation g is represented as $(1^{c_1}, 2^{c_2}, \ldots, n^{c_n})$, where $c_m \equiv c_{m,n}(g)$ is the multiplicity of cycles of length m in g, $m = 1, \ldots, n$. Cycles with null multiplicity are omitted in the notation. Table 5 shows the conjugacy classes $\mathcal{C}$, representatives of g^k, and the number $\gamma(g) = \sum c_{m,n}(g)$, for values of $k = 1, \ldots, 10$. We note that $\gamma(g)$ is the number of distinct orbits generated by g and its powers acting on $\{1, \ldots, n\}$ and that $|\mathcal{C}| = n!/\prod_{m=1}^{n} m^{c_m} c_m!$ (see also Appendix B). Table 6 shows values of tr g^k by conjugacy class $\mathcal{C}$, in S_5, for values of $k = 1, \ldots, 10$. Table 7 shows the probability distribution for $\tau_k = \text{tr } U^k$ and $E(\tau_k) = |D_{k,n}|$ in S_5, for values of $k = 1, \ldots, 10$ (also shown is $E(\tau_k^2)$). Table 8 shows values of $\tau_k = \text{tr } U^k = |D_{k,n}|$ in S_8 by conjugacy classes $\mathcal{C}$, for values of $k = 1, \ldots, 4$.

TABLE 5. Conjugacy classes $\mathcal{C}$, representatives of g^k, and number $\gamma(g)$ of $< g >$-orbits in S_5, for values of $k = 1, \ldots, 10$.

$\mathcal{C}$	$\gamma(g)$	$\lvert\mathcal{C}\rvert$	g	g^2	g^3	g^4	g^5	g^6	g^7	g^8	g^9	g^{10}
$\mathcal{C}_7$	1	24	5	5	5	5	1^5	5	5	5	5	1^5
$\mathcal{C}_6$	2	20	$2,3$	$1^2,3$	$2,1^3$	$1^2,3$	$2,3$	1^5	$2,3$	$1^2,3$	$2,1^3$	$1^2,3$
$\mathcal{C}_5$	2	30	$1,4$	$1,2^2$	$1,4$	1^5	$1,4$	$1,2^2$	$1,4$	1^5	$1,4$	$1,2^2$
$\mathcal{C}_4$	3	20	$1^2,3$	$1^2,3$	1^5	$1^2,3$	$1^2,3$	1^5	$1^2,3$	$1^2,3$	1^5	$1^2,3$
$\mathcal{C}_3$	3	15	$1,2^2$	1^5	$1,2^2$	1^5	$1,2^2$	1^5	$1,2^2$	1^5	$1,2^2$	1^5
$\mathcal{C}_2$	4	10	$1^3,2$	1^5	$1^3,2$	1^5	$1^3,2$	1^5	$1^3,2$	1^5	$1^3,2$	1^5
$\mathcal{C}_1$	5	1	1^5	1^5	1^5	1^5	1^5	1^5	1^5	1^5	1^5	1^5

TABLE 6. Values of $\tau_k = \text{tr } U^k$ by conjugacy class $\mathcal{C}$, in S_5, for values of $k = 1, \ldots, 10$.

$\mathcal{C}$	$\gamma(g)$	$\lvert\mathcal{C}\rvert$	τ_1	τ_2	τ_3	τ_4	τ_5	τ_6	τ_7	τ_8	τ_9	τ_{10}
$\mathcal{C}_7$	1	24	0	0	0	0	5	0	0	0	0	5
$\mathcal{C}_6$	2	20	0	2	3	2	0	5	0	2	3	2
$\mathcal{C}_5$	2	30	1	1	1	5	1	1	1	5	1	1
$\mathcal{C}_4$	3	20	2	2	5	2	2	5	2	2	5	2
$\mathcal{C}_3$	3	15	1	5	1	5	1	5	1	5	1	5
$\mathcal{C}_2$	4	10	3	5	3	5	3	5	3	5	3	5
$\mathcal{C}_1$	5	1	5	5	5	5	5	5	5	5	5	5

TABLE 7. Probability distribution for $\tau_k = \text{tr } U^k$, $E(\tau_k) = |D_{k,n}|$ and $E(\tau_k^2)$ when U is uniformly distributed in S_5, for values of $k = 1, \ldots, 10$.

τ	τ_1	τ_2	τ_3	τ_4	τ_5	τ_6	τ_7	τ_8	τ_9	τ_{10}
0	44	24	24	24	20	24	44	24	24	0
1	45	30	45	0	45	30	45	0	45	30
2	20	40	0	40	20	0	20	40	0	40
3	10	0	30	0	10	0	10	0	30	0
4	0	0	0	0	0	0	0	0	0	0
5	1	26	21	56	25	66	1	56	21	50
$E(\tau_k)$	1	2	2	3	2	3	1	3	2	3
$E(\tau_k^2)$	2	7	7	13	7	14	2	13	7	12

EXAMPLE 4.3. When $n = 2$, U^k is either constant and equal to the identity when k is even, or U^k is either $\begin{bmatrix} 0 & 1 \\ 1 & 0 \end{bmatrix}$ or the identity, each with probability 1/2, when k is odd. Also note that $2 - |D_{k,2}| = k \mod 2$. From Proposition 4.1, it then follows that tr $\text{Cov}(U^k) = k \mod 2$ and $\text{Cov}(U^k) = \frac{k \mod 2}{2} \begin{bmatrix} 1 & -1 \\ -1 & 1 \end{bmatrix}$.

4.2.2. *Derivation of the diagonal blocks of $E(\otimes^2 U^k)$.* We start with the derivation of the expected trace of $U_{ii,k}U^k$. In what follows we write *marginal sum* to indicate both row sum and column sum without distinction. Let $G_{ij,k}$ indicate the subset of G consisting of those permutation matrices g such that the ij-entry of g^k is equal to 1, that is $G_{ij,k} = \{g \in G; e_i' g^k e_j = 1\}$.

PROPOSITION 4.3. *The size of $|G_{ij,k}|$ is $(n-1)!|D_{k,n}|$ when $i = j$ and is $(n-2)!|D_{k,n}|$ when $i \neq j$.*

PROOF. In fact,

$$|G_{ii,k}| = |G|E(e_i' U^k e_i) = G\frac{|D_{k,n}|}{n} = |M_{n-1}||D_{k,n}| = (n-1)!|D_{k,n}|,$$

whereas, when $i \neq j$, $|G_{ij,k}| = |G|E(e_i' U^k e_j) = G\frac{n-|D_{k,n}|}{n(n-1)} = (n-2)!|D_{k,n}|$. $\square$

TABLE 8. Values of $\tau_k = \text{tr } U^k$ in S_8 by conjugacy classes $\mathcal{C}$, and $E(\tau_k) = |D_{k,n}|$, for values of $k = 1, \ldots, 4$.

$\gamma(g)$	g	$\lvert\mathcal{C}\rvert$	τ_1	g^2	τ_2	g^3	τ_3	g^4	τ_4
1	8	5040	0	4^2	0	8	0	2^4	0
2	$1,7$	5760	1	$1,7$	1	$1,7$	1	$1,7$	1
2	$2,6$	3360	0	$1^2,3^2$	2	$2,2^3$	0	$1^2,3^2$	2
2	$3,5$	2688	0	$3,5$	0	$1^3,5$	3	$3,5$	0
2	4^2	1260	0	2^4	0	4^2	0	1^8	8
3	$1^2,6$	3360	2	$1^2,3^2$	2	$1^2,2^3$	2	$1^2,3^2$	2
3	$1,2,5$	4032	1	$1^3,5$	3	$1,2,5$	1	$1^3,5$	3
3	$1,3,4$	3360	1	$1,3,2^2$	1	$1^4,4$	4	$1^5,3$	5
3	$2^2,4$	1260	0	$1^4,2^2$	4	$2^2,4$	0	1^8	8
3	$2,3^2$	1120	0	$1^2,3^2$	2	$1^6,2$	6	$1^2,3^2$	2
4	$1^3,5$	1344	3	$1^3,5$	3	$1^3,5$	3	$1^3,5$	3
4	$1^2,2,4$	2520	2	$1^4,2^2$	4	$1^2,2,4$	2	1^8	8
4	$1^2,3^2$	1120	2	$1^2,3^2$	2	1^8	8	$1^2,3^2$	2
4	$1,2^2,3$	1680	1	$1^5,3$	5	$1^4,2^2$	5	$1^5,3$	5
4	2^4	105	0	1^8	8	2^4	0	1^8	8
5	$1^4,4$	420	4	$1^4,2^2$	4	$1^4,4$	4	1^8	8
5	$1^3,2,3$	1120	3	$1^5,3$	5	$1^6,2$	6	$1^5,3$	5
5	$1^2,2^3$	420	2	1^8	8	$1^2,2^3$	2	1^8	8
6	$1^5,3$	112	5	$1^5,3$	5	1^8	8	$1^5,3$	5
6	$1^4,2^2$	210	4	1^8	8	$1^4,2^2$	4	1^8	8
7	$1^6,2$	28	6	1^8	8	$1^6,2$	6	1^8	8
8	1^8	1	8	1^8	8	1^8	8	1^8	8
$E(\tau_k)$			1		2		2		3

Note that $G_{ii,k}$ is not a subgroup of G whenever $|D_{k,n}|$ is not an integer divisor of n. In particular, $G_{ii,1}$ is a subgroup of G.

PROPOSITION 4.4. *If U is uniformly distributed in G, then $E(U_{ii,k}\ tr\, U^k) = \frac{|D_{k,n}|}{n}\sum_{m\in D_{k,n}} m\widehat{C}_{m,k}$, where $\widehat{C}_{m,k}$ is the average number of m-cycles in $G_{ii,k}$.*

PROOF.

$$\begin{aligned}
E(U_{ii,k}\text{tr } U^k) &= \sum_G e_i' g^k e_i \text{tr } g^k/|G| = \sum_{G_{ii,k}} \text{tr } g^k/|G| \\
&= \frac{|G_{ii,k}|}{|G|}\sum_{G_{ii,k}} \text{tr } g^k/|G_{ii,k}| = \frac{|D_{k,n}|}{n}\sum_{G_{ii,k}} \text{tr } g^k/|G_{ii,k}| \\
&= \frac{|D_{k,n}|}{n}\sum_{G_{ii,k}}\sum_{m\in D_{k,n}} mc_{m,n}(g)/|G_{ii,k}| \\
&= \frac{|D_{k,n}|}{n}\sum_{m\in D_{k,n}} m\sum_{G_{ii,k}} c_{m,n}(g)/|G_{ii,k}| \\
&= \frac{|D_{k,n}|}{n}\sum_{m\in D_{k,n}} m\widehat{C}_{m,k},
\end{aligned}$$

where $\widehat{C}_{m,k} = \sum_{G_{ii,k}} c_{m,n}(g)/|G_{ii,k}|, m = 1, \dots, n$ is the average number of m-cycles in $G_{ii,k}$. □

To obtain the main diagonal of $E(U_{ii,k}U^k)$ we start with the (ii,ii) entry. From (4.4),

$$E(U_{ii,k}^2) = E(U_{ii,k}) = \frac{|D_{k,n}|}{n}. \tag{4.5}$$

The $n-1$ remaining entries of the main diagonal are

$$E(U_{ii,k}U_{ss,k}) = \frac{E(U_{ii,k} \text{ tr } U^k) - |D_{k,n}|/n}{n-1}, \quad s \neq i. \tag{4.6}$$

PROPOSITION 4.5. *If* **U** *is uniformly distributed in G, then: (a) The entries $i \neq s$ of column t, and the entries $j \neq t$ of row s of $E(U_{ij,k}U^k)$ are equal to zero; (b) The marginal sums of $E(U_{ij,k}U^k)$ are constant and equal to $\frac{|D_{k,n}|}{n}$ for $i = j$, and $\frac{n-|D_{k,n}|}{(n-1)}$ for $i \neq j$.*

PROOF. Part (a) follows from the fact that each row (column) of U^k has exactly one entry equal to 1 and the remaining entries equal to 0; For (b), note that $E(\sum_s U_{ij,k}U_{st,k}) = E(U_{ij,k}\sum_s U_{st,k}) = E(U_{ij,k})$. Applying Propositions 4.2 and 4.1, the result follows (similarly for column sums). □

Proposition 4.5, and equations (4.5) and (4.6) directly determine all the entries at column $t = j$ and row $s = i$. Any other row (say row $s \neq i$) has the diagonal entry (s, s) given by (4.6), the (s, i) entry equal to zero and (for $n > 2$) the remaining n-2 entries given by

$$E(U_{ii,k}U_{st,k}) = \frac{|D_{k,n}|/n - E(U_{ii,k}U_{ss,k})}{n-2}, \quad t \neq i, s. \tag{4.7}$$

EXAMPLE 4.4. Evaluation of $E(U_{11,k}U^k \text{ tr } U^k)$ for n=3 and k=1,...,6. Table 9 summarizes the computational steps of Proposition 4.4. Consider the case k=2, to illustrate. First we evaluate $|G_{11,k}| = (n-1)!|D_{k,n}| = 4$. The corresponding permutations are marked with 1 on the box to the left of column $|G_{11,k}|$. Next, $\widehat{C}_{1,k}$, is obtained by summing $C_1(g)$ over those columns corresponding to permutations g marked with 1 (the first four when $k = 2$) and dividing by $|G_{11,k}|$. This leads to $\widehat{C}_{1,2} = 6/4$. Next, compute $W = \sum_{m \in D_{k,n}} m\widehat{C}_{m,k}$, or, is this case, $\sum_{m=1,2} m\widehat{C}_{m,2} = 6/4 + 2 \times 3/4 = 12/4$. Finally, $E(U_{11,2}\text{tr } U^2) = W|D_{2,3}|/3 = 2$. The derivations for other values of k are obtained similarly.

4.2.3. *Derivation of the off-diagonal blocks of $E(\otimes^2 U^k)$.* Let $[i\ j]$ indicate the permutation matrix representing the transposition (ij). Direct verification shows that $U_{ii,k}U^k \sim U_{jj,k}[i\ j]U^k[i\ j]'$, and that, for $i \neq j$, $U_{ij,k}U^k \sim U_{im,k}[j\ m]U^k[j\ m]'$. We note that in each of the cases (a) $s, t \in \{j, m\}$, (b) $s \in \{j, m\}$ and $t \notin \{j, m\}$, (c) $s \notin \{j, m\}$ and $t \in \{j, m\}$, and (d) $s, t \notin \{j, m\}$, it holds that

$$U_{ij,k}e_s'U^k e_t \sim U_{im,k}e_s'[j\ m]U^k[j\ m]'e_t.$$

PROPOSITION 4.6. *The following equalities hold:*

(1) $E(U_{ii,k} tr\ U^k) = \frac{1}{n}E(tr^2\ U^k)$,
(2) $E(U_{ij,k} tr\ U^k) = \frac{1}{n-1}E(tr\ U^k) - \frac{1}{n(n-1)}E(tr^2\ U^k), \quad i \neq j$.

TABLE 9. Values of $|G_{11,k}|$, $\hat{C}_{m,k}$, $W = \sum_{m \in D_{k,n}} m\widehat{C}_{m,k}$ and $E(U_{11,k}\text{tr } U^k) = W|D_{k,n}|/n$, for $n = 3$ and values of $k = 1, \dots, 6$.

k	$D_{k,n}$	1	(12)	(13)	(23)	(123)	(132)
1	$\{1\}$	1	0	0	1	0	0
2	$\{1,2\}$	1	1	1	1	0	0
3	$\{1,3\}$	1	0	0	1	1	1
4	$\{1,2\}$	1	1	1	1	0	0
5	$\{1\}$	1	0	0	1	0	0
6	$\{1,2,3\}$	1	1	1	1	1	1
	$C_1(g)$	3	1	1	1	0	0
	$C_2(g)$	0	1	1	1	0	0
	$C_3(g)$	0	0	0	0	1	1

k	$\lvert G_{11,k}\rvert$	$\hat{C}_{1,k}$	$\hat{C}_{2,k}$	$\widehat{C}_{3,k}$	W	$E(U_{11,k}\text{tr } U^k)$
1	2	4/2	1/2	0/2	4/2	2/3
2	4	6/4	3/4	0/4	12/4	2
3	4	4/4	1/4	2/4	10/4	5/3
4	4	6/4	3/4	0/4	12/4	2
5	2	4/2	1/2	0/2	4/2	2/3
6	6	6/6	3/6	2/6	18/3	3

PROOF. Write (the power notation is irrelevant and is omitted here), for part (1), $(\text{tr } U)U = U_{11}U + U_{22}U + \dots + U_{nn}U$. From Proposition 4.6, $(\text{tr } U)U \sim U_{11}U + U_{22}[1\ 2]U[1\ 2]' + \dots + U_{nn}[1\ n]U[1\ n]'$, so that, $(\text{tr } U)^2 \sim U_{11}\text{tr } U + U_{22}\text{tr } U + \dots + U_{nn}\text{tr } U \sim nU_{11}\text{tr } U$. Taking the expectation on both sides, the result follows. For part (2), again from Proposition 4.6, we write, for $m \neq i$,

$$\sum_{j \neq i} U_{im}[j\ m]U[j\ m]' \sim \sum_{j \neq i} U_{ij}U = (1 - U_{ii})U, \tag{4.8}$$

so that $(n-1)U_{im}\text{tr } U \sim \text{tr } U - \text{tr } U_{ii}U$. Taking expectations on both sides, and applying part (a), the result follows. □

EXAMPLE 4.5. Suppose n=4. The derivation of $Z_{12} = E(U_{12,k}U^k)$ is obtained as follows. First derive $M = E(1 - U_{11,k}U^k)$, following Example 4.4. Then apply Propositions 4.5 (b) and 4.6 (b), along with the estimating equations $M = Z_{12} + [2\ 3]Z_{12}[2\ 3]' + [2\ 4]Z_{12}[2\ 4]'$, to determine the components of Z_{12}. The remaining blocks on the first row are obtaining by conjugating Z_{12} (e.g., $Z_{13} = E(U_{13,k}U^k) = [2\ 3]Z_{12}[2\ 3]'$). The off-diagonal blocks on the remaining rows are complement-symmetric: $Z_{ij} = Z_{n-j,n-i}$ (e.g., $Z_{23} = Z_{12}$).

EXAMPLE 4.6. The following are the transition probability blocks,

$$P = E(C_1, [C_1', C_2', C_3'])$$

, and corresponding covariance matrices, $C = \mathrm{Cov}(C_1, [C_1', C_2', C_3'])$ for n=3 and k=1,...,6, and the complete 9×9 matrices P and C for $k = 4$ (lexicographic notation).

k=1

$$P = \frac{1}{6}\left[\begin{array}{ccc|ccc|ccc} 2 & 0 & 0 & 0 & 1 & 1 & 0 & 1 & 1 \\ 0 & 2 & 0 & 1 & 0 & 1 & 1 & 0 & 1 \\ 0 & 0 & 2 & 1 & 1 & 0 & 1 & 1 & 0 \end{array}\right],$$

$$C = \frac{1}{18}\left[\begin{array}{ccc|ccc|ccc} 4 & -2 & -2 & -2 & 1 & 1 & -2 & 1 & 1 \\ -2 & 4 & -2 & 1 & -2 & 1 & 1 & -2 & 1 \\ -2 & -2 & 4 & 1 & 1 & -2 & 1 & 1 & -2 \end{array}\right]$$

k=2

$$P = \frac{1}{6}\left[\begin{array}{ccc|ccc|ccc} 4 & 0 & 0 & 0 & 4 & 0 & 0 & 0 & 4 \\ 0 & 1 & 0 & 0 & 0 & 1 & 1 & 0 & 0 \\ 0 & 0 & 1 & 1 & 0 & 0 & 0 & 1 & 0 \end{array}\right],$$

$$C = \frac{1}{36}\left[\begin{array}{ccc|ccc|ccc} 8 & -4 & -4 & -4 & 8 & -4 & -4 & -4 & 8 \\ -4 & 5 & -1 & -1 & -4 & 5 & 5 & -1 & -4 \\ -4 & -1 & 5 & 5 & -4 & -1 & -1 & 5 & -4 \end{array}\right]$$

k=3

$$P = \frac{1}{6}\left[\begin{array}{ccc|ccc|ccc} 4 & 0 & 0 & 0 & 3 & 1 & 0 & 1 & 3 \\ 0 & 1 & 0 & 1 & 0 & 0 & 0 & 0 & 1 \\ 0 & 0 & 1 & 0 & 1 & 0 & 1 & 0 & 0 \end{array}\right],$$

$$C = \frac{1}{36}\left[\begin{array}{ccc|ccc|ccc} 8 & -4 & -4 & -4 & 2 & 2 & -4 & 2 & 2 \\ -4 & 5 & -1 & 5 & -4 & -1 & -1 & -1 & 2 \\ -4 & -1 & 5 & -1 & 2 & -1 & 5 & -1 & -4 \end{array}\right]$$

k=4

$$P = \frac{1}{6}\left[\begin{array}{ccc|ccc|ccc} 4 & 0 & 0 & 0 & 4 & 0 & 0 & 0 & 4 \\ 0 & 1 & 0 & 0 & 0 & 1 & 1 & 0 & 0 \\ 0 & 0 & 1 & 1 & 0 & 0 & 0 & 1 & 0 \end{array}\right],$$

$$C = \frac{1}{36}\left[\begin{array}{ccc|ccc|ccc} 8 & -4 & -4 & -4 & 8 & -4 & -4 & -4 & 8 \\ -4 & 5 & -1 & -1 & -4 & 5 & 5 & -1 & -4 \\ -4 & -1 & 5 & 5 & -4 & -1 & -1 & 5 & -4 \end{array}\right]$$

k=5

$$P = \frac{1}{6}\left[\begin{array}{ccc|ccc|ccc} 2 & 0 & 0 & 0 & 1 & 1 & 0 & 1 & 1 \\ 0 & 2 & 0 & 1 & 0 & 1 & 1 & 0 & 1 \\ 0 & 0 & 2 & 1 & 1 & 0 & 1 & 1 & 0 \end{array}\right],$$

$$C = \frac{1}{18}\left[\begin{array}{ccc|ccc|ccc} 4 & -2 & -2 & -2 & 1 & 1 & -2 & 1 & 1 \\ -2 & 4 & -2 & 1 & -2 & 1 & 1 & -2 & 1 \\ -2 & -2 & 4 & 1 & 1 & -2 & 1 & 1 & -2 \end{array}\right]$$

k=6

$$P=\frac{1}{6}\left[\begin{array}{ccc|ccc|ccc} 6&0&0&0&6&0&0&0&6\\ 0&0&0&0&0&0&0&0&0\\ 0&0&0&0&0&0&0&0&0 \end{array}\right],$$

$$C=\frac{1}{18}\left[\begin{array}{ccc|ccc|ccc} 0&0&0&0&0&0&0&0&0\\ 0&0&0&0&0&0&0&0&0\\ 0&0&0&0&0&0&0&0&0 \end{array}\right]$$

Complete transition probability and covariance matrices for $k = 4$, in lexicographic notation.

$$P=\frac{1}{6}\left[\begin{array}{ccc|ccc|ccc} 4&0&0&0&4&0&0&0&4\\ 0&1&0&0&0&1&1&0&0\\ 0&0&1&1&0&0&0&1&0\\ \hline 0&0&1&1&0&0&0&1&0\\ 4&0&0&0&4&0&0&0&4\\ 0&1&0&0&0&1&1&0&0\\ \hline 0&1&0&0&0&1&1&0&0\\ 0&0&1&1&0&0&0&1&0\\ 4&0&0&0&4&0&0&0&4 \end{array}\right],$$

$$\mathrm{Cov}(U^4)=\frac{1}{36}\left[\begin{array}{rrr|rrr|rrr} 8&-4&-4&-4&8&-4&-4&-4&8\\ -4&5&-1&-1&-4&5&5&-1&-4\\ -4&-1&5&5&-4&-1&-1&5&-4\\ \hline -4&-1&5&5&-4&-1&-1&5&-4\\ 8&-4&-4&-4&8&-4&-4&-4&8\\ -4&5&-1&-1&-4&5&5&-1&-4\\ \hline -4&5&-1&-1&-4&5&5&-1&-4\\ -4&-1&5&5&-4&-1&-1&5&-4\\ 8&-4&-4&-4&8&-4&-4&-4&8 \end{array}\right]$$

Complete transition probability and covariance matrices for $k = 4$, in $\otimes$-notation.

$$P_{\otimes}=\frac{1}{6}\left[\begin{array}{ccc|ccc|ccc} 4&0&0&0&1&0&0&0&1\\ 0&4&0&0&0&1&1&0&0\\ 0&0&4&1&0&0&0&1&0\\ \hline 0&0&1&4&0&0&0&1&0\\ 1&0&0&0&4&0&0&0&1\\ 0&1&0&0&0&4&1&0&0\\ \hline 0&1&0&0&0&1&4&0&0\\ 0&0&1&1&0&0&0&4&0\\ 1&0&0&0&1&0&0&0&4 \end{array}\right]$$

$$\text{Cov}(U^4)_{\otimes} = \frac{1}{36}\left[\begin{array}{ccc|ccc|ccc} 8 & -4 & -4 & -4 & 5 & -1 & -4 & -1 & 5 \\ -4 & 8 & -4 & -1 & -4 & 5 & 5 & -4 & -1 \\ -4 & -4 & 8 & 5 & -1 & -4 & -1 & 5 & -4 \\ \hline -4 & -1 & 5 & 8 & -4 & -4 & -4 & 5 & -1 \\ 5 & -4 & -1 & -4 & 8 & -4 & -1 & -4 & 5 \\ -1 & 5 & -4 & -4 & -4 & 8 & 5 & -1 & -4 \\ \hline -4 & 5 & -1 & -4 & -1 & 5 & 8 & -4 & -4 \\ -1 & -4 & 5 & 5 & -4 & -1 & -4 & 8 & -4 \\ 5 & -1 & -4 & -1 & 5 & -4 & -4 & -4 & 8 \end{array}\right]$$

5. Discussion and Summary

The covariance structure of U^k is naturally related to the structure of cycles in symmetric group. Here we will outline that connection from two perspectives: one based on simple combinatoric arguments and the other based on elementary character theory. First, note that Propositions 4.4 and 4.6 show that, in the uniform case,

$$E(tr^2\, U^k) = |D_{k,n}| \sum_{m \in D_{k,n}} m\widehat{C}_{m,k}, \tag{5.1}$$

where $\widehat{C}_{m,k}$ is the average number of m-cycles in $G_{ii,k} = \{g \in G; e_i' g^k e_i = 1\}$. In particular, when $k = 1$, one shows that

$$\sum_{g \in G_{ii,1}} \text{tr}\ g = (n-1)! + \sum_{h \in S_{n-1}} \text{tr}\ h. \tag{5.2}$$

Dividing both sides of equation (5.2) by $|G_{ii,1}| = (n-1)!$ (see Proposition 4.3), it follows that $\widehat{C}_{1,1} = 1 + E(\text{tr}\ Z)$, with Z uniformly in S_{n-1}. Because $E(\text{tr}\ Z) = 1$, we obtain $E(tr^2\ U) = \widehat{C}_{1,1} = 2$. As outlined below (Appendix A, Comment 3), one obtains the same result using Burnside's Lemma. The question of interest here is the determination of $E(tr^2\ U^k)$ for arbitrary k, when the distribution of U is uniform in G. Because

$$\text{tr}\ U^k = \sum_{m \in D_{k,n}} m c_{m,n}(U),$$

it follows that $E(tr^2\ U^k)$ is completely determined by the joint covariance structure of

$$c_n'(U) = (c_{1,n}(U), \dots, c_{n,n}(U)).$$

In fact, it is only necessary to determine the matrix $E(c_n(U)c_n'(U))$ of average cross-products.

5.1. The combinatorial approach. The joint probability law of $c_n(U)$ has the support the set K of points $c' = (c_1, \dots, c_n)$ with non-negative integer coordinates satisfying $\sum_{m=1}^{n} mc_m = n$. Clearly, $|K|$ equals the number of conjugacy classes in G and the law of $c_n(U)$ assigns the same probability to members within the same conjugacy class $|\mathcal{C}|$. Therefore $P(c_n(U) = c) = |\mathcal{C}|/n!$ for each

$c \in K$, where $\mathcal{C}$ is the conjugacy class with representative $(1^{c_1}, \ldots, m^{c_m})$. Standard combinatorics argument (e.g., [**Tak84**]) shows that $|\mathcal{C}| = n!/\prod_K m^{c_m} c_m!$, so that $P(c_n(U) = c) = 1/\prod_K m^{c_m} c_m!$. Given the law of $c_n(U)$, the matrix $E(c_n(U)c_n'(U))$ is then obtained by its own definition. The following illustrates the case n=3 and k=2.

EXAMPLE 5.1. Derivation of $E(tr^2\ U^2)$ in S_3. There are $|K| = 3$ conjugacy classes in S_3 and the points in the support set K are $c = (0,0,1)$, with probability 1/3, $c = (1,1,0)$ with probability 1/2 and $c = (3,0,0)$ with probability 1/6. The resulting marginals $c_{m,3}(U) \equiv c_m(U), m = 1,2,3$ have support the set $\{0,1,2,3\}$, with corresponding probabilities $(1/3, 1/2, 0, 1/6)$ for $c_1(U)$, $(1/2, 1/2, 0, 0)$ for $c_2(U)$ and $(2/3, 1/3, 0, 0)$ for $c_3(U)$. Note that $E(c_m(U)) = 1/m$, whereas $E(c_1^2(U)) = E(tr^2\ U) = 2$, as discussed above. In addition, $E(c_2^2(U)) = 1/2$ and $E(c_3^2(U)) = 1/3$. Similarly, the joint probabilities are determined. The law of $(c_1(U), c_2(U))$ assigns probabilities 1/3,1/2 and 1/6 to the points $(0,0),(1,1)$ and $(3,0)$, respectively, so that $E(c_1(U)c_2(U)) = 1/2$, whereas $(c_1(U), c_3(U))$ takes probabilities 1/3,1/2 and 1/6 at (0,1),(1,0) and (3,0), respectively, with $E(c_1(U)c_3(U)) = 0$. The law of $(c_1(U), c_2(U))$ assigns probabilities 1/6,1/3 and 1/2 to the points (0,0),(0,1) and (1,0), respectively, so that $E(c_2(U)c_3(U)) = 0$. In summary, we obtain,

$$E(c_3(U)c_3'(U)) = \begin{bmatrix} 2 & \frac{1}{2} & 0 \\ \frac{1}{2} & \frac{1}{2} & 0 \\ 0 & 0 & \frac{1}{3} \end{bmatrix}, \quad \mathrm{Cov}(c_3(U)) = \begin{bmatrix} 1 & 0 & -\frac{1}{3} \\ 0 & \frac{1}{4} & -\frac{1}{6} \\ -\frac{1}{3} & -\frac{1}{6} & \frac{1}{9} \end{bmatrix}.$$

Then, it follows that

$$\begin{aligned} E(tr^2\ U^2) &= E(c_{1,3}(U) + 2c_{2,3}(U))^2 \\ &= E(c_{1,3}^2(U)) + 4E(c_{1,3}(U)c_{2,3}(U)) + 4E(c_{2,3}^2(U)) = 6. \end{aligned}$$

5.2. The character theory approach. Let Θ indicate the $|K| \times |K|$ matrix of character values of G, with entries $\theta_{ij} = \chi_i(g_j)$, where g_j is a representative in the conjugacy class $\mathcal{C}_j$, $j = 1, \ldots, |K|$. Usually, χ_1 is the character of the trivial representation. Let T indicate the $n \times |K|$ matrix with entries

$$T_{mj} = \frac{|\mathcal{C}_j|}{n!} c_{m,n}(g_j), \quad j = 1, \ldots, |K|, \quad m = 1, \ldots, n.$$

One then shows that the components of m-th column of the matrix $\Theta T'$ are exactly the Fourier coefficients

$$< c_{m,n}, \chi_j >= \frac{1}{n!} \sum_G c_{m,n}(g)\bar{\chi}_j(g) = \frac{|\mathcal{C}_j|}{n!} \sum_{l=1}^{|K|} c_{m,n}(g_l)\bar{\chi}_j(g_l),$$

of the class function $c_{m,n}(g)$ relative to character χ_j, $j = 1, \ldots, |K|$. Moreover,

$$E(c_n(U)c_n'(U)) = T\Theta'\Theta T', \tag{5.3}$$

and $E(c_n(U)) = T\Theta' e_1$, so that $\mathrm{Cov}(c_n(U)) = T\Theta'\Theta T' - T\Theta' e_1 e_1' \Theta T'$. The following example illustrates the result:

EXAMPLE 5.2. We consider again the case $n = 3$, as in Example 5.1 above. We start with the matrix Θ of corresponding characters, where we define the conjugacy class $\mathcal{C}_1$ with representative 1^3, $\mathcal{C}_2$ with representative 2^1 and $\mathcal{C}_3$ with representative 3^1, so that

$$\Theta = \begin{bmatrix} 1 & 1 & 1 \\ 1 & -1 & 1 \\ 2 & 0 & -1 \end{bmatrix}.$$

From Example 5.1 we obtain the matrix T, as defined (note that the columns of T are indexed by the same labels defined by Θ),

$$T = \begin{bmatrix} \frac{1}{2} & \frac{1}{2} & 0 \\ 0 & \frac{1}{2} & 0 \\ 0 & 0 & \frac{1}{3} \end{bmatrix}.$$

From (5.3), we obtain

$$E(c_n(U)c_n'(U)) = T\Theta'\Theta T' = T \begin{bmatrix} 6 & 0 & 0 \\ 0 & 2 & 0 \\ 0 & 0 & 3 \end{bmatrix} T' = \begin{bmatrix} 2 & \frac{1}{2} & 0 \\ \frac{1}{2} & \frac{1}{2} & 0 \\ 0 & 0 & \frac{1}{3} \end{bmatrix},$$

which is as derived in Example 5.1.

Appendix A. The (n-1)-dimensional irreducible representation of G.

The existence and construction of the representation is outlined in [**Ser77**, p.17] and also in [**Dia88**, pp.6,134]. Let ρ indicate the permutation representation of the symmetric group ($G = S_n$), described in Section 1, and let Y indicate the product of p copies of $\{1, 2, \dots, n\}$. Let also $\chi(\pi) = \text{tr } \rho(\pi)$ indicate the character of representation ρ. The following hold:

(1) The character of the product representation (on Y) is χ^p;
(2) When $p = 1$ or $p = 2$, Y has p distinct orbits (the product action is n-transitive, e.g., [**Rob93**, Ch.7]). When $p = 3$, there are 5 distinct representatives, $(1,1,1)$, $(1,1,f)$, $(1,f,1)$, $(f,1,1)$, with $f \neq 1$ and (t,f,g), with $t \neq f, f \neq g, t \neq g$;
(3) In general, |G-orbits in Y| $= \sum_G \chi^p(g)/|G|$. This is essentially Burnside's Lemma [see [**Dia88**, p.134]]. This also gives all moments of the trace function of random permutations, and |G-orbits in Y| gives the first n moments of a Poisson (1) distribution. In particular, $E(\text{tr } U) = 1$ and $E(tr^2\ U) = 2$;
(4) Let 1 indicate the identity representation, and θ its complementary subspace. Let the corresponding characters be χ_1 and χ_θ. Burnside's Lemma implies that $1 = (\chi|\chi_1) = \sum \chi(g)/|G|$, so that 1 is contained in ρ exactly once. Write $\rho = 1 + \theta$;
(5) From 2 and 3 above, $2 = (\chi^2|1) = (\chi|\chi) = (1+\theta|1+\theta) = 1 + (\theta|\theta)$, so that $(\theta|\theta) = 1$ and hence θ is irreducible;
(6) The homotheties in V and W are the corresponding projections, given by $\sum_G \chi_1(g)\rho(g)\text{dim V}/|G|$ and $\sum_G \chi_\theta(g)\rho(g)\text{dim W}/|G|$, respectively. The homothety ratios are tr H_{11}/dim V and tr H_{22}/dim W, where H_{11} and H_{22} are the block partition of H corresponding to the dimensions of V and W, respectively. In the present case, the ratios are h_{11} and $(h_{22} + \dots + h_{nn})/(n-1)$. More generally, Schur's Lemma shows that

$$\sum_G (\theta_1(g) \vdots \theta_2(g)) H (\theta_1(g') \vdots \theta_2(g'))/|G| = diag\ (\frac{\text{tr } H_{11}}{d_1} I_1, \frac{\text{tr } H_{22}}{d_2} I_2),$$

for any two irreducible non-isomorphic representations θ_1 and θ_2 of dimensions d_1 and d_2 (and H of dimension $d_1 + d_2$).

Appendix B. The average number of cycles and m-cycles

Let $c_{m,n}(U)$ indicate the number of m-cycles in a random uniform permutation U in G, so that $\gamma_n(U) = \sum_{m=1}^{n} c_{m,n}(U)$ is the number of cycles in U. The average number $\bar{c}_{m,n}$ of cycles of length m (m-cycles) in G is $1/m$, for $m = 1, \dots, n$ and all n. This follows from Burnside's Lemma (e.g., [**DM96**, p.25]). The number of permutations in G with exactly k cycles is the Stirling number $S(n,k)$ [e.g., [**Tak84**]], so that (i) $\gamma_n(U) = k$ with probability $S(n,k)/n!$, $k = 1, \dots, n$, (ii) the mean number $\bar{\gamma}_n = E(\gamma_n(U))$ of cycles in G is

$$\bar{\gamma}_n = E(\gamma_n(U)) = \sum_{k=1}^{n} k\frac{S(n,k)}{n!} = \sum_{m=1}^{n} E(c_{m,n}(U)) = \sum_{m=1}^{n} \bar{c}_{m,n},$$

and (iii) $\bar{c}_{m,n}$ is the average number of m-cycles in G.

PROPOSITION B.1. *The mean number of cycles in G is $\bar{\gamma}_n = 1 + \frac{1}{2} + \ldots + \frac{1}{n}$.*

PROOF. This follows directly from the fact that $\bar{c}_{m,n} = 1/m$ for $m = 1, \dots, n$ and all n. A direct proof is based on the recursive relation $S(n+1,k) - nS(n,k) = S(n,k-1)$ for Stirling numbers. Simple calculation then shows that $\bar{\gamma}_{n+1} = \bar{\gamma}_n + \frac{1}{n+1}$, from which the result follows. □

Because $\gamma_n(U)$ is also the number of orbits generated by the multiplicative action of G on $\{1, \dots, n\}$, and $\gamma_n(U)$ is constant over conjugacy classes, we also obtain

$$\sum_j \gamma_j \frac{|C_j|}{n!} = \bar{\gamma}_n = 1 + \frac{1}{2} + \ldots + \frac{1}{n},$$

where the sum is indexed by the conjugacy classes and γ_j is the common value of $\gamma_n(U)$ for U in C_j.

References

[BE73] M. J. Buckley and G. K. Eagleson, *Assessing large sets of rank correlations*, Biometrika **73** (1973), no. 1, 151–7.

[Dan62] H. E. Daniels, *Processes generating permutation expansions*, Biometrika **49** (1962), 139–149.

[DF99] P. Diaconis and D. Freedman, *Iterated random functions*, SIAM Review **41** (1999), no. 1, 45–76.

[Dia88] P. Diaconis, *Group representation in probability and statistics*, IMS, Hayward, California, 1988.

[DM96] J. D. Dixon and B. Mortimer, *Permutation groups*, Springer, New York, NY, 1996.

[KW89] S. Klyce and S. Wilson, *Methods of analysis of corneal topography*, Refractive and Corneal Surgery **5** (1989), 368–371.

[LV99] H. Lee and M. A. G. Viana, *The joint covariance structure of ordered symmetrically dependent observations and their concomitants of order statistics*, Statistics and Probability Letters **43** (1999), 411–414.

[Nai82] M. A. Naimark, *Theory of group representations*, Springer-Verlag, New York, NY, 1982.

[OV95] I. Olkin and M. A. G. Viana, *Correlation analysis of extreme observations from a multivariate normal distribution*, Journal of the American Statistical Association **90** (1995), 1373–1379.

[PGL+88] V. Perkel, M. Gail, J. Lubin, D. Pee, R. Weinstein, E. Shore-Freedman, and A. Schneider, *Radiation-induced thyroid neoplasms: Evidence for familial susceptibility factors*, Journal of Clinical Endocrinology and Metabolism **66** (1988), no. 6, 1316–22.

[Rob93] Derek J.S. Robinson, *A course in the theory of groups*, Springer-Verlag, New York, 1993.

[Ser77] J-P. Serre, *Linear representations of finite groups*, Springer-Verlag, New York, 1977.

[Tak84] L. Takács, *Combinatorics*, Handbook of Statistics (New York, NY) (P. R. Krishnaiah and P. K. Sen, eds.), vol. 4, North-Holland, New York, NY, 1984, pp. 123–173.

[Via98] M. A. G. Viana, *Linear combinations of ordered symmetric observations with applications to visual acuity*, Order Statistics: Applications (Amsterdam) (N. Balakrishnan and C. R. Rao, eds.), vol. 17, Elsevier, Amsterdam, 1998, pp. 513–24.

[VO97] M. A. G. Viana and I. Olkin, *Correlation analysis of ordered observations from a block-equicorrelated multivariate normal distribution*, Advances in Statistical Decision Theory and Applications (Boston) (S. Panchapakesan and N. Balakrishnan, eds.), Birkhauser, Boston, 1997, pp. 305–322.

[VOM93] M. A. G. Viana, I. Olkin, and T. McMahon, *Multivariate assessment of computer analyzed corneal topographers*, Journal of the Optical Society of America - A **10** (1993), no. 8, 1826–1834.

The University of Illinois at Chicago, Eye Research Institute, Division of Biostatistics, and Department of Mathematics, Statistics and Computer Science, 1855 W. Taylor Street, Chicago, IL 60612

E-mail address: `viana@uic.edu`

Contemporary Mathematics
Volume **287**, 2001

The Extendibility of Statistical Models

Ernst Wit and Peter McCullagh

ABSTRACT. This paper introduces a new benchmark of statistical modelling, called "extendibility". Extendibility is defined as the unambiguity of the likelihood function under extensions of the statistical model. In this paper we discuss three forms of extendibility that should constrain the choice of a statistical model: (i) the addition or deletion of statistical units, (ii) the inclusion of alternative covariate levels and (iii) alternative measurement scales. Category theory is introduced to describe these extendibility conditions and create extendible statistical models.

1. Introduction

The platitude that all statistical models are wrong, but some are *useful*, has led to a certain pragmatism among statisticians. With a liberal use of an old proverb, statisticians tend to believe that 'the proof of the model is in the fitting.' There has been no attempt to formulate the principles underlying statistical modelling with any generality. It has been left to the intuition of each individual statistician to use only "sensible" statistical models.

In this paper we make an attempt to define a language that is capable of critiquing classical statistical models. Although our discussion does not consider Bayesian models, similar considerations can be brought to bear on them. We argue that a sensible model is one that is extendible consistently because a model is a tool that carries its own extension. A hammer would be quite absurd if it hit only nails of a particular color. Similarly, a statistical model would be silly if it were not extendible to a situation with additional statistical units or fewer covariate levels than in the actual situation.

Extendibility is the key word in this paper. Three forms of extensions we consider are the selection of statistical units, the selection of covariate values, and the transformation of the measurement scale. In turn, these extensions have consequences for the sample space, parameter space and the statistical model itself. A statistical model is understood to be a sequence of sets of probability measures on the sample space indexed by a parameter space. That is, for each sample space and parameter space that are induced by all possible sets of units and covariate levels a statistical model requires a different set of measures. We show how the structure of

1991 *Mathematics Subject Classification.* Primary 62A01, 18C10; Secondary 62A10.
Both authors were supported in part by NSF Grant DMS #97-05347.

the collections of statistical units, the covariate space and the measurement scale affect the ultimate structure of the statistical models that one can propose.

In Section 2 we describe the concept of extendibility in more detail. The notion of extendibility bears some similarity to Kolmogorov's *consistency conditions of finite dimensional processes* (Billingsley, 1995, p.483). In Section 3 we review basic notions of category theory with which help we formally define extendible statistical models for extensive and intensive response variables. Finally in Section 4 we explore the consequences of our new definitions by applying them to several practical situations.

2. Forms of extendibility

It is not uncommon that many choices have already been made before the statistician gets to see the data. Certain treatments were included or excluded in the study, counts are grouped in a certain way, centimetres rather than inches were used to present the data. The fact that these choices were made for the statistician does not mean that the consequences of these choices can be ignored. In particular, it seems reasonable to require that the model chosen to analyze the data can readily be extended to accommodate other sets of treatments, different measurement scales and more statistical subjects.

It is a legitimate question to ask at this point why one should be bothered at all by extendibility. First, why should one even consider models for a different number of units or covariate levels when one has a particular dataset at hand? And why should one consider different units of measurement, when the data are presented in a particular measurement scale? Secondly, even if these considerations are useful at all, how does that affect the kinds of models that can be used? The answer to these questions requires one to think more about the process than about the dataset. It lies in its very nature that a statistical model encompasses extensions to other statistical subjects. Sampling and prediction, two central activities of statistics, imply that the model has an extension to a larger collection of units. On the other hand, it depends on the ontological structure of the data whether the dataset observed could have been obtained by selection from a larger set of covariate levels than the ones observed. Frequently, this is the case. It is also almost always possible to present the data in another unit of measurement. Nevertheless, it remains a question what extendibility of a model really means and how desirable it really is. Even if we accept that we are at least implicitly responsible for the formulation of extensions of a statistical model, to what extent does that constrain our actual choice of a statistical model? The answer depends on the circumstances. For example, if a particular dataset could be considered a selection of a larger set of data and this selection does not affect the dynamics of the underlying stochastic process, then the model should be able to be written as a restriction of the larger model. An example of a model that does not satisfy this simple consistency constraint is the so-called type III model that we shall discuss later on. The failure of a proposed model to correspond to the restriction of the larger model implies that the likelihood of the data can be calculated in two conflicting ways. Therefore, an extendibility condition ultimately comes down to a form of unambiguity of the likelihood.

Almost every statistical design involves selection. At some point in the design process a decision is made regarding the factor levels and the number of units to be studied. For example, which drugs and dose levels will be included in the study, and what plots of land will be used for planting the varieties? Each of these decisions involves selection. For each of the selections it is possible to define a set of probability measures to describe the response of interest. Extendibility refers to a consistency relationship between these sets of probability measures. If the model proposed for a smaller set of levels (but with the same number of factors) corresponds to the restriction of a larger model to those levels, then this sequence of sets of probability measures is called "selection-sensible." The consistency property itself is sometimes referred to as "selection invariance." A model that lacks this property is called "selection-silly."

Grouping is another important operation, particularly for extensive response factors (Cox and Snell, 1981). Extensive response factors have levels that collect counts instead of recording a trait-value. Merging several levels of such factor aggregates the two counts in one super level, as in merging, e.g., *strongly disagree, disagree, neutral, agree, strongly agree* into *disagree, neutral, agree.* If the restriction of the model on the first set of levels to the levels of the second is consistent with the model on the second set of levels, then we call the model "merger-sensible" or invariant under the formation of subfactors. Otherwise it is called "merger-silly."

It is not uncommon that the unit of measurement has a large arbitrary component. Distances can be presented in metres, feet, inches or centimetres. It seems logical to require that the distributions for different response scales be related to one another by multiplicative constants. If temperature information is given in Centigrade or Fahrenheit and is to be included in the model, it is reasonable to require that the model be invariant not only under scale transformations but also origin translations, that is, under all affine transformations of the temperature scale. Circular and directional data can be represented by one or more angles and possibly the length of the vector. The reference frame with respect to which the angles are measured is generally completely arbitrary. Models for such data are frequently not preserved under change of measurement scale (cf. Anderson and Wu 1995, 1996). Models that are invariant under a change of measurement scale are called "scale-sensible," otherwise they are called "scale-silly."

3. Extendibility and category theory

In this section, we argue that category theory is precisely the right tool to express these forms of extendibility and their consequences for statistical modelling in a correct mathematical form. For our purposes, a category can be thought of as the collection of all possible extensions of the statistical units, the covariate space or the measurement scale. Using the language of category theory a statistical model can be defined on top of these three fundamental categories. We begin by introducing several important notions within category theory and then apply them in a statistical setting.

3.1. Some basic notions of category theory. Definitions of a category, a functor and a natural transformation can be found in MacLane's celebrated *Categories for the Working Mathematician* (1971). A more elementary book that introduces these concepts is *Arrows, Structures, and Functors: the Categorical Imperative* by Arbib and Manes (1975). The concept of a category representation was introduced into statistics by McCullagh (2000a, p.215).

Formally, a **category** is a collection of *objects* and *arrows*, such that (i) for each object Ω in the category the identity arrow 1_Ω is included in the same category and

(ii) for all arrows f and g such that $dom\ g = cod\ f$, the composition arrow $g \circ f$ is included in the category. Categories are the building blocks of category theory. A **contravariant** (resp. **covariant**) **functor** $T : \mathcal{C} \to \mathcal{K}$ is defined as a map between two categories $\mathcal{C}$ and $\mathcal{K}$, consisting of a object-map T_o and a arrow-map T_a, such that for every arrow $\varphi : \Omega \to \Omega' \in \mathcal{C}$, the arrow $T_a\varphi : T_o\Omega' \to T_o\Omega$ (resp. $T_a\varphi : T_o\Omega \to T_o\Omega'$) belongs to $\mathcal{K}$ in such a way that T_a maps identities to identities, $T_a 1_\Omega = 1_{T_o\Omega}$, and compositions to *reversed* (resp. non-reversed) compositions, $T_a(\varphi \circ \psi) = T_a\psi \circ T_a\varphi$ (resp. $T_a(\varphi \circ \psi) = T_a\varphi \circ T_a\psi$). With a useful abuse of language we sometimes refer to the collection of objects $\{T_o\Omega\}$ as the *functor*. The following picture describes two categories $\mathcal{C}$ and $\mathcal{K}$ and a contravariant functor T between the two.

$$\begin{array}{lccccc}
 & & \multicolumn{3}{c}{\overset{\varphi'\circ\varphi}{\longrightarrow}} & \\
\mathcal{C}: & \Omega & \overset{\varphi}{\longrightarrow} & \Omega' & \overset{\varphi'}{\longrightarrow} & \Omega'' \\
\downarrow T & & & & & \\
\mathcal{K}: & T_o\Omega & \overset{T_a\varphi}{\longleftarrow} & T_o\Omega' & \overset{T_a\varphi'}{\longleftarrow} & T_o\Omega'' \\
 & & \multicolumn{3}{c}{\underset{T_a\varphi\circ T_a\varphi'}{\longleftarrow}} &
\end{array}$$

A **category representation** on the category $\mathcal{C}$ is a contravariant functor $T : \mathcal{C} \to \mathcal{K}$ in which $\mathcal{K}$ is a category of linear transformations on vector spaces. Moreover, for a category representation the arrow-map T_a is fixed. For a $\varphi : \Omega \to \Omega'$ in $\mathcal{C}$ and for a $f : \Omega' \to \mathbb{R}$ in $\mathcal{X}_{\Omega'} \in \mathcal{K}$, the morphism $T_a\varphi(f)$ is defined as $f \circ \varphi$. This map $T_a\varphi := . \circ \varphi$ is known as the **pullback** map and generally written as φ^*.

Given two contravariant functors $\mathcal{X}, \mathcal{Y} : \mathcal{C} \to \mathcal{K}$, a **natural transformation** g from $\mathcal{X}$ to $\mathcal{Y}$ is a map which assigns to each object Ω of $\mathcal{C}$ an arrow $g_\Omega : \mathcal{X}_\Omega \to \mathcal{Y}_\Omega$, such that for every $\varphi : \Omega \to \Omega' \in \mathcal{C}$ the following diagram commutes:

$$\begin{array}{ccccc}
\Omega' & & \mathcal{X}_{\Omega'} & \overset{g_{\Omega'}}{\longrightarrow} & \mathcal{Y}_{\Omega'} \\
\varphi \uparrow & & \varphi^*_{\mathcal{X}} \downarrow & & \downarrow \varphi^*_{\mathcal{Y}} \\
\Omega & & \mathcal{X}_\Omega & \overset{g_\Omega}{\longrightarrow} & \mathcal{Y}_\Omega
\end{array}$$

Some examples of categories that frequently arise in statistical work are (i) the category $\mathcal{I}$ of injective maps on finite sets; (ii) the category $\mathcal{S}$ of surjective maps on finite sets; (iii) the general affine group $\mathcal{A}$; (iv) the category of linear transformations on vector spaces; (v) the category of surjective linear transformations on vector spaces. For example, the category $\mathcal{I}$ contains all finite sets and all injections between these sets. For instance, let $\varphi : \{1, 2\} \to \{a, b, c\}$ be defined as

$$\varphi(1) = c, \ \varphi(2) = b.$$

The map φ "selects" c and b from a larger set and "relabels" them as 1 and 2, respectively. It can be established easily that $\mathcal{I}$ is a category. It is intimately connected with the operations of selection and sampling. The category $\mathcal{S}$ consists of all finite sets and all surjections between those sets, such as for instance $\varphi : \{1, 2, 3\} \to \{a, b\}$, defined as

$$\varphi(1) = b, \ \varphi(2) = a, \ \varphi(3) = b,$$

which "merges" levels 1 and 3 into b. Again, it is easy to check that $\mathcal{S}$ is a category. It is related to the operation of amalgamation. The category $\mathcal{A}$ consists of the real line and all affine transformations of the real line. This constitutes a category that describes scale transformations of a linear measurement scale without a fixed origin.

It is possible to create new categories from old ones in several ways. For the purposes of this paper we introduce three such ways. The objects in the *product category* $\mathcal{C} \times \mathcal{K}$ are all product objects, typically product sets. The arrows in $\mathcal{C} \times \mathcal{K}$ are all order pairs $(\varphi_{\mathcal{C}}, \varphi_{\mathcal{K}})$, acting component-wise. From a single category $\mathcal{C}$ one can create $\mathcal{C}^n$, the *category of the order* n. This is a category whose objects are of the form Ω^n with $\Omega \in \mathcal{C}$ and whose morphisms come from $\mathcal{C}$ and act on all dimensions simultaneously. There is a distinction between $\mathcal{C}^2$ and $\mathcal{C} \times \mathcal{C}$. The objects in the former are all possible "squares," whereas objects in the latter are all possible "rectangles." A *subcategory* of a certain category $\mathcal{C}$ is a subcollection of objects and arrows from $\mathcal{C}$ that by itself satisfies the definition of a category. The category $\mathcal{C} \times \mathcal{C}$ contains $\mathcal{C}^2$ as a subcategory. Finally, the *opposite category* $\mathcal{C}^{\mathrm{op}}$ is constructed by the process of reversing all arrows in $\mathcal{C}$.

3.2. Definition of an extendible classical statistical model. In this section we translate the extendibility notions described in Section 2 into the language of category theory for a classical statistical model. A classical statistical model is conventionally thought of as a family of probability distributions on the sample space with parameters that depend on the treatments and other covariate levels. We instead stress that a statistical model is, *inter alia*, a model for a process rather than a single dataset.

Design category

Statistical units are at the basis of statistical theory. Generally, a statistical unit is defined as the smallest entity such that two distinct units may receive different treatments. One can think of plots of land in an agricultural trial or patients in a medical trial. The particular set of units observed is generally of little interest in itself: inference is usually concerned with plots or units other than those observed. From the point of view of statistical modelling, it is essential to aim for a universal statement applying to all sets of units. The observed set of units is then one of many such sets. It is naturally embedded in the sequence of all possible sets of statistical units. These sets are connected by injective maps describing selection. This effectively constitutes the category of injective maps $\mathcal{I}$ or a subcategory thereof. We refer to this category as $Cat_{\mathcal{U}}$.

To each statistical unit there corresponds a vector of distinguishing features, which we call a covariate vector. The set of all possible covariate vectors is denoted by Ω. The covariate space can also degenerate in a single point as is often the case in surveys where one is merely interested in the multivariate response. Again, the actual covariate space should be thought of as being embedded in a sequence of potential covariate spaces. More controls can be included or the covariates can be measured on a different measurement scale. Again, these sets are connected by morphisms that describe their logical relationship. We require that the sequence of quantitative or qualitative treatment and classification factors, i.e., the sequence of covariate spaces, constitutes a category. We denote this category by Cat_{Ω}.

A statistical design is a map $\psi : \mathcal{U} \to \Omega$ from the units into the covariate space. A morphism of design is a pair of injective maps $(\varphi_u, \varphi_c) : \psi \to \psi'$, in which $\varphi_u \in Cat_{\mathcal{U}}$ and $\varphi_c \in Cat_{\Omega}$ are maps such that $\varphi_c \psi = \psi' \varphi_u$. In other words a morphism of design is a pair of injective maps (φ_u, φ_c) such that the diagram shown below is commutative.

$$\text{Statistical Design}$$

$$\begin{array}{ccc} \mathcal{U} & \xrightarrow{\psi} & \Omega \\ \downarrow \varphi_u & & \downarrow \varphi_c \\ \mathcal{U}' & \xrightarrow{\psi'} & \Omega' \end{array} \tag{3.1}$$

The collection of statistical designs $\psi, \psi', \ldots$ and morphisms of design (φ_u, φ_c) is itself a category, Cat_{Design}.

Measurement scale category

The choice of measurement scale can have implications both for the sample space and the parameter space of a statistical model. In general, the collection of changes of a measurement scale that has a natural zero, such as length, volume, etc., constitute the groupoid of scale transformations. The measurement scale itself is the functor $\mathbb{R}^+$ on this category. For example, the measurement scale in an exponential model for a survival time affects both the observation and the survival parameter λ. If the measurement scale changes from weeks to days, then the morphism of sample spaces is a multiplicative constant 7 and the morphism of parameter spaces is a multiplicative constant $\frac{1}{7}$. The measurement scale category is a groupoid consisting of objects such as {years, weeks, days, ...}. The measurement scale itself is a functor $\mathbb{R}^+$ on this category.

For one-dimensional measurement scales without a natural origin the groupoid of all affine transformations constitutes the measurement category. The measurement scale is the functor $\mathbb{R}$ on this category. Typical examples are temperature and relative velocities.

Sample space functor

If $\mathcal{R}$ is the sample space for a single unit, then to each set of statistical units $\mathcal{U}$ there corresponds a sample space $\mathcal{R}^{\mathcal{U}}$. Also, to each morphism of units φ_u between two sets of units in the category $Cat_{\mathcal{U}}$ there corresponds a morphism φ_u^* between their associated sample spaces. The morphism φ_u^* is the pullback map, i.e., composition with φ_u. To a fixed set of units $\mathcal{U}$ and a change of measurement scale in the response scale $\rho : \mathcal{R} \to \mathcal{R}'$ there corresponds a change in sample space $\rho_{\mathcal{U}} : \mathcal{R}^{\mathcal{U}} \to \mathcal{R}'^{\mathcal{U}}$. In other words, the sample space is a covariant functor on the product category $Cat_{\mathcal{U}}^{\text{op}} \times Cat_{\mathcal{R}}$.

$$\begin{array}{ccccc} & & \mathcal{R} & \xrightarrow{\rho} & \mathcal{R}' \\ \mathcal{U} & & \mathcal{R}^{\mathcal{U}} & \xrightarrow{\rho_{\mathcal{U}}} & \mathcal{R}'^{\mathcal{U}} \\ \downarrow \varphi_u & & \uparrow \varphi_u^* & & \uparrow \varphi_u^* \\ \mathcal{U}' & & \mathcal{R}^{\mathcal{U}'} & \xrightarrow{\rho_{\mathcal{U}'}} & \mathcal{R}'^{\mathcal{U}'} \end{array}$$

Parameter space functor and subparameters

In a parametric model, there is a parameter space associated with the set of probability measures. This parameter space Θ_Ω depends on a given covariate space. An extendible parameter space sequence $\{\Theta_\Omega\}$ is a contravariant functor on the covariate space category Cat_Ω.

$$\begin{array}{ccc} \Omega & & \Theta_\Omega \\ \downarrow \varphi_c & & \uparrow \varphi'_c \\ \Omega' & & \Theta_{\Omega'} \end{array}$$

The map φ'_c is normally an extension of the pullback map φ^*_c. For instance, for a one-way model with unknown variance the parameter space is $\mathbb{R}^\Omega \times \mathbb{R}^+$ for the mean value parameters and the variance. The maps φ'_c in this case are defined as (φ^*_c, id). Linear models enjoy a lot of attention due to their mathematical convenience and wide applicability. In our terms, a linear model is a statement about the structure of the sets Θ_Ω, namely, that they are linear function spaces on the sets Ω. An extendible linear model Θ is defined as a (surjective) representation on the covariate space category Cat_Ω. In Section 4.1 we show how restricting linear models to category representations can act as a normative principle that rejects certain models and suggests others.

Although the parameter space may be extendible, a subparameter may not be. This is nothing new to the statistician who frequently defines models that need arbitrary constraints to make certain parameters identifiable. But the issue goes deeper than that of identifiability. Even if a parameter is identifiable it does not necessary mean that it is extendible. For a parameter to make "sense" it has to sustain a constancy of interpretation in different extensions of the model. For example, one can ask which parameters in the two-way additive model $\mu_{ij} = \alpha_i + \beta_j$ are extendible, if one of the blocks j' were eliminated. We call parameters that are extendible *natural subparameters*. A natural subparameter is defined as a natural transformation $g : \Theta \to \Xi$ of functors on the category Cat_Ω. To each object Ω in Cat_Ω, the natural transformation g associates a map $g_\Omega : \Theta_\Omega \to \Xi_\Omega$ in such a way that the diagram below commutes. A natural transformation whose components $g_\Omega : \Theta_\Omega \to \Xi_\Omega$ are linear, is called a homomorphism of representations.

$$\begin{array}{ccccc} \Omega & & \Theta_\Omega & \xrightarrow{g_\Omega} & \Xi_\Omega \\ \downarrow \varphi_c & & \uparrow \varphi'_c & & \uparrow \varphi''_c \\ \Omega' & & \Theta_{\Omega'} & \xrightarrow{g_{\Omega'}} & \Xi_{\Omega'} \end{array}$$

For example, in a two-way additive model $A + B$ the treatment effect A is not a natural subparameter; it is impossible to find a natural transformation that can distinguish between an additive constant in A and B. However, the quotient projection $A + B \to A/1$ is a natural subparameter. More dramatically, it can be shown that the mean value parameter in the Box-Cox model is not a natural subparameter. A practical consequence of this fact is that a confidence interval for the mean value parameter alone is meaningless—although computable—since it depends intrinsically on the transformation parameter.

Statistical model

Conventionally, a classical statistical model is defined as a family of probability distributions on the sample space depending on an associated parameter space. In other words, a conventional statistical model can be thought of as a map from the parameter space Θ_Ω to a set of measures $\mathcal{P}(\mathcal{R}^\mathcal{U})$ on the sample space. We have seen how extendibility considerations expand the concepts of a sample space and a parameter space. As a consequence, a statistical model is a more elaborated object, namely, a sequence of families of probability distributions.

An extendible statistical model is a contravariant functor on the design category Cat_{Design} (for a fixed response scale $\mathcal{R}$). This functor associates with each design $\psi : \mathcal{U} \to \Omega$ a model object P_ψ. This model object is a map P_ψ from the parameter space Θ_Ω into a set of distributions on the sample space $\mathcal{P}(\mathcal{R}^{\mathcal{U}})$. This form of extendibility guarantees that the restriction of the set of probability distributions to a smaller design corresponds to the set of distributions actually proposed for that design. It is furthermore reasonable to require that the parameters in both models sustain a constant interpretation, i.e., that the operations "putting the parameters in the model" and "taking marginals" are commutative. For example, if μ_1 is the mean value parameter for the placebo in a trial with four treatments, it should remain the mean value parameter for the placebo if one of the other treatments is eliminated. This extendibility condition on a statistical model corresponds to requiring that the map P_ψ be a natural transformation. This completes the definition of a statistical model. The pieces of this definition can be put together in the following diagram, where all the squares are commutative.

$$\begin{array}{ccccccc} \text{Design} & & & \text{Sample space} & \text{Model} & & \\ \mathcal{U} & \xrightarrow{\psi} & \Omega & S = \mathcal{R}^{\mathcal{U}} & \mathcal{P}(S) & \xleftarrow{P_\psi} & \Theta_\Omega \\ \downarrow \varphi_u & & \downarrow \varphi_c & \uparrow \varphi_u^* & \uparrow \varphi_u^\dagger & & \uparrow \varphi_c' \\ \mathcal{U}' & \xrightarrow{\psi'} & \Omega' & S' = \mathcal{R}^{\mathcal{U}'} & \mathcal{P}(S') & \xleftarrow{P_{\psi'}} & \Theta_{\Omega'} \end{array}$$

One can put further restrictions on the map P_ψ so that the parameter Θ is identifiable. This means that distinct parameter values $\theta_1 \neq \theta_2$ should lead to distinct distributions on the sample space, i.e., $P_\psi\theta_1 \neq P_\psi\theta_2$; in other words, the map P_ψ should be injective for the design $\psi \in Cat_{Design}$ to make the parameter Θ identifiable at ψ.

This general framework defines an extendible classical statistical model. In Section 4 we explore the consequences of this definition. In particular, we show that it can act as a helpful tool to limit the search for sensible models as well as an arbiter in rejecting some commonly used models.

3.3. Definition of an extendible contingency table. Two common interpretations of a contingency table are that (i) it represents a multivariate response or that (ii) it contains a univariate response and that the rest of the dimensions are covariates that explain the response. In data analysis one frequently switches back and forth between these two views. From an extendibility point of view these two interpretations are entirely different. Whereas selection is a typical operation for explanatory variables, amalgamation of levels and aggregation of counts is the corresponding operation for extensive response variables. We give a description of the extendibility considerations for a multivariate response table. The covariate case is discussed in Wit (2000b).

Table category: raw contingency table

In general, the table category $Cat_{\mathcal{T}}$ is a subcategory of all surjective maps between all finite tables $\mathcal{S}$. Which surjective maps are included in $Cat_{\mathcal{T}}$ depends on the particular structure of the marginal response variables. Under certain circumstances merging response levels is unnatural. However, for tables with ordered response levels $Cat_{\mathcal{T}}$ is the category of weakly monotone surjective maps on non-empty, finite tables. This is a natural extension of the underlying category in the concept of "palindromic invariance" (McCullagh 1978). Moreover, merging adjacent levels is a natural operation for ordinal response variables with arbitrary cut-offs.

Sample space functor: counts in the table

The sample space is the collection of all possible, finite counts on a table D. A count on a table D can be represented by a non-negative, discrete measure M_D on D,

$$M_D : \sigma(D) \to \mathbb{N}_0^+,$$

where $\sigma(D)$ is a sigma algebra on D. For finite tables $\sigma(D)$ is always the powerset and the measure M_D is completely characterized by its density. For example, if table $D = \{1, 2, 3\}$ is collapsed into the table $D' = \{1, \{2, 3\}\}$ as a result of map $\varphi \in Cat_{\mathcal{T}}$, then the counts (m_1, m_2, m_3) get mapped into $(m_a = m_1, m_{\{2,3\}} = m_2 + m_3)$. This operation $\varphi^\dagger$ between the non-negative, discrete measures on D and those on D', i.e., $\varphi^\dagger : \mathcal{M}_D \longrightarrow \mathcal{M}_{D'}$, can be defined as

$$\varphi^\dagger(M_D) = M_D \circ \varphi^{-1}.$$

This yields indeed the desired result:

$$m_{\{2,3\}} = M_{D'}(\{2,3\}) = M_D \circ \varphi^{-1}(\{2,3\}) = M_D(\{2,3\}) = m_2 + m_3.$$

In other words, an extendible sequence of sample spaces of non-negative measures $\mathcal{M}_D$ is a covariant functor on the table category $Cat_{\mathcal{T}}$.

Model functor: probabilities of counts

A model for a contingency table is a set of probability measures $\mathcal{F}(\mathcal{M}_D)$ on all possible counts in the table. An extendible statistical model is one where the proposed set of measures for a table relates naturally to the proposed set of measures for a collapsed table. For instance, if in the previous example the total count $m_1 + m_2 + m_3$ were fixed and the increments were independent, then the probability distribution $P_D \in \mathcal{F}(\mathcal{M}_D)$ is determined by the probabilities (π_1, π_2, π_3) (Agresti, 1990). Given this probability measure P_D on $\mathcal{M}_D$ and certain counts $(m_1, m_{\{2,3\}})$ in table D' the probability of obtaining $(m_1, m_{\{2,3\}})$ is

$$\begin{aligned} Pr(\{m_1, m_{\{2,3\}}\}) &= P_D\left(\{m_1, m_2, m_3\} \in \mathcal{M}_D \mid m_1 = m_1, m_2 + m_3 = m_{\{2,3\}}\}\right) \\ &= P_D\left((\varphi^\dagger)^{-1}(m_1, m_{\{2,3\}})\right). \end{aligned}$$

In general, the probability measure P_D in $\mathcal{F}(\mathcal{M}_D)$ is transformed to a measure in $\mathcal{F}(\mathcal{M}_{D'})$ by the map $\varphi^{\dagger\dagger}$, where $\varphi^{\dagger\dagger} : \mathcal{F}(\mathcal{M}_D) \longrightarrow \mathcal{F}(\mathcal{M}_{D'})$ is defined as

$$\varphi^{\dagger\dagger}(P_D) = P_D \circ (\varphi^\dagger)^{-1}.$$

Therefore, a model $\mathcal{F}$ is *extendible*, if it is a covariant functor on the table category $Cat_{\mathcal{T}}$. Diagram (3.2) summarizes this idea.

$$\text{(3.2)} \qquad \begin{array}{ccc} \text{Table} & \text{Sample Space} & \text{Model Space} \\ D & \mathcal{M}_D & \mathcal{F}(\mathcal{M}_D) \\ \downarrow \varphi & \downarrow \varphi^\dagger & \downarrow \varphi^{\dagger\dagger} \\ D' & \mathcal{M}_{D'} & \mathcal{F}(\mathcal{M}_{D'}) \end{array}$$

Many log linear models are not extendible, i.e., not merger-sensible. This may not be as dramatic as it seems, since it is dubious whether invariance under merging discordant classes is really desirable. Other examples will be discussed in sections 4.2 and 4.3.

4. Consequences of extendibility considerations for actual models

The real test for our new definitions of statistical models is (i) whether it has actual consequences for the models that we are able to propose or only introduces unnecessary clutter and (ii) whether the models that arise from the definition make good statistical sense. This section is dedicated to exploring the consequences of our categorical efforts. In Section 4.1 we try to dispel one more time the "Type III ghost" by showing that the model is selection-silly. In Section 4.2 we analyze two association models for contingency tables with ordered factors and show that the most common one is merger-silly. Finally in Section 4.3 we propose a merger-sensible model for a contingency table with cyclicly ordered factors.

4.1. The selection-silly type III model. In 1977 Nelder expressed his dissatisfaction with the conventional algebraic formulation and practical analysis of factorial models, a dissatisfaction with what he termed *the neglect of marginality.* According to Nelder, effects should be viewed in the light of nested vector spaces. He argued that the arbitrary imposition of constraints on effects can obscure the underlying algebraic structure of the statistical model.

The type III model as well as the *weak heredity principle* (Hamada and Wu, 1992) promote the use of certain statistical models that neglect marginality. Recently Venables (1998) has criticized such models. Nevertheless, the U.S. Food and Drug Administration recommends the use of type III models when the design is unbalanced. According to Yandell (1997, p.172) "the type III hypotheses represent the correct approach for assessing importance of model effects (main effects and interactions) when there are unequal numbers of observations per factor combination." In this section we show that the marginality principle is a result of selection invariance, thereby demonstrating that a type III model is selection-silly.

The two-way type III model defines a sequence of parameter spaces for all $a, b \in \mathbb{N}$,

$$III_{ab} = \{\mu \in \mathbb{R}^{ab} \mid \mu_{1.} = \mu_{2.} = \ldots = \mu_{a.}\}.$$

The idea behind this linear model is to test for the presence of an A-effect. According to the definition in Section 3.2, a linear model is extendible if it is a representation of the underlying covariate space category Cat_Ω. Under mild conditions, the category Cat_Ω is the category of all injective maps on all possible two-way layouts, $\mathcal{I}\times\mathcal{I}$. It is easy to check that the sequence $\{III_{ab}\}$ is not a representation on $\mathcal{I}\times\mathcal{I}$. For example, if we restrict an element in III_{23} by selecting the first two columns only, then the result does not lie in III_{22}.

	1	2	3	$\bar{\mu}_{i.}$
1	10	12	17	13
2	13	15	12	13

selection $\varphi_c^{\bullet}$ ⟶

	1	2	$\bar{\mu}_{i.}$
1	10	12	11
2	13	15	14

More specifically, there exists no model for any number of columns and rows that has the type III model for three columns and two rows as its restriction. So what was the reason for using the type III model in the first place? McCullagh (2000a) and Wit (2000b) establish that the hierarchical factorial models are the only representations on the category of injective maps. Nelder's marginality principle is equivalent with the requirement of extendibility under selection.

4.2. Correlation and association models. When the rows and columns of a two-way contingency table are of an ordinal nature, Goodman (1979) proposes an association model that can be presented as

$$\pi_{ij} = \alpha_i \beta_j \exp(\mu_i \nu_j). \tag{4.1}$$

The scores μ_i and ν_j are considered parameters monotone in i and j and capture the "distance" between the levels of the rows and columns, respectively.

Frequently, the grouping of levels of ordinal factors is arbitrary and it is quite likely that two or more consecutive levels have already been merged into one. Therefore, in many circumstances the table category includes all weakly monotone, surjective maps. If the first two levels are merged into a single level c, the association model can, in general, not be written in the form of equation (4.1):

$$\begin{aligned} \pi_{cj} &= \pi_{1j} + \pi_{2j} = \alpha_1 \beta_j \exp(\mu_1 \nu_j) + \alpha_2 \beta_j \exp(\mu_2 \nu_j) \\ &= \beta_j \{\alpha_1 \exp(\mu_1 \nu_j) + \alpha_2 \exp(\mu_2 \nu_j)\}. \end{aligned}$$

The Goodman association model is not extendible under the action of the category of weakly monotone, surjective maps.

Another model that has been proposed to study the interdependence of the variables of a contingency table is the correlation model. This model can be written as:

$$\pi_{ij} = \alpha_i \beta_j (1 + \mu_i \nu_j), \tag{4.2}$$

where the parameters μ_i and ν_j are monotone in i and j respectively. Notice that if the levels i and $i+1$ are merged into a single level, the result is of the same functional form as equation (4.2):

$$\begin{aligned} \pi_{ij} + \pi_{i+1\,j} &= \alpha_i \beta_j (1 + \mu_i \nu_j) + \alpha_{i+1} \beta_j (1 + \mu_{i+1} \nu_j) \\ &= (\alpha_i + \alpha_{i+1}) \beta_j \left(1 + \frac{\alpha_i \mu_i + \alpha_{i+1} \mu_{i+1}}{\alpha_i + \alpha_{i+1}} \nu_j\right). \end{aligned}$$

Since $\frac{\alpha_i \mu_i + \alpha_{i+1} \mu_{i+1}}{\alpha_i + \alpha_{i+1}}$ is simply a weighted average of adjacent scores, monotonicity is preserved by merging levels. In contrast to the association model, the canonical correlation model is extendible under the action of surjective order preserving and reversing maps. This is an important conclusion because a lot of the traditional sociological literature seems to favor Goodman's association model.

4.3. A sensible model for a contingency table with cyclic factors. In this section we propose a model for tables with cyclic row and column factors. The model is a variation on the canonical correlation model that was presented in Section 4.2 and the reduced rank models that were introduced by De Leeuw and Van der Heijden (1991). This model has the advantage of being closed under any cyclic relabelling of the factors. Consider the following example.

EXAMPLE 4.1. **Frequency of mining disasters**
Jarrett (1979) describes a data set of 191 major coal mining disasters in terms of daily frequencies. It is possible that the data were further cross-classified by season

in which the accidents happened. Table 1 presents an imaginary cross-classified contingency table where the marginal day totals corresponds to Jarrett's.

TABLE 1. Frequency of mining disasters between 1851-1962.

	Mon	Tue	Wed	Thu	Fri	Sat	Sun	
Autumn	7	10	5	5	6	7	1	41
Winter	5	9	10	10	11	7	0	52
Spring	3	7	10	12	13	9	2	56
Summer	4	8	8	9	5	6	2	42
	19	34	33	36	35	29	5	

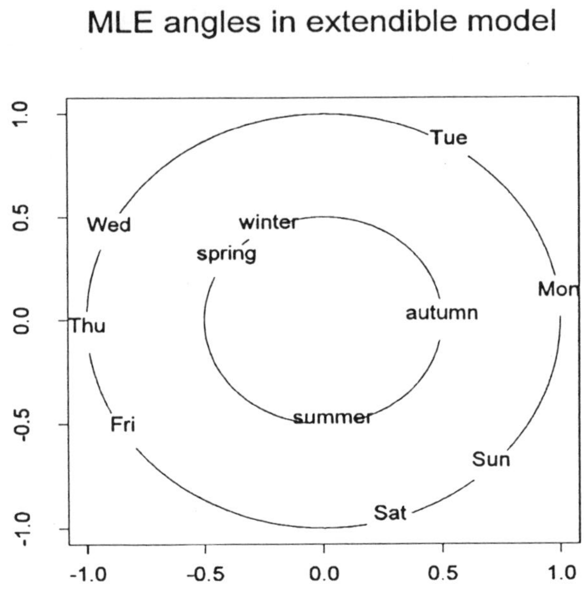

FIGURE 1. The estimated angles of the seasons and days in the extendible models for the mining disaster data.

The underlying table category Cat_T is a subcategory of the surjective maps. Only surjective maps that preserve the cyclic order of the covariates are considered; this category is the category of all cycle-preserving surjections for both rows and columns, $\mathcal{S}_C \times \mathcal{S}_C$. We propose a model that is extendible under $\mathcal{S}_C \times \mathcal{S}_C$:

$$\pi_{ij} = \alpha_i\beta_j(1+\lambda\cos\theta_i\cos\varphi_j), \tag{4.3}$$

where θ, φ are ordered angles on $[0, 2\pi)$, α_i, β_j, and $\lambda \geq 0$ and $\sum_i \alpha_i = \sum_j \beta_j = 1$. The model describes particular deviations from independence. It is obvious that the model is closed under cyclic relabelling. Moreover, if two neighbouring levels

— in a cyclic sense — are merged, say i and $i+1$ into level c, then the result is of the same functional form as equation (4.3):

$$\begin{aligned}\pi_{cj} &= \pi_{ij} + \pi_{i+1,j} \\ &= \beta_j(\alpha_i + \alpha_{i+1})(1 + \lambda \cos\varphi_j \frac{\alpha_i \cos\theta_i + \alpha_{i+1}\cos\theta_{i+1}}{\alpha_i + \alpha_{i+1}}) \\ &= \beta_j \alpha_c (1 + \lambda \cos\varphi_j \cos\theta_c),\end{aligned}$$

where $\alpha_c = \alpha_i + \alpha_{i+1}$. The angle θ_c is between θ_i and θ_{i+1} because the cosine is a continuous function.

The mining disaster data in table 1 can be fitted to the two-way model (4.3) by means of maximum likelihood. Particularly interesting are the angles of the days and seasons as fitted in this model. Figure 1 shows the two sets of angles on the unit circle and a circle with radius 0.5 respectively. To assess the significance of this model we notice that the independence model is nested in model (4.3), so that we can compare the log likelihood of these models. The difference in residual deviances of the independence model and the merger-sensible model is $2(-602.2 + 615.9) = 27.4$, which, under the null hypothesis of no effect, is approximately a χ^2 distribution with $21 - 10 = 11$ degrees of freedom. This suggests the merger-sensible model is a significant improvement (p-value $= 0.0040$) over the independence model.

5. Conclusions

This paper has introduced a new benchmark of statistical modelling called "extendibility". Every statistical model has natural extensions by omitting one or more factors, by merging several classes, or by changing the measurement scale. We have argued that closure under natural extensions is a form of logical consistency and that models that lack this property are suspect. Without extendibility statistical models are unsuitable for statistical inference.

With the help of category theory we have given descriptions of a classical statistical model and a contingency table that incorporate their extensions. Not only does this put statistical modelling on a firm formal basis, but also suggest ways of determining actual models in practical situations. As a result of our algebraic analysis of statistical modelling, Goodman's association model and the type III model, including the weak heredity principle, have been discredited, whereas, on the other hand, several new models have been proposed.

References

[1] Agresti, A. (1984) *Analysis of Ordinal Categorical Data*, New York: John Wiley & Sons.

[2] Agresti, A. (1990) *Categorical Data Analysis*, New York: John Wiley & Sons.

[3] Anderson, C. M., Wu, C. F. J. (1995), "Measuring location effects from factorial experiments with a directional response," *International Statistical Review*, **63**, 3, 345–63.

[4] Anderson, C. M., Wu, C. F. J. (1996), "Disperion measures and analysis for factorial directional data with replicates," *Applied Statistics*, **45**, 1, 47–61.

[5] Arbib, M. A., Manes, E. G. (1975), *Arrows, Structures, and Functors: the Categorical Imperative*, London: Academic Press.

[6] Billingsley, P. (1995), *Probability and Measure*, 3rd ed., New York: John Wiley & Sons.

[7] Cox, D. R., (1958) *Planning of Experiments*, London: Wiley, Chapman & Hall.

[8] Cox, D. R., Snell, E. J., (1981) *Applied Statistics*, London: Chapman and Hall.

[9] Diaconis, P., (1988) *Group Representations in Probability and Statistics*, Hayward, CA: Institute of Mathematical Statistics, Lecture Notes — Monograph series, Vol. 11.

[10] Goodman, L. A., (1979) "Simple models for the analysis of association in cross-classifications having ordered categories," *J. of the American Statistical Association*, **74**, 367, 537–552.
[11] Goodman, L. A. (1981) "Association models and canonical correlation in the analysis of cross-classification having ordered categories," *J. of the American Statistical Association*, **76**, 374, 320–334.
[12] Goodman, L. A. (1985) "The analysis of cross-classified data having ordered and/or unordered categories," *The Annals of Statistics*, Vol. **13**, 1, 10–69.
[13] Halmos, P. R. (1948) *Finite Dimensional Vector Spaces*, Princeton, N.J.: Princeton University Press.
[14] Hamada, M. and Wu, C. F. J. (1992) "Analysis of designed experiments with complex aliasing," *J. Qual. Technology*, **24**, 130–7.
[15] Hannan, E.J. (1965), "Group representations and applied probability," *Journal of Applied Probability* **2**, 1–68.
[16] Hand, D. J. *et al.* (1994), *A Handbook of Small Data Sets*, London: Chapman and Hall.
[17] Jarrett, R. G. (1979), "A note on the intervals between coal mining disasters," *Biometrika*, **66**, 191–93.
[18] De Leeuw, J., Van der Heijden, P. G. A. (1991), "Reduced rank models for contingency tables," *Biometrika*, **78**, 1, 229–32.
[19] MacLane, S. (1971), *Categories for the Working Mathematician*, New York: Springer-Verlag.
[20] McCullagh, P. (1978) "A class of parametric models for the analysis of square contingency tables with ordered categories," *Biometrika*, **65**, 2, 413–8.
[21] McCullagh, P (1999) "Quotient spaces and statistical models," *The Canadian Journal of Statistics*, **Vol. 27**, No. 3, 1–10.
[22] McCullagh, P. (2000a) "Invariance and factorial models" (with discussion), *J. R. Statist. Soc. B*, **62**, Part 2, 209–256.
[23] McCullagh, P. (2000b) "What is a statistical model," Technical Report No. 499, University of Chicago.
[24] Nelder, J. A. (1977) "A reformulation of statistical models," (with discusion), J. R. Statist. Soc. A, **140**, 48–77.
[25] Nelder, J. A. (1998) "The Selection of Terms in Response-Surface Models: How Strong is the Weak-Heredity Principle?", *The American Statistician*, Vol. **52**, Nr. 4, 315–18.
[26] Serre, J-P. (1977) *Linear Representations of Finite Groups*, New York: Springer-Verlag.
[27] Van der Heijden, P. G. M., De Falguerolles, A., De Leeuw, J. (1989), "A Combined Approach to Contingency Table Analysis Using Correspondence Analysis and Log-Linear Analysis," *Applied Statistics*, Vol. **38**, No. 2, 249–292.
[28] Venables, W. N. (1998) "Exegeses on Linear Models," *Proceedings 1998 International S-PLUS User Conference.*
[29] Wit, E. C. (2000a), *J. R. Statist. Soc. B*, Discussion on the Paper by McCullagh, **62**, Part 2, 251–2.
[30] Wit, E. C. (2000b), *The Categorical Imperative: Extendibility Considerations for Statistical Models*, Ph.D. Thesis, University of Chicago.
[31] Yandell, B. S. (1997) *Practical Data Analysis for Designed Experiments*, London: Chapman and Hall.

DEPARTMENT OF STATISTICS, UNIVERSITY OF GLASGOW, GLASGOW, G12 8QW, U.K.
E-mail address: ernst@stats.gla.ac.uk

DEPARTMENT OF STATISTICS, UNIVERSITY OF CHICAGO, CHICAGO, ILLINOIS 60637, U.S.A.
E-mail address: pmcc@galton.uchicago.edu

Selected Titles in This Series

(*Continued from the front of this publication*)

262 **Paul Igodt, Herbert Abels, Yves Félix, and Fritz Grunewald, Editors,** Crystallographic groups and their generalizations, 2000

261 **Gregory Budzban, Philip Feinsilver, and Arun Mukherjea, Editors,** Probability on algebraic structures, 2000

260 **Salvador Pérez-Esteva and Carlos Villegas-Blas, Editors,** First summer school in analysis and mathematical physics: Quantization, the Segal-Bargmann transform and semiclassical analysis, 2000

259 **D. V. Huynh, S. K. Jain, and S. R. López-Permouth, Editors,** Algebra and its applications, 2000

258 **Karsten Grove, Ib Henning Madsen, and Erik Kjær Pedersen, Editors,** Geometry and topology: Aarhus, 2000

257 **Peter A. Cholak, Steffen Lempp, Manuel Lerman, and Richard A. Shore, Editors,** Computability theory and its applications: Current trends and open problems, 2000

256 **Irwin Kra and Bernard Maskit, Editors,** In the tradition of Ahlfors and Bers: Proceedings of the first Ahlfors-Bers colloquium, 2000

255 **Jerry Bona, Katarzyna Saxton, and Ralph Saxton, Editors,** Nonlinear PDE's, dynamics and continuum physics, 2000

254 **Mourad E. H. Ismail and Dennis W. Stanton, Editors,** q-series from a contemporary perspective, 2000

253 **Charles N. Delzell and James J. Madden, Editors,** Real algebraic geometry and ordered structures, 2000

252 **Nathaniel Dean, Cassandra M. McZeal, and Pamela J. Williams, Editors,** African Americans in Mathematics II, 1999

251 **Eric L. Grinberg, Shiferaw Berhanu, Marvin I. Knopp, Gerardo A. Mendoza, and Eric Todd Quinto, Editors,** Analysis, geometry, number theory: The Mathematics of Leon Ehrenpreis, 2000

250 **Robert H. Gilman, Editor,** Groups, languages and geometry, 1999

249 **Myung-Hwan Kim, John S. Hsia, Yoshiyuki Kitaoka, and Rainer Schulze-Pillot, Editors,** Integral quadratic forms and lattices, 1999

248 **Naihuan Jing and Kailash C. Misra, Editors,** Recent developments in quantum affine algebras and related topics, 1999

247 **Lawrence Wasson Baggett and David Royal Larson, Editors,** The functional and harmonic analysis of wavelets and frames, 1999

246 **Marcy Barge and Krystyna Kuperberg, Editors,** Geometry and topology in dynamics, 1999

245 **Michael D. Fried, Editor,** Applications of curves over finite fields, 1999

244 **Leovigildo Alonso Tarrío, Ana Jeremías López, and Joseph Lipman,** Studies in duality on noetherian formal schemes and non-noetherian ordinary schemes, 1999

243 **Tsit Yuan Lam and Andy R. Magid, Editors,** Algebra, K-theory, groups, and education, 1999

242 **Bernhelm Booss-Bavnbek and Krzysztof Wojciechowski, Editors,** Geometric aspects of partial differential equations, 1999

241 **Piotr Pragacz, Michał Szurek, and Jarosław Wiśniewski, Editors,** Algebraic geometry: Hirzebruch 70, 1999

240 **Angel Carocca, Víctor González-Aguilera, and Rubí E. Rodríguez, Editors,** Complex geometry of groups, 1999

For a complete list of titles in this series, visit the AMS Bookstore at **www.ams.org/bookstore/**.